To my father *John Henry Rollins*
who introduced me to the joys of walking in wild places
and to my daughter *Emily* in the hope that she, too,
will enjoy the thrill of original exploration.

COMPLIMENTARY COPY
NOT FOR RE-SALE

Disclaimer

This book is intended for experienced cavers: the information in it is incomplete and may be inaccurate.

Caving can be dangerous. Most caves contain drops and loose rock. Accidents occurring underground are compounded by the inability for conventional rescue to take place. An injured person could wait several days for help to arrive. At present there is no cave rescue organization based in the Canadian Rockies.

It is also assumed that the reader is fully conversant with alpine travel and the use of climbing equipment needed to access the caves.

Rocky Mountain Books
Calgary–Victoria–Vancouver

Above: The Straw Gallery in Arctomys Cave, Mount Robson Provincial Park. Photo Ian Drummond.

Right: Fang Cave, McGregor Range. Photo Dave Thomson.

CAVES

of the Canadian Rockies and Columbia Mountains

Jon Rollins

Above: Camp in Castleguard Cave entrance, Banff National Park. Photo Dave Thomson.

Right: Chas Yonge on the third pitch in Shorty's Cave, Crowsnest Pass. Photo Dave Thomson.

Above: Squeeze in Rat's Nest Cave, Canmore. Photo Chas Yonge.

Left: Terminal passage in Serendipity, Crowsnest Pass. Photo Ian McKenzie.

Above: Drying out after a winter trip to the Small River karst. Photo Dave Thomson.

Left: Randy Spahl descending through the ice floor in Serendipity, Crowsnest Pass. Photo Ian McKenzie.

Tony Bennnett in Bloodstone Passage, Yorkshire Pot, Crowsnest Pass.
Photo Dave Thomson.

Above: Waiting at the top of the pitch, Supplies Passage, Castleguard Cave, Banff National Park. Photo Chas Yonge.

Right: The Coliseum Middle Entrance, Fang Cave, McGregor Range. Photo Dave Thomson.

Laure Morel and Bill MacDonald in the entrance to Gargantua Cave, Crowsnest Pass. Photo Dave Thomson.

We acknowledge the financial support of the Government of Canada through the Book Publishing Industry Development Program (BPIDP) and the support of the Alberta Foundation for the Arts for our publishing program.

All rights reserved. No part of this work may be reproduced or transmitted in any form or by any means, electronic or mechanical, including photocopying and recording, or by any information storage or retrieval system, except as may be expressly permitted in writing from the publisher.

© Copyright Jonathan Rollins 2004

Printed in Canada

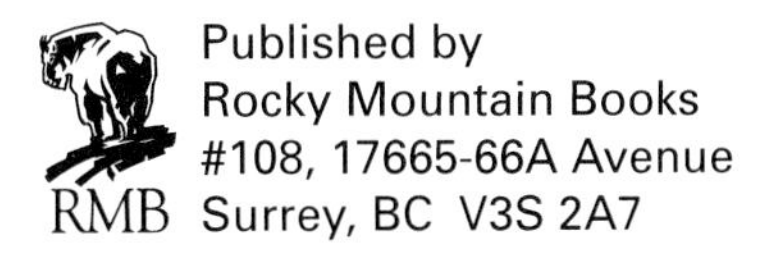

Published by
Rocky Mountain Books
#108, 17665-66A Avenue
Surrey, BC V3S 2A7

National Library of Canada Cataloguing in Publication

Rollins, Jon, 1957-
Caves of the Canadian Rockies and Columbia Mountains / Jon Rollins.

Includes bibliographical references and index.
ISBN 0-921102-94-1

1. Caves--Rocky Mountains, Canadian (B.C. and Alta.)--Guidebooks. 2. Caves--British Columbia--Columbia Mountains--Guidebooks. 3. Rocky Mountains, Canadian (B.C. and Alta.)--Guidebooks. 4. Columbia Mountains (B.C.)--Guidebooks. I. Title.

GV200.66.C3R64 2004 551.44'7'09711 C2004-900733-5

Above: John Donovan. Photo Dave Thomson.

Opposite far top: Looking out of the entrance of Caveat, Roche Miette caves. Photo Jon Rollins.

Opposite far bottom: Jim McPhail in Shorty's Cave entrance, Crowsnest Pass. Photo Dave Thomson.

Opposite: Surface karst on the Andy Good Plateau. Photo Jon Rollins.

Front Cover: Ian Mckenzie descending by Dracula's Tongue, The Pellet Factory, Fernie Area. Photo Henry Bruns.

Back cover: The big shaft , Close to the Edge, Dezaiko Range. Photo Ian McKenzie.
Back cover top inset: Wedge Cave, Top of the World Provincial Park. Photo Ian Drummond.
Back cover bottom inset: Formations in Bloodstone Passage, Yorkshire Pot, Crowsnest Pass. Photo Ian Drummond.

Top left: Ron Lacelle, first recipient of the ASS Golden Garbage Can Award. Photo Ian McKenzie.

Top right: Peter Thompson. Photo Peter Thompson Collection.

Bottom right: Chas Yonge in Porcupine Cave. Photo Jon Rollins.

Contents

Cave Descriptions 35

Canadian Rockies

Columbia Mountains

Northern Alberta

Natural History 303

Geology — The Rock that contains the Caves 304

The Cave Environment 315

Cave Fauna and Flora 318

Appendices 323

Glossary of Caving Terms 325

Sources & Resources 330

Index 332

Author's Foreword

I can still remember my first caving trip and probably will for the rest of my life — the cool darkness, the moist bedrock surfaces, and my absolute reliance on the experienced cavers who took me along. The overwhelming memories of that first trip are not so much of the cave environment, but rather of the physical nature of caving that left little time for casual observation. The initial downclimbs and use of ropes to descend a shaft were familiar, my sport at the time being rock climbing, but as the passages got smaller and smaller I found myself having to control feelings of panic. Following a wallow through a flooded passage, nose pressed against the ceiling, I realized I could cope, and surrendered myself to the intense satisfaction generated by extreme physical activity.

The next few trips were similar, getting into seemingly impossible situations in deep cave systems, but relying on experienced colleagues and physical endurance to see me through. And always there was the immense feeling of satisfaction following a hard trip. I was hooked. It wasn't until later trips that I began to appreciate the beauty and subtleties of cave environments like the sculpting of bedrock surfaces by running water into fantastic shapes, the dense impenetrable blackness when you turned your light out, and a silence so complete you could hear your heart beating. Within a year of my first trip I experienced one of the ultimate thrills in caving: that of exploring a newly discovered cave system. One of the advantages of being a caver in the Canadian Rockies (and believe me, there are few) is the opportunity to find and explore new caves, to go where no one has been before. Along with cave exploration came an introduction to the multiple facets of speleology, the science of caves and caving: learning to survey caves and draft cave maps, identify mineral deposits and cave fauna, and interpret topographical and geological maps. Unlike sports such as rock climbing and skiing where the physical activity is paramount, caving can be a perfect blend of hands-on activity and scientific theory.

The discovery and exploration of every cave in this book was accompanied by excitement and euphoria, sometimes followed by disappointment when the cave ended short of the explorer's dreams of being the longest or deepest. For a few who take up caving it becomes a lifelong pursuit, and although the body may complain owing to countless treks up steep mountains hauling huge packs, the enthusiasm remains undiminished. Always there is the hope of finding that kilometre-deep cave system just over the next mountain ridge. In using this guide I hope you, too, may experience some of the thrill of discovery. Caving is still a relatively new activity in the Rockies, and it could be that the "golden age" of discovery is still to come.

I apologize for inaccuracies or omissions. If you have any comments, please contact me at 1114 Larch Place, Canmore, Alberta T1W 1S8, jrollins@telus.net, so that this guide can be made as accurate as possible.

Jon Rollins 2004

Opposite: Digging into Yorkshire Pot in the winter of 1973. Photo Ian Drummond.

Introduction

A BRIEF HISTORY OF CAVE DISCOVERY

"Stoney Indians associate caves with a race of immortal little people called makutibi. These little people are males only. Hunters, on occasion, may see traces of these creatures, their tiny footprints or that of their tiny dogs. At night, they may hear these dogs barking. Little people normally shun human contact, but from time to time they will speak to people. They are credited with being instructors to the Indians, and they live in villages beneath the mountains. These villages may be reached through the caves. According to the Stoney, not all caves lead to underground villages, but because they did not know which caves did and which did not, they treated all caves with respect. Before approaching a cave, Stoneys traditionally conducted a brief ritual to assure the dwellers within that no harm was intended. This was wise since there was the risk of captivity among the little people." from *The David Thompson Highway: A Hiking Guide* by Jane Ross & Dan Kyba.

Arrowheads, Pelican Lake style (2000-3000 years old), have been found in the entrance of Rat's Nest Cave. Two other cave sites, Devona Cave and Crowsnest Spring, have been documented as containing pictographs. For the main part it appears Canadian Rockies caves were too cold to be suitable for human habitation. However, it is probable that some cave entrances have been used by aboriginal peoples for shelter, as a place out of a storm to dress a kill, to build a fire, or even as destinations for vision quests. However, no records or artifacts other than those mentioned have been found.

The first documented record of cave discovery involving Old World emigrants began with the Cave and Basin in 1875, when local Stoney Indians directed Peter Younge and Benjamin Pease to the site. People would have also visited Crowsnest Spring in the 1860s when the Dewdney Trail, the first wagon road through the Rockies, was constructed. Henry Cody, a prospector in the Selkirks, discovered Cody Caves in the 1880s, and in 1904 C. H. Deutschmann was credited with the discovery of Nakimu Caves, which were surveyed in 1907 by A. O. Wheeler. Certainly known to aboriginal people because of its obvious entrance, Canyon Creek Ice Cave was visited by Stan Fullerton in 1905. A description of a cave discovered in 1910 by hunter Russ McFall near Nordegg is almost certainly Wapiabi Cave. In 1911, the Rev. G. B. Kinney and Conrad Kain descended Arctomys Cave to a depth of 76 m, and in 1921, Banff guide Cecil Smith discovered Castleguard Cave while searching for his horses. A photograph taken by Byron Harmon looking out the Hole-in-the-Wall dates from the early 1900s. It is likely Goat Cave above Banff was also visited around this time.

There is no further documented evidence of cave exploration until W. L. Brigg and R. S. Taylor surveyed Cadomin Cave in 1959.

Between 1965 and 1976, students of Derek Ford — British and Canadian cavers working out of a karst research group in the Department of Geography and Geology at McMaster University — scoured the southern Canadian Rockies for caves. They re-surveyed Nakimu Caves in the Selkirk Mountains, tripling the known length, and in 1967 explored Castleguard Cave to "The Crutch." Two years later, in 1969, many of the caves of Crowsnest Pass were discovered and subsequently explored. That same year caving activities were documented in the first issue of the magazine "Canadian Caver," initiated by Julian Coward and published at McMaster. In 1973 Arctomys Cave was explored to its current depth. In 1976 Peter Thompson produced "Cave Exploration in Canada," the first (and only) attempt at a complete inventory of Canadian caves.

Lewis Freeman, Soapy Smith and Ulysses LaCasse in Castleguard Cave entrance, 1924. Photo Byron Harmon, courtesy Whyte Museum of the Canadian Rockies.

Since 1970 nearly all Rockies caves have been explored by members of the Alberta Speleological Society (ASS), with some notable exceptions in the northern Rockies as exploration shifted farther afield. A look at the cave location maps on pages 11 and 12 would seem to indicate a lack of caves and karst areas in the northern Rockies. This is far from the case. What it does indicate is the high cost of hiring planes and helicopters, and the degree of organization necessary to get groups of cavers and equipment into remote areas. The "three cavers and a rope" exploration mode used to good effect in the southern Rockies doesn't work well in the more remote north, and organization being anathema to most Canadian Rockies cavers, there have been only a few caving expeditions to date. Formerly much exploration was carried out by Jasper members of the ASS. Lately, more northerly based communities of cavers are becoming established. In Prince George the Devil's Club and Caledonia Ramblers hiking club have found a number of caves, and with the advent of the University of Northern British Columbia, the UNBC Caving Club often pokes around the Dezaiko and McGregor ranges.

Members of the British Columbia Speleological Federation (BCSF) have also been known to venture as far east as the Rockies, resulting in a friendly rivalry between cavers from Alberta and British Columbia.

Expeditions have often had an international theme, with the usual opportunistic British cavers getting in on the act. Members of the 1983-1984 Anglo-Canadian Rocky Mountain Speleological Expeditions (ACRMSE) were instrumental in exploring the Small River and Bocock karst areas, and in 1986 the Imperial College Caving Club (ICCC) had an exciting time wrestling with canoes and thrashing around in the B.C. bush. More recently, the Lancaster University Speleological Society (LUSS) explored some connections in Yorkshire Pot and American cavers were the first to penetrate "The Crack" and make it to the 430 m depth in Close to the Edge.

The southern Rockies continue to yield discoveries, not just in the Crowsnest Pass, which is a veritable caving cornucopia, but also in the canyons and on the high mountain ridges of the national parks.

On the other hand, exploration in the northern Rockies has barely scratched the surface. Several of the karst areas in the northern Rockies are substantial: the Bastille and Dezaiko plateaus each encompass 175 sq. km of karst with potential for kilometre-deep caves. An invaluable tool has been aerial reconnaissance and the use of aerial photographs. However, as the 1986 Caribou Mountain Expedition discovered, the existence of well-developed surface karst is no guarantee of finding caves, and there is no substitute for ground-walking an area before organizing a full-scale expedition. Follow-up expeditions to any of the remarkable karst areas of the northern Rockies are long overdue, and it is hoped that this guide will stimulate further activity. Let's organize some more caving expeditions soon!

Byron Harmon in Castleguard Cave in 1924.
Photo Byron Harmon, courtesy Whyte Museum of the Canadian Rockies.

Looking out of the entrance to Hole-in-the-Wall.
Photo Byron Harmon, courtesy Whyte Museum of the Canadian Rockies.

WHERE THE CAVES ARE

When the five founding members of the Alberta Speleological Society (ASS) first met in 1968 to contemplate the formation and aims of the caving club, there were only two caves on the cave list: Canyon Creek Ice Cave and Cadomin Cave. (And of the two only Canyon Creek Ice Cave had been visited by any of the members.) Thirty-five years on, this guide documents over 200 caves discovered to date in the Canadian Rockies — a mountain range 1450 km long and 150 km wide, covering an area of 180,000 km^2. It stretches from the Yukon border in the north to the United States (Montana) border in the south, and from the Rocky Mountain Trench in British Columbia to the eastern foothills in Alberta.

For completeness, I have included northern Alberta and the few caves and karst areas in the Columbia Mountains of BC west of the Trench.

Aside from the gypsum caves in Wood Buffalo National Park, the majority of caves are located above the 2000 m elevation in the main ranges of the Rocky Mountains, in areas of limestone karst straddling the Continental Divide. With no respect for jurisdictional boundaries, caves are located in both federal and provincial parks, and on Crown land, some even passing through mountains between Alberta and British Columbia.

Because carbonates are the dominant bedrock, it has long puzzled cavers why relatively few caves have been discovered in the Canadian Rockies. Although systematic exploration has only been carried out since the late 1960s, it is apparent that large numbers of caves, or more importantly cave entrances, just aren't there to be found. In contrast, Vancouver Island has possibly as many as a thousand caves located in scattered lenses of limestone. Certainly, sequential periods of heavy glaciation have been responsible for removing caves and burying cave entrances, so that digging, both into caves and at the end of plugged passages, is a common caving activity. Despite this major obstacle, the apparent paucity of enterable caves, the isolated locations of karst areas and the small numbers of local cavers, some impressive cave systems have been explored over the years.

Of the caves that have so far been discovered, 11 are about 250 m deep and 23 are more than 500 m long. Outstanding caves include Yorkshire Pot with 200 m of entrance pitches leading to 12 km of passage, 20 km-long Castleguard Cave that ends in an ice-plugged passage beneath the Columbia Icefield, the 536 m-deep Arctomys Cave that is the deepest cave north of Mexico and the more recently discovered 472 m-deep Close to the Edge with its huge 255 m-deep entrance shaft — the deepest single shaft north of Mexico.

Approximately half of the caves in this guide are located in just ten karst plateaus: Mount Bocock, Moon River, Bastille, Small River, Snaring, Castleguard, Mount Ball, Top of the World, Andy Good and Ptarmigan. These are high alpine plateaus, sometimes mantled but more often covered with large areas of bare limestone pavement dotted with fissures and shafts. Meltwater from glaciers and snowmelt sink into these karstic plateaus, often emerging as valley springs a kilometre lower. The high altitude of these sites provides only two or three months of snow-free caving.

Another common location for caves are major east-west gaps through the mountains. This is the case with the Roche Miette caves, Wapiabi Cave, Rat's Nest Cave and some of the caves of the Crowsnest Pass.

Seeming incongruous are the occasional voluminous cave passages and chambers discovered high in precipitous mountain ridges. Examples include Fang, Cadomin, Wapiabi, Block Lakes, Gargantua and Cleft caves.

Having said all this, if there is one defining rule for locating enterable caves in the Canadian mountains, it is the lack of predictability. Despite the pontificating of numerous karst academics over the years, nowhere is the adage "caves are where you find them" more true than in the Rockies.

"Once away from the few main roads and gravel logging routes, the country becomes a steep and densely vegetated wilderness, with raging streams of glacial melt water, precipitous cliffs and potentially dangerous wild animals. Gaining access to the fascinating and largely unexplored areas of high alpine karst on foot is a time consuming, injury-liable and often frustrating experience, particularly when back-packing heavy loads." Deej Lowe.

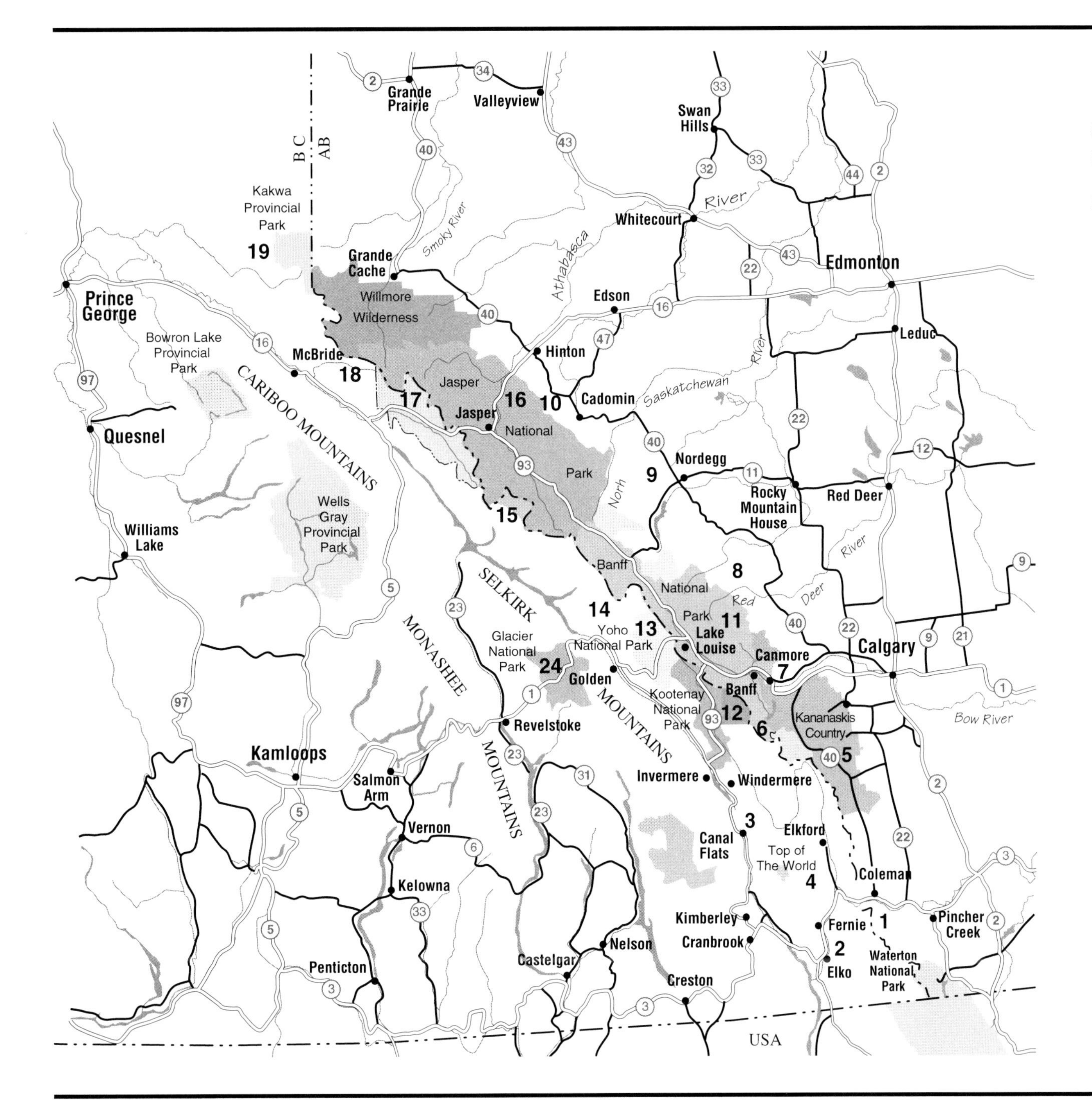

Southern Rockies & Columbia Mountains

KEY

1 Crowsnest Pass
2 Fernie Area
3 Whiteswan Provincial Park
4 Top of the World Provincial Park
5 Kananaskis Country
6 Assiniboine Provincial Park
7 Canmore Area
8 Red Deer River
9 Nordegg Area
10 Cadomin Area
11 Banff National Park
12 Kootenay National Park
13 Yoho National Park
14 Golden Area
15 Hamber Provincial Park
16 Jasper National Park
17 Mount Robson Provincial Park
18 Small River Karst
19 Kakwa
24 Rogers Pass

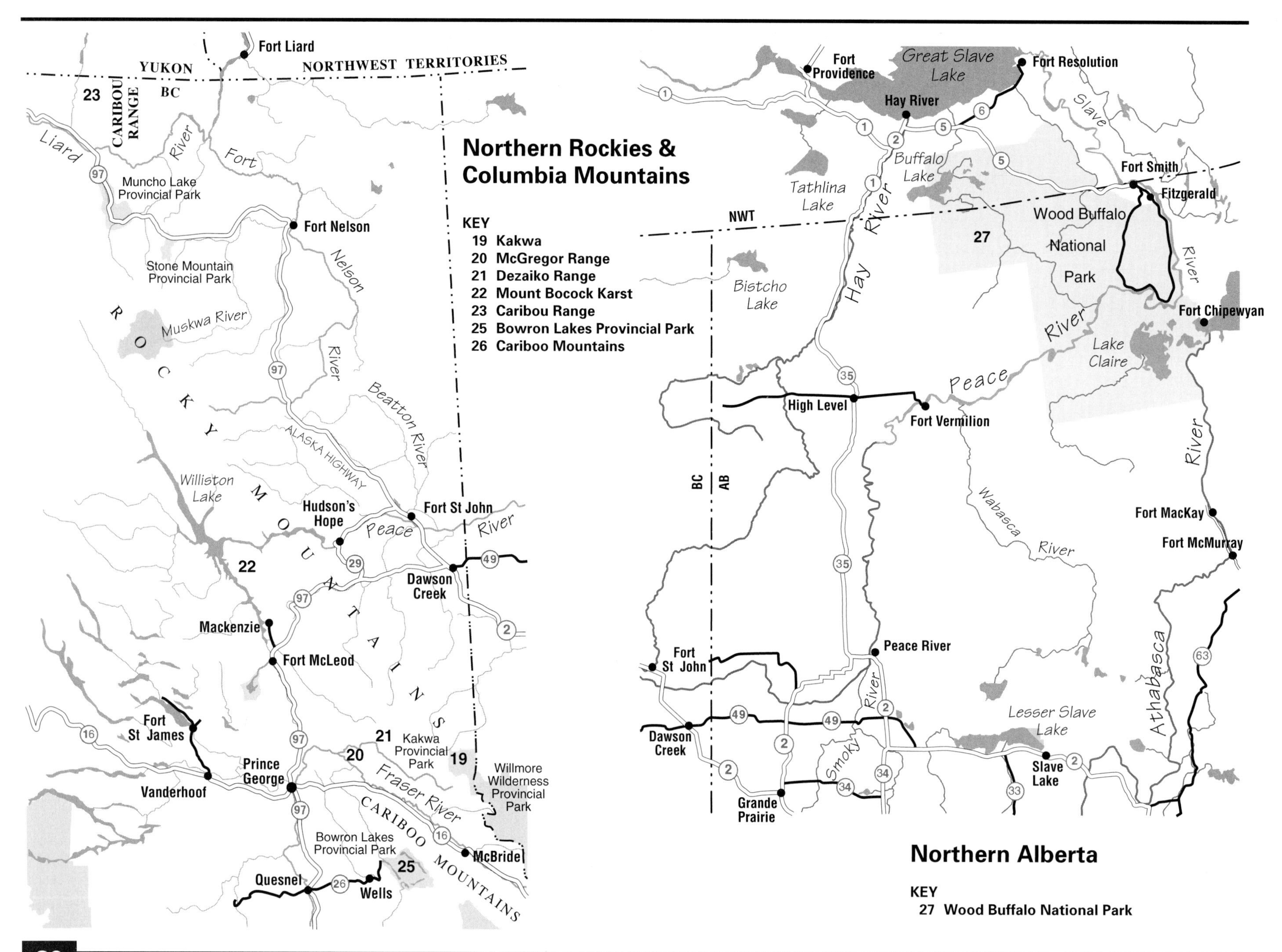

Northern Rockies & Columbia Mountains
KEY
19 Kakwa
20 McGregor Range
21 Dezaiko Range
22 Mount Bocock Karst
23 Caribou Range
25 Bowron Lakes Provincial Park
26 Cariboo Mountains
YUKON
NORTHWEST TERRITORIES
BC
Fort Liard
Caribou Range
Liard River
Fort River
Muncho Lake Provincial Park
Fort Nelson
Nelson
Stone Mountain Provincial Park
Muskwa River
Rocky Mountains
River
Beatton River
Alaska Highway
Williston Lake
Hudson's Hope
Fort St John
Peace River
Dawson Creek
Mackenzie
Fort McLeod
Fort St James
Vanderhoof
Prince George
Kakwa Provincial Park
Fraser River
Willmore Wilderness Provincial Park
Cariboo Mountains
Bowron Lakes Provincial Park
McBride
Quesnel
Wells
Northern Alberta
KEY
27 Wood Buffalo National Park
Fort Providence
Great Slave Lake
Fort Resolution
Hay River
Slave
Buffalo Lake
Tathlina Lake
Fort Smith
Fitzgerald
NWT
Wood Buffalo National Park
River
Bistcho Lake
Hay River
Fort Chipewyan
Lake Claire
Peace River
High Level
Fort Vermilion
BC
AB
Wabasca River
Fort MacKay
Fort McMurray
Peace River
Fort St John
Smoky River
Lesser Slave Lake
Athabasca
Dawson Creek
Slave Lake
Grande Prairie

GETTING AROUND

Travel

Arriving by Air
Edmonton and Calgary both have international airports with connecting flights to Prince George, Kamloops, Cranbrook, Fort St. John, Fort Nelson and Peace River.

Travelling by Bus
Greyhound Bus Lines operate a reliable bus service to the mountain centres of Crowsnest Pass, Canmore, Banff, Jasper, Field, Golden and Prince George.

You need a Vehicle
If coming from overseas, you need to rent a vehicle to get to a trailhead. It is extremely difficult to go caving in the Canadian Rockies without one. During the summer months (July to October) a two-wheel drive vehicle with normal ground clearance is usually adequate. Make sure you have a good spare wheel — a space saver spare might not be durable enough for some of the gravel roads. A four-wheel drive vehicle is strongly recommended for attempting reconnaisance trips into new areas, or if you want to go caving during the rest of the year.

Most of the roads used to reach cave trailheads are frequented by logging or gravel trucks, so drive with caution.

Where to Stay

All towns and communities have hotels, motels and B &Bs. From information centres pick up free copies of information booklets and phone ahead.

Roadside Campgrounds
Vehicle accessible campgrounds abound throughout the national parks, provincial parks and provincial forests. There are also private campgrounds and municipal campgrounds. Some have showers and a small store attached. In the summer, campsites fill up quickly, particularly those in the national parks.

Hostels
Many hostels are conveniently close to trailheads.

In Banff National Park: Tunnel Mountain (Banff), Castle Mountain (Hwy. 1A), Lake Louise (townsite), and along the Icefields Parkway (Hwy. 93): Mosquito Creek, Rampart Creek, Hilda Creek, Beauty Creek and Athabasca Falls.

In Jasper National Park: the townsite, The Whistlers (toward the Jasper Tramway), Maligne Canyon (off Hwy.16) and at Mount Edith Cavell.

In Yoho National Park: Whiskey Jack near Takakkaw Falls.

In Kananaskis Country, at Ribbon Creek off Hwy. 40.

Close to Nordegg on Hwy. 11: Shunda Creek.

On Hwy. 3 in the Crowsnest: the Grand Union International Hostel in Coleman.

In Fernie: town centre off Hwy. 3.

For booking hostels: from Calgary to the Columbia Icefield call (403) 283-5551, and north to Jasper (780) 432-7798. For the hostel in Fernie call (604) 684-7111 and for Coleman call (403) 563-3433.

Getting Gas & Supplies

All towns and communities have grocery stores, banks, laundromats and eating places. Often gas stations have a small grocery store attached. In north-central BC, gas stations and stores are almost non-existent outside of major highways.

Most outdoor stores only carry gear for hunters, anglers and roadside campers. Only in the major cities and in smaller mountain communities of Banff, Lake Louise and Jasper will you find stores carrying gear for hikers, backpackers and climbers. Caving gear is not available at the present time.

Getting Information

Information Centres
Banff National Park (403) 762-1550
Jasper National Park (780) 852-6176
Kootenay National Park (250) 347-9505
Yoho National Park (250) 343-6783
Rogers Pass (250) 837-7500
Canmore (403) 678-5277
Kananaskis Country (403) 673-3985
Crowsnest 1-800-661-8888
Golden 1-800-622-GOLD
Mount Robson Provincial Park (250) 566-4325
Valemount (250) 566-4846
McBride (250) 569-3366
Prince George (250) 562-3700
Northern BC (250) 561-0432
Hudson's Hope (250) 783-9901
Wells 1-877-451-9355

Jon's Recommendations

Cavers have traditionally been a poor, penny pinching lot (many of them coming from Britain) — nursing a beer until chucking out time, and then choosing to sleep in their car or a ditch. However, should you be looking for a little more in the way of comfort before or after a caving trip, a few of the caving areas offer the following hostelries, inns, hotsprings and restaurants which have been used, and get the thumbs up from cavers (a sometimes dubious accolade).

Crowsnest Pass The Kozy Knest Kabins (403) 563-5155 offers motel accommodation right at the pass on Highway 3. You won't get any closer to the caves other than by camping, and just west is the Inn On The Border (403) 563-3101, a former caver's hangout which has moved upscale, but is still a great place for a pint. For those just out of the caves, you won't feel conspicuous at the Greenhill Hotel (403) 562-2232 on the main street in Blairmore.

Canmore Cave guides and clients retire after a hard trip to the Rose and Crown (403) 678-5168. The patio is great on a summer afternoon. Otherwise, try the Drake (403) 678-5131 across the road for good food and beer. The Bolo Ranch House (403) 678-5211 on Main Street has good food and does breakfast at 8:00 am.

Of the numerous hotels and motels in Canmore, the most reasonable place to stay is the Alpine Club of Canada Clubhouse (403) 678-3200.

Banff The Magpie and Stump (403) 762-4067 on Caribou Street provides a good plate of calories at a reasonable price, and a few of their Dos Equis beers may bring back memories of warm southern caves. Whether you've just spent five days underground in Castleguard, or just returned from a short hike to Goat Cave, the Upper Hot Springs feels wonderful (please shower on your way in!). Call 1-800-767-1611 for hours of operation.

Jasper All good Central Rockies cave trips start and end at the Atha. B. — the Athabasca Hotel (780) 865-2120, the usual meeting location for caving expeditions on their way up the Yellowhead Highway into British Columbia. The Miette Hotsprings is an hour's drive east towards Edmonton. Call 1-800-767-1611 for hours of operation.

Rogers Pass The Glacier Park Lodge (250) 837-2126 is very comfortable and has a hot pool to soak away those weary caving muscles. Two summer campgrounds are provided by Parks Canada: the Illecillewaet and Loop Brook (first come, first served).

Hinton On your way to Cadomin Cave there are numerous hotels, motels and eateries in Hinton. Notable is Tokyo Sushi (780) 865-2120.

Golden The Mad Trapper Saloon (250) 344-6661 has long been a popular rendezvous for cavers.

CAVING IN THE CANADIAN ROCKIES

The Caving Season

The length of the caving season depends on the altitude and aspect of the cave. The majority of caves are alpine, which means access is limited to the summer months of July to October. In heavy snow years some caves may never open, and many cave trips have consisted of digging very deep holes in a wide expanse of snow with no results. Winter caving trips do take place, often in the hope that water levels in flooded sections of a cave may have dropped, thus making more passage accessible. A few low elevation caves remain accessible all year round: Crowsnest Spring, Eagle Cave and the caves low on Sentry Mountain, Canyon Creek Ice Cave, Rat's Nest Cave, Goat Cave and Tokumm Cave. Some caves, such as Cadomin Cave, should not be visited in the winter owing to hibernating bats.

Only a few caves are subject to flooding, although spring melt accompanied by heavy rain can make things exciting. The entrance series of Castleguard Cave, for example, floods on a regular basis in the summer, so trips usually occur at Easter. Read the individual cave descriptions for further information.

Getting to the Caves

From the trailhead most modes of getting to a cave or karst area — helicopters, float planes, 4x4 vehicles, ATVs, mountain bikes, skidoos, skis and on foot — have been used singly or in combination.

The traditional route, to hire an outfitter and have gear carried in by pack train, was tried in concert with other modes of transport in the Caribou Range, and may be a viable transportation mode for some of the more isolated areas in the northern Rockies.

Mountain bikes are useful when a road of some description exists, but motorized vehicles are not permitted i.e. Amos Cave, Plateau Mountain Ice Cave, Forgetmenot Pot, Burstall Pots, Canyon Creek Cave, the Vroom Closet. With few exceptions, mountain biking is not allowed in national parks.

Many caves in this guide can be reached on foot from the trailhead. A few are located close to a highway. Others are close to a logging or resource road or seismic line accessible by a 4x4 or mountain bike. Assuming you know exactly where your cave is, in this scenario cavers can hike up in daylight, cave overnight and hike down the next day. i.e. Crowsnest Pass caves.

Many caves, though, are more remote and located at high altitude (typically over 2000 m), which means that most trips begin with a long drive, followed by a long hike, possibly on a trail, followed by a climb or scramble up a mountainside — and all this before the caving even starts! Sure, your pack may be huge — in excess of 20 kg with overnight gear — and the hike horrendously steep, but this is all part of the sport of caving in the Rockies. It helps ensure only fit, capable cavers venture into Rockies caves.

Apart from difficult terrain, two other hazards of accessing Rockies caves are tricky river crossings and bears. Both these potential obstacles can be overcome provided you are not alone. Holding of hands and the use of long sticks will get you across the shallower parts of most rivers, although the author has on occasion ended up swimming in both the Moose and Whirlpool rivers. Make sure you can shed your pack in a hurry if you have to. Much has been written on bears, particularly grizzlies, and how to avoid them. The simple solution seems to be to travel in a group of three or more. The national park information centres hand out free literature on this topic.

Helicopters have an impact. They are a major imposition on the peaceful mountain environment. You are not likely to see any of those charismatic carnivores or ungulates for a few days after flying in to a cave, and hikers in the area are not going to be friendly. However, in really isolated areas (i.e. Bocock, Bastille, Dezaiko, Moon River) helicopters may be the only way in. Though as logging proceeds inexorably into the wilderness it may be that trailheads will become close enough to the caves to make foot access feasible (this may now be the case with Bocock and Dezaiko Plateau caves).

Cost wise, helicopters may be reasonable if they are already working in the area and you don't have to pay the staging costs. The larger helicopters are way out of the price range for most cavers. However the 'A' Star and the Bell Jet Ranger can carry four to five passengers at a time. Passengers are not carried when caving equipment is slung in (carried in a net underneath), so you are looking at at least two flights. For thirty minutes flying time with no staging costs you are looking at between $500-600. The price can be considerably less if the pilot can fit you in between jobs.

Remember that mountain weather is fickle and helicopters can only fly during the day when the cloud base is high. Also, flying in mountains with no established landing areas can be hazardous. Follow the pilot's instructions carefully — don't slam the doors and don't walk into the tail rotor!

It is tempting to bring everything when flying, but make sure everything either gets flown out or carried out at the end of the trip.

Know that helicopters are not usually allowed in national parks, though Parks Canada uses them extensively.

For the northern Rockies contact Yellowhead Helicopters at Valemount, (250) 566-4401. Alpine Helicopters fly out of Golden (250) 344-7444 and service Mistaya Lodge. They also make regular flights from Canmore (403) 678-4802 into Mount Assiniboine Provincial Park.

Overnighting

For most caving areas camping in the alpine is part of the experience, and allows the caving to take place at a more leisurely pace. Often two or three days are required to reach and explore the caves and karst areas.

Wilderness Camping

For caves located in Provincial Forest lands, there are few restrictions on camping. However, provincial parks and national parks tend to have a stronger environmental protection mandate, and usually require you to camp in designated camping areas. They may issue you a bivouac permit if you are camping above the tree line. See the section on jurisdictions.

The majority of caves are located in the alpine, just above the tree line, a fragile ecological zone that does not easily recover from repeated use. Use a stove rather than an open fire, bury your fecal waste and burn or carry out your toilet paper. In short, use minimal impact camping techniques. One of the joys of caving in the Canadian Rockies is the pristine scenery, often with no sign of human impact. Please keep it this way. Never camp in the caves or at cave entrances.

To avoid attracting bears do not cook in the tent you sleep in and store food some distance away.

Mountain Huts

Occasionally, mountain huts are located conveniently close to caves or cave areas. This includes the Fish Lake Cabin (Top of the World) (250) 422-4200, the Naiset Cabins and Hind Hut (Assiniboine) (250) 422-4200. For the Elizabeth Parker Hut (Lake O'Hara), Fryatt Creek Hut, Wates-Gibson Hut (Chrome Lake), Canmore Club House and Wheeler Hut (Rogers Pass) call the Alpine Club of Canada at (403) 678-3200.

Avoiding Problems

- If you are a novice caver contact the Alberta Speleological Society or Canadian Rockies Cave Guiding and go on caving trips with experienced cavers before attempting any caves involving drops. Practise self-rescue techniques and contact BCCR for information on rescue courses. Get all possible information on the cave you are visiting (including a map) before you go. Note that very few Rockies caves are suitable for the inexperienced. See the appendices for a list of caves suitable for novice cavers.

- Cavers should be familiar with rugged mountain terrain and have good navigation skills.

- A group of three works well as a minimum. In the event of an accident, one person can get help while the other stays with the injured person. Maximum group size depends on the nature of the cave and experience of the cavers. However, more than five tends to move slowly, especially if pitches are involved as they are in most Rockies caves.

- Any caver who tells you they've never been lost underground is lying. The trick is not to panic and have sufficient light and energy to deal with unexpected delays.

- Always tell someone where you are going (name of cave and route you intend to take) and when you expect to be back. Include details regarding your vehicle: where it will be parked and the number plate.

- Carry a cell phone. While this can be a life-saving device, bear in mind it will not work underground or from many surface locations in the Rockies. Satellite phones would be useful equipment for expeditions to more isolated areas and would speed up cave rescue callout in the event of a serious accident.

In the Event of an Accident

British Columbia Cave Rescue (BCCR) coordinate all major cave rescues in Western Canada. Their callout number is 1-800-663-3456. Keep this telephone number inside your helmet. The cave rescue seminars and regional orientations held by BCCR are highly recommended. For further information check out their web site at www.cancaver.ca/bccr.

Because it will be at least 12 hours from the time of callout before the BCCR arrive at the cave entrance, it is best to get yourself out if at all possible. Self-rescue is the name of the game. Organizing a mechanical advantage hauling system is central to self-rescue. There is no point in carrying the equipment unless you know how to use it. Practise before you go underground.

Cave rescue is an incredibly slow and arduous process, and incapacitating accidents occurring only a short distance into a cave can involve large numbers of rescue personnel and many hours of waiting. Serious injuries deep underground will probably result in death, either through trauma or hypothermia.

Remember — there are no local cave rescue organizations for the Canadian Rockies.

BCCR: 1-800-663-3456

EQUIPMENT

Caves in the Canadian Rockies are colder than most U.S. caves; we can only look with envy at the T-shirt and shorts-clad TAG (Tennessee, Alabama, Georgia) cavers. On the other hand, caves are not as wet as those in Britain and the full wetsuits traditionally used by Brits would be horribly hot and constrictive over here. PVC-type suits worn by many European cavers, although adequate, also have shortcomings (see below). Consequently, Canadian cavers have developed their own unique style of cave wear.

Preparation is the most important item on your equipment checklist. Play with your caving gear at home, and become familiar with the construction of your caving lights. Next time they go wrong underground, you will be able to fix them more quickly.

Likewise, maintenance is very important. Wash all your gear after every caving trip. This avoids transferring mud from one cave to another, and also ensures everything will work on your next trip. I use the high-pressure sprayers at a car wash; this is fast and avoids domestic disputes caused by blocking your drains at home. Various rope-washing devices can be easily constructed or purchased. Store everything by hanging in a dry place out of direct sunlight. Metal equipment, especially ladders and carabiners, are prone to corrosion if stored damp. Never store ladders in garbage bags, and always unclip carabiners of different metals to prevent corrosion through electrolysis.

Other than items manufactured at home by a few industrious cavers, purpose-made caving equipment is not currently available in Canada and must be purchased from the U.S. or Europe.

Coveralls

Because Canadian Rockies caves are cold (-1 to +4°C), damp and often muddy, wearing warm, protective clothing is important. Purpose-made caving coveralls are an essential item and are usually made of a Cordura-type fabric that is water resistant but not waterproof. The waterproof PVC-type suits are not suitable for Rockies caving: they tear easily and don't breathe, making them hot and wet inside. Although Rockies caves are damp, it is rare that you actually have to immerse in water. Some people use cotton coveralls, which are adequate for short dry trips where most of the caving is walking.

Undersuit or "Furry Suit"

For longer trips many cavers wear a one-piece pile suit under their coveralls. This helps keep you warm and prevents cold spots developing at the small of the back when clothing creeps up. They also provide some extra padding, much appreciated when moving through small passages. I have found the thinner variety best, because overheating accompanied by dehydration can be a common problem on longer trips. For many years I used a Farmer John-style pile outfit and found it a good compromise that allowed the upper body to breathe while keeping the lower body warm. A useful addition to the undersuit is foot loops that prevent the legs from pulling up under your coveralls during crawls.

Underclothing

Under your coveralls you should make sure all skin surfaces are covered. As with cross-country skiing, thin layers of synthetics (polar fleece or Lycra blends) or wool are best. Too much clothing can lead to overheating and dehydration. Experience will tell you what works best for you. Jeans, cotton clothing and bulky padded jackets do not work well.

Gloves

These protect your hands and keep them warm. If you are going into a particularly wet cave, they may need to be waterproof. If the gloves are too cumbersome they will make handling equipment difficult and you will have to take them off repeatedly and consequently may lose them. Insulated winter gloves or leather gloves are not suitable; they quickly become wet and unwieldy. I use orange "spun" gloves made of cotton with a rubber compound beaded on them. These are available for just a few dollars from most hardware stores. For wet caves these can be combined with green rubber waterproof gloves that reach over your wrists.

Kneepads

Many cave trips require crawling, and although not essential, kneepads will make caving far more pleasant. I find hockey kneepads work well, although purpose-made neoprene kneepads last longer, stay in place better and do not absorb water. If a lot of crawling is necessary, or if you have bad knees, buy heavy-duty roofing or skateboarding pads that have plastic or leather cups to protect your kneecaps. They do, however, restrict leg flexibility.

Boots

Good quality leather hiking boots with a stiff Vibram sole are essential both for caving and for getting to the cave.

Helmets

The helmet is an essential item for any caving trip. It protects your head, not so much from falling rock — a common occurrence on pitches — but from banging your head on the ceiling in low passages. It also provides a place to mount your light. Climbing helmets are suitable; some already have metal hooks and elastic necessary for mounting Petzl lights. The helmet should fit snugly and not move around, even with the light mounted. A secure chinstrap is essential.

Lights

More than one light source per person is essential. I suggest having three and mounting both your primary and secondary lights on your helmet. Cavers used to use carbide lamps, but these are messy and hard on the cave environment. Nowadays, electric lights are more common. If you do much caving, investing in a rechargeable electric is a good idea. I have used FX Speleotechnics lights, which give between 7 and 14 hours of light depending on the model. The lighter, cheaper Petzl lights, although not very bright, are adequate for easier caving or as a second light source, and are locally available. With a flat 4.5v battery and regular bulb they will run for approximately 11 hours; with a halogen bulb they last for less than 3 hours. Make sure you carry spare bulbs and batteries.

A third light source can be a small flashlight carried on a cord around your neck; the metal Maglites are good for this. Recent innovations in white LED (light emitting diode) lights for caving are promising. Models are now on the market that provide useable light for more than 24 hours. Although not bright enough to use as a primary light, their longevity and compactness make then ideal back-up lights.

Backpack

Most caving trips will require at least a small pack in which to carry spare lights and batteries, food and drink, and a first aid kit. Longer, harder trips will require a larger pack to carry ropes and SRT gear. Whatever the size, it is worth investing in purpose-made caving packs constructed from heavy duty, smooth, abrasion resistant materials similar to those made for hauling on big-wall climbs. They should have straps for wearing on your back, and a handle for carrying through low passages. Extra loops or compression straps should be avoided as they catch when the pack is being dragged. Purpose-made cave packs have a smooth cylindrical shape and a hole in the bottom to let out the water on waterfall pitches. They come in a wide variety of sizes. For delicate equipment such as cameras or scientific monitoring gear, Pelican boxes, modified by the addition of a carrying strap, work well.

Ropes

Stretchy dynamic climbing ropes are not suitable: they are delicate, have little abrasion resistance and absorb water. Static ropes such as Bluewater are designed for caving, and come in 11 mm or 9 mm diameter. The 9 mm is lighter, but less abrasion resistant and should only be used by experienced cavers familiar with rebelay techniques. Rope pads or rope protectors are occasionally used to protect a rope where it runs over an edge. These are useful as a temporary measure: for example, when descending a shaft for the first time to check it out. If a pitch is to receive much traffic, rebelays should be arranged. The rigging of ropes in caves, as with anchors, is a skilled activity that requires knowledge and experience.

Ladders

Ladders designed for caving are made of steel and aluminum and come in 8 m lengths that can be joined together using C links. They should always be used in conjunction with a rope to lifeline persons up and down. More popular in the past, ladders have been superseded by Single Rope Technique (SRT), whereby a rope is climbed using ascenders (see ascenders).

Harnesses

A sit harness is essential if you are doing any vertical caving. Climbing harnesses are adequate, but tend to wear out quickly unless worn under your coveralls. Caving harnesses are more robust — especially the leg loop supports — but are not available locally. Many cavers also use a chest harness when climbing ropes using SRT technique.

Descenders and Ascenders

There are three main types of descending gear used in rappelling. The Figure Eight is light and simple, but tends to twist the rope, is tricky to lock off and wears out quickly. It has inadequate heat dispersal properties for longer pitches. Many U.S. cavers use the Rack or Brake Bar. These take practice to control properly and are heavy and tricky to lock off. The Bobbin or Petzl Stop is probably the best available descender for caving, although it takes practice to attach it correctly to the rope. They are relatively light, and merely releasing a handle can lock off the Stop version. None of the devices used by climbers for belaying and descending work well for caving.

Climbing back up ropes is an essential element of Single Rope Technique (SRT). Ascenders attached to your harness and foot loops with slings allow you to climb the rope. Numerous devices are used: Jumars, Petzl Jammers, Crolls and Shunts. I use a Jumar for my leg loops and a Petzl Shunt for my chest ascender. Practice and perfect equipment set-up is very important. An inch slack in the system can turn a straightforward 100 m-high rope climb into an exhausting experience. For pitches over 100 m it is worth considering a rope walking system that incorporates an extra ascender, usually worn on the ankle.

For a review of devices check out *On Rope*, an NSS publication on caving equipment and techniques.

Anchors

Most Rockies caves contain drops that require ropes or ladders. Attachment points (anchors or belays) can be natural, such as threads, or artificial, such as bolts. The placement and use of anchors requires knowledge and experience beyond the scope of this book. The author suggests a course on climbing anchors together with a review of vertical caving techniques. In order to avoid the proliferation of bolts such as has occurred at pitch heads in many European caves, Rockies cavers are starting to install glue in anchors in more popular caves. These are durable, and can be replaced without drilling further holes. Please don't place any bolts underground without conferring with the caving community.

Emergency Equipment

Much can be done by experienced cavers to assist injured cavers out of a cave. The basic equipment of pulleys and prussics should be carried on all caving trips involving pitches. I strongly recommend you attend a crevasse or cave rescue course in order to learn the basics of mechanical advantage hauling systems.

A large, thick plastic bag will help reduce heat loss and takes up little space in the bottom of your pack. A small first aid kit is all that realistically can be carried underground. It should be packed in a durable waterproof container or it will be useless after one trip.

Cavers usually keep a stash of rescue equipment — known as a "bash kit" — in a vehicle at the trailhead. The bash kit contains supplies to stabilize an injured caver during the long wait for rescue.

JURISDICTIONS & PERMITS

Caving is a relatively new activity among the list of recreational pursuits land mangers have to deal with. For many jurisdictions, protocols for access to caves have not yet been established. For the harder caves, you may be asked for affiliation to a caving club, or proof of caving experience. Some land managers may wish to accompany you to learn more about the caves — and to make sure you don't damage the caves. Be patient, apply well before your intended trip date and make sure you don't sour relations for future cavers.

The caves fall mainly into three jurisdictions: national parks, provincial parks and provincial forests.

National Parks

The National Parks General Regulations, section 8, state that except where it is indicated by a notice posted by the superintendent, it is illegal to enter any cave in a national park without permission from the park superintendent. This applies to Glacier, Yoho, Kootenay, Banff and Jasper national parks. Currently, caves in Wood Buffalo National Park are closed to the public.

Along with a permit, Parks Canada will give you a UTM grid reference of the entrance, assuming the cave is known to them.

In order to help streamline which caves are open to public visitation, a three-tiered access classification is being developed:

Class 1 Caves Access by Application.
These caves are not for recreational purposes. The caves may contain significant resources that cannot handle unlimited or even moderate use, or there may be special surface resources near their entrances or along the approach route that cannot tolerate significant human use. Any visit to this class of cave must contribute to the knowledge of the cave or give a net benefit to the cave. Only a small number of national park caves in the Rockies are in this category.

Class 2 Caves Access by Special Activity Permit.
The majority of the national park caves will fall into this class. There are some management concerns regarding the surface or in-cave resources, or the consequences of an accident. There may be seasonal use restrictions to protect bat hibernacula and other wildlife, or for public safety (closure/restriction due to avalanche control, bear activity etc.). Users may be advised of resources that require extra care, or what additional information a park may want them to gather during their visit.

Class 3 Caves Unrestricted Public Access.
These caves have few or no resource management concerns and/or low accident potential. In order to comply with the spirit of the regulations, a list of these caves will be posted and kept at the park administration office, warden office and trail office. These caves will not be advertised or marketed. The list will be available to the public upon request. No permit is required.

Obtaining a Caving Permit

A permit to enter a cave is required for all Class1 and 2 caves.

The few designated Class 1 caves require a written application detailing the purpose of your visit and what information will be gained from the visit. Reference checks and cave specific management guidelines may apply. These caves are for serious exploration and/or research, not recreation. Applications must be made at least two months prior to an intended visit.

Class 2 caves are suitable for recreation. A permit can be obtained from the relevant national park. Apply in advance; some parks may require more time, anything from one day to two weeks. The applicant will be the trip leader whose qualifications may be assessed. You will be asked for your name, address and phone number, vehicle make, model, colour and license plate number, the name of the cave, the number of people in the party and the date of visit.

Specific Regulations

There are rules specific for Nakimu Caves in Glacier National Park:

- You must have a permit to enter the closed area (same as cave permit).
- Minimum party size of six (to minimize chance of a bear encounter).
- Group must stay together while in the closed area.
- Access is limited to Balu Pass route.
- Cavers are encouraged to use camping facilities near closed area access (Balu Pass).
- No camping is permitted in the cave area.

Permits may take two weeks to process. Applicants are judged on cave conservation ethics. Placing fixtures or altering the cave in any way (e.g. digging) requires permission. Quotas have been established and may apply.

Other Permits Required

All national park users are required to pay a "Personal User Fee." The cost per person varies depending upon age and the specific park to be visited. Fees are also charged for wilderness camping. Campsite or bivouac locations may be restricted. Check with the park trail office for details.

Jasper National Park
tel: (780) 852-6155
fax: (780) 852-2369
e-mail: jasper_dispatch@pch.gc.ca

Banff, Kootenay and Yoho National Park
tel: (403) 762-1470
fax: (403) 762-3240
e-mail: banff_dispatch@pch.gc.ca

Glacier National Park
Box 350
Revelstoke, BC.
V0E 2S0
tel: (250) 837-7500
fax: (250) 837-7536
e-mail: revglacierreception@pch.gc.ca

Wood Buffalo National Park
For up-to-date information on closures, contact Mike Keizer, Client & Heritage Services Manager, tel: (867) 872-7956.

Provincial Parks

Provincial Parks allow caving, but each park has particular protocols that need to be observed. Some of the new provincial parks are still formulating cave management policies. Contact the relevant land manager before embarking on a cave trip.

Provincial Parks (BC)

For current information go to www.elp.gov.bc.ca/bcparks/explore/regions.htm.

- For Kakwa and Hamber provincial parks, and Close to the Edge and Evanoff protected areas contact the Prince George Area Office at (250) 565-7086.
- For Bocock Peak and Bowron Lakes provincial parks contact the Cariboo District Office at (250) 398-4414.
- For Mount Robson and Small River Caves provincial parks contact the Mount Robson Area Office at (250) 566-4325.
- For Top of the World, Whiteswan Lake and Mount Assiniboine provincial parks contact the Kootenay District Office at (250) 422-4200.

Provincial Parks (Alberta)

*For Peter Lougheed and Spray Valley provincial parks call (403) 591-6309.

Other Parks (Alberta)

- For Whitehorse Wildland Park call (780) 723-8554.
- For Plateau Mountain Ecological Reserve, Don Getty Wildland Park and Canyon Creek caves call Elbow District Ranger Station at (403) 949-3754.
- For Rat's Nest Cave Provincial Historic Site call 403-678-3522.

Provincial Forests

There are currently no restrictions on caving in Provincial Forests. However, contact the regional forestry office for information on access and camping.

BC Forests

- Icefall Brook contact the Recreation Officer, Columbia Region at (250) 344-7504.
- For Fernie area, Crowsnest Pass, Amos Cave (Elkford) contact Cranbrook Forest District (250) 426-1700.
- For the Cariboo Mountains contact the Peace-Liard District Office at (250) 787-3407.

Alberta Forests

- For Crowsnest Pass contact the Blairmore Office at (403) 562-3210.
- For Pinto Lake Cave and Nordegg area caves contact the Nordegg Area Office at (403) 721-3743.
- For caves in the Cadomin area other than those in Whitehorse Wildland call the Hinton area office at (780) 865-8267.

CAVE MANAGEMENT

"Visitors to Canadian caves will note the complete absence of any rubbish and minimal damage to not only speleothems but also to sedimentary deposits and breakdown areas. The Canadians take great pride in their unspoilt landscape and caves and it is the responsibility of visiting cavers to follow their example."
Deej Lowe et al., 1983.

For the majority of Canadian Rockies caves isolation continues to be the most significant factor in their preservation. A few of the more accessible caves in the foothills and front ranges have been adversely impacted by human visitation, i.e. Cadomin, Wapiabi, Rat's Nest and Canyon Creek Ice Cave, which have seen formation breakage, garbage and sewage pollution, spray paint and mud distribution. The bat population has been severely impacted in Cadomin Cave by thoughtless visitors lighting fires and partying underground. Caving is slowly increasing in popularity, especially in the Crowsnest area where Ptolemy Plateau is showing signs of heavy use. Volunteers from the Alberta Speleology Society are working with land managers to try and find management solutions to these problems.

Before humans came bumbling in, caves were essentially pristine sterile environments. Unlike the surface world where damaged vegetation will recover given time, and weathering processes will eventually soften the worst impacts, caves do not recover. This is especially true for most Canadian Rockies caves, which are low energy environments, long abandoned by the glacial meltwater streams that formed them. We can, however, minimize our impacts by following certain practices. Essentially, the rules for minimal impact caving are simple: do not leave any sign of your passing in the cave and do not alter the cave in any way. In practice this can be hard to achieve, but let me pass on some suggestions that cavers have found useful over the years.

Group Size
Larger groups tend to create greater impacts.

Route Marking
Experienced cavers do not leave spray paint arrows on the walls, build cairns or leave marker tape. They make note of memorable features at passage junctions, and treat temporary diversions as part of most caving trips. When mapping caves it used to be common practice to use a carbide lamp to mark survey stations. The smoked dots and numbers can still be seen in many caves. This is no longer considered acceptable, and old survey marks are slowly being removed.

Camping Underground
Don't do it. Camping in cave entrances or farther underground is extremely hard on the cave environment. It compacts floor sediments, pollutes cave water and leads to the dumping of human waste, grey water, food particles and other garbage, no matter how careful you are. Castleguard Cave is the only cave in the Canadian Rockies that is long enough to warrant camping underground.

Human Waste
If you have to defecate or urinate in a cave, carry it out with you. Use zip lock bags or commercially available products such as WagBag or ResStop for feces and toilet paper, and an old water bottle clearly marked for urine. Devices used by Himalayan climbers are available that enable women to peee into a bottle (my wife recommends you practise at home first). There is no excuse for polluting a cave with human waste. It is damaging to cave biota, and disgusting and hazardous for other cavers. Remember, there are no weathering processes or sunlight to mitigate sewage problems underground.

Garbage
Carry it all out with you, as well as any you see left by others. Be careful with any food you consume underground; avoid leaving food particles on the floor. Take snacks that don't crumble and scatter.

Respect Cave Formations
These include mineral formations as well as mud formations and sediments. Avoid touching or walking on any cave formation. Walking in a line and following previous footsteps can minimize damage. If you can't continue without trampling or breaking formations, don't go.

Cave Fauna
Caves containing bat roosts should not be disturbed during the winter months of October to April. This includes Guano Sauna, Cadomin Cave, Procrastination Pot, Wapiabi Cave and Bisaro But Beautiful. Avoid disturbing pack rats and harvesters (daddy long-legs), common in many cave entrances. When examined closely, many pools in Rockies caves have been found to contain isopods, amphipods and occasionally larger biota. Some of these life forms may be unique to cave environments. Avoid walking through pools underground and never place any waste in cave water.

Bones
Avoid disturbing animal bones and bone fragments. Never remove them from the cave. Report any large accumulations or unusual bones, and send pictures to the Provincial Paleontologist at Old St. Stephen's College, 8820-112 Street, Edmonton, Alberta T6G 2P8.

USING THIS GUIDE

Criteria for including a Cave

All caves in this guide are natural, formed by the dissolving of bedrock by water, the exception being a moulin explored in a glacier in the Small River karst. Moulins are temporary features, but are of interest to cavers because it is thought they may lead to passages in bedrock beneath glaciers.

In order to be included, a cave has to go out of daylight and contain some horizontal passage. This precludes hundreds of shafts found on karst plateaus which descend up to 30 m, only to end in glacial rubble or in a plug of snow and ice. The exceptions are a few short caves with historical significance i.e. Hole in the Wall in Mount Cory and Goat Cave in Sulphur Mountain. Caves such as these have provided enormous pleasure as hiking destinations over many years.

Other short, not particularly noteworthy caves in remote locations are also included. Justification is the possibility of more caves being found in the area. It is amazing how many caves have been discovered while searching for one already known, one example being Procrastination Pot, an impressive cave found while searching for the relatively diminutive Disaster Point Caves.

Organization of Caves

Wherever possible, caves are organized south to north by national or provincial park or recreation area. Caves in the Rockies are documented first, followed by caves in the Columbia Mountains and in Northern Alberta. Within each area the caves are listed according to trailhead. Occasionally, caves are listed according to the jurisdiction they are accessed from, rather than the jurisdiction they are located in. i.e. Pinto Lake Cave,

Map

Most cave descriptions include a grid reference for use with 1:50,000 topographic maps. However, in some cases this may be a little optimistic, and caution should be used as to their accuracy. A snow-covered plateau rarely provides easily identifiable features on which to get a fix, and Canadian Rockies topographic maps are unreliable concerning the position of treelines and glaciers. Nevertheless, the use of a GPS greatly aids the locating of cave entrances. National Parks policy is not to reveal the grid reference. It will be given to you when you apply for a caving permit.

A few caves in this guide have no map information. These are the "lost caves." They are definitely out there somewhere, but the original discoverers cannot remember where they are exactly, or in a few cases, choose not to divulge the information. I have included them anyway in the hope they will be rediscovered and we can fill in the blanks.

Jurisdiction

Refer to the introductory section titled "Regulations & Permits" for a breakdown on regulations for each area.

Number of Entrances

The number of entrances, unless otherwise specified, refers to those large enough to permit human access (debatable in some cases). Yorkshire Pot at the last count has eight entrances, but because each entrance has long been identified as a separate cave with its own exploration history, they have been listed as separate caves

Length, Depth

Some caves have passages both higher and lower than the datum point at the entrance. These are combined in the cave depth figure, but may be broken down into + (above the entrance) and – (below the entrance).

Discovered

Crediting certain persons or an expedition with discovering a cave is fraught with pitfalls. Aboriginal peoples, European settlers and explorers may have known about some caves, but without documentation credit must be given elsewhere. Non-cavers play a large part in pointing out possible cave entrances, but discovery goes to those who actually establish the existence of an enterable cave. Despite potential inaccuracies, I decided to include discovery credits because they provide an information source for caves that may only have been visited the one time. It also helps to have a name to curse when staggering around on a steep mountainside in a blizzard looking for a small hole. It is the discoverer's prerogative to name a cave, and these have not been subject to editing.

Footprints in a cave are a sure sign a cave is not a new discovery. However, even experienced cavers have surveyed a cave and subsequently found a map already exists. This is the sure way to get on the list of Golden Garbage Can recipients.

Access

Descriptions and access maps showing road access to the trailhead, trail access and approximate cave locations are found under each section. Where campsites, or in a few instances, huts are within reach of the caves, it has been noted.

Location

Once you are in the general area, this gives specific information on finding a cave. However, if neither you nor any member of your party have visited a particular cave before, your chances of finding it on your first trip are probably less than 50%. Experienced cavers know they are going to be spending many hours traipsing up and down mountainsides.

Warning

While nearly all caves have dangers such as loose rocks and drops, some caves and/or their approaches have particular hazards the caver should know about.

Exploration & Description

The accompanying description often describes the cave as observed during a particular visit. Water and ice levels can fluctuate significantly depending on the time of year. For example, in many Crowsnest caves the ice has melted dramatically in the last decade. Cave exploration in the Canadian Rockies has a relatively short history, and this guide is just a beginning that will hopefully get fleshed out with time. Information was often obtained from other cavers and has been accepted as correct.

Cave Softly

All caves and karst areas should be treated as sensitive to human impact. Those particularly vulnerable in some way have been noted.

Notes on Geology, Geomorphology, Hydrology, Speleogenesis and Fauna

Some karst areas and specific caves include much information on these subjects. Other caves have just a brief description incorporated with the cave description.

Surveys

Mapping caves is accomplished by using a compass, clinometer and tape measure. The data is then fed into a dedicated computer program (previously accomplished using log tables and ruler) that produces a "line plot" showing the cave location as a line in three-dimensional space. Passage detail is then added freehand. Some cave surveys include both plan and elevation views. A largely horizontal cave is best shown in plan view, whereas a vertical cave containing lots of pitches is best shown by elevation.

The British Cave Research Association (BCRA) has a grading system for cave surveys using six subdivided grades. Grade One is a sketch from memory, and Grade Six X involves the use of sophisticated laser-sighted survey equipment. Most of the surveys in this guide, unless otherwise indicated, are somewhere in-between, usually Grade Three, which is accurate to +/- 2.5° and +/- 50 cm. The differences in appearance of the many cave surveys in this guide are indicative of the different styles and abilities of the many cavers who drafted them. These range from tweed-clad A. O. Wheeler with his brass tripod-mounted theodolite in Nakimu Caves, to Joe Blow who forgot his survey book and pencil, and is scratching numbers Flintstone-style using rock fragments. Cave surveys take practice to interpret and are not as useful underground as one might suppose. Check whether pitch depths are given in feet or metres before packing rope for your trip.

NOTE: For the purposes of this book, text on some of the surveys has been enlarged for readability.

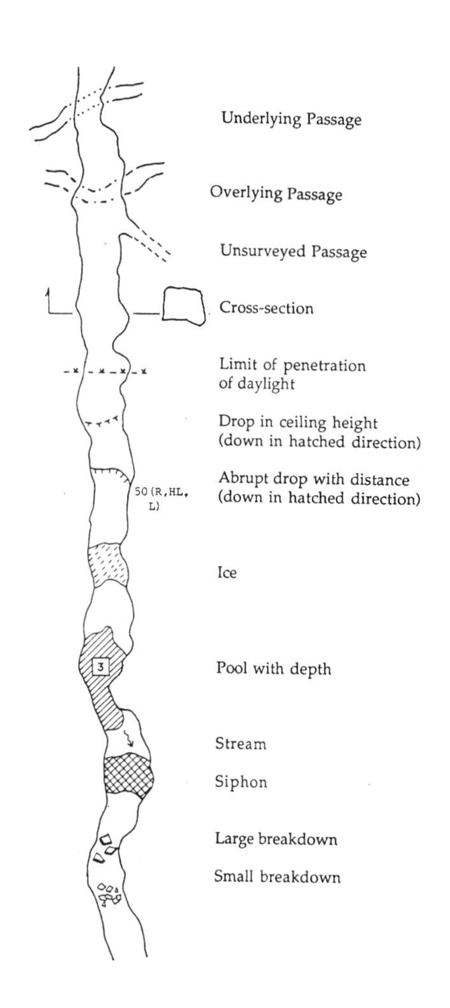

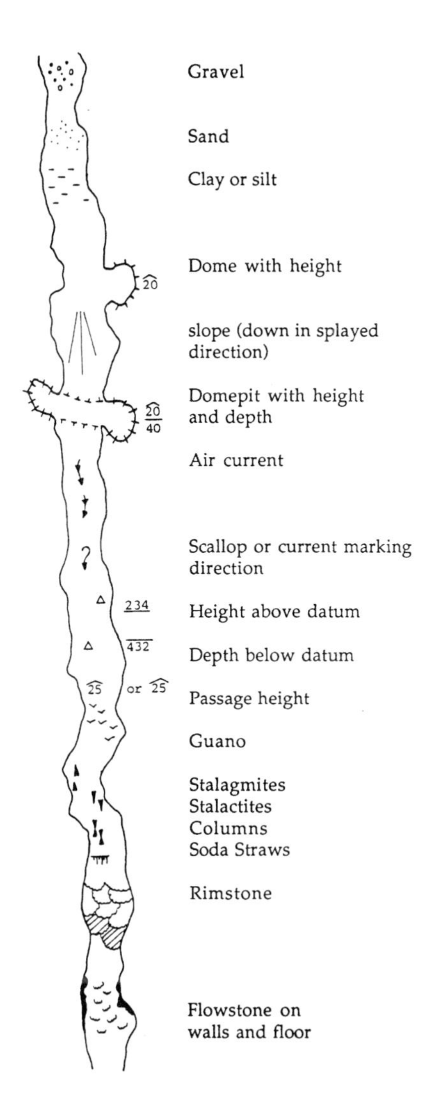

Cave Difficulty

In Europe, caves are often graded according to difficulty, much like rock climbs. I have not attempted to grade Canadian caves for a number of reasons. Access often involves hazards as great as bottoming the caves themselves. Many caves have only been visited a few times, making it hard to compare difficulty. And lastly, any cave involving pitches for which ropes are required are beyond the ability of the casual visitor, and the majority of caves in this book contain pitches.

Abbreviations used in text

Js^{-1}	joule (watt power)
l/m	litres per minute
l/s	litres per second
ka	thousand years
Ma	million years
m^3/sec	cubic metres per second
ACC	Alpine Club of Canada
ACRMSE	Anglo-Canadian Rocky Mountain Speleological Expeditions (1983/84)
ACRO	Alberta Cave Rescue Organisation
AFS	Alberta Forestry Service
ASS	Alberta Speleological Society
AWA	Alberta Wilderness Association
BCSF	British Columbia Speleological Federation
BCCR	British Columbia Cave Rescue
BCRA	British Cave Research Association
BP	Before Present (1950)
CAJ	Canadian Alpine Journal
CC	Canadian Caver periodical
CPS	Canadian Parks Service
GPS	Global Positioning System
GSC	Geological Survey of Canada
ICCC	Imperial College Caving Club
JOSM	Journal of Subterranean Metaphysics (ASS newsletter)
LUSS	Lancaster University Speleological Society
MUCCC	McMaster University Climbing and Caving Club
NSS	National Speleology Society
PEP	Provincial Emergency Program
SAR	Search and Rescue
SRT	Single Rope Technique
VICEG	Vancouver Island Cave Exploration Group
WLH	Working Level Hours

References used in text

Alberta Wilderness Association, *Eastern Slopes Wildlands Our Living Heritage.* A.W.A., 1986

Angerilli, N. & Holmberg R., *Harvestmen of the Twilight Zone.* Canadian Caver 17, 1984

Atkinson, T. C., Smart, P. L. & Wigley, T. M. L., *Climate and Natural Radon Levels in Castleguard Cave, Columbia Icefields, Alberta, Canada.* Arctic and Alpine Res. 15, 1983

Atkinson, T., *Chimney Effect Winds in Caves.* Canadian Caver 17. 1984

Balch, E. S., *Glaciers or Freezing Caverns.* Lane and Scott, 1900.

Barton, T., *Crowsnest Spring Dives.* Canadian Caver 13(2), 1981

Barton, T., *Rat's Nest Dives.* Canadian Caver, 13(2),1981

Beckmann H., *Fire Hose Pit.* JOSM 5 May, 1988

Beers, D., *The Magic of Lake O'Hara.* Rocky Mountain Books, 1981

Bellemare, Y., *Yves' Drop, Wilcox Col.* JOSM 119, 1995

Boles, G. W., Putman, W. L. & Laurilla, R. W., *Place Names of the Canadian Alps.* Footprint Publishing, 1990

Boon, J. M., *Down to a Sunless Sea.* Stalactite Press, 1977

Boon, J. M., *Toad River.* Canadian Caver 12(2), 1980

Boon, J. M., *Diving in Crowsnest Spring.* Canadian Caver, 5 (1), 1973

Borneuf, D., *Springs of Alberta.* Earth Sciences Report 82-3, Alberta Research Council, 1983

Brophy, J. A., *Sinkholes, Swallow Holes and Sunken Valleys: The Gypsum Karstlands of Wood Buffalo National Park., The Natural History of Canada's North: Current Research.* Occasional Papers of the Prince of Wales Northern Heritage Centre No. 3, 1988

Brown, M. C., *Karst Geomorphology and Hydrology of the Lower Maligne Basin, Jasper, Alberta.* Unpublished PhD thesis, McMaster University, 1970

Brown, M. C. & Goodchild, M. F., *A Brief History of Exploration and Commercial Developments of Nakimu Caves, Glacier National Park, British Columbia.* McMaster Report to the National Parks, 1971

Burns, J. A., *Fossil Vertebrates from Rats Nest Cave, Alberta.* Canadian Caver 21(1) 1989

Burns, J. A., *Mid-Wisconsian Vertebrates and their Environment from January Cave, Alberta, Canada.* Quartenary Research 35 130-143 1991.

Burns, P., *Vertical Development of Castleguard Cave.* Canadian Caver 12(1), 1980

Coward, J., *Caves in the Ptolemy Creek Area.* Canadian Caver, 1, 1969

Crystalline Group, The, *Calgary In Winter.* The Faculty of Environmental Design, Univ. of Calgary, 1988

Drake, J. J., *Hydrology and Karst Solution in the Southern Canadian Rockies.* Unpublished PhD thesis, McMaster University, 1974

Drake, J. J., Stein, R. & Lewis, R., *The Gypsum Karst of Wood Buffalo National Park.* Canadian Caver 11(2), 1978

Drake, J. J., *The Geomorphic Implications of the Geohydrology of Gypsum Karst Areas.* Unpublished Masters thesis, McMaster University, 1971

Drummond, I. & Coward, J., *The ASS Cave Radio.* Canadian Caver 13 (2) 1981

Drummond, I., *Icefall Brook.* Canadian Caver 9(2), 1977

Ellis, B., *Surveying Caves.* British Cave Research Association, Somerset, England, 1976

Evans, M. & Worthington, S., *A New Survey of Yorkshire Pot.* Canadian Caver, 21(1), 1989

Forbes, J., *Heave Ho!* Canadian Caver, 20(1), 1988

Ford, D. C. & Williams P., *Karst Geomorphology and Hydrology.* Unwin Hyman, 1989.

Ford, D. C., *Alpine karst in the Mount Castleguard - Columbia Icefield Area, Canadian Rocky Mountains.* Artic and Alpine Research, 3, 1971

Ford, D. C., Smart, P. L. & Ewers, R. O., *Physiography and Speleogenesis of Castleguard Cave, Columbia Icefields, Alberta, Canada.* Arctic and Alpine Res., 15, 1983

Gadd, B., *Snaring Karst Exploration, Jasper National Park.* An unpublished report for Parks Canada by the ASS, 1983

Gadd, B., *Handbook of the Canadian Rockies.* Corax Press, 1983; Second Edition 1986.

Gascoyne, M., Latham, A. G., Harmon, R. S. & Ford, D. C. *The antiquity of Castleguard Cave, Columbia Icefields, Alberta, Canada.* Arctic and Alpine Res., 1983

Gillieson, D., *Caves Processes, Development, Management.* Blackwell, 1996

Grundy S. & Hatherley, P., *Caribou Mountains Expedition.* Canadian Caver 18(2), 1986

Harmon, R.S. & P. Thompson., *Late Pleistocene Paleoclimates of North America as Inferred from Stable Isotope Studies of Speleothems.* Quaternary Research 6, 1978

Harris, S. A., *Ice Caves and Permafrost Zones in Southwest Alberta.* Erkundi 33, 1979

Hill, C.A. & Forti, P., *Cave Minerals of the World.* Natl. Spel. Soc., Huntsville, 1986

Holland, H. D., Kirsipu, T. W., Huebner, J. S. & Oxburgh, U. M., *On Some Aspects of the Chemical Evolution of Cave Waters.* J. Geol. 72, 1964

Holsinger, J.R., Mort, J.S. & Recklies, A.D., *The subterranean Crustacean Fauna of Castleguard Cave, Columbia Icefields, Alberta, Canada, and its Zoogeographic Significance.* Arctic and Alpine Res. 15, 1983

Humphrey, S. R., *Disturbances and Bats.* Oklahoma Underground 2, 1972

Jennings, J. N., *Karst Geomorphology.* Basil Blackwell, 1985.

Kendall, A. C., *Continental and Supratidal Evaporites. Facies Models.* Geological Association of Canada, 1985

Kinney, G. & Kain, C., *Arctomys Cave.* Canadian Alpine Journal Vol. IV, 1912

Lewis, R., *Cave Report*, submitted to Wood Buffalo National Park by R. Lewis, seasonal naturalist, 1974-75.

Li, W. X., Lundberg, J., Dickin, A. P., Ford, D. C., Schwarcz, H. P., McNutt, R. & Williams, D., *High-precision Mass-spectrometric Uranium-series dating of Cave Deposits and implications for Palaeoclimate Studies.* Nature 339, 1989

Lock, Harry et al. *Imperial College Caving Club Canada Expedition 1986.* ICCC, 1986

Lowe, D.J., *The Origin of Limestone Caverns: An Inception Horizon Hypothesis.* Unpublished PhD thesis, Manchester Polytechnic, 1992

Lowe, D.J. & C. J. Yonge., *The Anglo-Canadian Rocky Mountains Speleological Expeditions 1983 and 1984.* British Geological Survey, 1983/4

Lyon, B., *ACA Canadian Lynx - Geomorphological Report*, unpublished, British Army caving expedition to the Rockies, 1983

MacDonald, B., *Bastille Karst Expedition 1988.* Canadian Caver 21(1), 1989

MacDonald, W. D., *Holocene Paleoclimate of the Bow Corridor Determined from Speleothems.* Paper Submitted for course work, University of Calgary, 1991

MacDonald, W. D., *Stable Isotopes of Ice in Three Southern Alberta Ice Caves.*, Paper submitted for course work, University of Calgary, 1989

Marsh, J., A taped interview with Tommy Frame. The Archives of the Whyte Museum of the Canadian Rockies, July 5, 1980

Marshall, P., *Limestone Caverns.* Canadian Caver 5(2),1973

McEachren, M., *Current Research in the Upper Livingstone River Valley.* The Alberta Speleologist Vol. 1, 1972

McKenzie, I., *Heli-Caving in the Rockies.* Canadian Caver 15(1), 1983

McKenzie, I., *Close to the Edge.* Canadian Caver 18(2), 1986

McKenzie, I., *A New Survey of Yorkshire Pot.* Canadian Caver, 19(2), 1987

McKenzie, I., *Explorations near Crowsnest Pass.* Canadian Caver, 21(1), 1989

McKenzie, I., *Further Explorations in Snowslope Pot.* Canadian Caver, 22(1), 1990

McKenzie, I., *Hoodoo Creek.* JOSM 122, 1996

McKenzie, I., *Men in the Moon.* Canadian Caver 20(1), 1988

McKenzie, I., *The Backdoor Resurvey.* Canadian Caver, 21(1), 1989

McKenzie, I., *The Lost Caves of Sentry Mountain.* JOSM 127, 1997

McKinnon, N. A. & Stuart, G. S. L., *Man and the Mid-Holocene Climatic Optimum.* Proc. 17th Annual Chacmool Conf., Dept. of Archaeology, University of Calgary, 1987

McManus, J. J., *Activity and Thermal Preference of the Little Brown Bat, Myotis velifer.* Unpublished PhD dissertation, University of Arizona, 1974

Meinke, S., *Halfway Caves Revisited.* Canadian Caver, 18(2), 1986

Middleton, J. & Waltham, T., *The Underground Atlas.* St. Martin's Press, 1986

Miller, T., *Storming the Bastille.* Canadian Caver 21(1), 1989

Ministry of Forests, British Columbia. *Cave Management Handbook.* B.C. Forest Service, 1990

Mohr, C. E., *The World of the Bat.* Lippincott Company, 1976

Moore, G. W. & Sullivan ? ., *Speleology - The Study of Caves.* Clarendon, 1978

Morris, T., *Caving in the Canadian Rockies.* Canadian Caver 3, 1970

Mort, J. S. & Recklies, A. D., *The Biology of Castleguard Cave*. Canadian Caver 12(1), 1980

Mountjoy, E. W. & Price R. A., *Geology of Jasper, Alberta*. Geological Survey of Canada, map 1611A, scale 1:50,000, 1985

Muir, D., & Ford D. C., *Castleguard*. Canadian Govt. Publ. Centre, 1985

Nelson, J. S. & Paetz M. J., *Evidence for Underground Movement of Fishes in Wood Buffalo National Park*. Canadian Field Naturalist 88(2), 19??

Parks Canada, Glacier National Park. *Cougar Valley - Nakimu Caves*. Park Management Planning, Western Region. 1983

Peck, S. B., *A Review of the Cave Fauna of Canada, and the Composition and Ecology of the Invertebrate Fauna of Caves and Mines in Ontario*. Can. J. Zool. Vol.66, 1988

Pollack, J., *Two Dives in Bluebell Cave*. Canadian Caver 18(2), 1986

Prosser, D. W., *Cadomin Cave, Alberta*. Canadian Caver, 4(1), 1972

Pybus, M., *Bats of Alberta*. Alberta Forestry, Lands and Wildlife, 1988

Ridley, A., *Flop Pot*. Canadian Caver, 6(2), 1974

Rollins, J. L. & Yonge C., *Recent Developments in Castleguard Cave*. Proceedings of 9th Congreso Internacional de Espeleologia, Espana, 1986

Rollins, J. L., *Mount Ball Karst*. Canadian Caver 20(1),1989

Sawatsky K., *Pinto Lake Cave*. Canadian Caver 11(2), 1979

Sawatsky, K. D., *Diving in Castleguard*. Canadian Caver 19(2), 1987

Schroeder, J., *Inside the Glaciers - Svalbard, Norway*. Canadian Caver 22(1), 1990

Schroeder, J. & Ford, D. C., *Clastic Sediments in Castleguard Cave, Columbia Icefields, Alberta, Canada*. Arctic and Alpine Res. 15, 1983

Seale, R., A taped interview with Cecil Smith. The Archives of the Whyte Museum of the Canadian Rockies, July 5, 1980

Shaw, P., *Aquatic Cave Bugs*. Canadian Caver 19(2), 1987

Shreeve, J., *The Dating Game*. Discover, September 1992, p.82

Smart, C. C., *Quantitative Tracing of the Maligne Karst System, Alberta, Canada*. J. Hydrol., 98. 1988

Smart, C. C., *The Hydrology of the Castleguard Karst, Columbia Icefields, Alberta, Canada*. Unpublished PhD thesis, McMaster University, 1983

Smart, C. C., *The Maligne System*. Canadian Caver 4(3), 1971

Smart, C. C. & Huntley, D., *Glacier-Karst Research at Small River, B.C.* Canadian Caver 21(1), 1989

Smart, P., *Rat's Hole Cave*. Canadian Caver 5, 1971

Sweeting, M. M., *Karst Landforms*. Columbia University Press, 1973

Thompson, P., *?...from Alberta*. Canadian Caver 11(2), 1979

Thompson, P., ed., *Cave Exploration in Canada*. Canadian Caver Magazine special issue, Department of Geography, University of Alberta, 1976

Thomson, D., *Caving in the Rockies*. Canadian Geographic, Oct/Nov 1986

Thomson, D. ,*Mistaya Cave*. B. C. Caver 5(6), 1991

Tsui, P.C., Deformation, *Ground Subsidence and Slope Movements along the Salt River Escarpment in Wood Buffalo National Park*. Unpublished MSc thesis, University of Alberta, 1982

van Everdingen, R. O., *Physical, Chemical, and Distributional Aspects of Canadian Springs*. The Arctic Institute of North America, University of Calgary, 1991

van Everdingen, R. O., *Thermal and Mineral Springs in the Southern Rocky Mountains of Canada*. Environment Canada, Water Management service, 1972

Waltham, A., *Karst and Caves*. British Cave Research Assn., 1987

Waltham, A., *The World of Caves*. Orbis Publishing, 1976

Whitten, D. G. & J. R. V. Brooks., *The Penguin Dictionary of Geology*. Penguin Books, 1983

Whyte Collected letters, 1980. Whyte Museum of the Canadian Rockies

Wigley, T. M. L. & Brown, M. C., *The Science of Speleology*. Academic Press, 1976

Wigley, T. M. L. & Plummer, L. N., *Mixing of Carbonate Waters*. Geochim, Cosmochim Acta 40, 989-95, 1976

Wood, K., *The Hidden Caves* The Albertan, December 3, 1955

Worthington S. R. H., *Karst Hydrology of the Canadian Rocky Mountains*. Unpublished PhD thesis, McMaster University, 1991

Yonge C. J., *Which Way to the North Pole?* Canadian Caver 18(2), 1986

Yonge, C. J., *Shorty's Cave is Getting Longer*. Canadian Caver 18(1), 1986

Yonge, C. J., *The 1984/1985 Castleguard Expeditions*. Canadian Caver 17, 1985

Yonge, C. J. & Worthington, S. R. H., *Recent Explorations around Mount Robson*. Canadian Caver 17, 1984

Yonge, C. J., *Rat's Nest Cave: Provincial Historic Resource, 1991*. Unpubished report.

Yonge, C. J., *Shorty's Cave: The Connection to Yorkshire Pot*. Canadian Caver 18(2), 1986

Yonge, C.J., *Studies at Rats Nest Cave: Potential for an Underground Laboratory in the Canadian Rocky Mountains*. Cave Science Vol. 18 no.3, December 1991

Zacharda, M. & Pugsley, C. W., *Robustocheles occulta sp.n., a new troglobitic mite (Acari: Prostigmata: Rhagidiidae) from North American Caves*. Can. J. Zool. Vol. 66, 1988

Opposite: A well-defined frost line in Coulthard Cave, a classic cold-trap cave with icicles.
Photo Dave Thomson.

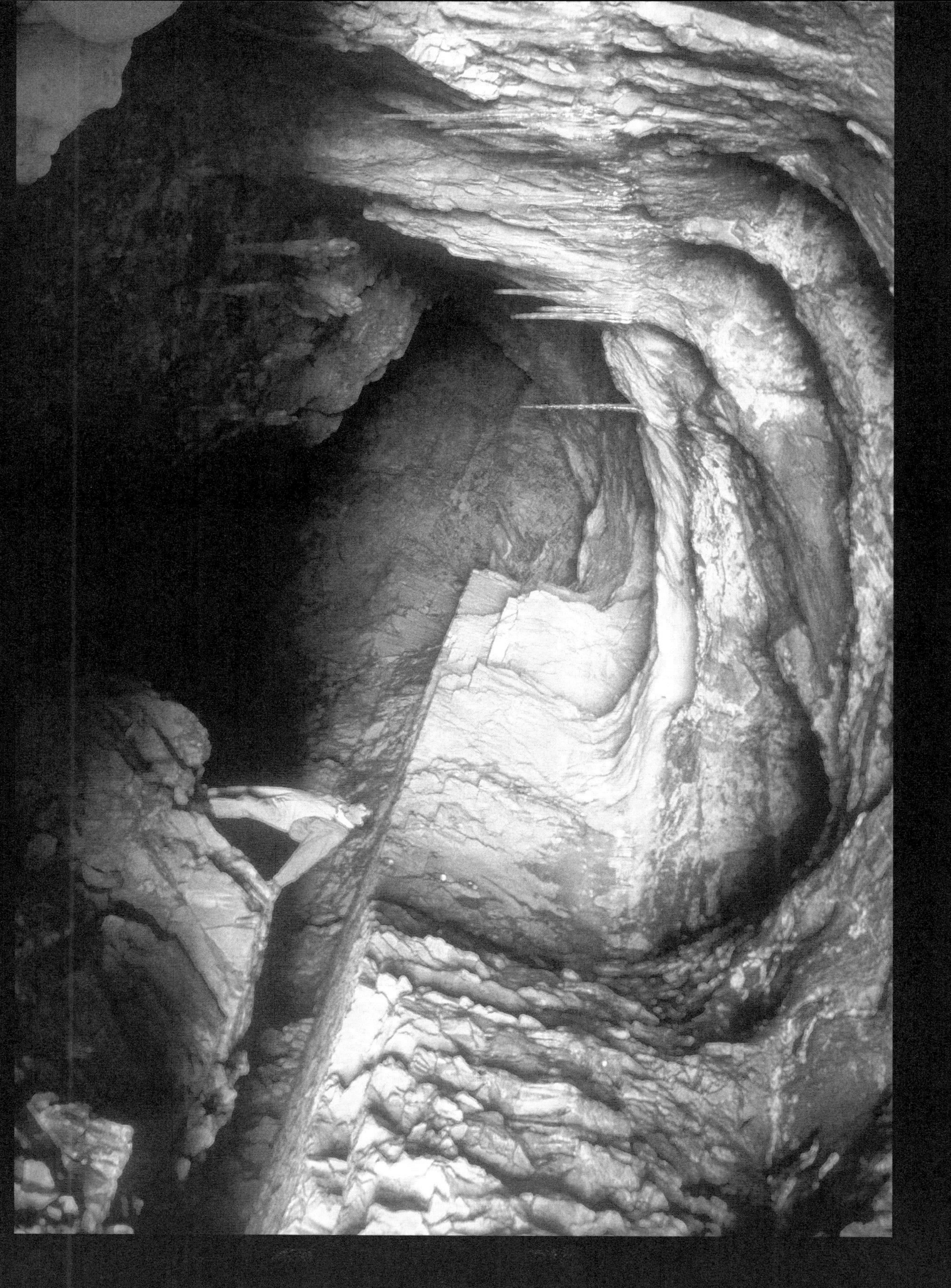

Cave Descriptions

CROWSNEST PASS

Crowsnest Pass, one of the finest areas of alpine karst in North America, contains the greatest known concentration of subterranean passages anywhere in Canada. It has now seen 30 years of cave exploration, with a current total of 39 caves discovered. Caves are located in five main areas: above North York Creek, on either side of Hwy. 3, above lower Ptolemy Creek, in and above the Ptolemy Plateau, and — farther to the south — in and above the Andy Good Plateau.

The caving season for Crowsnest Pass is from late July to October. The majority of entrances are hard to find and buried under snow for most of the year.

Most of the Crowsnest caves have a large vertical component making them accessible only to experienced cavers with equipment. There are, however, two caves that are accessible to inexperienced novices: Cleft Cave and the upper section of Gargantua.

Access for all Caves From Hwy 3 between the Municipality of Crowsnest Pass (Blairmore and Coleman) and the Alberta/B.C. boundary at Summit Lake. For details see area accesses.

Area Exploration McMaster cavers first visited the pass in 1965. A year later the McMaster University Caving and Climbing Club (MUCCC) was formed in Hamilton, Ontario. Mainly composed of ex-Brits, these experienced cavers were students of Derek Ford, a faculty member in the Geography Department since 1959. For over a decade up to the mid 1970s this group dominated cave exploration, not just at Crowsnest Pass but all over the Canadian Rockies.

1965 marked the first of many trips by the McMaster cavers to the Crowsnest Pass, during which several entrances were noted and return trips planned. At first, owing to obvious resurgence of Crowsnest Spring visible from the highway, reconnaisance focussed on the north side of the pass. In 1967 not much was found other than Pinnacle Pot, a blocked shaft just below Philipps Peak. However, when Mike Boon went for a

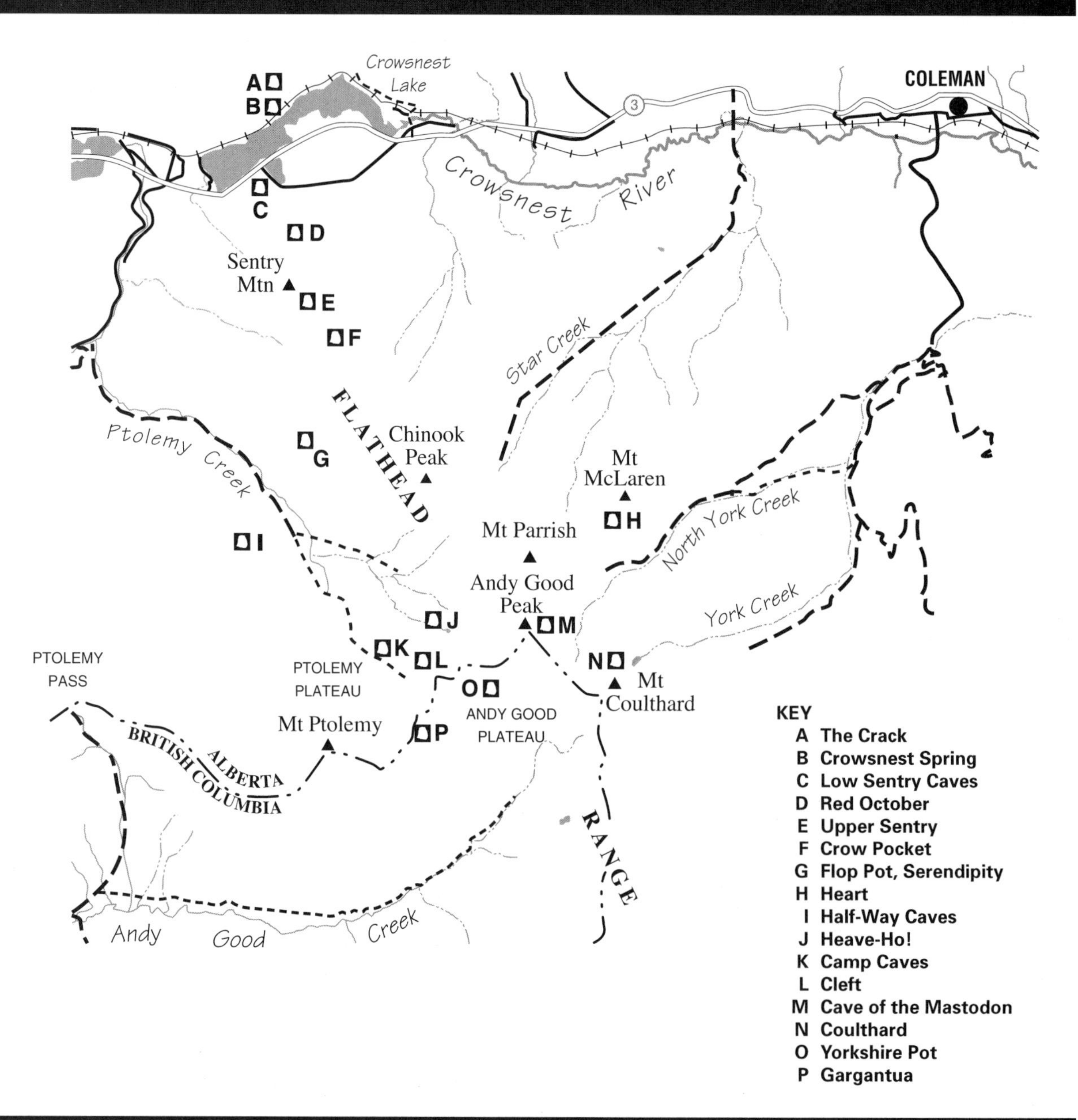

stiff walk on the south side of the pass he discovered High Sentry Cave high up on Sentry Mountain. The following year some of the road-visible entrances were checked, including Eagle Cave just above Crowsnest Spring and Low and Middle Sentry caves. The most promising discovery was Coulthard Cave, the entrance of which is visible from Blairmore.

It wasn't until 1969 that Tich Morris and Gary Pilkington finally walked up Ptolemy Creek to Camp Caves and "The Promised Land." Discoveries on the Andy Good Plateau included Halfway Caves, Ice Hall, Mendips Cave to the 7 m pitch, Yorkshire Pot (YP) to the 40 m pitch, Derbyshire Pot and Snowslope Pot to the bottom of the second (side) shaft.

In 1970 Gargantua was discovered, and all the major passages were explored with the exception of Winter Warren Series and Back of Beyond Series. YP was pushed via the Roller Coaster to the Tight White Way, and the Upper and much of the Lower Alberta Avenue were explored. YP was now 2026 m long and 344 m deep. This same year ASS members pushed Mendips Cave to a depth of 200 m.

In 1971 the Green Pool Series in YP was surveyed to a sump and the Rat Route pushed for 400 m. A winter trip to Gargantua resulted in 600 m of passage being surveyed. The inlets off Interprovincial Way were pushed in the hope of getting more depth, and the Winter Warren series was explored. Other caves discovered included Rat's Hole Cave and the short Double Pots Cave.

"We strode quickly down from the col in the light of a half-moon; the evening was cold but rudely pleasant, with only the occasional gust whipping snow crystals up to sting our faces. We were perhaps a little underdressed for the weather, but were in good spirits; we had had a good day, accomplished what we had come to do, and were still fairly fresh. As I gritted my teeth and blinked into another gust of snow, I felt as strong and alive as I ever have; one of those great days in the mountains. Caving has given me a lot of moments like that." Ian McKenzie, 1991.

In 1973 a YP winter (April) expedition was organised using a helicopter. After digging through 8 m of snow, 12 cavers camped underground for four days in the Green pool Series. The result was 0.3 m added to the total cave depth! YP had now been pushed to its current 384 m depth. The bypass to the Seven Steps was discovered, the P and F surveys were pushed and the drop below Chocolate Chamber was descended to The Horror Show that was surveyed to the Muddy Gulch Series. Shortly after the expedition ended, on May 4, an accident occurred when a group of Calgary cavers, unrelated to the ASS, had an accident in YP. One of the group fell down the second entrance pitch after the rope slipped off an inadequate belay point. A helicopter rescue was organised involving a team of miners from Blairmore. The same year Shorty's Cave was discovered.

The next year, 1974, ASS members discovered Flop Pot. In Gargantua, the Caterpillar Tube was surveyed and the pitches in the Middle Entrance dropped. YP passages and drops around the Horror Show and Muddy Gulch Series were surveyed. Shorty's Cave was pushed to the bottom of an 80 m shaft and believed finished. 1975 marked the end of the McMaster group domination of Crowsnest Pass cave exploration, most subsequent exploration being carried out by the ASS together with a few individual cavers from the original McMaster group.

In 1978 the Green Pool Sump in YP was dived for 13 m to a constriction, and the Muddy Gulch Series connected with Alberta Avenue. The next year Serendipity and Ice Chest were discovered and in Gargantua the Dye Streamway and Lower Streamway were surveyed. In 1980 the Backdoor was discovered and connected to Mendips and YP at Leprechaun's Leap — unfortunately, the survey notes were lost. In 1981 Crowsnest Spring was dived by University of Calgary geology student Tom Barton, who was an important motivational force behind many new cave discoveries at Crowsnest Pass in the late 1970s and '80s.

In 1983 a lead in Gargantua was pushed for 200 m off B.C. Chamber (not shown on survey), and in 1984 an unsuccessful attempt at recording music in Boggle Alley provided a helicopter to carry out rope and garbage left behind by the 1973 YP expedition. In 1984 Betelgeuse was discovered.

In 1986 Gargantua's Waterfall Exit was discovered by using a cave radio, and the first "pull-down" through trip was made. An exit to Winter Warren series was discovered from the surface, a small hole in the cliffs above Waterfall Exit. Gargantua now had five entrances. Shorty's Cave was extended and eventually joined to YP at the F survey and through trips were made. Elsewhere in YP, the Crowbar Extension and P survey were pushed, and the Backdoor to YP through trip made. That same year Upper Halfway Cave was extended. In 1987 a Speleofest was held at Crowsnest Pass, with 60 cavers from many countries attending. During this event Little Moscow and Heave-Ho! caves were discovered and YP was completely de-tackled for the first time since 1973. The F survey was pushed to a loose conclusion.

In 1988 Merlin Cave was discovered and Mendips Cave entrance was found to be free of ice for the first time since 1971. Ian McKenzie — motivational force behind much of the exploration at the time — finally surveyed the cave to a connection with the Backdoor and thus to YP. Ian also discovered Snowslope and in 1991 connected it to F survey in YP. Also in 1991 a second speleofest was held at Crowsnest Pass, with 30 cavers attending. Quinta Penta Pot was discovered at this time.

In 1993 there was an accident in Gargantua at the top of the first (56') pit. A falling rock caused a back injury and necessitated a stretcher recovery to the entrance. The injured person was flown out the following morning. In 1994, with helicopter support from the BC Forest Service, garbage was removed from the old 1973 winter camp in YP. Quinta Penta Pot was connected with the Horror Show and Surprise Streamway in YP by Taco van Ieperen, who later hauled a novice caver, unable to SRT, up the 180 m of entrance pitches out of YP. LUSS cavers connected the No Bar extension in YP with Shorty's Cave via the Yorkshire Gripper. The Crack was excavated above Crowsnest Spring. A sign and logbook were placed at the head of Ptolemy Creek and caver counters were installed in Cleft Cave and Gargantua.

In 1995 a lead in Mendips Cave was connected with Quinta Penta Pot. The road up Ptolemy Creek was rendered temporarily impassable due to extensive flood damage. In 1996 Exhibition Way in YP was connected with Inhibition Way in Snowslope Pot. In 1997 Heave Ho! was connected via a dig with YP. Low Sentry and Middle Sentry caves were pushed and surveyed. Red October was discovered and surveyed. Gargantua Cave was equipped with stainless steel eco-hangers and the Rat Route pushed in YP. In 1998, during the third Speleofest, through trips were made from YP to Heave-Ho! and from Heave-Ho! to YP (potential Golden Garbage Can recipients). An attempted connection between Little Moscow and YP, thought to be very close, was unsuccessful. Jason Morgan, a recent ASS member and a dynamo of energy, discovered several new caves.

Although the focus of exploration has switched to the Flathead Range near Fernie, the Andy Good Plateau continues to be one of the best places in the Canadian Rockies to find new cave passages. As Ian McKenzie, an ASS caver who has masterminded many of the recent finds commented in 1989, "There are perhaps a thousand surface leads and digs nearby, and about a hundred unsurveyed leads in the cave itself [Yorkshire Pot]."

Jon Rollins, Randy Spahl & John Donovan on the Andy Good Plateau.
Photo Dave Thomson.

Area Geology Mts. Andy Good and Ptolemy are part of the High Rock Range. This range was formed by the Lewis thrust fault, which resulted in a northeasterly displacement of Paleozoic limestones, shales and dolomites over younger Mesozoic rocks. As a result of the thrust, the massively bedded cavernous limestones dip at an angle of 20-30° to the southwest. At Yorkshire Pot the total thickness of the limestone is approximately 600 m. The upper member of this section is the Livingstone Formation (180 m of Mississippian age), a massively bedded calcarenitic limestone with thin chert interbeds near the base. This resistant formation forms the steep cliffs that are evident along Ptolemy Valley. The Livingstone conformally overlies the Banff Formation (180 m thick), a crystalline limestone that downgrades into cherty calcareous shale. Beneath the Banff Formation is the Exshaw Formation, an aquitard composed of black, non-calcareous shale of varying thickness. The lowest member of the series is the Palliser Formation (213 m thick), another resistant, cliff-forming limestone. It is thickly bedded, fine-grained limestone that has been extensively altered to dolomite, particularly in the basal sections. The Palliser Formation unconformally overlies Cretaceous sandstones and shales. Yorkshire Pot is developed in the Livingstone and Banff formations. (Adapted from Peter Thompson and Julian Coward, *Canadian Caver*, 1973.)

Area Hydrology Steve Worthington (formerly of McMaster University) spent the summers of 1985-1987 studying the groundwater flow at Crowsnest Pass for his PhD. He identified five principal springs in the area, including Ptolemy Spring, located at 1600 m just west of Ptolemy Creek. It can be reached by following the resurgence stream up from the third creek crossing on the Ptolemy seismic road. The water emerges from a bank of gravel and boulders. Dye traces carried out by Steve and others has confirmed that Gargantua and Yorkshire Pot, as well as other locations on the plateau, drain to this spring. An exception is the east side of the plateau, which possibly drains to Parish Spring in the north cirque. Flow-through times from Yorkshire Pot and Gargantua to the Ptolemy Spring vary from one day at peak discharge to months during the winter.

Top: Steve Worthington. Photo Dave Thomson.

Bottom: Ian Mckenzie, Gille Roy and Jon Rollins on Ptolemy Plateau. Photo Dave Thomson.

Cave Fauna Huge volumes of pack rat droppings have been found in the entrances of some caves, notably Rat's Hole Cave. Small bundles of plant material can be found on ledges close to cave entrances in the summer, the pack rats drying out their summer forage. A salamander was seen in the Green Pool Series near the bottom of Yorkshire Pot. In 2000 Heidi Macklin collected and identified some cave invertebrates in Crowsnest Spring and Low Sentry caves.

Cave Softly In 1979 Summit Lime Works, who operate two quarries in the Crowsnest Pass, filed an application with the Alberta government to operate a quarry in the Ptolemy basin. Their application was to quarry away the tip of the prominent ridge that separates the north and south cirques at the head of Ptolemy valley, barely 600 m away from Ice Hall and the Ptolemy benches. Following vigorous protests by cavers from all over the World, the application was denied. Cavers also opposed suggestions that the Great Divide Trail be routed through the Ptolemy Plateau on the grounds that it would compromise the isolation of the caves.

Fortunately, Cleft Cave and the accessible upper chamber of Gargantua contain little in the way of delicate mineral formations and are populated only by pack rats and swallows. Apart from some trail erosion and garbage (including human waste) around the entrance to Gargantua — please don't camp in the cave or at the cave entrance — and spray paint on the walls of Cleft, little damage has occurred to the caves. There are, however, signs of heavy visitation in the alpine and sub-alpine meadows below Andy Good Col, specifically trail erosion, camp fire damage and tree cutting in popular camping areas below Camp Caves. Please use minimal impact camping techniques in this area, and use a stove.

In 1994 Alberta Forestry, Lands and Wildlife and the ASS erected a sign at the trailhead for neophyte cavers providing information on equipment and caving etiquette. A useful addition to this sign would be a line or two discouraging camping in Gargantua.

Despite this area being one of the most notable areas of alpine karst in Canada, plans to get the Andy Good Plateau declared a provincial park under the B.C. Special Places 2000 program have been unsuccessful.

Warning There have been several cave rescues at Crowsnest Pass over the years, two serious enough to require the use of helicopters and rescue personnel. A caver fell down the first pitch in Yorkshire Pot when an inadequate belay failed, and a caver was injured at the top of the first pitch (the '56') in Gargantua, struck in the back by a falling rock. The approaches to both Gargantua and Cleft involve crossing steep slopes, prone to rock fall in the summer and avalanches in the spring. Use extreme caution when accessing either of these caves. Many of the other caves also involve steep approaches and may require climbing gear.

Since a bottom entrance has been discovered to both Yorkshire Pot and Gargantua, several rescues have occurred. A pull-down trip, where you rappel and pull the rope down after you, is a committed undertaking, and several parties have become trapped. Fortunately, in past incidents other cavers have been around to help extricate trapped parties. This may not always be so!

Randy Spahl, ASS Caver involved in exploration all over the Rockies for the last 20 years. Photo Dave Thomson.

North York Caves

Access for all Caves Start from Coleman. South of the railway tracks follow 81 Street that turns into the gravelled York Creek road. In about 4.5 km, just before the road crosses York Creek, park at the entrance to a logging road coming in from the right.

Hike the logging road to the river. A little way upstream is a log bridge that crosses the creek to the resumption of the logging road on the left bank. At the junction go right and recross the creek. Road climbs up northwest bank of North York Creek. Keep right at all junctions.

KEY
A Coulthard
B Cave of the Mastodon
C Heart Cave

Coulthard Cave

Map 82 G/10 755925
Jurisdiction Alberta Forests (Flathead)
Entrance elevation 2440 m
Number of entrances 1
Length, Depth 370 m, 61.2 m (-27.5 m +33.7 m)
Discovered 1968, by Monica & Tammy Morris

Location When under Mt. McLaren the logging road crosses to the other bank. The cave — a large 7 m diameter entrance — is now visible low on the north face of Mt. Coulthard. It has a large tongue of snow leading up to it for much of the year. Take a rope and ice axe as the final climb to the entrance may require crossing a bergschrund.

Description Because its large entrance is visible from Hwy. 3, Coulthard was the first cave visited in the area.

From a snow bank, a large passage leads downward to a point 10 m below the entrance. Here, a well-defined frost line can be seen circling the walls of the cave just below the ceiling. The passage ends after 122 m in a rising boulder choke that was unsuccessfully pushed in 1968 using explosives. A prominent side lead accessed by a loose 6 m climb leads to a spectacular wall of polished black ice blocking a large passage. In the other direction, the same passage heads down a rift to end in a 7 m blind pit. Two other high side passages near the entrance also end in ice and boulder blockages.

Coulthard Cave entrance.
Photo Dave Thomson.

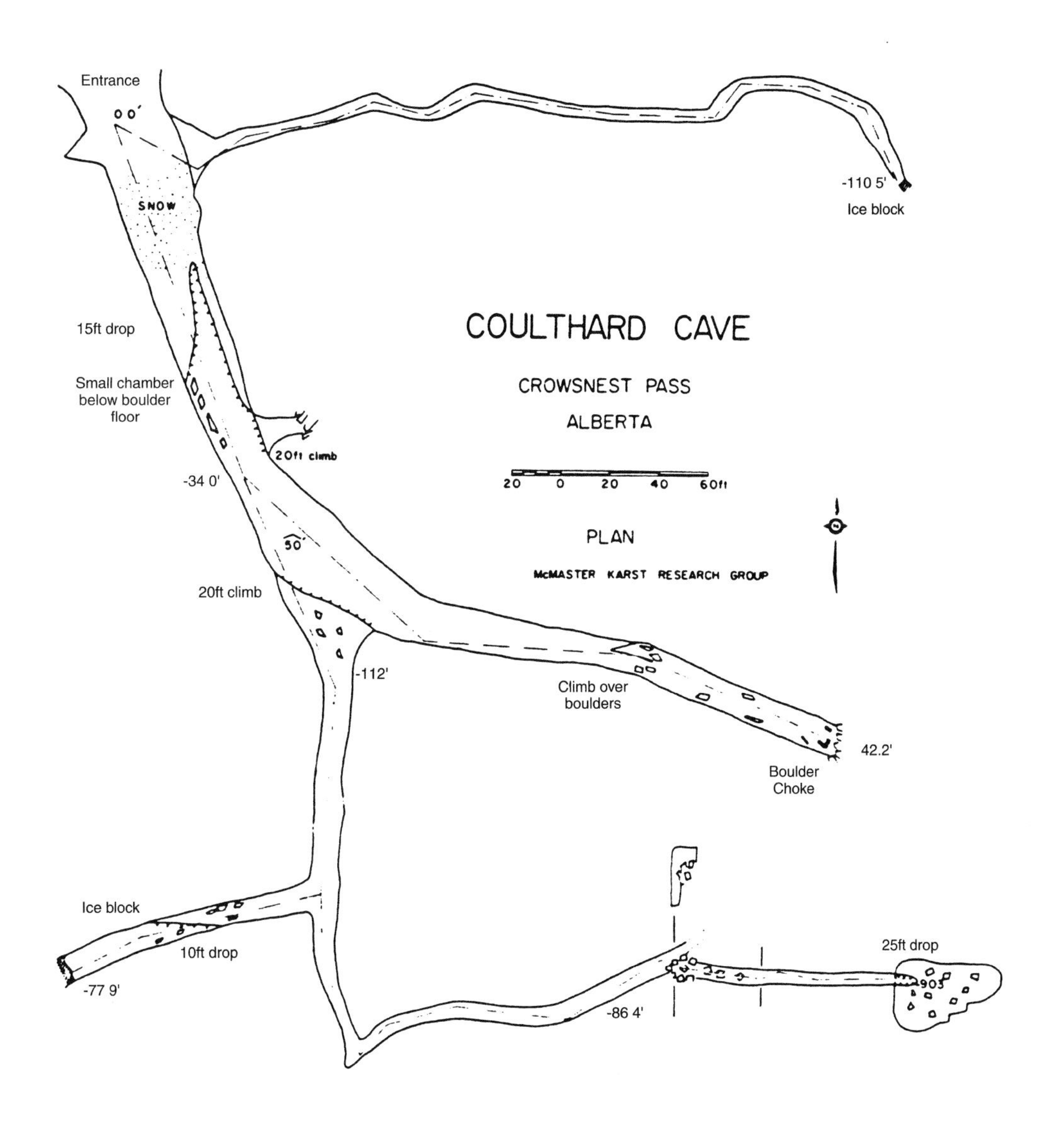

Cave of the Mastodon

Map 82 G/10 747928
Jurisdiction Alberta Forests (Flathead)
Entrance elevation 2400 m
Number of entrances 1
Length c. 200 m
Discovered 1998, by Jason Morgan

Location Hike to the end of the logging road. Cross the creek and follow a trail up the west side of the creek into the cirque between Andy Good Peak and Mt. Coulthard. The cave is located high in the east face of Andy Good Peak and can only be accessed using climbing gear.

Description This cave is a large ice-floored chamber with a side passage blocked by ice. This cave has been visited only once and detailed information is lacking.

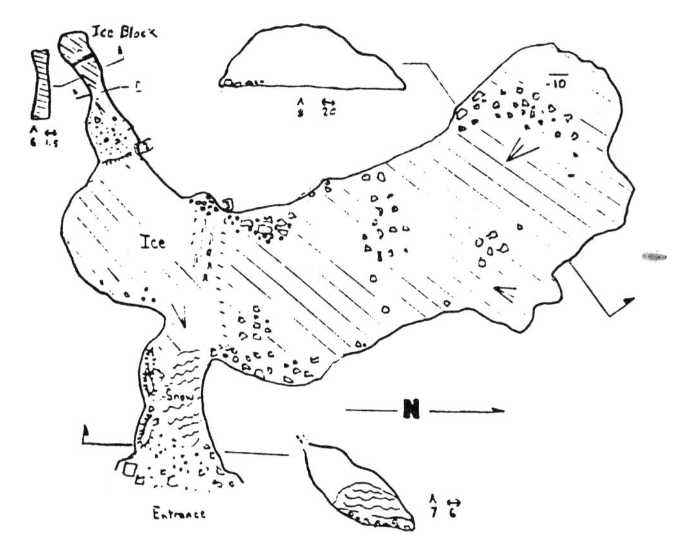

CAVE OF THE MASTODON
Andy Good Pk, Crowsnest Pass
Grade 1
Sketch by Jason Morgan
Approx length 200m

Heart Cave

Map 82 G/10 759946
Jurisdiction Alberta Forests (Flathead)
Entrance elevation 2100 m
Number of entrances 1
Length, Depth 90.5 m, 12 m
Discovered 1998, by Jason Morgan

Location When under Mt. McLaren, scramble steeply up its southeast slope. A rope and climbing gear is required.

Description Located on the southeast side of Mt. McLaren are numerous frost pockets. Heart Cave is an exception, actually having some passage. This cave has been visited only once, so detailed information is lacking. However, it is known that its name comes from the heart-shaped cross-section of a passage. There is a promising dig in the lower chamber.

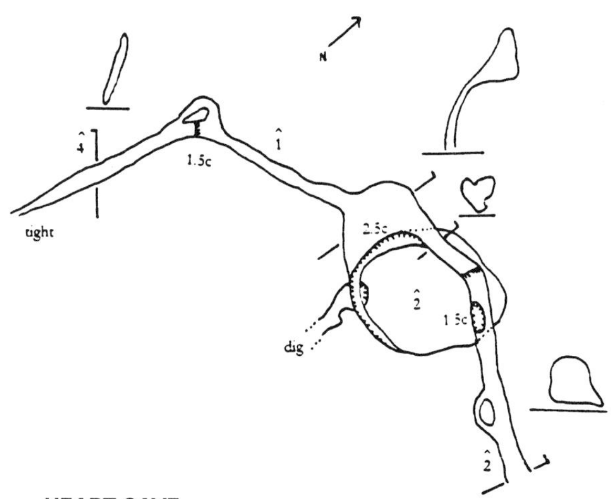

HEART CAVE
Crowsnest Pass
Grade 5 survey with memory detail
Jason Morgan & Rej Desjardins, 1998
Length: 90.5 m, Depth: 12 m
Alberta Speleological Society

Crow Pocket

Map 82 G/10 715965
Jurisdiction Alberta Forests (Flathead)
Entrance elevation 2072 m
Number of entrances 1
Length 21 m
Discovered 1974, by John Donovan

Access & Location Just east of the Travel Alberta Information Centre near Sentinel (at the east end of Crowsnest Lake) turn south and follow an old mining road up a small valley to the mine site. The road is driveable for a short distance depending on spring floods. From the mine site, although the road continues on towards Chinook Peak, head west to where a large entrance is visible high on the east side of the unnamed mountain that lies between Chinook and Sentry mountains.

Description Crow Pocket is a frost pocket, 6 m high, 15 m wide and 21 m deep. Close by are three other possible cave entrances that have not been checked, as they are hard to reach.

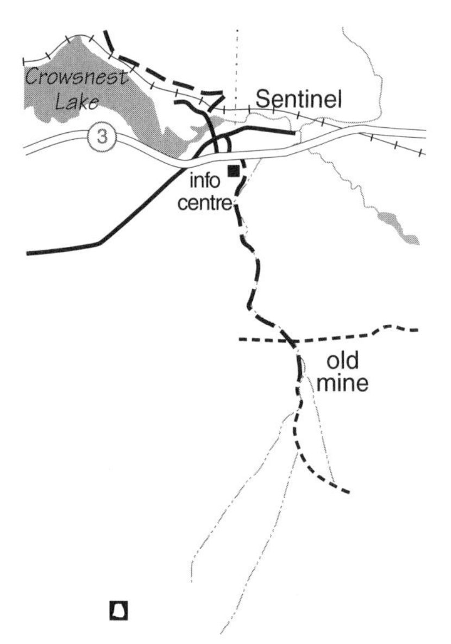

Crowsnest Lake Area

Owing to the prominent spring on the north side of Crowsnest lake, initial exploration in the Crowsnest Pass focussed on the Mt. Tecumseh and Phillips Peak area, but nothing significant was found other than Pinnacle Pot, a 25 m shaft plugged with snow. The following caves are all in the vicinity of the spring and visible from Hwy. 3, except for The Crack.

Access to all Caves There are two ways of getting there from Hwy. 3, both involving the railroad. Keep off the tracks; you may not hear a train approaching around a corner!

From the west: Park at the nformation kiosk at the west end of Crowsnest Lake. Follow the trail along the west end of the lake and climb up to the railroad tracks. Turn right and in about 1 km reach Crowsnest Spring.

Crowsnest Spring

Map 82 G/10 704 003
Jurisdiction Municipality of Crowsnest Pass
Entrance elevation 1388 m
Number of entrances 3: The Cave Spring, Lower Spring and The Crack
Length, Depth 275.9 m, 136.3 m (including The Crack)
Discovered Aboriginal peoples

Location A short scramble above the railroad tracks to two outlets known as the Lower Spring and the Cave Spring.

Warning Do not attempt to dive the cave unless you are an experienced cave diver. The sport of cave diving has one of the highest fatality rates of any activity (on a par with Himalayan climbing) and it takes years of experience to become proficient, if you survive! Even a casual visit to the entrance requires caution. The rocks are slippery, even in the summer. Move cautiously to avoid falling in.

Exploration & Description At least six attempts have been made to dive the Cave Spring: Peter Thompson in 1969, Dave Drew in 1970, Mike Boon in 1973, Tom Barton in 1981, Greg Resche in 1992/1993 and David Sawatzky and Eric Anderson in 1999. In summer the large amount of water coming out of the cave makes entry impossible, the preferred diving times being late fall or winter.

The first serious attempt was by Mike Boon who bottom-walked using two 72 cubic foot bottles. "After about 36 m at a depth of 12-15 m I reached a slot, beyond which the passage rose more steeply at 25-30° as a tube. This tube broke into the side of a large flooded rift that descended and rose vertically beyond my line of sight. Having no fins I returned, in poor visibility."

Tom Barton dived twice in 1981, pushing beyond the point Mike Boon had reached. "It was not too hard to reach the bottom of the first drop. The passage narrowed down at the bottom of the dogleg, but I got past the constriction with very little difficulty and swam to the top of the lookout where Mike had stopped. At this point, I decided to check out the upward-trending lead. Hoping to find a dry passage carrying on, I was happy

"The luckless Peter Thompson was the first to investigate the cave, being pushed in one summer in a dry suit weighted down with rocks. Fortunately the turbulence was sufficient to eject him." Mike Boon, *Canadian Caver*, 1976.

Crowsnest Spring entrance.
Photo Dave Thomson.

SURVEY: GRADE 1 / GRADE 3
SURVEYED: NOV., 1992
APR., 1993
BY: GREG RESCHKE
JIM SARGENIA
DRAWN BY: GREG RESCHKE

TOTAL SURVEYED LENGTH = 580 FT. (176.7m)
TOTAL SURVEYED DEPTH = 195 FT. (59.4m)

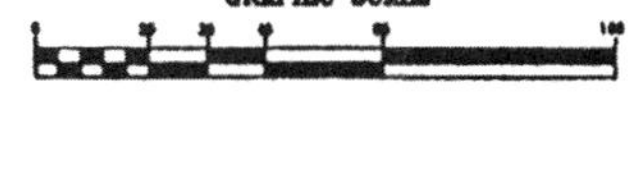

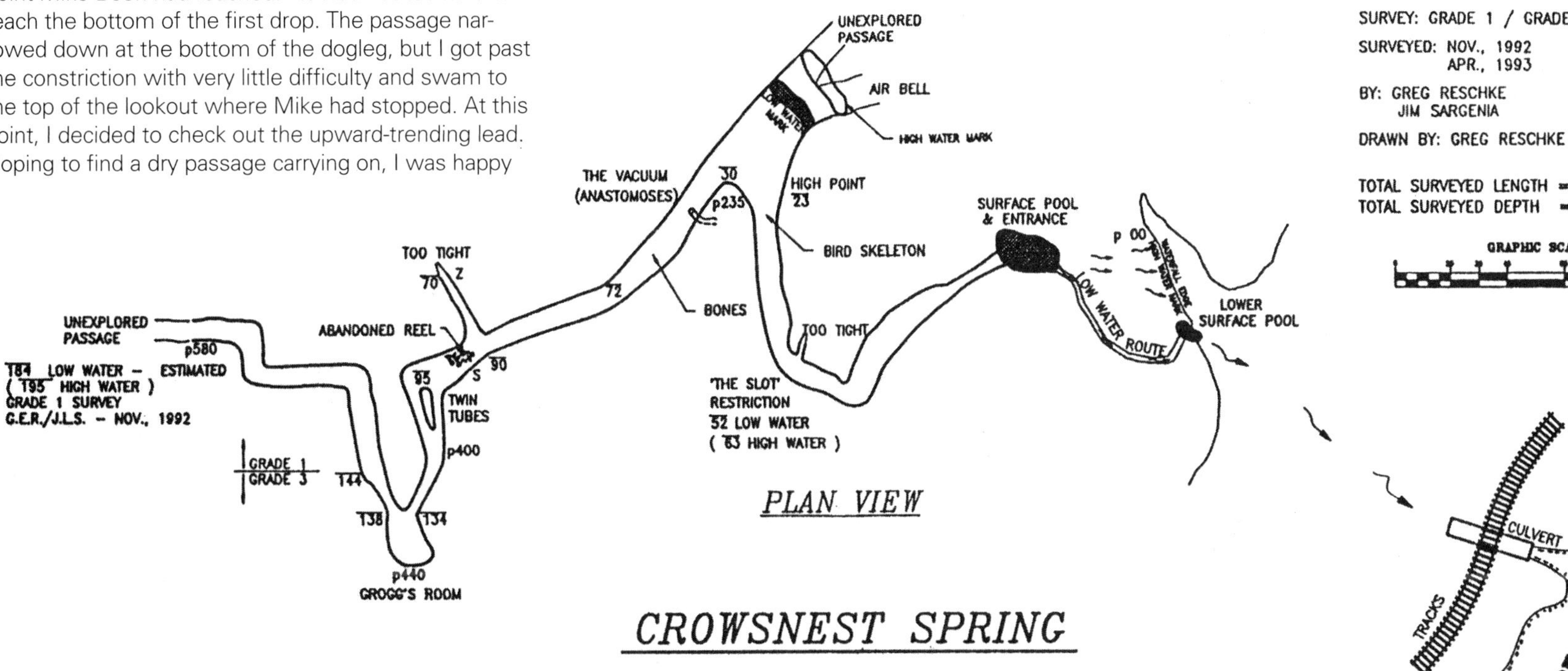

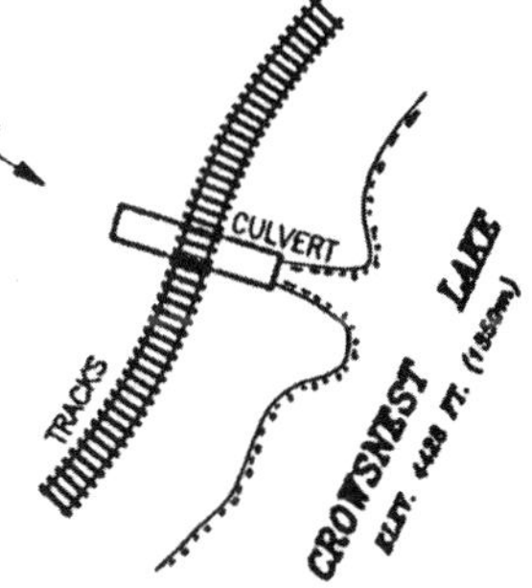

there was an air-filled bell chamber [The Crack]. The chamber is 1.5 m wide and six to nine metres long, however it is not possible to get out of the water since the walls rise at a 60-70° angle. The chamber follows a bedding plane and appears to break into smaller passages about six metres above the water. This lead is not too promising as it is heading back towards the cliff above the spring entrance. It probably joins Eagle Cave located about 30 m above the spring [actually it joins The Crack]. From the bell chamber I started to descend down past the lookout. The passage continued to drop at a 60° angle. I began to have pressure equalization problems but soon reached the bottom of the drop, where the cave trended horizontally. I equalized the pressure in my ears but found I was very cold. This seemed strange, since I had been deeper in lakes without getting nearly as cold. The trip out was uneventful."

A month later Tom returned with an extra 80 cubic-foot bottle. Reaching a short distance farther, at the head of twin descending shafts, he discovered he had misread his depth-gauge on his previous dive and that he was actually at –39 m rather than the –21 m he had believed! Tying his line to a rock he headed back out with two decompression stops.

In November, 1992 and April ,1993, Greg Resche dived a short distance beyond where Tom's line was tied off, attaining a depth of 59.4 m. David Sawatzky and Eric Anderson made three dives in October of 1999, surveying to a depth of 61.6 m — Greg's limit on air — and then on trimix laid line. They removed all the old line and Tom's abandoned reel, and commented on the strong current and coldness of the dives. David commented, "A return visit is required but until trimix computers and warmer drysuits are available, there is no rush. The cave is going horizontal and even if it headed back up, we could not surface with the current technology."

The "warm" nature of the spring suggests an input of thermally heated water, and so it may be a lot deeper than the dives to date suggest.

Hydrology Crowsnest Spring drains karst to the north of the pass. It is a thermal spring and flows year round with "warm" (up to 5.5°C) sulphate-rich water. In the winter, the water temperature is 1.5° warmer than in the summer when snowmelt enters the system. "In terms of heat fluxes, Crowsnest Spring produces 2200 Js^{-1}, which makes it the fifth in importance in the southern Rockies, after the commercialized hot springs at Banff, Fairmont, Radium and Miette." (Steve Worthington)

Formally incorrectly known as the Old Man River Spring, Crowsnest Spring actually feeds the Crowsnest River. The output was monitored during the summers of 1968 and 1969 and was found to have a maximum discharge of about 400 l/s, the overflow Cave Spring handling the bulk of the flow (Steve Worthington). In the winter, the cave spring ceases to flow and the water level drops by 4 m to the base of a small beach. The lower spring is active year round.

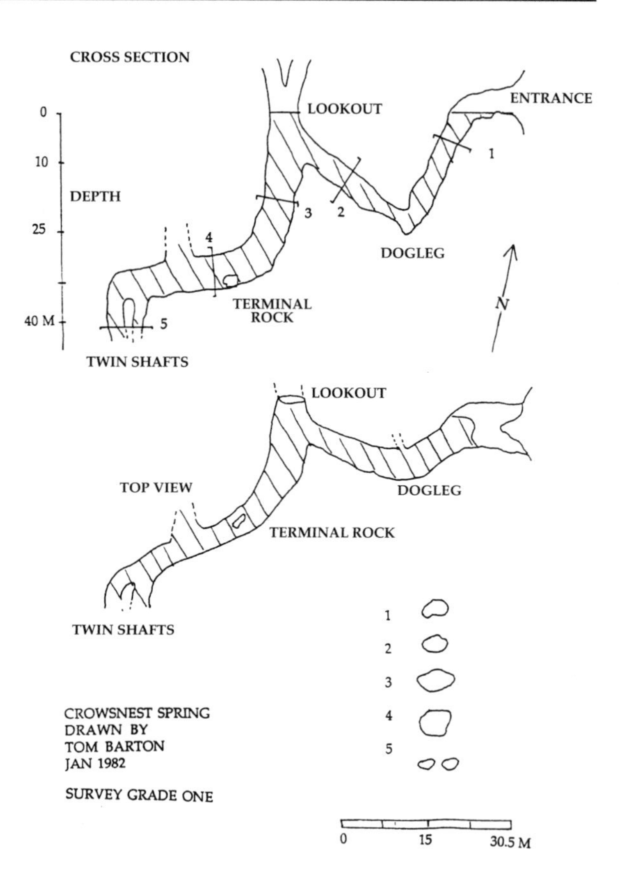

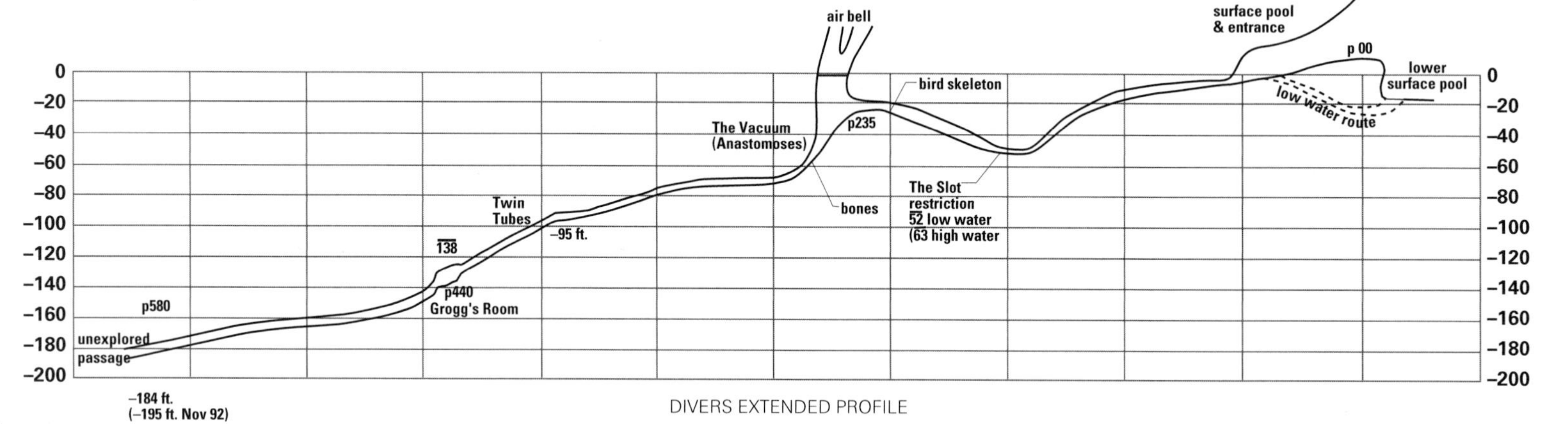

DIVERS EXTENDED PROFILE

The Crack

Map 82 G/10 704003
Jurisdiction Municipality of Crowsnest Pass
Entrance elevation 1462.7 m
Number of entrances 1
Length, Depth 97.2 m, 74.7 m
Discovered 1994, by Cam Hutchinson-Maclean, Jason Stone and Tony Young

Location The Crack is located directly above Crowsnest Spring, but is not visible from below. It is best reached by scrambling (a rope is useful) up a scree slope and rock buttress, starting about 50 m along the railroad tracks to the west of Crowsnest Spring.

Exploration & Description The Crack is a fossil overflow for Crowsnest Spring, connecting at the air bell. Gaining access to The Crack involved the excavation of a considerable amount of material and required the construction of a wooden headworks in the cave entrance. Beyond the initial 9 m drop, a splendid phreatic tube descends at a constant 60° for 60 m into the air-bell sump, across from which a possible lead heads back up, but is difficult to reach.

Warning The 60 m rope pitch descends directly into the air-bell sump (The Wishing Well) without a ledge. Take care not to rappel into the sump; it could be very awkward to climb out.

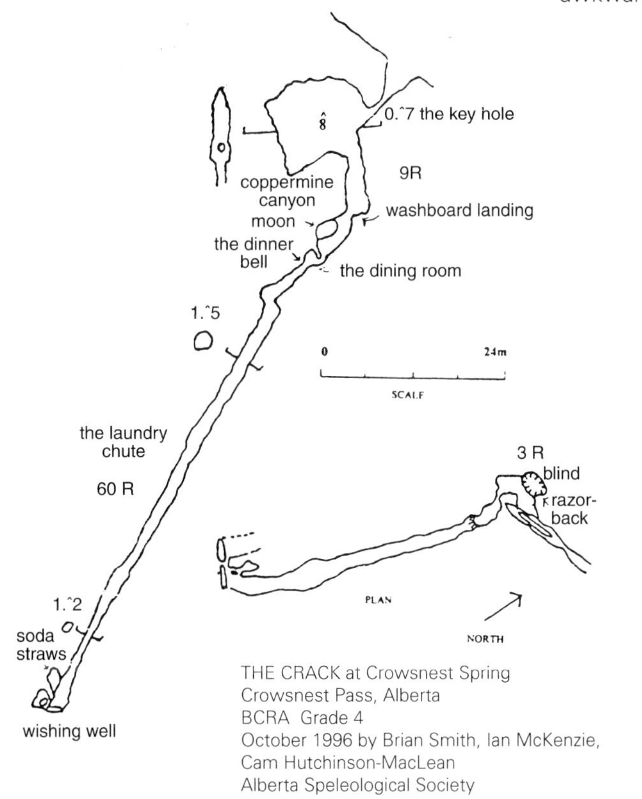

THE CRACK at Crowsnest Spring
Crowsnest Pass, Alberta
BCRA Grade 4
October 1996 by Brian Smith, Ian McKenzie, Cam Hutchinson-MacLean
Alberta Speleological Society

Eagle Cave

Map 82 G/10 705007
Jurisdiction Municipality of Crowsnest Pass
Entrance elevation 1418 m
Number of entrances 1
Length c. 60 m
Discovered Unknown

Location Eagle Cave is located above and slightly to the right (east) of Crowsnest Spring. A short section of exposed scrambling is required to reach the entrance (a rope is useful).

Exploration & Description The cave was the site of an archaeological dig by the University of Alberta in 1968. They excavated a low crawl that ended after about 60 m in a small room. Further attempts at digging have failed to produce much more passage.

Bill MacDonald in Eagle Cave. Photo Dave Thomson.

Sentry Mountain Caves

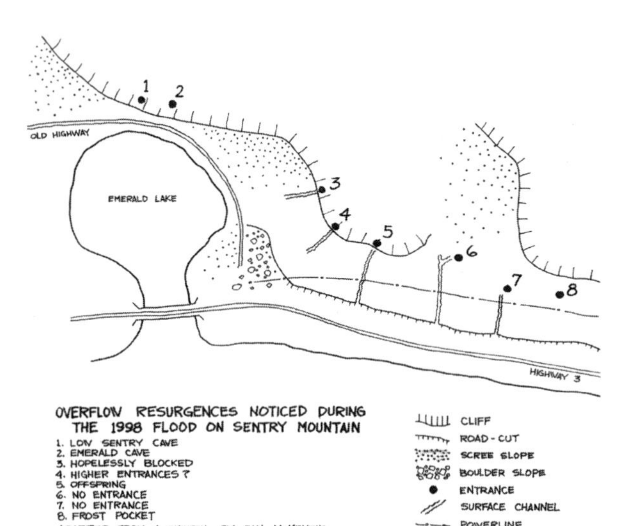

ROAD-VISIBLE ENTRANCES ON SENTRY MOUNTAIN

1. FROST POCKET MARKS START OF ROUTE UP TO MIDDLE SENTRY, RED OCTOBER, AND UPPER SENTRY
2. FROST POCKET (MIDDLE SENTRY CAVE JUST BELOW)
3. DEEP FROST POCKET. RED OCTOBER BETWEEN 2 AND 3
4. LOW SENTRY CAVE
5. EMERALD CAVE
6. FROST POCKET

ADAPTED FROM A SKETCH BY IAN McKENZIE

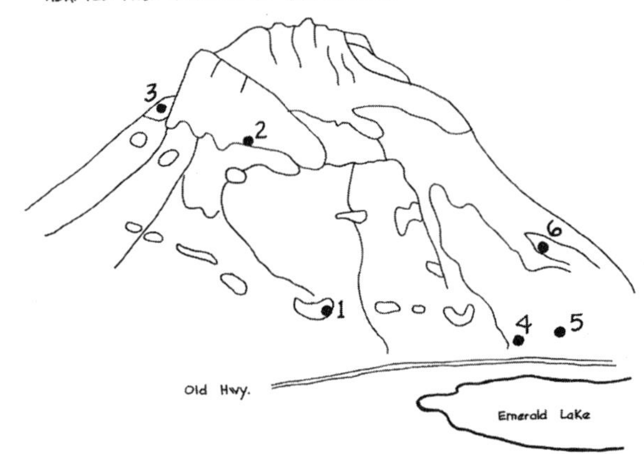

Low Sentry Cave

Map 82 G/10 704994
Jurisdiction Municipality of Crowsnest Pass
Entrance elevation 1380 m
Number of entrances 1
Length, Depth 113.9 m + 19.1 m
Discovered Unknown

Access Start from Hwy. 3 at the Travel Alberta Centre just west of Sentinel at the east end of Crowsnest Lake. Head west on a gravel road (the old highway) to where it skirts the south edge of Emerald Lake.

Low Sentry Cave is the lowest entrance at 10 m above the road and is accessed by a short scramble.

Exploration and Description In the late 1960s, McMaster cavers documented two caves (Low Sentry and Middle Sentry) in Sentry Mountain above Emerald Lake. Unclear as to which of several visible holes these were, Ian McKenzie, accompanied by various ASS cavers, attempted to sort out the mystery in 1997 he determined that of the two prominent entrances located just above the old highway, the more accessible was Low Sentry Cave.

It starts very tight, but opens up a bit further in. In 1998 this cave was pushed through a sump to a second sump and surveyed.

Warning The cave pours water in wet weather and should not be attempted during this time.

LOW SENTRY CAVE
Sentry Mountain, Crowsnest Pass
Grade 5
Survey by Rej Desjardins and Jason Moore
January/February 1998
Alberta Speleological Society
Surveyed Length 113.9 m +19.1m

0m
25m
Scale (metres)
Entrance
1.5
0.6
Dig
Spider Clusters
0.3
1st Duck
0.9
0.95
Soda Straws
Flow Stone
Soda Straws and Stals
1.15
Steep Climb
Moon Milk
Too Tight
1.25
Moon Milk
0.65
0.5
2nd Duck
Moon Milk
Too Tight
Draughting
North

Emerald Cave

Map 82 G/10 704994
Jurisdiction Municipality of Crowsnest Pass
Entrance elevation 1400 m
Number of entrances 1
Length 15 m
Discovered Jason Morgan

Access Start from Hwy. 3 at the Travel Alberta Centre just west of Sentinel at the east end of Crowsnest Lake. Head west on a gravel road (the old highway) to where it skirts the south edge of Emerald Lake.

The cave is located slightly to the west of and slightly higher than Low Sentry Cave. A short climb past a bolt accesses the entrance.

Description A well-scalloped 1.2 m diameter tube has been excavated for 15 m.

Warning The cave pours water in wet weather and should not be attempted during this time.

Hopelessly Blocked

Map 82 G/10 703995
Jurisdiction Municipality of Crowsnest Pass
Entrance elevation 1500 m
Number of entrances 1
Length c. 13 m
Discovered No one has claimed responsibility

Access Start from Hwy. 3 at the Travel Alberta Centre just west of Sentinel at the east end of Crowsnest Lake. Head west on a gravel road (the old highway) to where it skirts the south edge of Emerald Lake.

The cave is located at the base of cliffs, above and close to the barricaded end of the old highway.

Description This cave has recently undergone a series of transformations, from hopelessly sumped to hopelessly muddy to hopelessly tight — with air!

Warning The cave pours water in wet weather and should not be attempted during this time.

Middle Sentry Cave

Map 82 G/10 711985
Jurisdiction Alberta Forests (Flathead)
Entrance elevation 1860 m
Number of entrances 2
Length, Depth 75 m, 10 m
Discovered 1968, by Derek Ford and Gary Pilkington

Access Start from Hwy. 3 at the Travel Alberta Centre just west of Sentinel at the east end of Crowsnest Lake. Head west on a gravel road (the old highway). Park at the first junction where a minor spur road heads off to the right (north). To the south, a large frost pocket is visible in a cliff band at the bottom of a gully.

A steep two-hour hike adjacent to, then up the gully (no trail) and over to a subsidiary gully farther east accesses the cave entrance.

Exploration & Description Checking out three large "entrances" above Low Sentry Cave produced three frost pockets. However, below frost pocket #2 was a small cave leading to a large lookout with a view onto Crowsnest Mountain. This was presumed to be Middle Sentry Cave.

"An elegant little cave that quickly chokes." Derek Ford, 1968, referring to either Middle or Low Sentry caves.

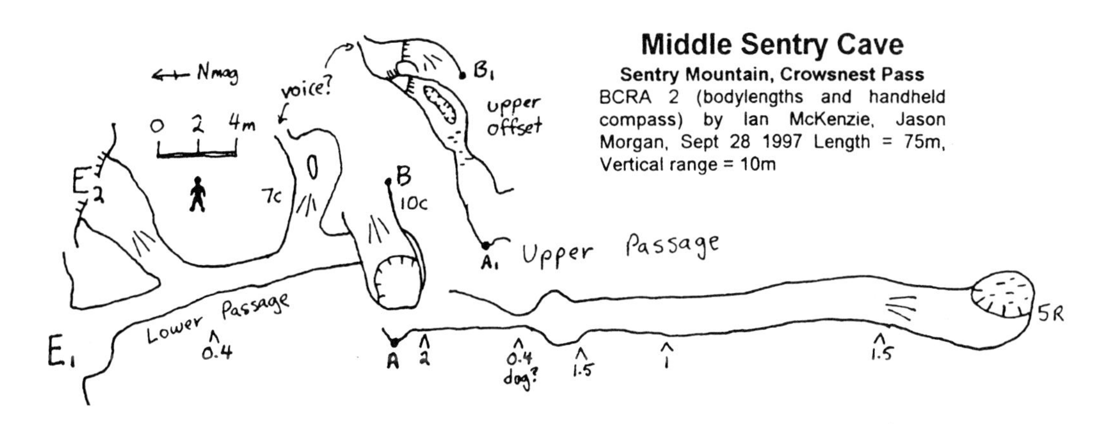

Offspring

Map 82 G/10 700995
Jurisdiction Municipality of Crowsnest Pass
Entrance elevation c. 1520 m
Number of entrances 1
Length, Depth 29 m, 11 m
Discovered June 1999, by Lynnette Jessop & Ian McKenzie

Access Start from Hwy. 3 at the Travel Alberta Centre just west of Sentinel at the east end of Crowsnest Lake. Head west on a gravel road (the old highway) to where it skirts the south edge of Emerald Lake.

From the end of the old highway climb up to the power line and follow it west to where it crosses a small stream channel. Follow the channel up to the base of the cliff to where the normally dry creek bed emerges from a small crawl-down entrance.

Exploration & Description This cave was excavated in the course of several visits from various ASS cavers. A short straightforward crawl ends at an awkward bend in the passage where it heads upwards into a loose rift full of cobbles, sand and bones. Excavation beyond this point was terminated when a large slab looked set to fall and block the access passage beneath.

Warning Material in the rift at the end of the cave is very unstable. In wet weather the cave pours water and should not be attempted during this time.

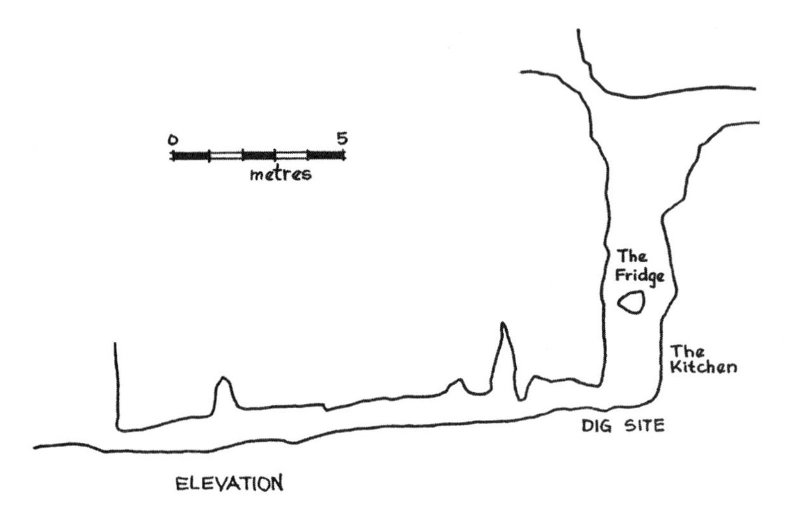

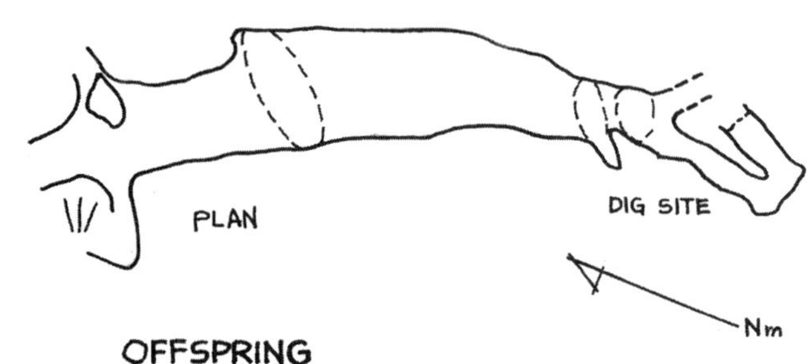

OFFSPRING
SENTRY MOUNTAIN, CROWSNEST, ALBERTA

SUUNTOS AND TAPE SURVEY (BCRA GRADE 5)
APRIL 21 & 23, 2000
DAVID BRANDRETH, DAN GREEN, GEORGE HANUS, IAN McKENZIE
ADAPTED FROM SURVEY BY: DAN GREEN

Red October

Map 82 G/10 712984
Jurisdiction Alberta Forests (Flathead)
Entrance elevation 1920 m
Number of entrances 1
Length, Depth 341.9 m, 85 m
Discovered 1997, by Jason Morgan and Ian McKenzie

Access Start from Hwy. 3 at the Travel Alberta Centre just west of Sentinel at the east end of Crowsnest Lake. Head west on a gravel road — the old highway. Park at the first junction where a minor spur road heads off to the right (north). To the south, a large frost pocket is visible in a cliff band at the bottom of a gully.

A steep two-hour hike adjacent to, then up the gully (no trail) and over to a subsidiary gully further east accesses the general area of the cave. Expect to do considerable searching. The slot entrance (bolt and a piton) is well hidden at the back of a small ledge. Look for a prominent cairn.

Exploration & Description On the way to check out Entrance #3 (Hopelessly Blocked), Jason Morgan found a cairn next to a pit entrance. A return visit with tackle revealed a 25 m drop into a huge (12 m wide, 30 m long, 15 m high) beautifully decorated passage. "We followed [a] walking-sized tube under natural bridges and up a few easy climbs past flowstone cascades. A second tube was completely coated with calcite and decorated with red blobs and stripes like bleeding ulcers inside an intestine; the calcite causes amazing reverberation for deep notes sung down the passage. We stopped at a flowstone free climb that needed a chock to protect. There were no traces of a previous visit." The climb led to further walking passage and a large well-decorated chamber.

The mystery of the cairn, and why the cave was not previously explored remains unsolved.

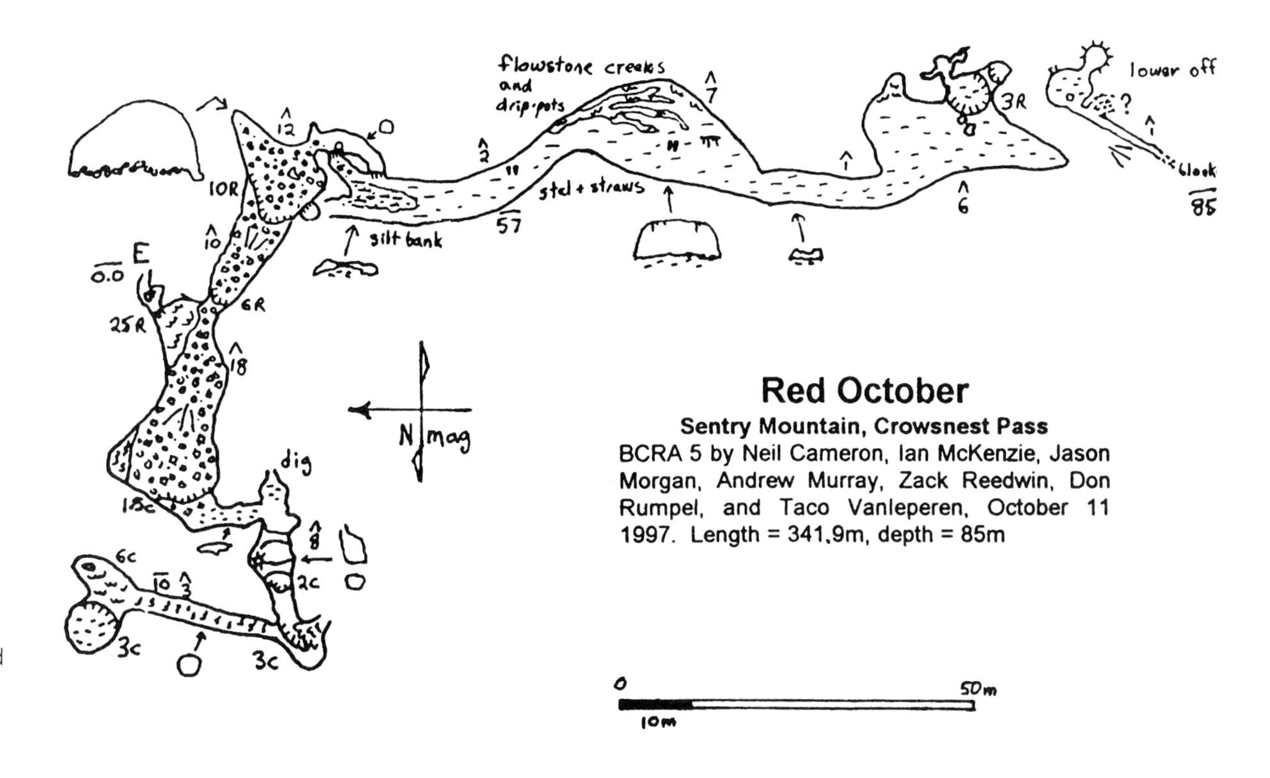

"Red October is by far the largest cave remnant in Sentry Mountain and gives hope that further 'big cave' will be found in the mountain." Ian McKenzie.

Upper Sentry Cave (Boon's Glittering Ice Palace)

Map 83 G/10 710974
Jurisdiction Alberta Forests (Flathead)
Entrance elevation 2226 m
Number of entrances Several skylights
Length, Depth 286 m, 84 m (-66 m +18 m)
Discovered 1967, by Mike Boon

Access Start from Hwy. 3 at the Travel Alberta Centre just west of Sentinel at the east end of Crowsnest Lake. Head west on a gravel road (the old highway). Park at the first junction where a minor spur road heads off to the right (north). To the south, a large frost pocket is visible in a cliff band at the bottom of a gully.

A steep two-hour hike adjacent to, then up the gully (no trail) and over to a subsidiary gully further east accesses Middle Sentry Cave. Continue up and contour below cliffs to the left and southeast of Red October until approaching the ridge line. Then scramble steeply up grassy ledges where the large frost-rimmed slot entrance lies hidden.

It may also be possible to reach the cave by following the west ridge towards the summit of Sentry Mountain (see "Scrambles in the Canadian Rockies" by Alan Kain), then heading back along the south ridge.

Exploration & Description The cave is composed of three passages. The Gothic Passage ends in an ice-decorated room at a boulder choke. The Skylight Passage, an upper section of the cave, climbs steeply to the bottom of a pit with a roof skylight. The Scree Run is a steeply descending ice slope that reaches a wall at the low point of the cave (sheep skeleton). From here a right turn leads up steep ice ending beneath two avens.

When revisited in 1990, the entrance to this cave had changed drastically since its discovery in 1967. Most of the upper ice had melted including the ledge to The Skylight Passage.

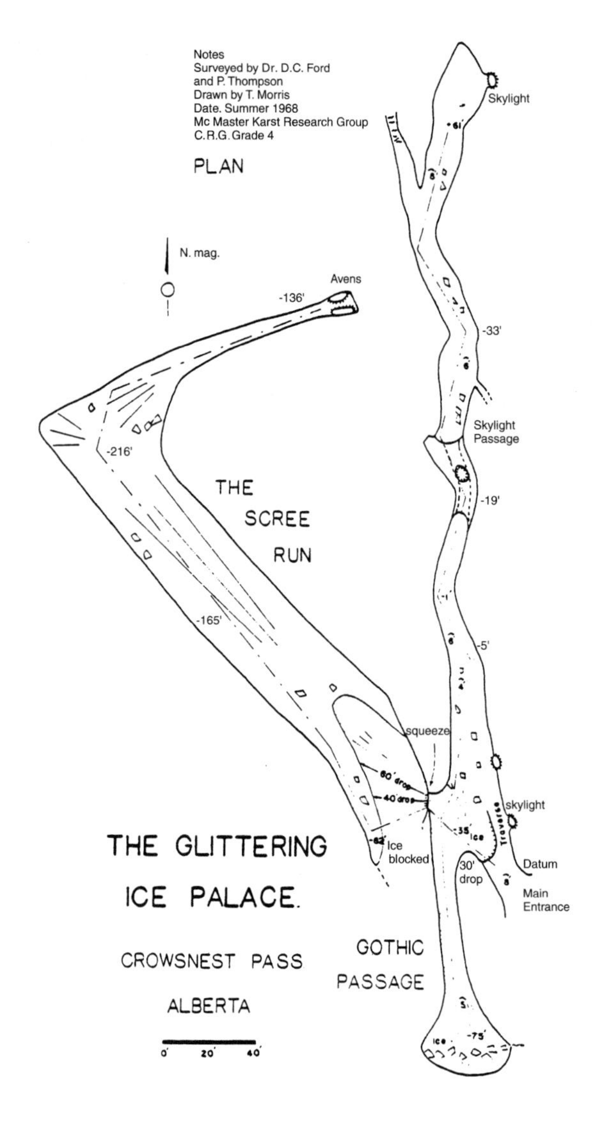

"It's probably very beautiful down there, but I don't really think I should go down." Kevin, aged 10, 1974.

Ice Chest

Map 82 G/10 703984
Jurisdiction Alberta Forests (Flathead)
Entrance elevation 2250 m
Number of entrances 2
Length, Depth 381 m, 30 m (-2.0 m +28 m)
Discovered 1979, by Tom Wilson and Bob Isaac

Access Hwy 3 at the motel at the west end of Crowsnest Lake. Not easy to find, the cave is located just to the east of the steep gully that splits the northwest face of Sentry Mountain.

Climb the gully starting from the south side of the highway opposite the motel. Just above where the gully begins to merge into the mountainside, a short, leftward traverse brings you to the 1 m by 1.5 m entrance. A second entrance 30 m farther to the east is blocked, but drafts strongly (it connects to Ice Chest).

Exploration & Description The first entrance leads to a 12 m wide passage that after 73 m intersects a large 20 m by 122 m chamber. When first discovered almost the whole floor of the cave was covered with ice. In 1991 the ice had retreated to the large chamber and many of the spectacular ice crystals formerly found in the Crystal Room had melted. This cave is rarely visited because of the steep approach and illusive entrance, so it is unclear whether the melting of ice formations is permanent.

Several smaller caves and entrances have been located across the gully 200 m to the west.

Cave Softly Avoid breaking ice formations, and keep your visit short to avoid raising the cave temperature.

ICE CHEST

Notes
1. Brunton and tape survey September 1979 by Tom Wilson, Bob Isaak & Tom Barton.
2. Drawn by Tom Barton.
3. Total length 250 ft.
4. Entrance location 49° 36' 40" W114° 38' 45"

Entrance to Ice Chest. Photo Jon Rollins.

Ptolemy Valley Caves

Access for all caves Hwy. 3, 4.8 km west of the Travel Alberta Information Centre. Just after crossing Crowsnest Creek, turn left (south) onto the gravelled Chinook Coal Road. Follow the road alongside Crowsnest Creek for about 3 km to a ford (sometimes with bridge) immediately adjacent to the left side of the road. Park here if you have little ground clearance. Otherwise cross the ford and follow a muddy track up into a meadow and park. From here on, a 4x4 vehicle is required to continue driving up the Ptolemy Creek seismic road.

Otherwise, hike the seismic road up Ptolemy Creek valley. Caves are accessed shortly after the seventh and final creek crossing at 3.4 km from the Chinook Coal Road.

"Ptolemy is a pleasant valley. Snow-capped mountains surround us; the stream was swollen with the spring run-off. A city lass, I grew ecstatic over the wild flowers growing in profusion there. Idly I wondered why John's head was tilted up so often; he must be looking at birds or admiring the view. I looked about me for frost pockets. Suddenly Wes and John babbled excitedly and pointed upwards. My eyes followed their trembling fingers upwards. To my horror there were holes in the rock at least two thousand feet above us — frost pockets." Anne Ridley.

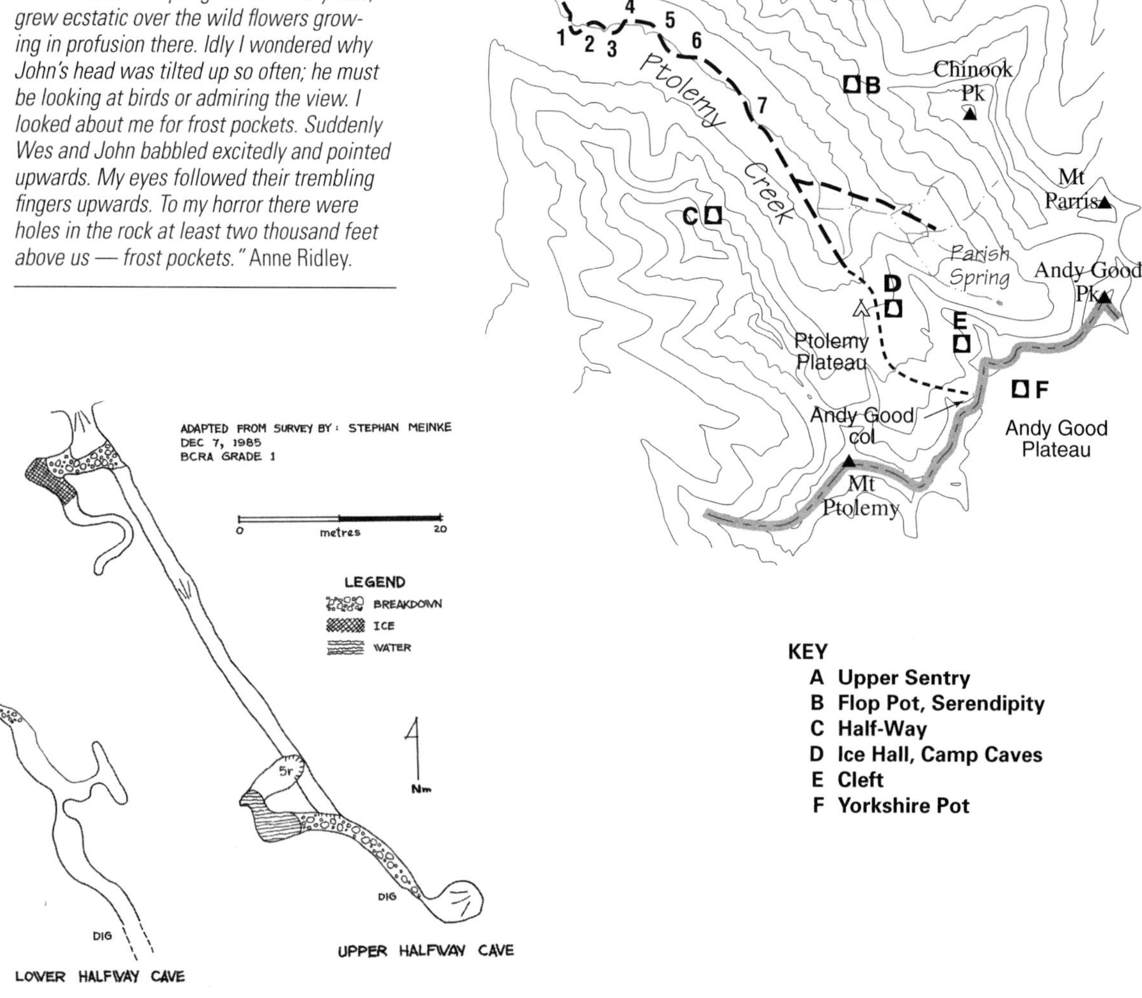

Half-Way Caves

Map 82 G/10 699938
Jurisdiction Alberta Forests (Flathead)
Entrance elevation 2135 m
Number of entrances 2 caves, each with 1 entrance
Length c. 60 m & 120 m
Discovered 1969, by Tich Morris

Location The entrances, located on the long ridge to the west, are fairly easy to locate with binoculars. Shortly after the seventh and final creek crossing hike up over benches and through small cliff bands.

Exploration & Description The upper of these two caves was one of the first finds in the Ptolemy valley, the lower cave being dug into shortly afterward. Both end in chokes and were thought to connect, but digging has so far proved unsuccessful.

Flop Pot

Map 82 G/10 715956
Jurisdiction Alberta Forests (Flathead)
Entrance elevation 2165 m
Number of entrances 1
Length, Depth 200 m, 90 m
Discovered 1974, by Wes Davies, John Donovan & Anne Ridley

Location Flop Pot is located high on the east side of the valley. Shortly after the seventh and final creek crossing on the seismic road, recross the creek and climb steep slopes towards the distinctive bleached trunks of old pine trees. Make for a low point on the ridge connecting Sentry Mountain and Chinook Peak. The large 6 m-high and 10 m-wide entrance is situated below an overhang near the top of the ridge.

Exploration & Description One of these "frost pockets" turned out to be Flop Pot, named after Anne's dog Flop.

A steep snow and scree slope leads to an ice-floored chamber beyond which, in quick succession, a series of further pitches lead to the end of the cave, a mud choke. Some nice ice columns decorate a small chamber near the bottom. Flop Pot probably connects with Serendipity, located just above; leads in both caves are drafting through a stone-choked connection.

During a visit in 1993 the cave was completely blocked by ice not far from the entrance.

Randy Spahl at the entrance to Flop Pot (left) and Serendipity. Photo Ian Mckenzie.

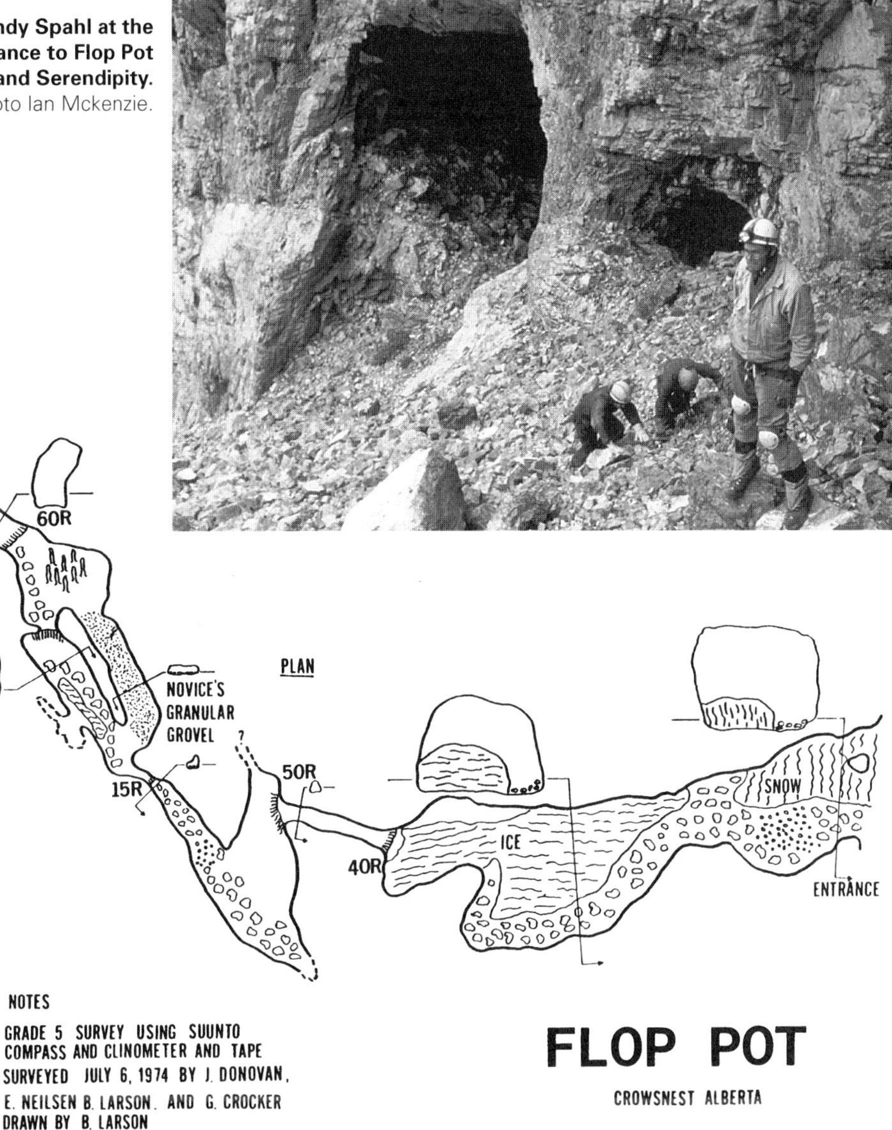

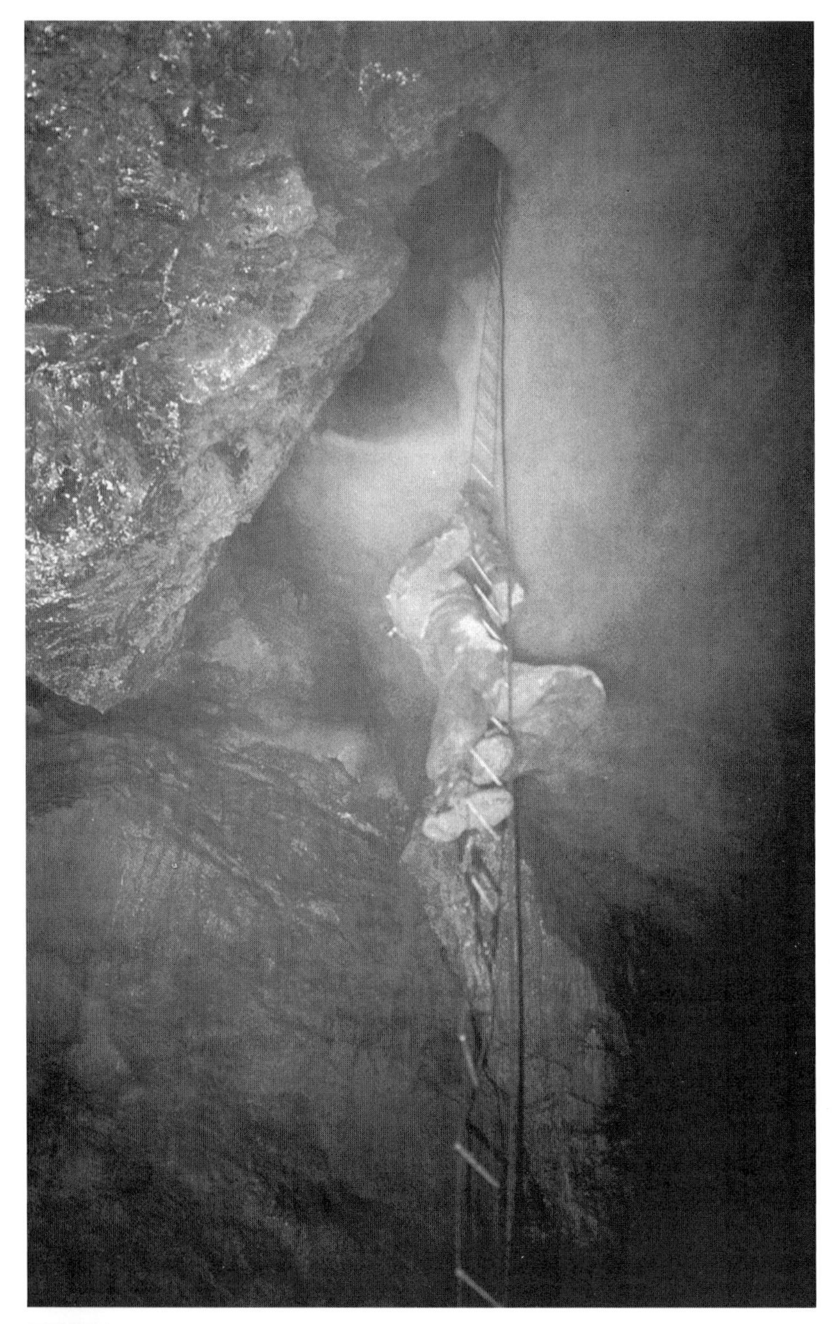

Entrance to Serendipity. Photo Jon Rollins.

Randy Spahl descends through the ice floor in Serendipity. Photo Ian McKenzie.

Serendipity

Map 82 G/10 715956
Jurisdiction Alberta Forests (Flathead)
Entrance elevation 2165 m
Number of Entrances 2 (second entrance closed to reduce ice melt)
Length, Depth 473 m, 20 m
Discovered 1979, by Tom Barton

Location Just above Flop Pot beneath an impressive overhang.

Exploration & Description The entrance narrows down to a crawl-in bedding plane slot before opening out into the upper section of the cave, remarkable for a large chamber almost completely filled with ice. Down to the left is a narrow boulder-blocked rift leading to the probable connection with Flop Pot.

The glacier-like ice mass forms the floor of the upper chamber, one wall and part of the ceiling of the chamber below. The placement of two ice screws allows the rigging of a 15 m ladder through a small gap between the ice and the chamber ceiling. The massive prow of ice rearing over your head as you descend is impressive and contrasts strangely with the surrounding limestone. As in a painting by Margrit, where an apple fills a room, one wonders how this massive curved ice mass with its alternating bands of white, light blue and grey, came to fill the chamber. Long ice flutings reminiscent of rillen karren run the height of one wall, probably formed by melt water when the ice was in close contact with the wall. The floor is covered with large boulders. An arch at the back of the chamber marks the continuation of the cave passage, an ice-free rift, which eventually becomes too narrow to continue along.

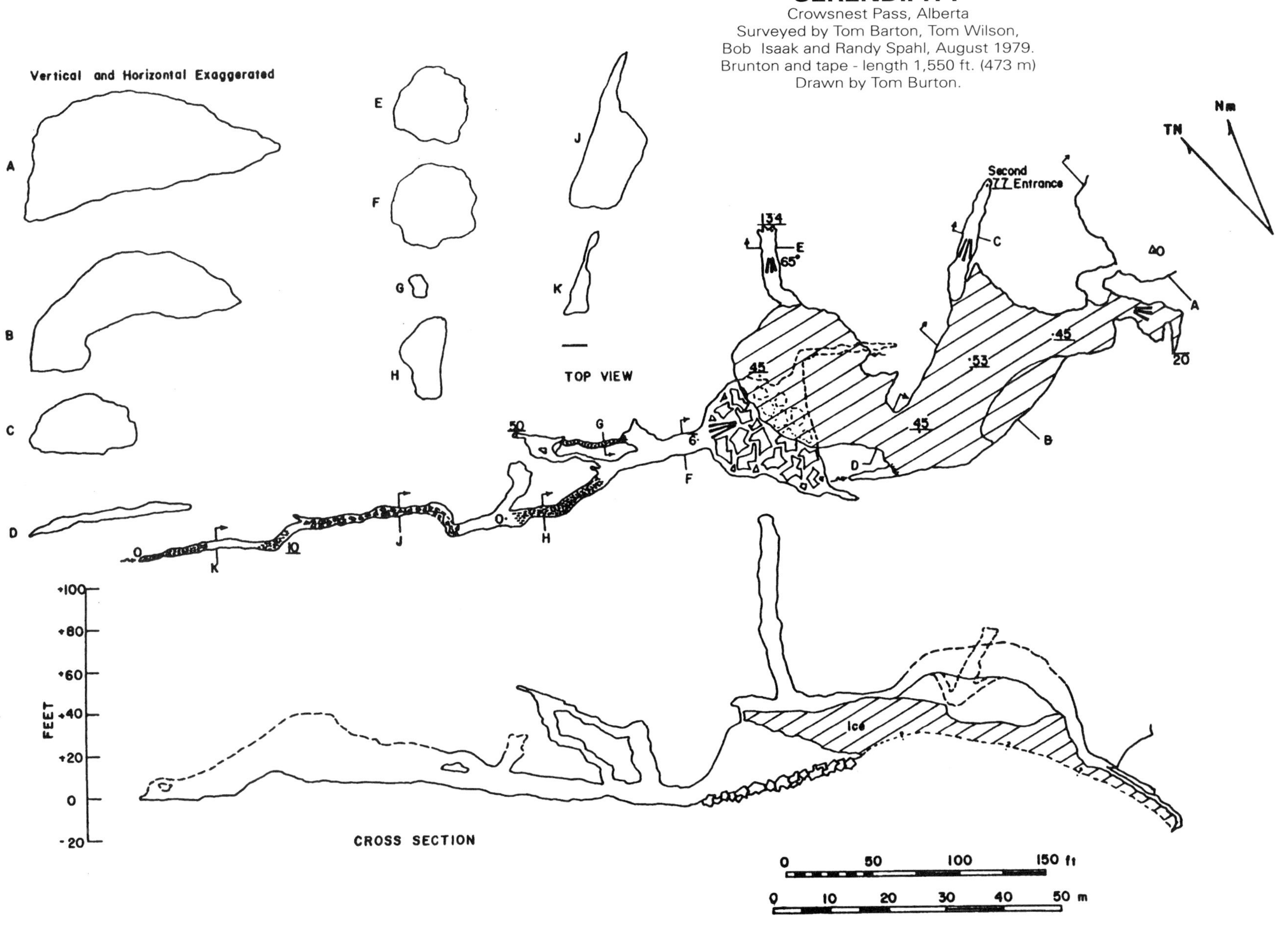
SERENDIPITY
Crowsnest Pass, Alberta
Surveyed by Tom Barton, Tom Wilson,
Bob Isaak and Randy Spahl, August 1979.
Brunton and tape - length 1,550 ft. (473 m)
Drawn by Tom Burton.
Vertical and Horizontal Exaggerated
A
B
C
D
E
F
G
H
J
K
TOP VIEW
Second Entrance
77
134
65°
45
53
40
20
50
6
10
0
TN
Nm
Ice
FEET
+100
+80
+60
+40
+20
0
-20
CROSS SECTION
0 50 100 150 ft
0 10 20 30 40 50 m

Heave-Ho!

Map 82 G/10 733927
Jurisdiction Alberta Forests (Flathead)
Entrance elevation 2104 m
Number of entrances 1 (connects with Yorkshire Pot)
Length 326.6 m
Discovered August 1987, by Martin Davis

Location Heave-Ho! and Hi-Ho! caves are located in the north cirque under Mt. Parish and Andy Good Peak.

From the seventh and final creek crossing on the seismic road, continue up the road for a further 1.5 km to a junction. Turn left. Follow this road as it descends to cross Ptolemy Creek, then winds steeply upwards (abandoning your 4 X 4 vehicle as it steepens) towards the cirque. When the road ends, hike to the small lake fed by Parish Spring at 733928.

The entrance to Heave-Ho! lies just above the small lake, below the base of large cliffs to the west. These same cliffs contain the two lookouts of Cleft Cave.

Exploration & Description Heave-Ho! was named following the combined effort needed to remove a 200 kg boulder blocking the entrance. It was discovered during a cave radio trip to find the surface location of the bottom of Yorkshire Pot.

A 4 m down-climb leads to a small chamber from which a dig through cobbles and pack rat debris accesses a 2 m diameter phreatic tube. Beyond, another dig was eventually connected to Yorkshire Pot below the Seven Steps in 1997. This dig connection is usually left blocked to prevent an increase in airflow drying out YP's lower passages.

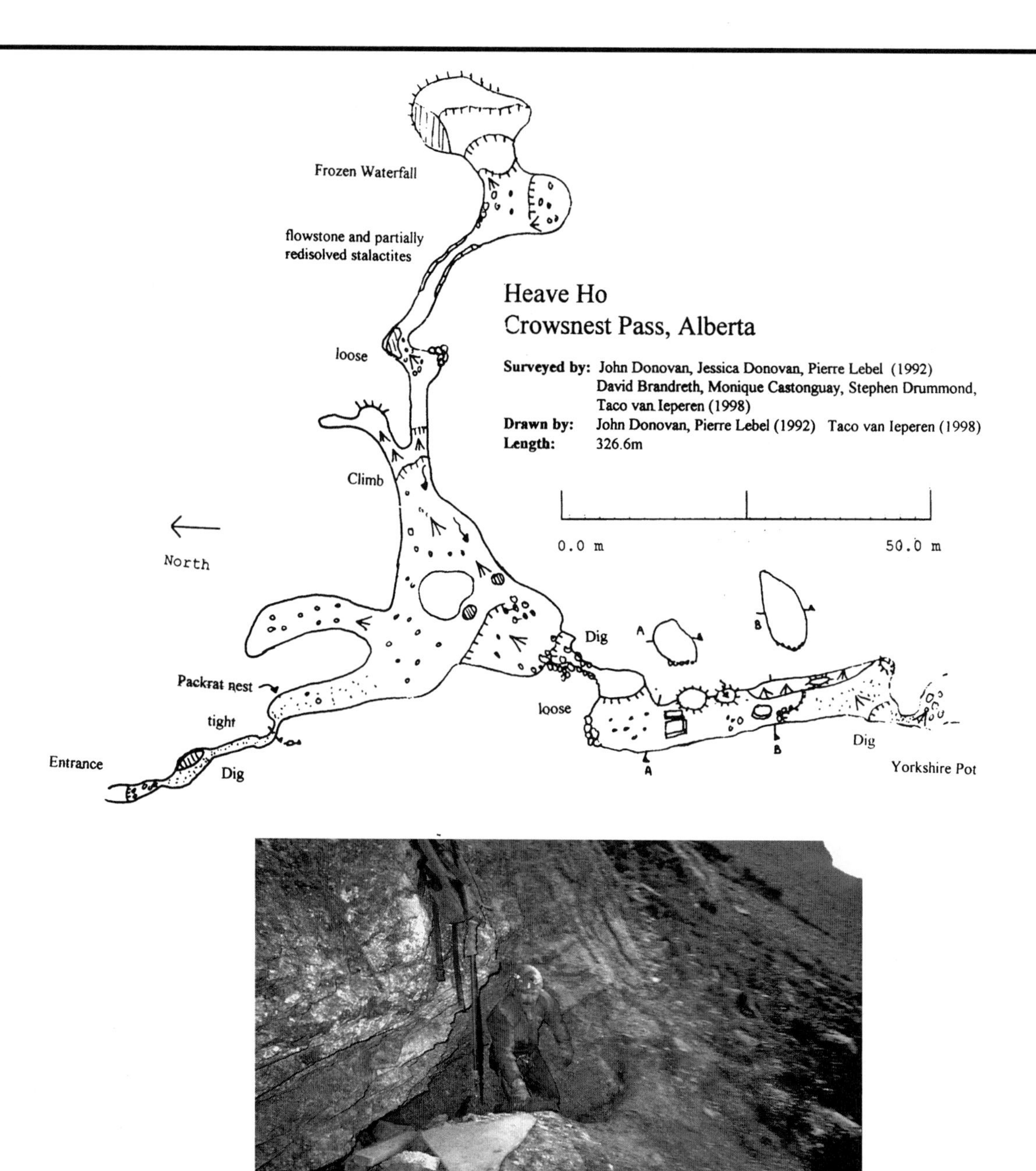

Entrance to Heave-Ho!
Photo Chas Yonge.

Hi-Ho!

Map 82 G/10 733927
Jurisdiction Alberta Forests (Flathead)
Entrance elevation 2104 m
Number of entrances 1 (Connects with Heave Ho!)
Length 326.6 m
Discovered 1999, by the ASS

Location Hi-Ho! is located about 100 m above Heave-Ho! The small, excavated entrance lies at the top of a scree slope at the base of the headwall.

Description Hi-Ho connects with Heave-Ho! by way of a frozen waterfall pitch first pushed from Heave-Ho! before the entrance to Hi-Ho! was found and the connection to YP made. Attempts included ice climbing and the use of a maypole. Hi-Ho! has not been surveyed. However, a description of a through trip made between the two caves indicates two pitches of at least 10 m were encountered, bolts being placed at the head of the first pitch.

Ptolemy Plateau

The so-called Ptolemy Plateau is actually a series of alpine meadows at treeline, located about 500 m above the end of the Ptolemy Creek seismic road.

Access to all caves After the seventh and final crossing of Ptolemy Creek on the seismic road, follow the road to a junction. Keep straight and reach the end of the road at 6 km from the Chinook Coal Road. The final turn-around area is sometimes marked by a caving information sign and log book.

Hidden in the willows is an ongoing trail that climbs steeply through scree to some large boulders (half-way rocks), and then winds up through dense trees to the commonly-used wild camping areas on Ptolemy benches adjacent to the Ice Hall doline and below Camp Caves (1½ hours from the road). See access map on page 52.

If camping here, use minimal impact camping techniques by carrying a stove. Water can be obtained from melting snow patches higher towards the Andy Good Col early in the season. In summer, get water from a year-round spring located at the nose of the impressive cliff that dominates the view to the west.

Camp Caves

Map 82 G/10 725923
Jurisdiction Alberta Forests (Flathead)
Entrance elevation 2165 m
Number of entrances Numerous passage remnants
Length, Depth Not surveyed
Discovered 1969, by Gary Pilkington & Tich Morris

Location In the meadows above the impressive Ice-Hall doline are dotted at least four entrances to the passage remnants of Camp Caves.

Description These cave passage remnants have been chopped up, and in places de-roofed by glacial activity. They provide a handy shelter for tentless visitors in stormy weather, complete with all-night entertainment from a large population of pack rats.

Camp Caves, a series of cave remnants deroofed by glaciation. Photo Gillean Daffern.

Ice Hall

Map 82 G/10 725923
Jurisdiction Alberta Forests (Flathead)
Entrance elevation 2180 m
Number of entrances 1
Length, Depth c. 130 m, 97 m
Discovered 1970, by Pete Fuller, Tich Morris & Peter Thompson

Ice Hall, frozen shut for many years.
Photo Gillean Daffern.

Location Adjacent to the camping areas is a large doline backed by an overhanging cliff. The pond in the bottom rests on the frozen entrance of Ice Hall Cave.

Exploration & Description Never easy to enter, this cave has been shut since 1987 owing to the the collapse and build-up of ice. When opened in 1970 by explosives, it was explored through two impressive chambers to a tight tube ending in a drafting dig. Large amounts of redissolved flowstone were found in the lower chamber.

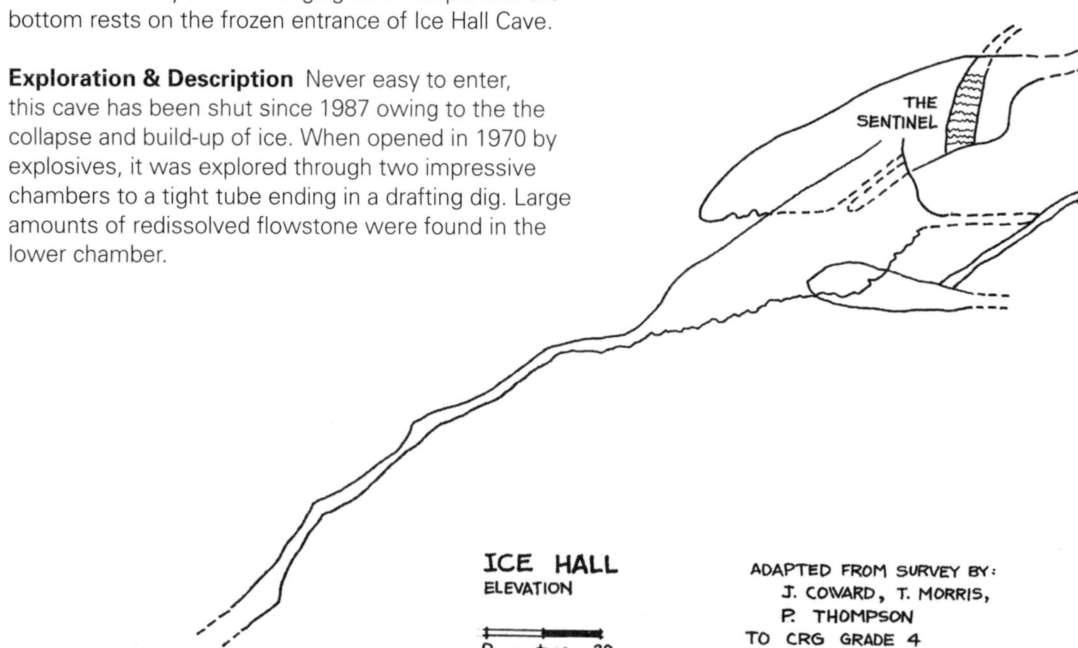

Cleft Cave

Map 83 G/10 730924
Jurisdiction Alberta Forests (Flathead)
Entrance elevation 2379 m
Number of entrances 3
Length 412 m
Discovered Unknown

View of the Cleft cirque from Camp Caves. The Andy Good Col to right accesses Yorkshire Pot, Gargantua and the Andy Good Plateau. Photo Gillean Daffern.

Location Cleft Cave is situated 200 m above Camp Caves in the axis of the synclinal bowl to the southeast. The 18 m-high, 3 m-wide slot-like entrance is clearly visible from the camping area.

The entrance is best approached from the right (south). Follow a rough trail up scree on the right-hand side of the cirque, then after a short climb up outcropping rock, traverse left along ledges to below the entrance. The steep entrance gully is best climbed one at a time to avoid rockfall.

Description Cleft Cave, which passes straight through a ridge to two lookout points across from Andy Good Peak, is a popular destination for weekend hikers.

The first 100 m of passage is a spacious canyon with large scallops on the walls. When the passage gets low and large breakdown blocks obstruct the way on, the Lower Loop passage leads off to the left. This passage contains some very pretty ice crystals, and following a short squeeze and climb, leads to Lofty Lookout with a spectacular view towards Mt. Parish and Andy Good Peak. (Move cautiously to avoid falling out of the lookout.)

The main passage, called the Wind Tunnel owing to a strong breeze, is a tight winding fissure leading to another lookout. A delightful phreatic crawl tube, Ice Crystal Tube, connects the two passages. Some redissolved flowstone and stalactites can be seen in the roof of the Wind Tunnel.

A 15 m pit just off the Lower Loop is blind. However, a draft from a boulder choke above the other side indicates the possibility of more passage. Cleft Cave is obviously a remnant of a much larger cave system that was removed by glacial activity in the two bounding cirques.

Lofty Lookout in Cleft Cave. Photo Dave Thomson.

Warning Reaching the cave entrance requires crossing steep avalanche-prone slopes. There has been more than one accident when visitors fell or were hit by falling rocks immediately below the entrance.

Cave Softly Spray paint arrows have been placed on the walls in several locations. These are unnecessary, unsightly and very hard to remove. Please remove any garbage you see in the cave.

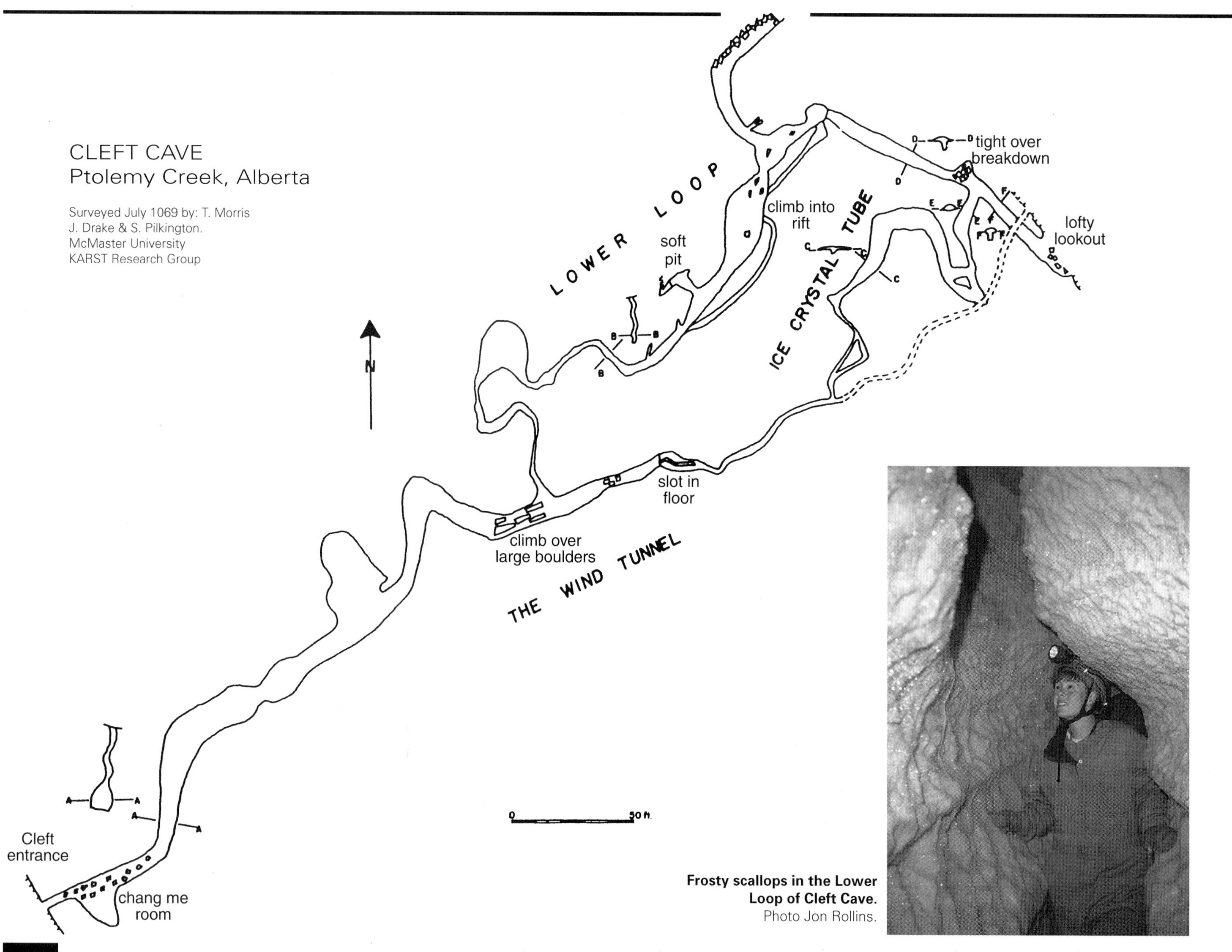

Frosty scallops in the Lower Loop of Cleft Cave.
Photo Jon Rollins.

Andy Good Plateau

The large area of bedrock and rubble-strewn benches south of the Andy Good Col in British Columbia is known as the Andy Good Plateau. The plateau karst area is dominated by two major contrasting cave systems: Yorkshire Pot with 200 m of entrance shafts leading to 12 km of largely horizontal phreatic passages, and Gargantua with its massive entrance chamber and large lower phreatic and vadose canyon passages. Although the entrances lie only 1 km apart, the two cave systems have not been connected and have probably been separate since glacial activity removed much of the rock from above the present level of the plateau surface. Both caves drain to Ptolemy Spring, and so share a hydrological catchment. In recent years seven caves have been connected to Yorkshire Pot: The Backdoor, Mendips Cave, Shorty's Cave, Snowslope Pot, Quinta Penta, Heave Ho! and Hi Ho!. Although now technically part of the Yorkshire Pot system, these caves all have their own exploration histories, and so have been listed separately.

Steep cliffs dropping down to the headwaters of Andy Good Creek delineate the plateau's southern extent. From the Andy Good Col a cairned trail winds up to the entrance of Gargantua Cave, hidden high in Mt. Ptolemy's southeast face. To the northeast is Rat's Hole Cave. The remainder of the caves, mainly the numerous entrances to Yorkshore Pot lie directly below the col (the Backdoor), and to the northeast where the small dome of limestone emerging from the scree marks the main entrance to YP — Canada's second longest cave and fifth deepest.

Snowslope, Mendips, Derbyshire, Quinta Penta, Shorty's and Little Moscow caves are all hard to find, and depending on snow levels, may be buried or blocked by ice for a good part of the year. Even the original discoverers have a hard time relocating the entrances, and often a close examination of the survey is necessary to see if a cave is a new discovery or not.

Usually all of the caves on Andy Good Plateau require overnight camping down on Ptolemy Plateau.

Access to all caves From the camping area on the Ptolemy benches, the way over to the Andy Good Plateau is via the obvious Andy Good Col directly to the southeast. The easiest way is via an indistinct trail that traverses grassy knolls to the right (south) before traversing back left across scree to the col itself. The well-worn trail traversing scree to the left (north) is the down trail. See access map on page 52.

Chas Yonge on the Andy Good Plateau. Photo Dave Thomson.

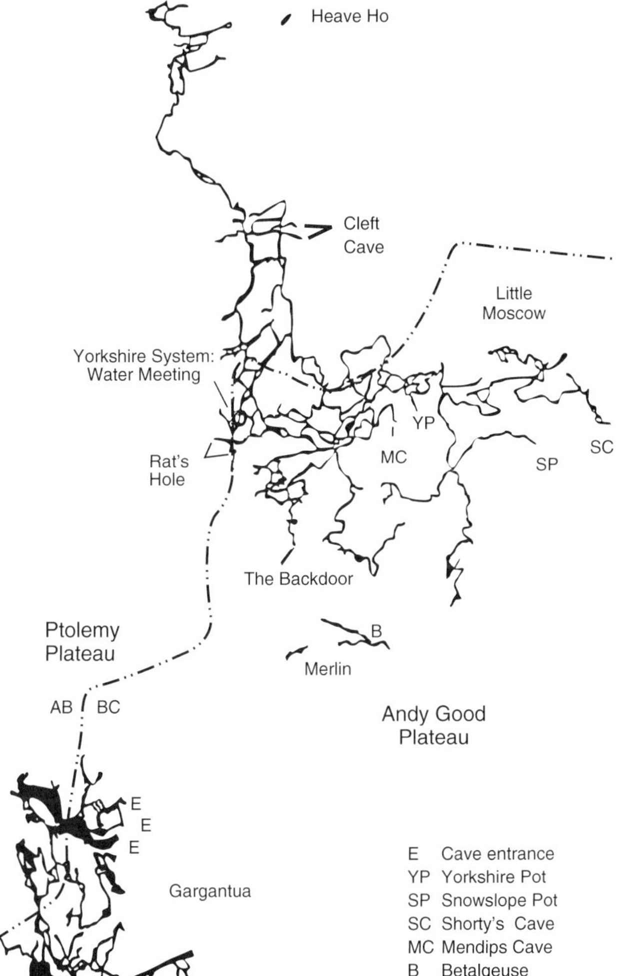

The Caves of the Ptolemy and Andy Good plateaus. (Mike Evans & Steve Worthington, 1989) .

Rat's Hole

Map 82 G/10 732922
Jurisdiction Alberta Forests (Flathead) & BC Forests (Elk)
Entrance elevation 2455 m
Number of entrances 2
Length, Depth 130 m, 19.2 m
Discovered 1971, by Mike Shawcross

Location The higher B.C. entrance lies close to the top of the ridge adjacent to the east side of the Andy Good Col. From the col scramble steeply up to your left. Shortly before the top of the ridge, the low crawl-in entrance is located at the base of a small bluff.

The lower Alberta entrance looks out over the trail leading up to the col from Ptolemy Plateau.

"Mike Shawcross got excited by the huge draft blowing from the entrance. A short dig led to large passage and a three metre drop. He returned to camp convinced the way to the deepest cave on the plateau lay open. Next day he returned with Pete Smart laden down with all the team's gear. They found out where the draft came from; the second lower entrance." Pete Smart.

Exploration & Description The higher crawl-in entrance over scree quickly accesses a large chamber and then a rift passage that leads to the only drop in the cave. Beneath the 3 m drop, a large chamber leads to a junction. To the left, a large beautifully sculptured keyhole passage with extensive scalloping on the walls heads out of the mountainside (Alberta entrance). Straight on are the remains of a large ice mass that used to block the passage. In 1990 this had melted back, allowing access to a short section of canyon passage ending at more ice and a dig with a fluctuating draft, probably from another entrance.

Rat's Hole is named for the massive piles of pack rat debris that block roof tubes in the cave.

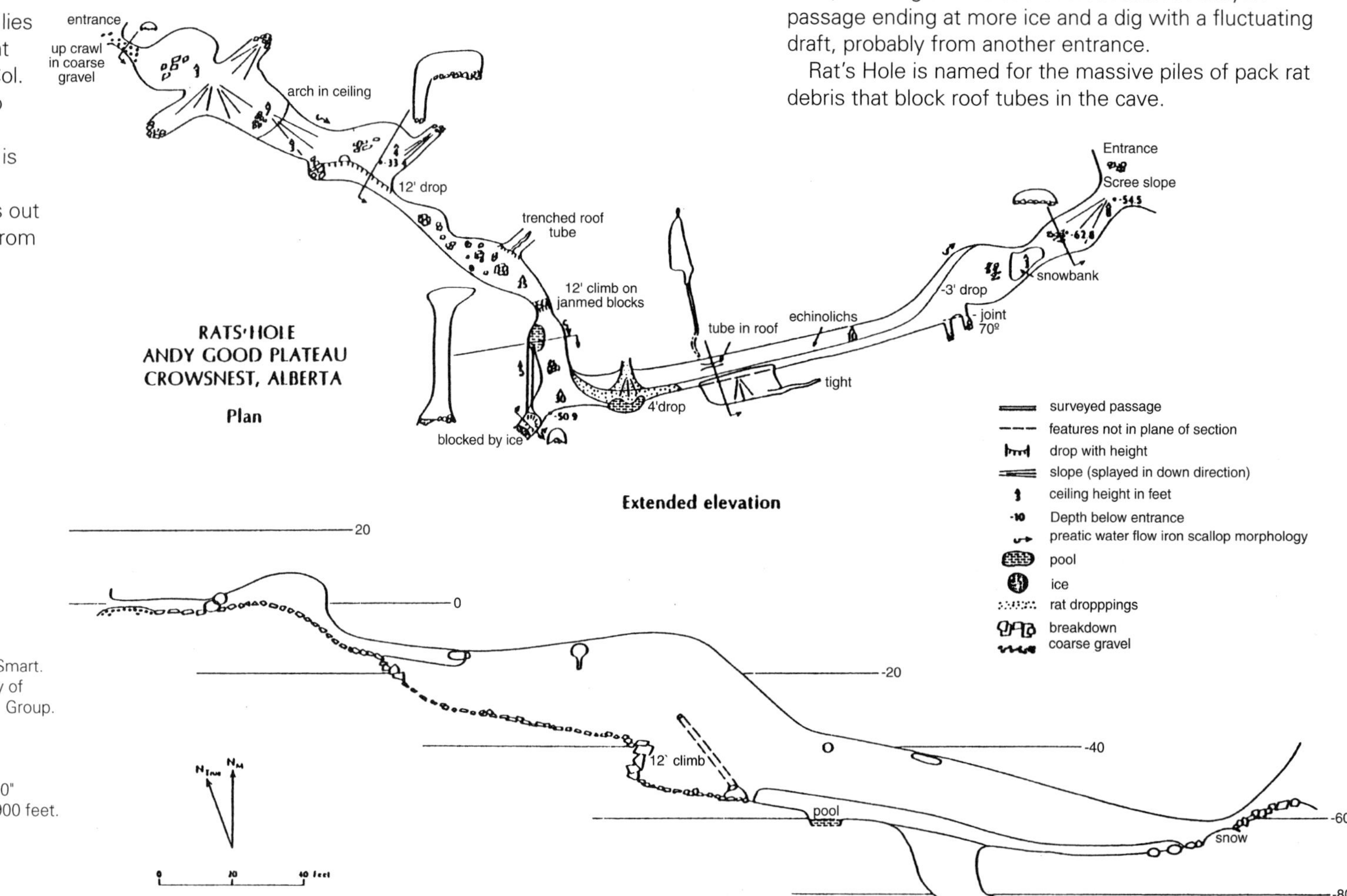

Notes
1. Surveyed 1971 by R. Bignell & P. I. Smart.
2. Compiled by P. I. Smart at University of Alberta for Canadian Karst Research Group.
3. Survey CRC gade 5 d
4. Total surveyed length 357.2 feet, total depth 63 feet.
5. Grid reference at entrance 49° 33' 40" 114° 35' 25" entrance elevation 7,900 feet.
6. Magnetic declination 21° L.

"Heading out of the cave, Pete Livesey and Peter Thompson only had one pair of jumars between them. Livesey apparently was unaware of this and left Thompson behind. Thompson cut some rope from the bottom of the 40 m pitch and prussicked the 200 m of pitches out of the cave, arriving back late at camp. No one had noticed he was missing." Ian Drummond.

Chas Yonge rapelling in Yorkshire Pot.
Photo Dave Thomson.

Yorkshire Pot

Map 82 G/10 732921J
Jurisdiction BC Forests (Elk)
Entrance elevation 2412 m
Number of entrances 8 (connects with The Backdoor, Shorty's Cave, Mendips Cave, Snowslope Pot, Quinta Penta, Heave-Ho! & Hi Ho!)
Length, Depth 13.812 km (including all connecting caves), 389 m
Discovered July 1969, by Derek Ford, Owen Ford & John Drake

Main entrance location The dome of limestone that protects the entrance, projects up above the frost-shattered surrounding karren and can be seen from the Andy Good Col at a similar elevation a few hundred metres to the southeast. Look for an intermittent trail, The entrance fissure forms a 3 m vertical slot.

Exploration & Discovery From the entrance slot drop into a small chamber from which a narrow, 3 m-high canyon leads to the head of Lover's Leap, an awkward 6 m down-climb (most parties use a rope). Following this is an obvious T-junction. To the right lies No Bar Extension and a connection with Shorty's Cave.

The way on, to the left, is an awkward narrow canyon that accesses the head of the first pitch, a 13 m drop. A short passage then leads to the second 17 m pitch, scene of Al Schaffer's fall and rescue in 1973. This is followed by an unpleasant squeeze at the head of the third 7 m pitch. These first three pitches are known collectively as Rubble Pots.

After the fourth 28 m pitch, you enter Confluence Rift. From a floor of jammed boulders a short scramble and down-climb along the rift leads to the Horrible Holdless Squeeze at the top of the 40 m free-fall drop. One more 15 m pitch leads to a sandy crawl and the Chocolate Chamber, the start of the horizontal passage in Yorkshire Pot.

Once rigged, these entrance pitches are relatively straightforward to descend, although parties have been known to miss the turn into the canyon after Lover's Leap. On the way out after a typical 12 hours of hard caving (prior to the connection of Heave Ho!), this seemingly endless series of pitches can be very exhausting. With a large pack, the squeeze at the top of the 40 m pitch and the final climb up Lover's Leap are especially memorable. As Ian McKenzie, mastermind behind much recent exploration noted: "Despite twenty years of cave discoveries elsewhere, it's still one of the most serious bottoming trips in the country, requiring two teams over two days. There are more pitches in this cave than in any other in Canada [now surpassed by Close to the Edge]."

From Chocolate Chamber the Roller Coaster Run — a large phreatic series — heads to the west and is crossed by several slippery vadose trenches. Just before the first of these, The Greasy Traverse, a passage to the left leads down to the Muddy Gulch Series and Leprechaun's Leap — the junction with The Backdoor. Heading back up from Leprechaun's Leap, the F Survey leads to connections with Snowslope Pot and Shorty's Cave, and via the Horror Show and Surprise Streamway to a connection with Quinta Penta Pot. In one direction the Horror Show terminates beneath the Chocolate Chamber (43 m connecting pitch) and in the other connects with Lower Alberta Avenue.

The next junction on the Roller Coaster Run, known as Second Look Junction, leads to Upper Alberta Avenue that runs southwest at an average downward dip of 20°. Scallop formations indicate water flow was uphill, turning downhill again to the west into the High Level Duck (dry in winter). Alberta Avenue drops to a point where two streams join and form The Water Meeting.

A narrow sinuous stream passage leads off to the right via several small drops to a sump. The sump at –283 m has a bedrock floor and was the lowest point reached in 1970.

From The Water Meeting, Lower Alberta Avenue rises steeply heading east along a large 3 m by 3 m walking passage that eventually connects with the Horror Show. Two of the right-hand passages, the P Survey and the Rat Route, head out toward the middle of the Andy Good Plateau. The P Survey has been pushed via Exhibition Way to a series of chambers and loose canyons and a second connection with Snowslope Pot at Inhibition Way. Rat Route ends in some tight upward crawls.

Back at Roller Coaster Run, the passage starts to rise and at its highest point takes an abrupt left turn and plunges downhill in a 1.5 m by 1 m passage with a calcite floor and a stalactite-covered roof (The Tight White Way). The passage is drafting but gets increasingly tight. The way on is a down-dip side passage that following an awkward down-climb gets larger before ending abruptly at the Green Pool Sump.

Backtracking to the first junction, you come to a side passage leading to a well-decorated bedding-plane room, after which it opens up again (site of 1973 camp) and continues to descend (Blood Stone Passage) to the head of a series of beautifully decorated drops totalling 40 m called The Seven Steps. Beyond the Seven Steps a large passage floored with chert heads steeply down (30°–40°) to a mud sump at –384 m that at the time of discovery in 1972 established a new North American depth record. This is still the deepest point in the cave, and although numerous side passages have been explored, no way round the terminal sump has been found. The 1998 connection of this part of the cave with Heave Ho! has stimulated exploration in these lower passages.

Yorkshire Pot has now been connected with seven other caves on the Andy Good Plateau, making it the second longest cave in Canada. As with Gargantua and Rat's Hole Cave, Yorkshire Pot is an interprovincial cave, the entrance shafts being in BC and much of the horizontal passage (including Heave Ho!) in Alberta. There are still hundreds of leads to be checked and even more to be surveyed. However, exploration is often motivated by competition and as the next contending cave (Thanksgiving Cave) is 5 km shorter, it will probably be a while before all the gaps are filled in.

Upper Alberta Avenue, Yorkshire Pot. Photo Ian Drummond.

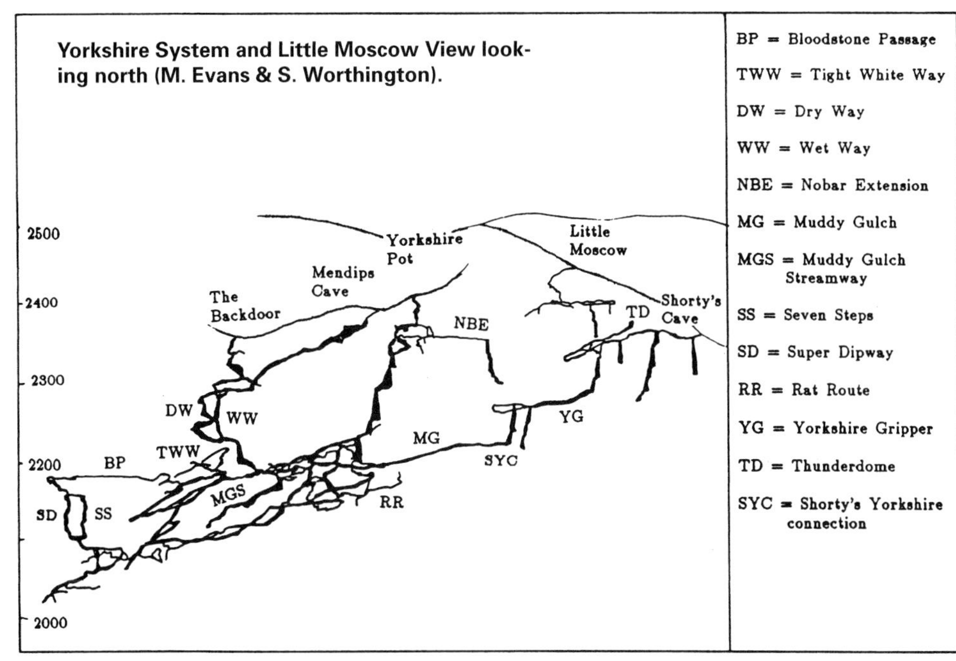

Yorkshire System and Little Moscow View looking north (M. Evans & S. Worthington).

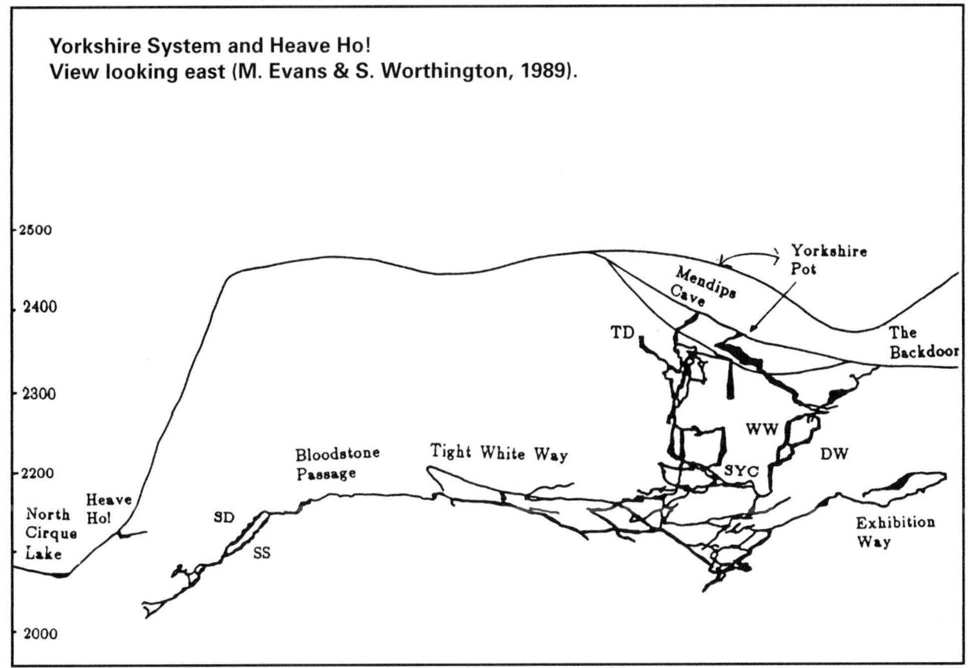

Yorkshire System and Heave Ho!
View looking east (M. Evans & S. Worthington, 1989).

Above: Camp in Yorkshire Pot. Photo Ian Drummond.

Warning Yorkshire Pot is a serious caving trip. An accident in the cave could easily become a fatality. Gain experience in easier caves before attempting Yorkshire Pot.

Cave Softly Most of the heavily decorated passage in Yorkshire Pot lies in the area of Bloodstone Passage and The Seven Steps. Superb examples of many types of formations can be found here in pristine condition, including cave pearls. Prior to discovery of the exit through Heave Ho!, Yorkshire Pot was only visited by experienced cavers, and the formations toward the end of the cave could be reached only after descending many pitches followed by complex route finding along extensive horizontal passages. It remains to be seen whether increased traffic from through trips and easier access to the lower passages adversely impacts the cave. A strong wind through the cave generated by the opening of the bottom entrance has been identified as a potential threat to speleothem formation, and the connection with Hi Ho! is usually left blocked with rubble.

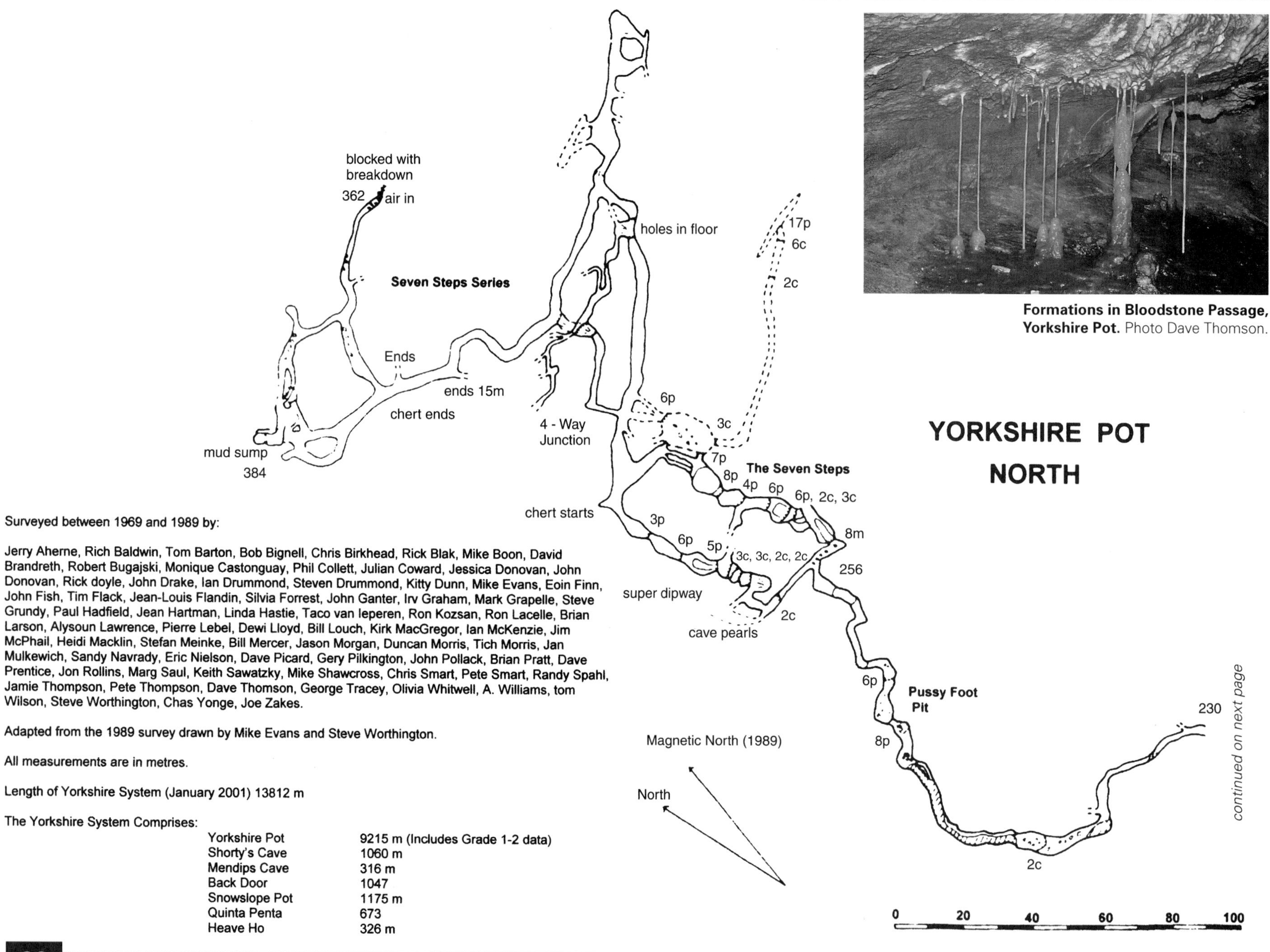

Formations in Bloodstone Passage, Yorkshire Pot. Photo Dave Thomson.

continued on next page

Surveyed between 1969 and 1989 by:

Jerry Aherne, Rich Baldwin, Tom Barton, Bob Bignell, Chris Birkhead, Rick Blak, Mike Boon, David Brandreth, Robert Bugajski, Monique Castonguay, Phil Collett, Julian Coward, Jessica Donovan, John Donovan, Rick doyle, John Drake, Ian Drummond, Steven Drummond, Kitty Dunn, Mike Evans, Eoin Finn, John Fish, Tim Flack, Jean-Louis Flandin, Silvia Forrest, John Ganter, Irv Graham, Mark Grapelle, Steve Grundy, Paul Hadfield, Jean Hartman, Linda Hastie, Taco van Ieperen, Ron Kozsan, Ron Lacelle, Brian Larson, Alysoun Lawrence, Pierre Lebel, Dewi Lloyd, Bill Louch, Kirk MacGregor, Ian McKenzie, Jim McPhail, Heidi Macklin, Stefan Meinke, Bill Mercer, Jason Morgan, Duncan Morris, Tich Morris, Jan Mulkewich, Sandy Navrady, Eric Nielson, Dave Picard, Gery Pilkington, John Pollack, Brian Pratt, Dave Prentice, Jon Rollins, Marg Saul, Keith Sawatzky, Mike Shawcross, Chris Smart, Pete Smart, Randy Spahl, Jamie Thompson, Pete Thompson, Dave Thomson, George Tracey, Olivia Whitwell, A. Williams, tom Wilson, Steve Worthington, Chas Yonge, Joe Zakes.

Adapted from the 1989 survey drawn by Mike Evans and Steve Worthington.

All measurements are in metres.

Length of Yorkshire System (January 2001) 13812 m

The Yorkshire System Comprises:

Yorkshire Pot	9215 m (Includes Grade 1-2 data)
Shorty's Cave	1060 m
Mendips Cave	316 m
Back Door	1047
Snowslope Pot	1175 m
Quinta Penta	673
Heave Ho	326 m

Tight White Way

? Continues

Blocked with flowstone

Tight

3c

The Roller Coaster

Tight

231

Second Look Junction

Low crawl

3c

1973 Camp

Frozen Pools

Blood Stone Passage

236

continued from previous page

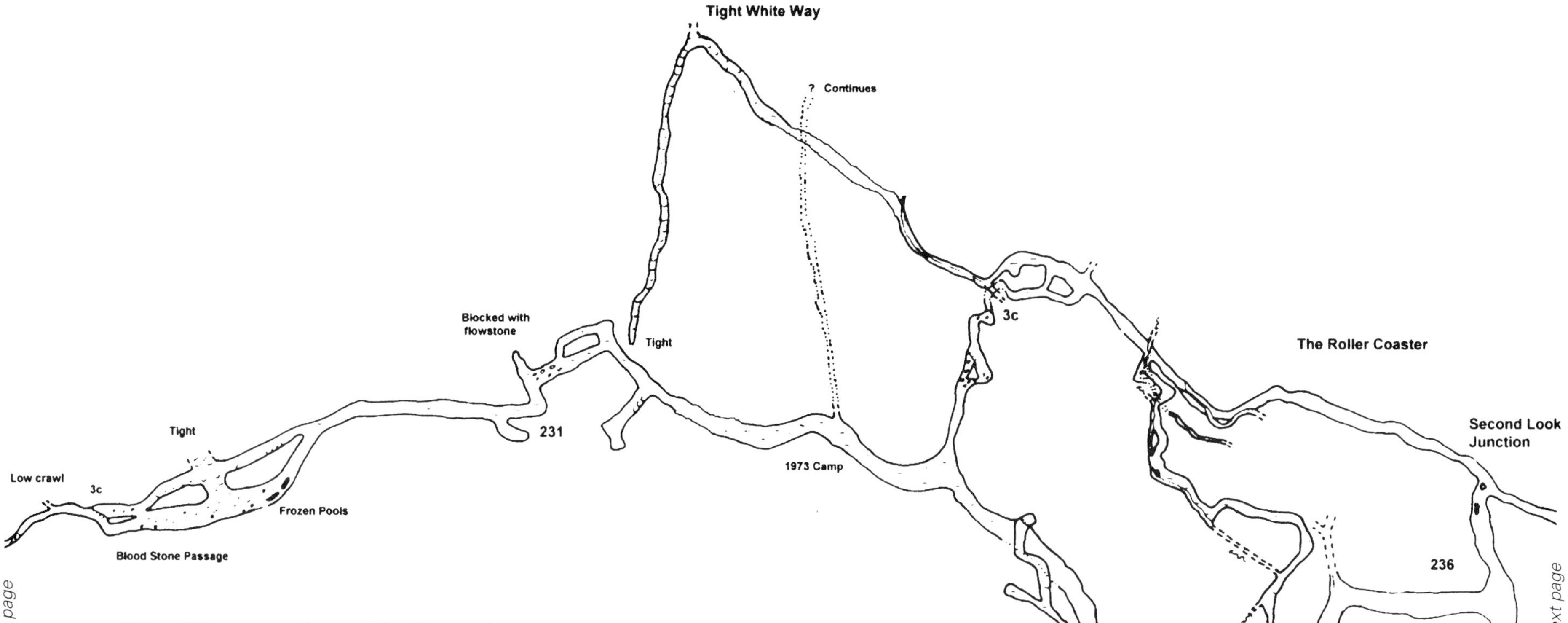

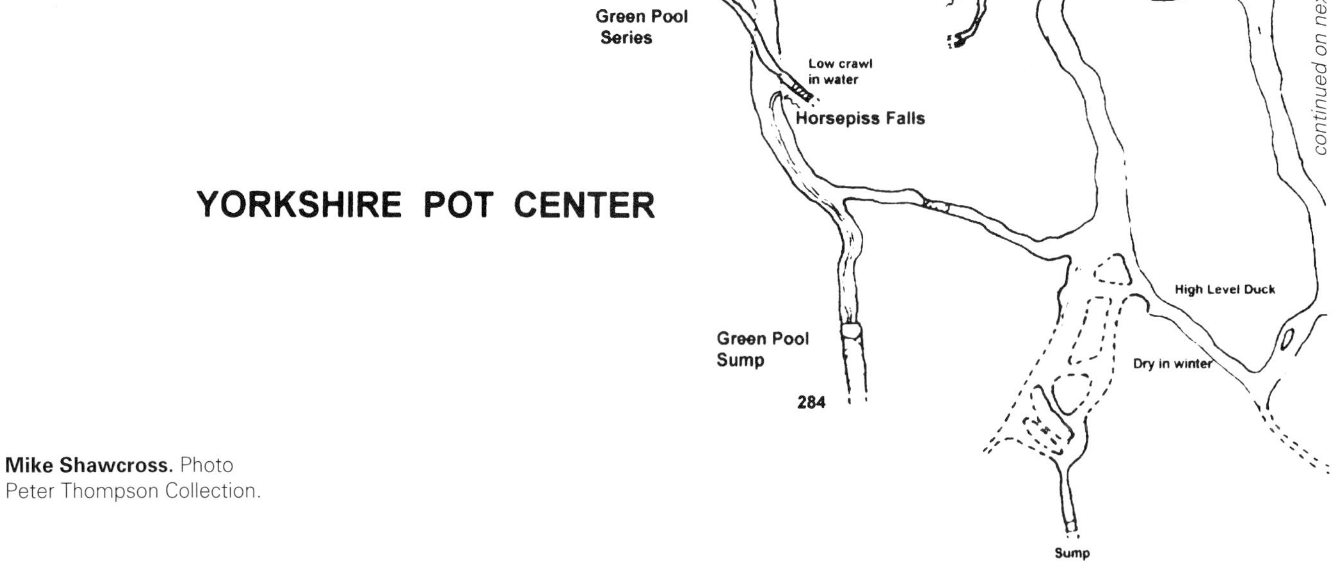

Green Pool Series

continued on next page

Low crawl in water

Horsepiss Falls

YORKSHIRE POT CENTER

High Level Duck

Green Pool Sump

Dry in winter

284

Mike Shawcross. Photo Peter Thompson Collection.

Sump

continued at bottom of next page

continued from previous page

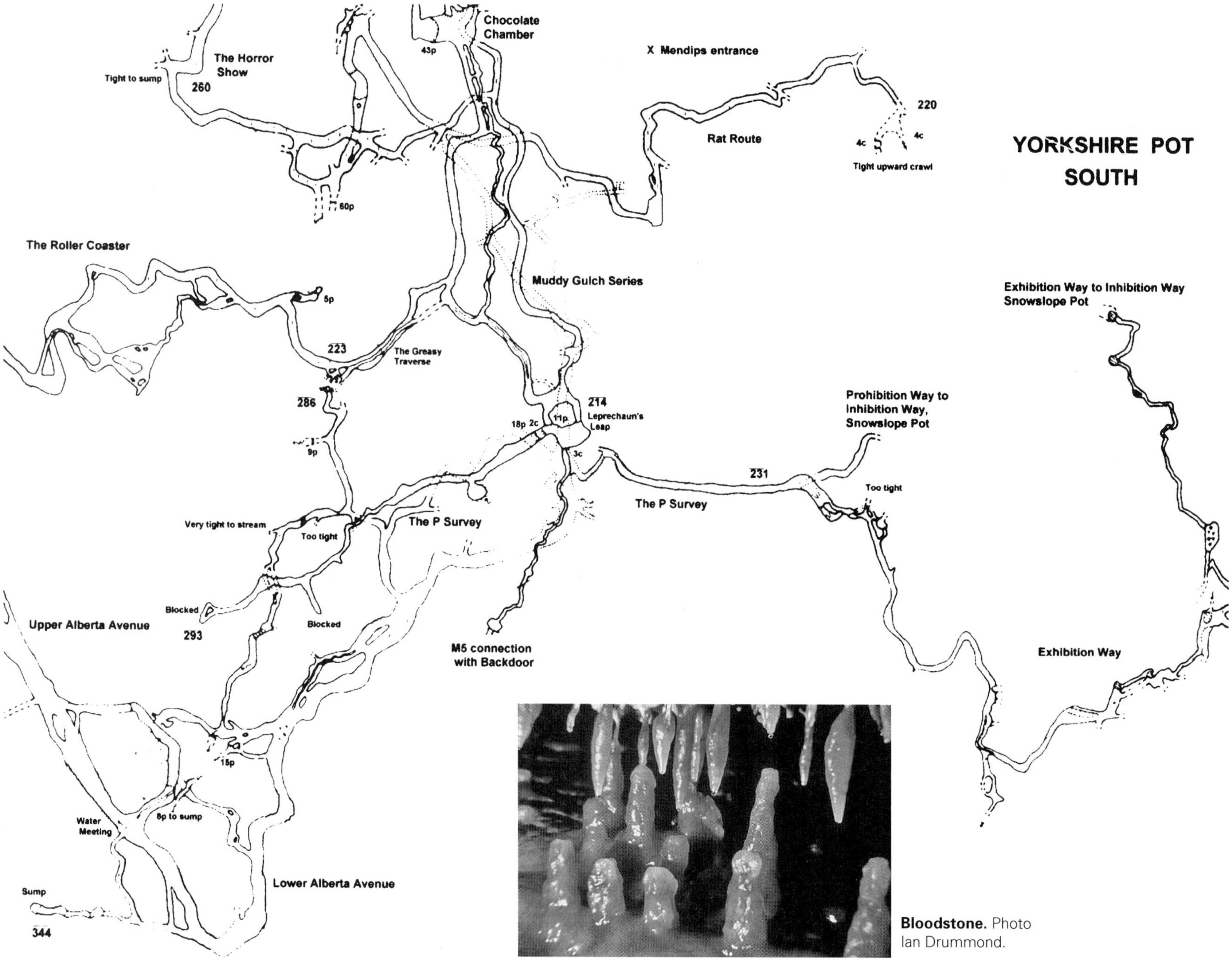

YORKSHIRE POT SOUTH

Chocolate Chamber
43p
X Mendips entrance
The Horror Show
Tight to sump
260
220
Rat Route
4c
4c
Tight upward crawl
60p
The Roller Coaster
Muddy Gulch Series
5p
Exhibition Way to Inhibition Way Snowslope Pot
223
The Greasy Traverse
286
214
Leprechaun's Leap
11p
18p 2c
Prohibition Way to Inhibition Way, Snowslope Pot
9p
3c
231
Too tight
The P Survey
Very tight to stream
Too tight
The P Survey
Blocked
Upper Alberta Avenue
293
Blocked
M5 connection with Backdoor
Exhibition Way
15p
Water Meeting
8p to sump
Lower Alberta Avenue
Sump
344

Bloodstone. Photo Ian Drummond.

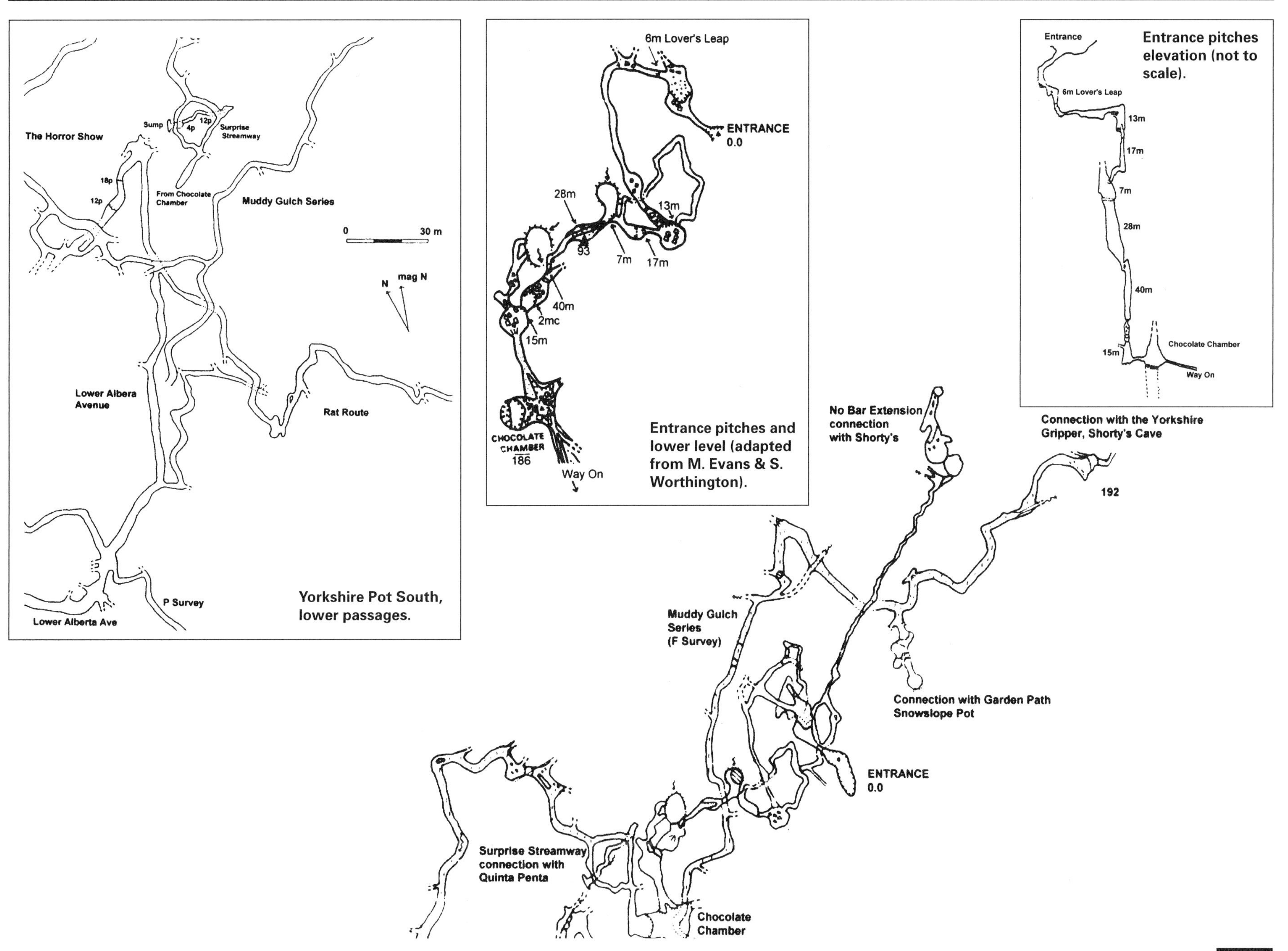

Yorkshire Pot South, lower passages.

Entrance pitches and lower level (adapted from M. Evans & S. Worthington).

Entrance pitches elevation (not to scale).

continued from top of previous page

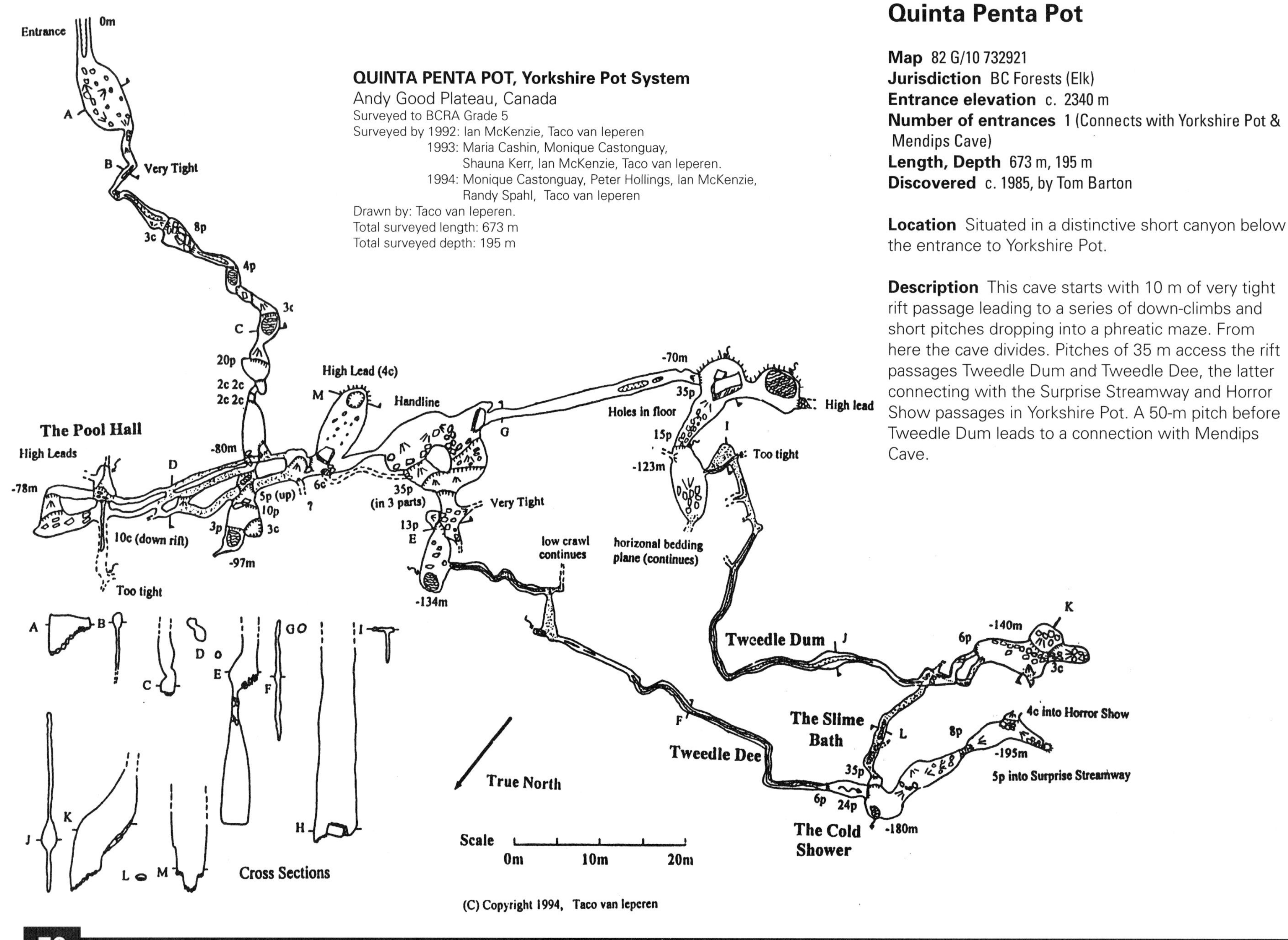

Quinta Penta Pot

Map 82 G/10 732921
Jurisdiction BC Forests (Elk)
Entrance elevation c. 2340 m
Number of entrances 1 (Connects with Yorkshire Pot & Mendips Cave)
Length, Depth 673 m, 195 m
Discovered c. 1985, by Tom Barton

Location Situated in a distinctive short canyon below the entrance to Yorkshire Pot.

Description This cave starts with 10 m of very tight rift passage leading to a series of down-climbs and short pitches dropping into a phreatic maze. From here the cave divides. Pitches of 35 m access the rift passages Tweedle Dum and Tweedle Dee, the latter connecting with the Surprise Streamway and Horror Show passages in Yorkshire Pot. A 50-m pitch before Tweedle Dum leads to a connection with Mendips Cave.

continued at right

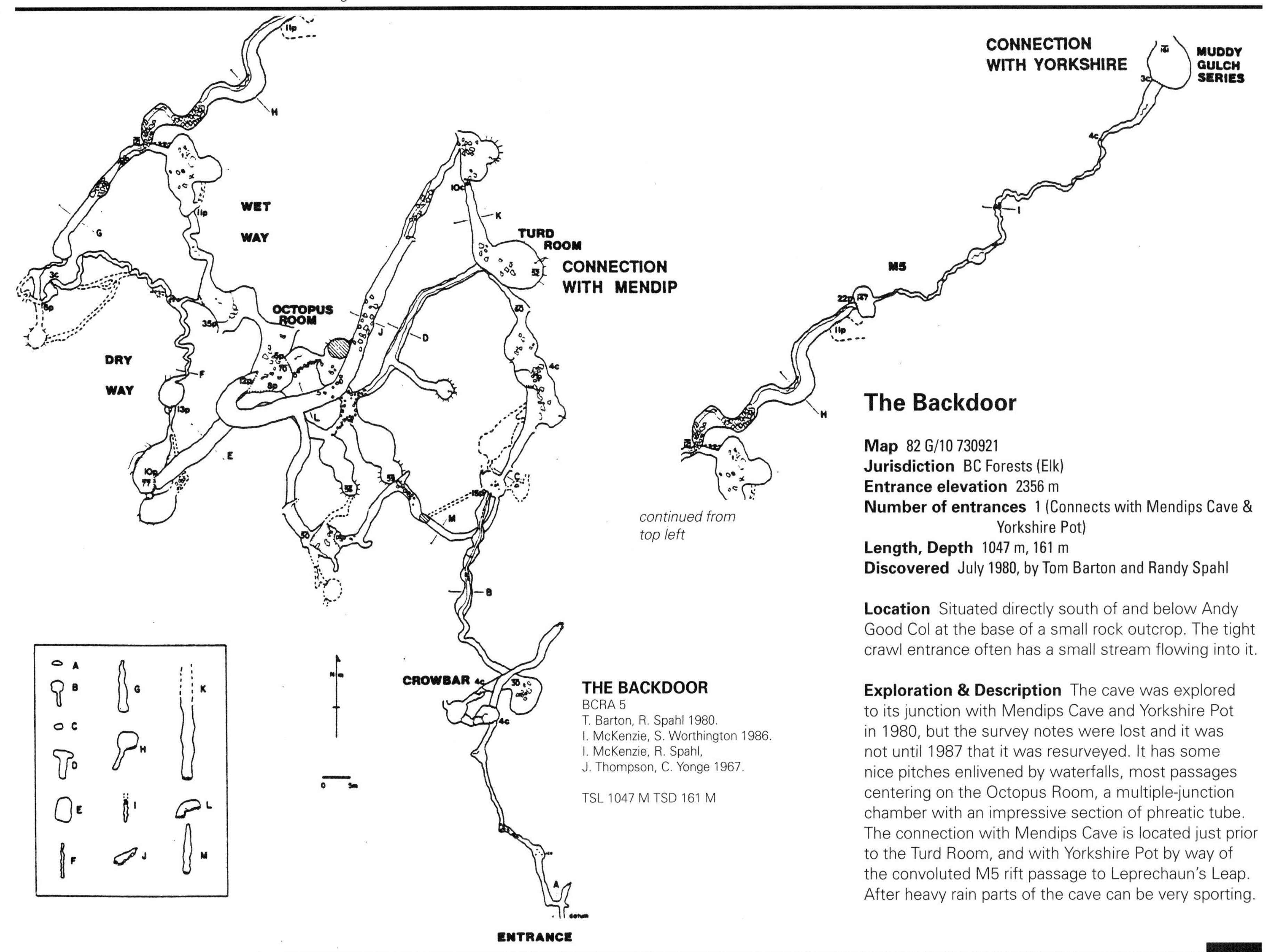

The Backdoor

Map 82 G/10 730921
Jurisdiction BC Forests (Elk)
Entrance elevation 2356 m
Number of entrances 1 (Connects with Mendips Cave & Yorkshire Pot)
Length, Depth 1047 m, 161 m
Discovered July 1980, by Tom Barton and Randy Spahl

Location Situated directly south of and below Andy Good Col at the base of a small rock outcrop. The tight crawl entrance often has a small stream flowing into it.

Exploration & Description The cave was explored to its junction with Mendips Cave and Yorkshire Pot in 1980, but the survey notes were lost and it was not until 1987 that it was resurveyed. It has some nice pitches enlivened by waterfalls, most passages centering on the Octopus Room, a multiple-junction chamber with an impressive section of phreatic tube. The connection with Mendips Cave is located just prior to the Turd Room, and with Yorkshire Pot by way of the convoluted M5 rift passage to Leprechaun's Leap. After heavy rain parts of the cave can be very sporting.

Mendips Cave

Map 82 G/10 732920
Jurisdiction BC Forests (Elk)
Entrance elevation 2386 m
Number of entrances 1 (Connects with The Backdoor & Quinta Penta)
Length, Depth 316 m, 85 m
Discovered August 1969, by Tich Morris

Location About 25 m lower and 45 m to the south of Yorkshire Pot. The entrance is a steep ice slope that was frozen shut shortly after its initial discovery in 1971, and not found open again until 1988.

Trevor "Tich" Morris. Photo Peter Thompson Collection.

Exploration & Description According to Tich Morris, the original exploration in 1969 was interrupted a short distance in by "Two scruffy looking characters who ran off with our ladder yelling something about a 1000' shaft [Derbyshire Pot]."

In 1988 the cave was surveyed for 316 m to a connection with The Backdoor at the Octopus Room. Most of the cave is a single passage with travel at various levels. It provides a more comfortable and interesting alternative to The Backdoor entrance, which has a lot of crawling and tends to be wet.

"Mendips at last." Logo on a T-shirt worn by Ian McKenzie following many years of trying to get into Mendips Cave.

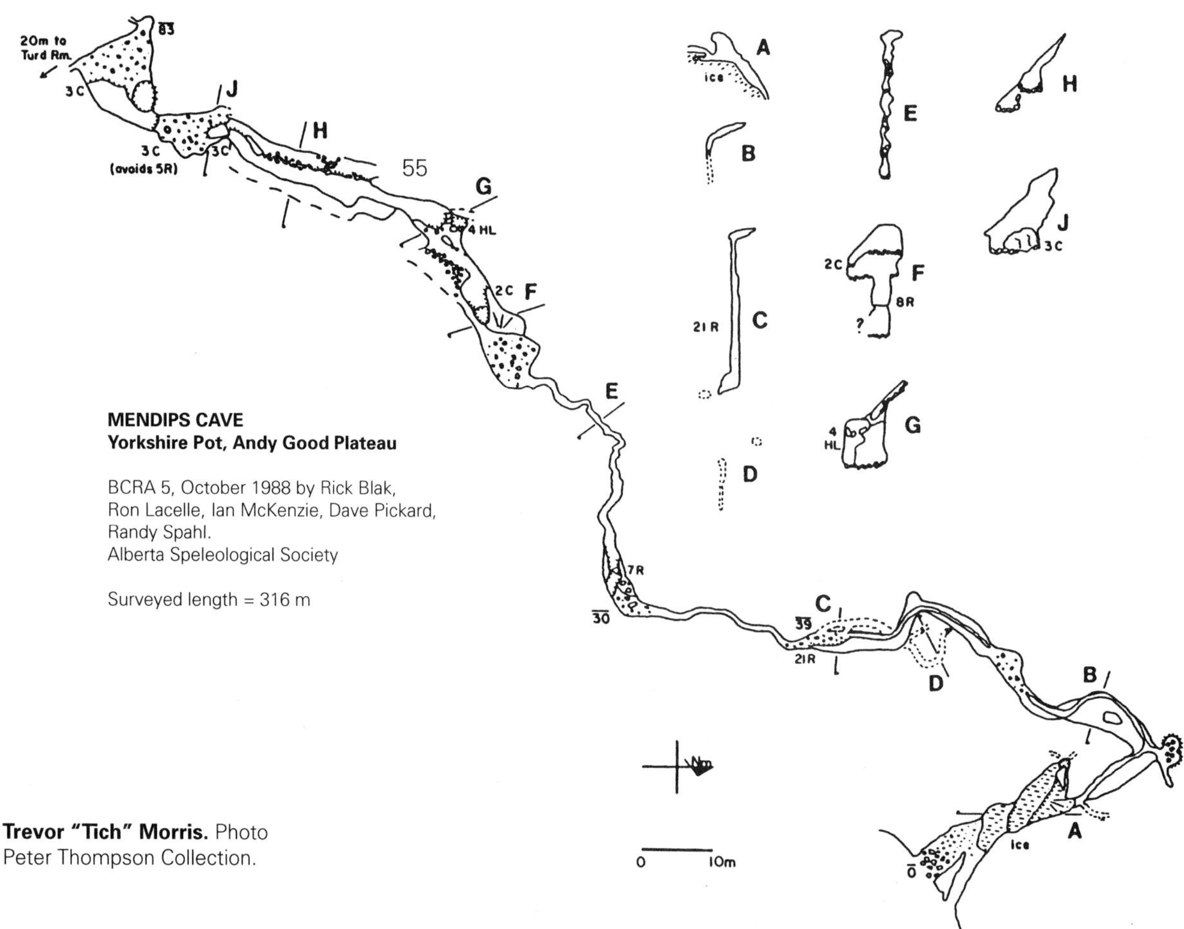

Snowslope Pot

Map 83 G/10 735920
Jurisdiction BC Forests (Elk)
Entrance elevation 2379 m
Number of entrances 1 (Connects with Yorkshire Pot)
Length, Depth 1175 m, 196 m
Discovered August 1969, by Julian Coward, John Drake & Peter Thompson

Location 250 m southeast of YP and 150 m west of Shorty's Cave. The entrance is in a snow-filled depression sloping down against a headwall with a horizontal slot entrance.

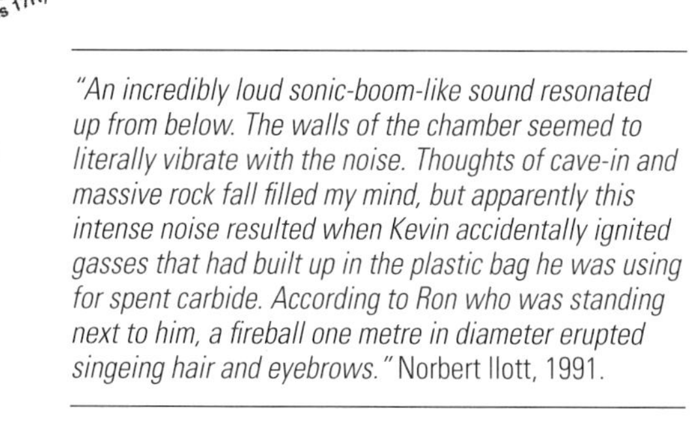

"An incredibly loud sonic-boom-like sound resonated up from below. The walls of the chamber seemed to literally vibrate with the noise. Thoughts of cave-in and massive rock fall filled my mind, but apparently this intense noise resulted when Kevin accidentally ignited gasses that had built up in the plastic bag he was using for spent carbide. According to Ron who was standing next to him, a fireball one metre in diameter erupted singeing hair and eyebrows." Norbert Ilott, 1991.

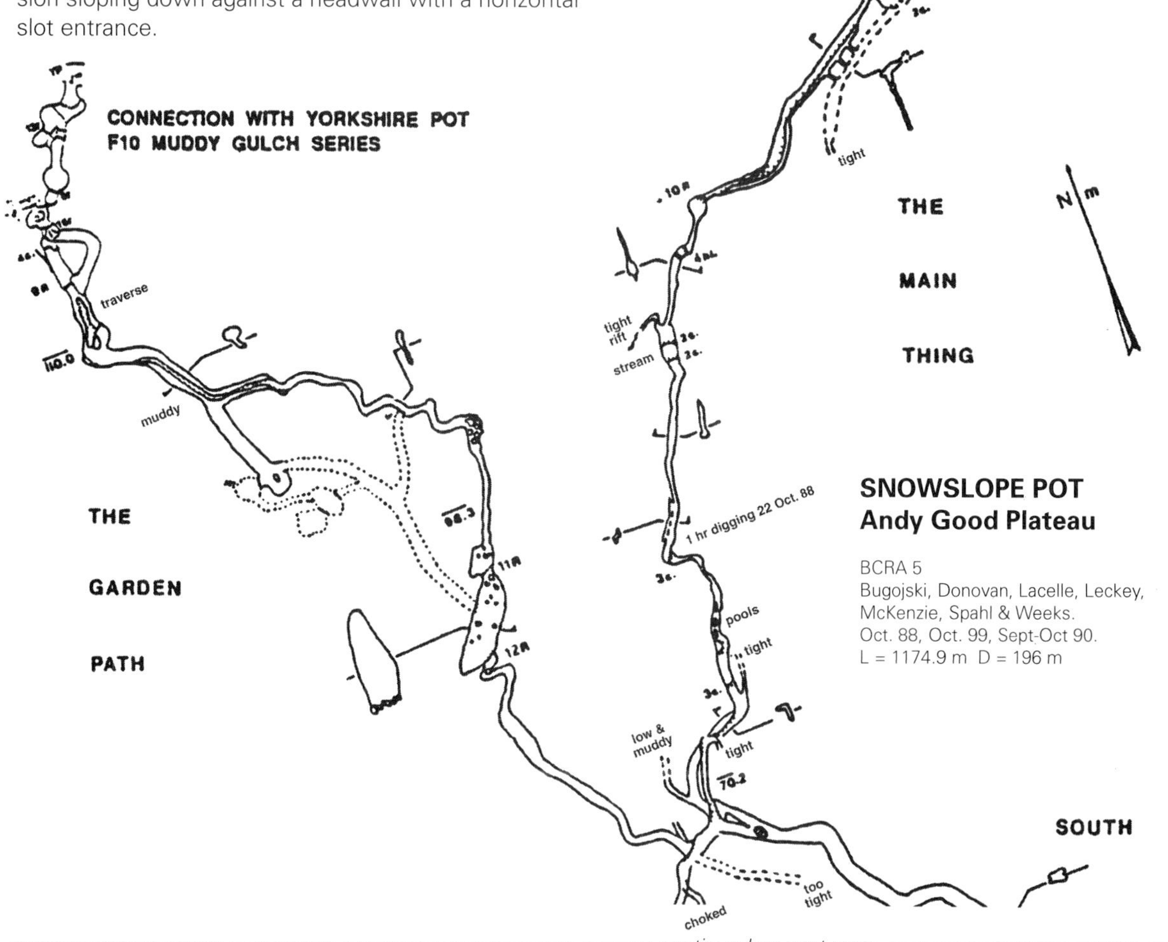

Exploration & Description Snowslope Pot was rediscovered in 1988 and named Stonehenge. When reading the first issue of "The Canadian Caver," Ian McKenzie noticed a reference to a natural rock bridge, a feature just below the entrance pitch, and realized the mistake.

The cave was written off in 1969 as a loose, blind 70 m pot, the original exploration being fairly brief with better prospects such as Yorkshire Pot having just been discovered. Ian McKenzie describes how he found the way on: "While surveying out I had Randy belay me round the pitch mouth into a chamber with a rift leading to a pitch. At the bottom the passage closed down and zigzagged to a choke, something neither of us would have dug five years ago. We did; it went."

With an unusual overall X shape, Snowslope Pot has grown to a respectable length, much of it hard caving along muddy crawls and loose, shattered rifts. The northern axis of Snowslope Pot, The Garden Path, was connected with the F Survey in Yorkshire Pot in 1991 and the western arm, Inhibition Way, was joined with Exhibition Way in 1996. The southern arm, the South Alternative, is heading into an area of the Andy Good Plateau previously thought devoid of caves, and so is of great interest.

continued on next page

continued from previous page

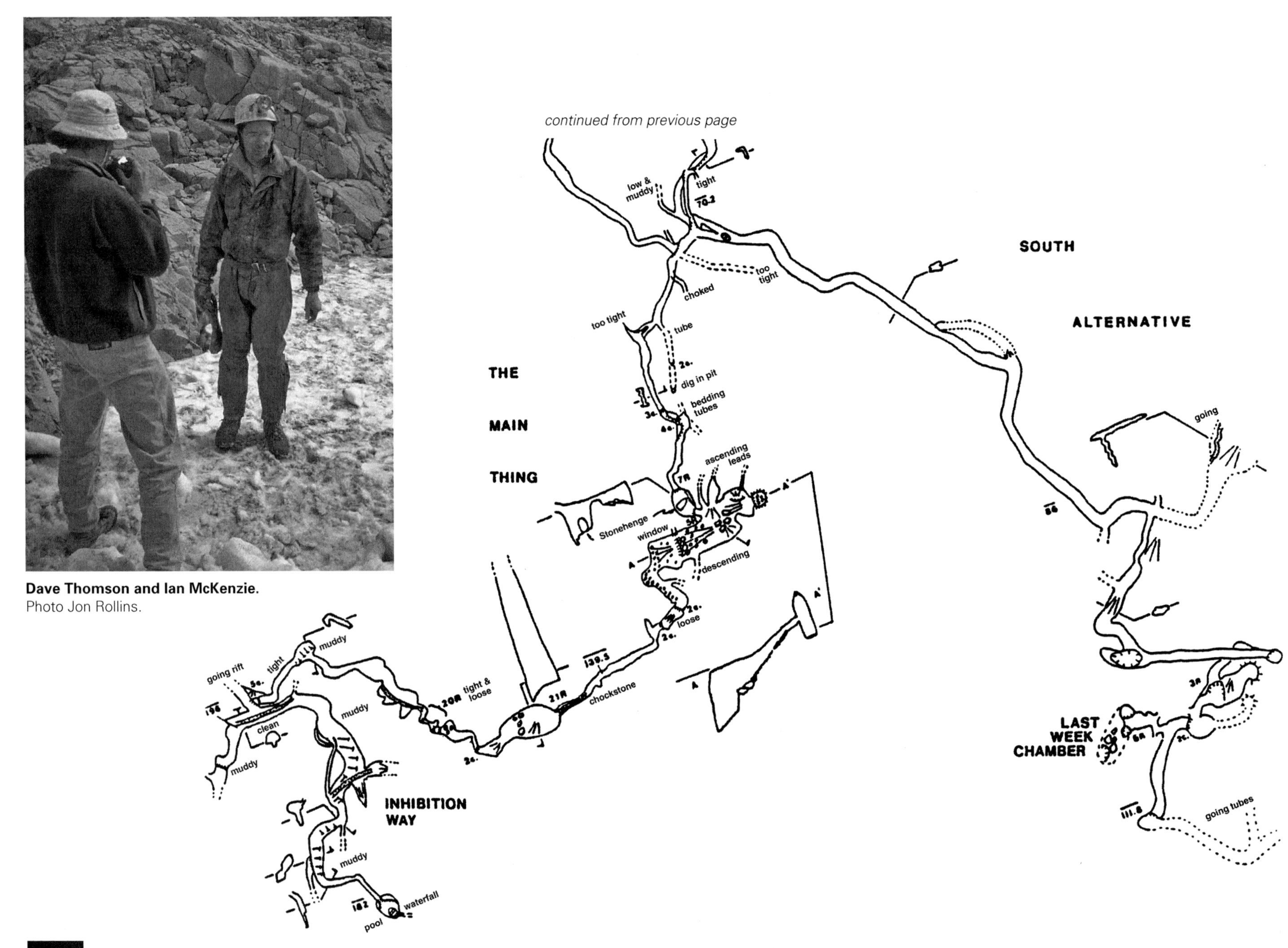

Dave Thomson and Ian McKenzie.
Photo Jon Rollins.

Shorty's Cave

Map 83 G/10 737921
Jurisdiction BC Forests (Elk)
Entrance elevation 2378 m
Number of entrances 1 (Connects with Yorkshire Pot)
Length, Depth 1060 m, 180 m
Discovered 1973, by Mike Shawcross

Location The easternmost entrance to the Yorkshire Pot system. Look for a low arch with a strong fluctuating draft.

Exploration & Description This cave is extremely challenging, with lots of complex route finding and rope work, and with a habit of dropping large rocks on people. It has seen little traffic since the visiting ICCC team did an exchange trip via Yorkshire Pot and detackled the cave in 1986.

A short way in, a 48 m blind pit is skirted after which the phreatic passage intersects a canyon and another series of drops totalling 82 m to a dead-end. This was the extent of exploration for ten years until Tom Barton rediscovered the cave in 1983.

A rift was found crossing the first of the drops, more phreatic passage leading to another impressive chasm 30 m deep and 20 m across. Some straightforward aid climbing enabled the head of this shaft to be crossed (Overpass Pit), the large phreatic passage continuing northward to the Thunderdome, a possible area for a connection with Little Moscow Cave. However, no way on could be found. In 1993 a possible voice connection was made in the vicinity of the Thunderdome.

Backtracking revealed a canyon leading to the site of Chas's mishap, the Snuff Box. From here it became clear that the cave was heading for the F Survey of Yorkshire Pot with which it was eventually connected via the Yorkshire Gripper Passage. In 1994 the Lancaster University caving expedition connected the Nobar extension in Yorkshire Pot with Shorty's Cave, again via the F Survey.

Shorty's Cave is named after the landlord of the Summit Inn, the caver's local watering hole on the summit of Crowsnest Pass.

Surveyed by: P. Thompson, B. Zakes (1974).
T. Barton, M. Evans, S. Grundy, I. McKenzie, J. McPhail, D. Prentice, K. Sawatzky, O. Whitwell, C. Yonge (1988).

Drawn by: C. Yonge.

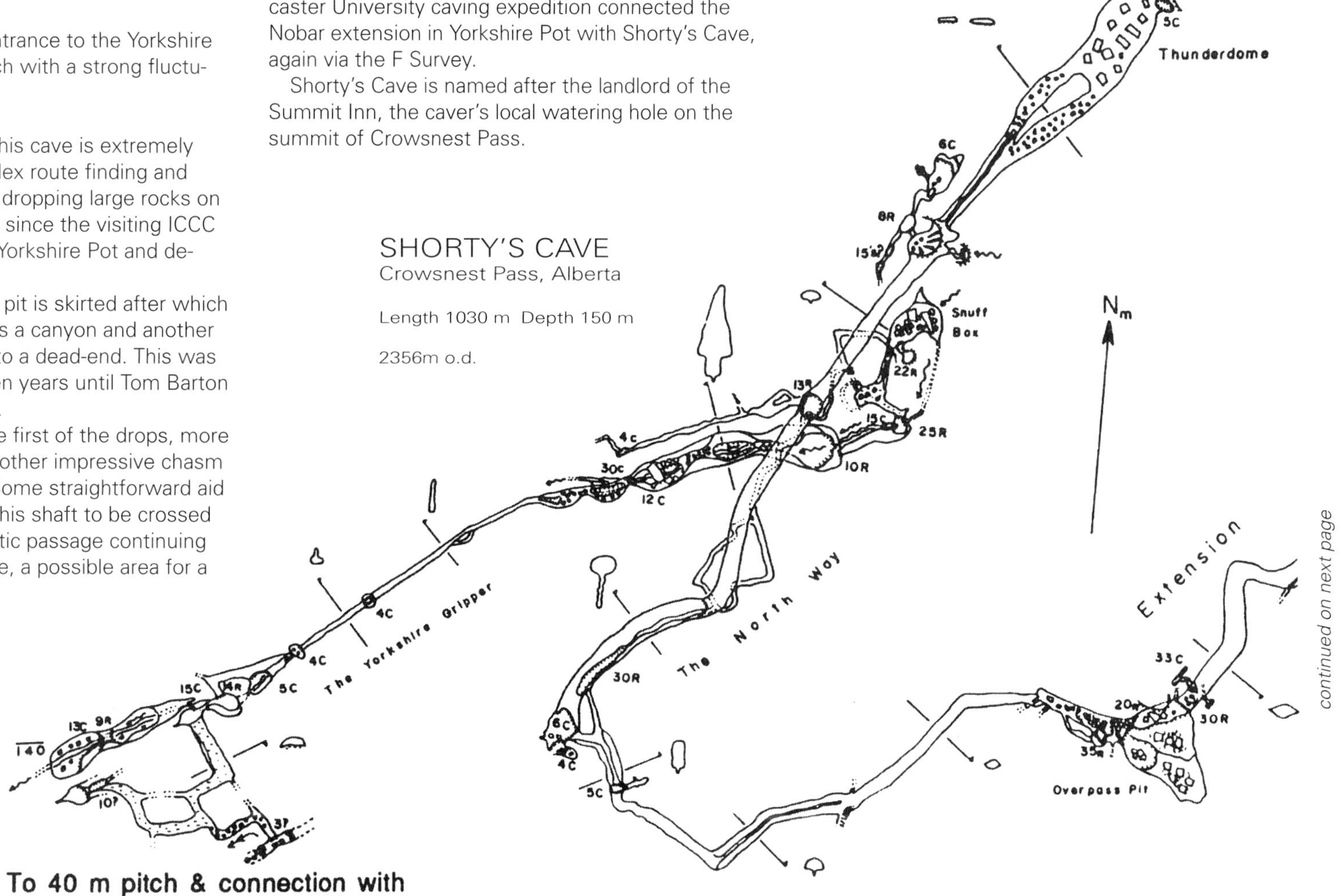

continued on next page

"Cathy Barton was three metres down the rope when a large rock peeled off the head of the pitch. It brushed her back and continued on towards me at the bottom. I heard a shout and an increasing rush of air and dived into a small side rift. My heel was crimped momentarily by the boulder followed by a deafening explosion behind as the boulder disintegrated. An acrid sulphurous smell ensued. On gathering my wits, I returned to my cave bag only to find that the survey book was severed neatly in three portions as if guillotined and my Teknalite consisted of two ends only — the middle bit, batteries and all, had been ground to dust. Incidents such as these serve to remind us of the seriousness of initial exploration in caves."
Chas Yonge.

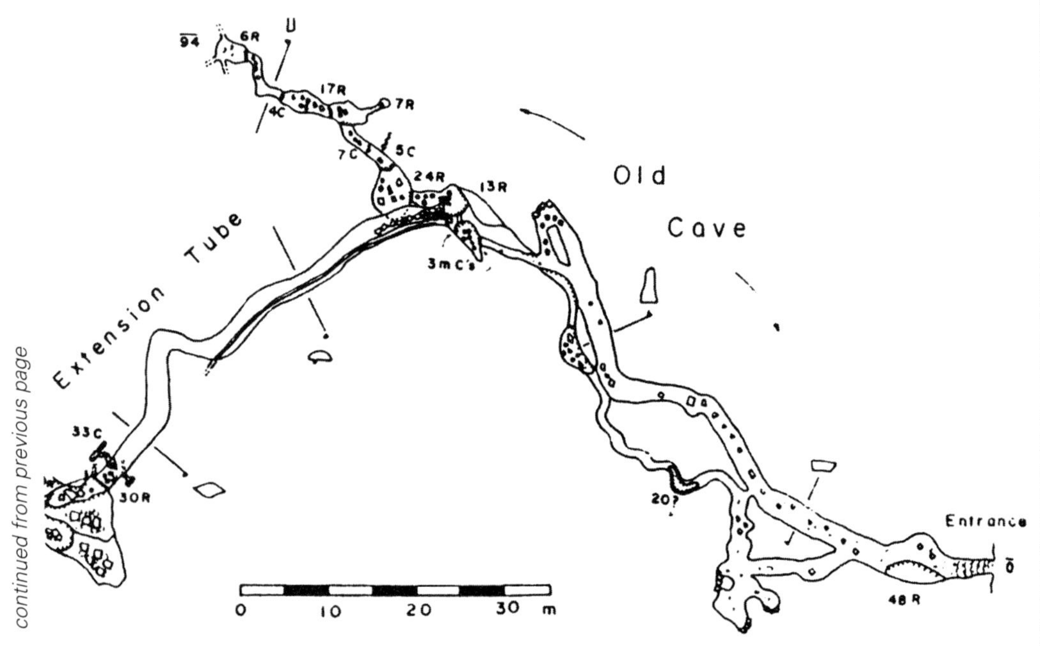

Shattered passage in Shorty's Cave. Photo Dave Thomson.

continued from previous page

Little Moscow

Map 83 C/10 734923
Jurisdiction BC Forests (Elk)
Entrance elevation 2425 m
Number of entrances 1
Length, Depth 540 m, 96 m
Discovered 1987, by Tom Barton & Jon Rollins

Location The entrance is small and hard to locate, being buried by snow for most of the year.

Exploration & Description The small drafting entrance was dug into and the cave explored during the 1987 Speleofest. It has some nice dip-tubes and horizontal maze passage. The strong fluctuating draft present in much of the cave probably originates from a large snow-plugged doline just to the east of the entrance. The bottom of the 31 m pit, Oblast Pot, should be very close to the Thunderdome area of Shorty's Cave. If ever connected, Little Moscow will be the highest entrance to the Yorkshire Pot system.

Little Moscow was a derisive name for Blairmore when it was a communist party stronghold in the 1930s.

LITTLE MOSCOW
Andy Good Plateau B. C.

I. Drummond, J. Ganter, M. Grapelle,
D. Hartman, R. Kozsan, D. Morris,
T. Morris, J. Rollins, C. Yonge.
Drawn by J. Rollins 1988.

Length 540 m Depth 96 m

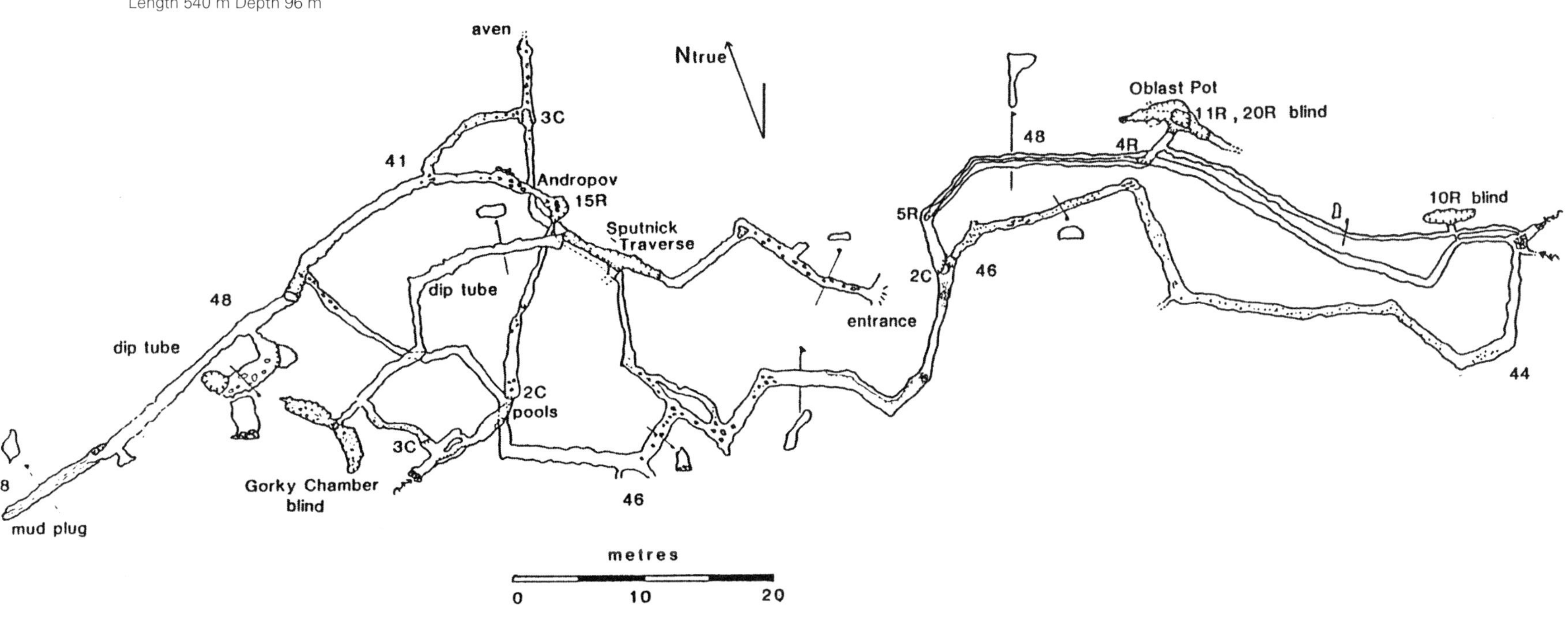

Derbyshire Pot

Map 83 G/10 734920
Jurisdiction BC Forests (Elk)
Entrance elevation 2378 m
Number of entrances 1
Depth c. 55 m
Discovered 1969, by Gary Pilkington

Location Reputedly located about 110 m to the southeast of Yorkshire Pot and 34 m lower; a "long stone's throw from Shorty's."

Exploration & Description This cave may have been entered only once since its discovery.

The entrance consists of a series of pitches and steep, loose talus slopes followed by a short series of muddy crawls that have yet to be fully explored. If connected to Yorkshire Pot, this will provide one of the least attractive entrances to the system.

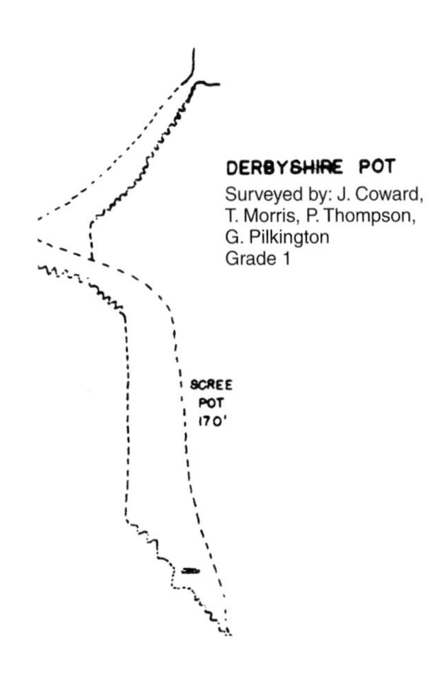

Gary Pilkington. Photo John Donovan.

"'Hey, this one has a 60 second drop!' shouted Gary, as we were busy heaving stones into likely looking holes on the Andy Good Plateau. 'About 50,000 feet deep' I said nonchalantly and strolled up and heaved another stone in. A rumble followed — and continued — and continued — and continued! 'Yes — it sounds a trifle loose,' I commented." Julian Coward.

Double Pots

Map 83 G/10 737921
Jurisdiction BC Forests (Elk)
Entrance elevation 2394 m
Number of entrances 1
Length, Depth 71 m, 39 m
Discovered 1971, by Pete Smart

Location This is the southeasternmost cave on the plateau, situated a short distance beyond Shorty's Cave. Look for two shafts.

Description The shorter of two shafts connects with the other after 8 m, at a squeeze between rock and snow. Below this is a chamber with a boulder run-in at one end and a short, icy shaft at the other. An ice-floored canyon then leads to a terminal chamber with a high crawling lead that chokes.

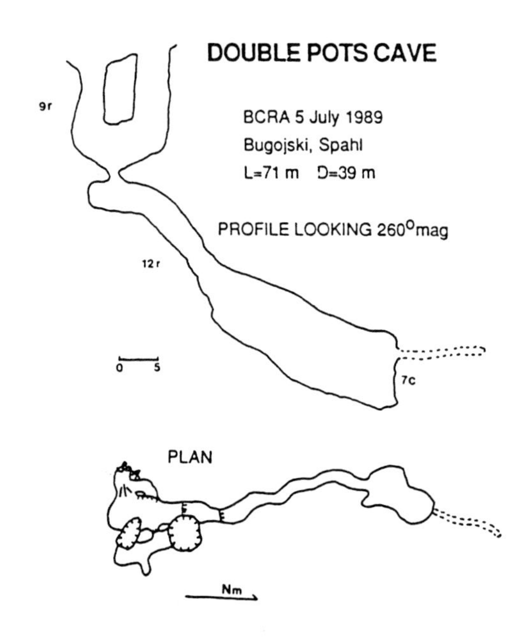

The Big Dipper in Gargantua.
Photo Ian Drummond.

"Frost-Pocket: A cave-like entrance caused by frost shattering because of water seepage through limestone beds. All frost pockets have the following characteristics: 3,000 feet above a road, 50 feet above the thickest timbered mountain side in the area, never go beyond daylight and cannot be checked out with binoculars. It takes a glory-seeking idiot to check out more than five in a lifetime." Tich Morris.

Gargantua

Map 83 G/10 727915
Jurisdiction BC Forests (Elk)
Entrance elevation 2501 m
Number of entrances 5
Length, Depth 5940 m, 286 m (-271, +15)
Discovered 1970, by Trevor Morris & Mike Shawcross

Location Gargantua's three entrances are clustered on an east-facing ledge high on the side of Mt. Ptolemy. From the Andy Good Col turn right and follow the rough trail up the northeast ridge of the mountain. Upon reaching a pond on a flat area of the ridge, continue up, but angle round to the left (south) side of the ridge (cairns). From the end of a karst bench a trail traverses a steep scree slope (sometimes snow-covered) before climbing up to the distinctive jug-handle main entrance.

The lower entrance is located just around the corner on the same ledge system.

The Waterfall Exit is reached by descending from the Andy Good Col and traversing southwest around the edge of the Andy Good Plateau. Then head west into the scree bowl below Mt. Ptolemy. The exit is located about 30 m above the screes, beneath a small overhang, or, when the stream is running, behind a waterfall. If snow is present you can assume the exit is closed.

Exploration & Description The cave was not discovered until 1970. From a distance the main entrance recess looks like a frost pocket, and no one except the "glory seeking" Tich Morris was motivated enough to make the long hike up to check it.

True to its name, the upper chamber of Gargantua is huge. Taking the form of a passage 290 m long, 30 m wide and around 25 m high, and known as Boggle Alley, it is believed to be the largest natural subterranean cavern discovered in Canada. The lower entrance permits daylight to penetrate 150 m into the chamber.

The easy round trip from the upper to the lower

Waterfall Exit at centre of photo. Photo Jon Rollins.

Top: Main entrance to Gargantua.
Photo Jon Rollins.

entrances is popular with casual visitors. The through trip to Waterfall Exit is for experienced cavers only. See warning.

From Boggle Alley two shafts descend, the 56' Pitch (17 m), and toward the end of the chamber, a 26 m drop (Storm Pot). From the 56' a series of dip-tubes named the 30°-50° Dipway (because of the dip on the limestone beds) lead to a chamber known as B.C. Chamber. Turning right leads to a blind 40 m shaft, Picnic Pot. Turning left leads to another 6 m pitch, the Window Pitch, which drops into Alberta Chamber. A small strongly drafting crawl over boulders at the far end of Alberta Chamber accesses Interprovincial Way, and the lower, generally horizontal section of Gargantua.

Interprovincial Way leads to Two Day Junction, the point reached on day two of initial exploration. Turning right accesses some superb, gently rolling, large phreatic tubes known as The Big Dipper Series. Issuing from a small rift, the Dye Streamway intersects this passage, and during a winter trip was followed to the deepest point in the cave at -271 m.

Turning left at Two Day Junction leads to The Canyon (30 m high and 4.5 m wide), which after two pitches ends abruptly close to the edge of the mountainside. A tight series of passages beneath the second pitch in The Canyon (not surveyed) lead to Waterfall Exit low on the east-facing side of Mt. Ptolemy (often buried beneath a snowslope).

Initial interest in the cave quickly waned when no passage could be found descending below the 270 m level. As a result many of the side passages that lead up from Interprovincial Way have been left incompletely explored or surveyed, despite the passing of 20 years since the original exploration.

Warning Use care on approaching the main entrance. When snow is present, use an ice-axe and be aware the slope could avalanche.

Do not attempt a through trip from the main entrance to Waterfall Exit unless you have first checked that the exit is open (1 hour return trip to check from the Andy Good Col). A pull-down trip, where you rappel and pull the rope down after you, is a committing undertaking, and several parties have become trapped when they could not find or use the exit. On October 16, 1999, a guided group was trapped for 22 hours until rescued by a group of cavers from Chilliwack. Exactly the same situation occurred on October 21, 2002, when a school party of six spent 28 hours waiting for rescue. Both groups failed to check the exit was open before entering the cave. If in any doubt as to the exit route, or the state of the exit, leave the pitches rigged and carry SRT gear.

Cave Softly Please do not camp in Gargantua Cave. The entrance has been polluted with human feces, urine and garbage. Pick up any garbage you find in the cave.

Opposite top left: Ron Lacelle at the Waterfall Exit. Photo Jon Rollins.

Opposite bottom left: Laure Morel in The Big Dipper series. Photo Dave Thomson.

Opposite far right: The 56' pitch in Gargantua. Photo Ian Drummond.

GARGANTUA

Crowsnest, B. C.

NOTES

1. Surveyed by: R. Bignell, F. Binney, M. Brown, P. Charkiw, J. Donovan, I. Drummond, E. Finn, L. Hastie, J. Johnson, S. Jones, B. Larson, W. Louch, P. Marshall, N. Montgomery, T. Morris, E. Neilsen, A. Nixon, G. Pilkington, M. Shawcross, W. Skinner, C. Smart, P. Smart, P. Thompson and D. Vincent.
2. Total surveyed length = 17,277 feet (5266·0m)
3. Surveyed depth = 888 feet (270·7m)
4. Elevation of Upper entrance approximately 8300 feet (2530m) a.s.l.
5. Surveyed using hand-held Brunton & Suunto compass/clinometer and fabric tape. Bearings read to ±2° on average, elevations to ±1°.
6. Loop closure errors:

	misclosure			
	horizontal	vertical	traverse length	accuracy ratio*
Interprovincial Way - Big Dipper	4·9	65·3	2006·5	1:30·6
Boggle Alley - Storm Pot	20·8	9·4	1178·5	1:51·7
Interprovincial Way - Gary's Lost Passage	11·4	8·8	1100·7	1:76·4

* 1:(traverse length ÷ radial error of closure)

Surveyed July 1970 - October 1975. Drawn by T. Morris and P. Thompson.

Note: All measurements on survey converted to metres by J. Rollins, 2002
Not shown on the survey are passages leading from The Canyon to the Waterfall Exit and a complex of passages leading from Two-Day Junction to a higher fifth exit

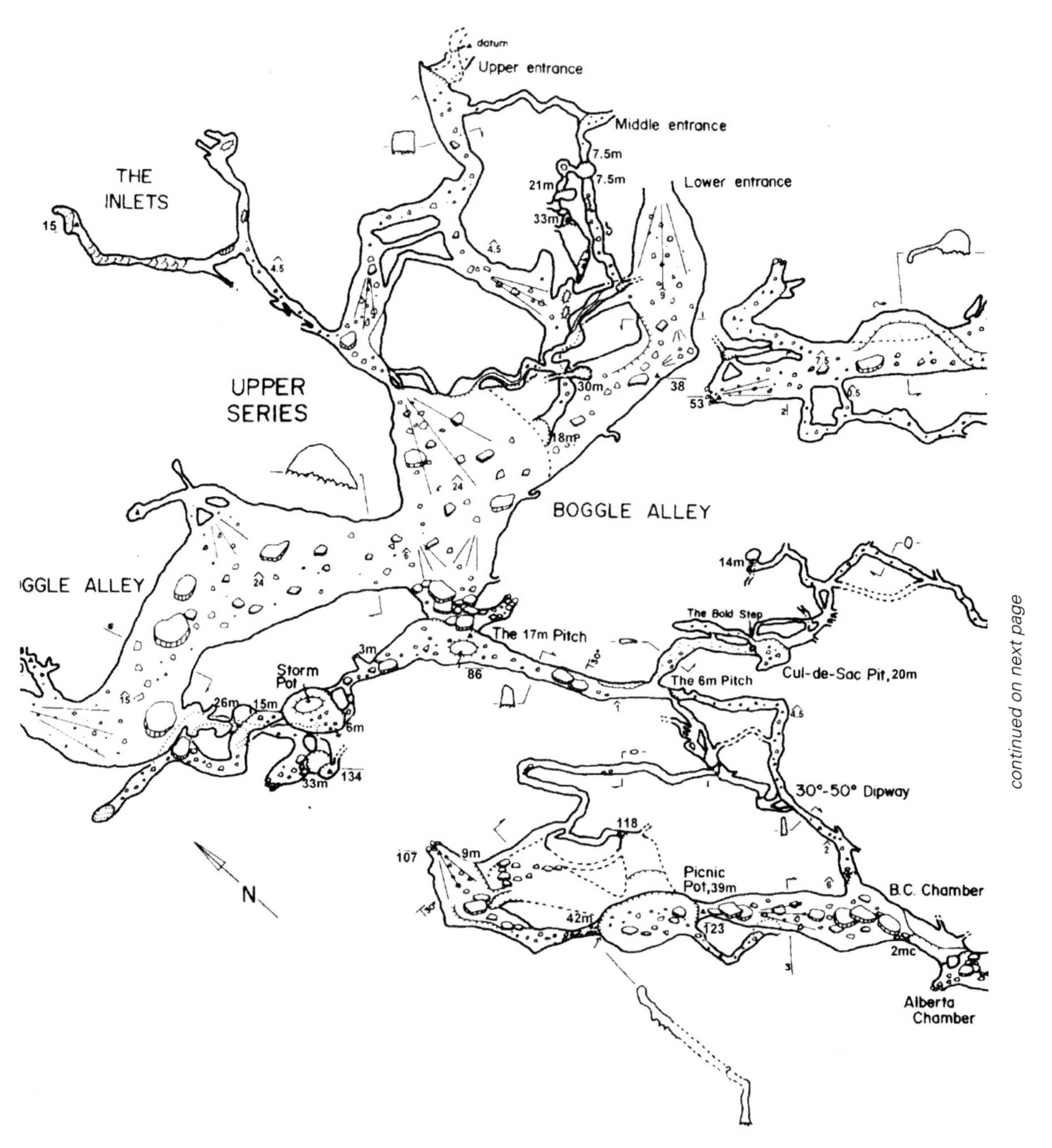

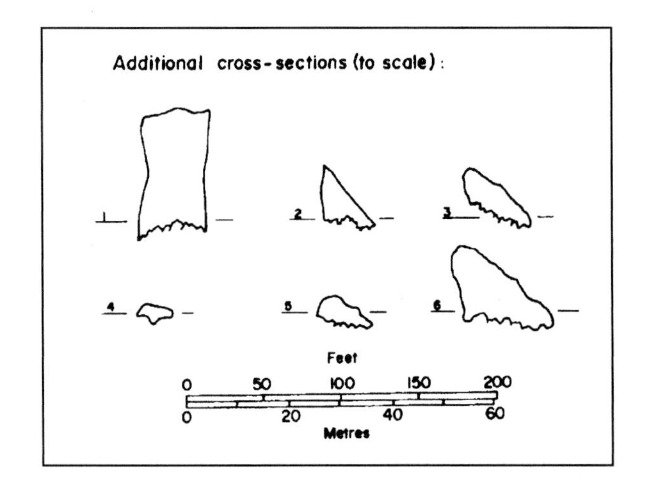

continued on next page

continued at right

continued from previous page

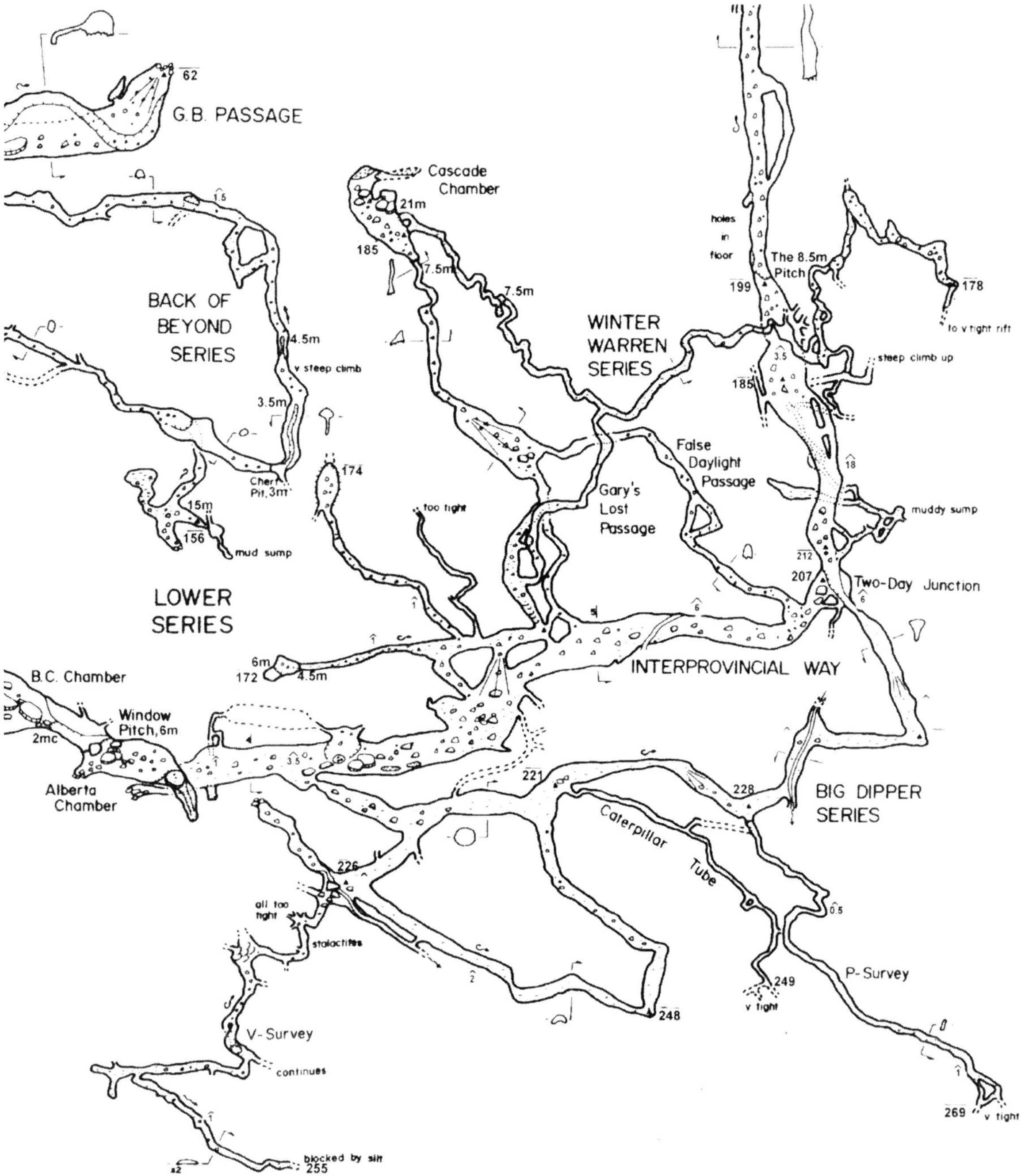

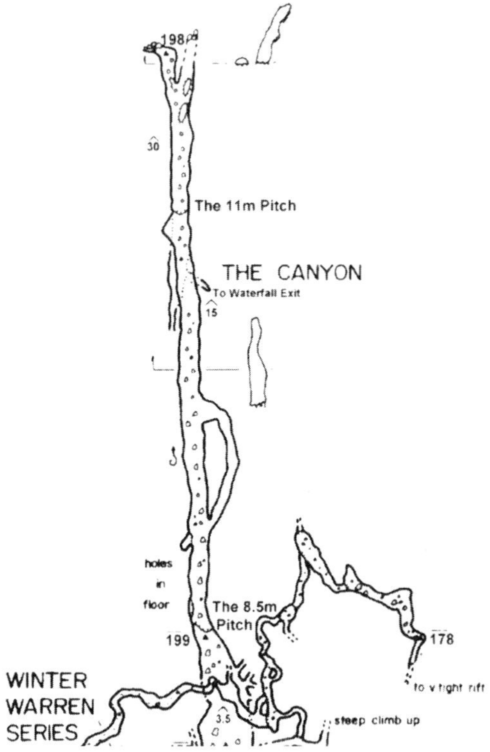

continued from top left

Speleogenesis Tich Morris provides a colourful, but brief speleogenesis of Gargantua: "Left by the rending chomp of by-gone glaciers, Gargantua stands as a relic of the past. Passages that once swallowed the jokulhaups from glacial snouts now see only the crisscross piracy of summer melt-water streams from snow patches."

Coral Corridor

Map 82 G/10 729905
Jurisdiction BC Forests (Elk)
Entrance elevation 2318 m
Number of entrances 1
Length 57 m
Discovered 1987, by Chas Yonge & Jon Rollins

Location Coral Corridor lies just south of the Waterfall Exit to Gargantua.

Description This short section of small phreatic passage has some nice cave coral formations. It ends in a drafting calcited tube that doesn't look a good prospect for extension.

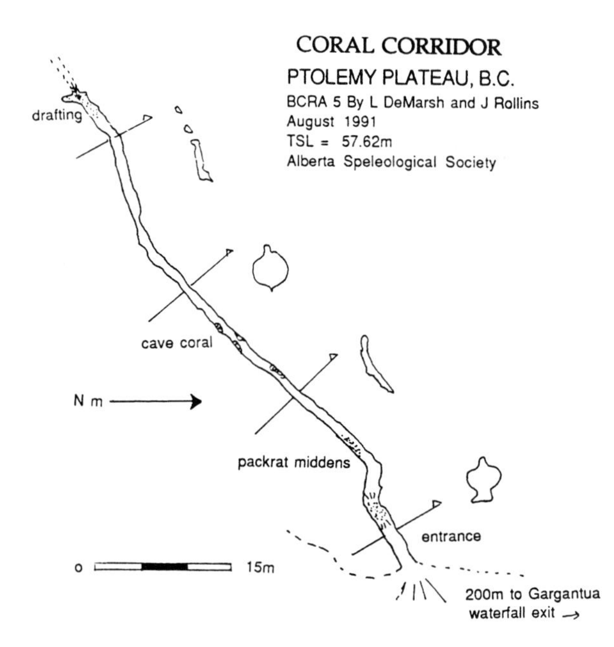

Merlin

Map 82 G/10 731916
Jurisdiction BC Forests (Elk)
Entrance elevation 2379 m
Number of entrances 1
Length, Depth 104 m, 65.2 m
Discovered 1988, by Tom Barton

Location 200 m south of The Backdoor.

Exploration and Description Named after Tom Barton's long-time feline companion, Merlin is a short but superb vertical cave with a tight squeeze at the head of the first 6 m drop.

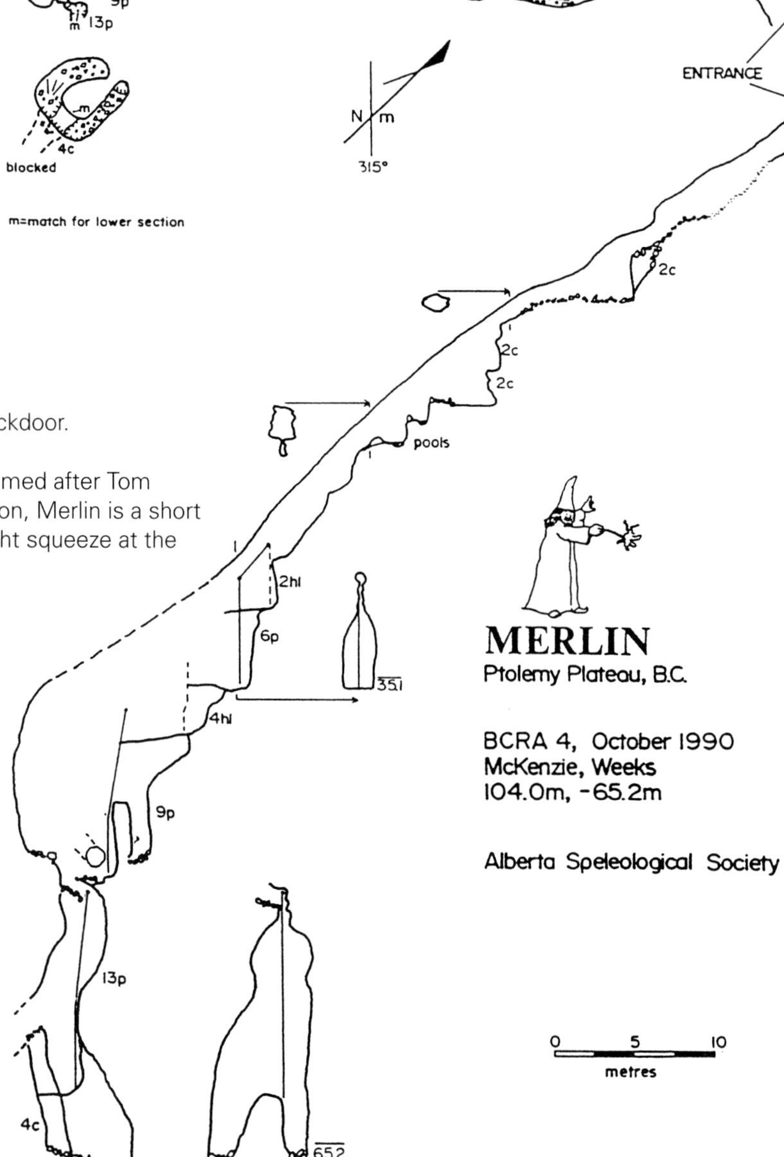

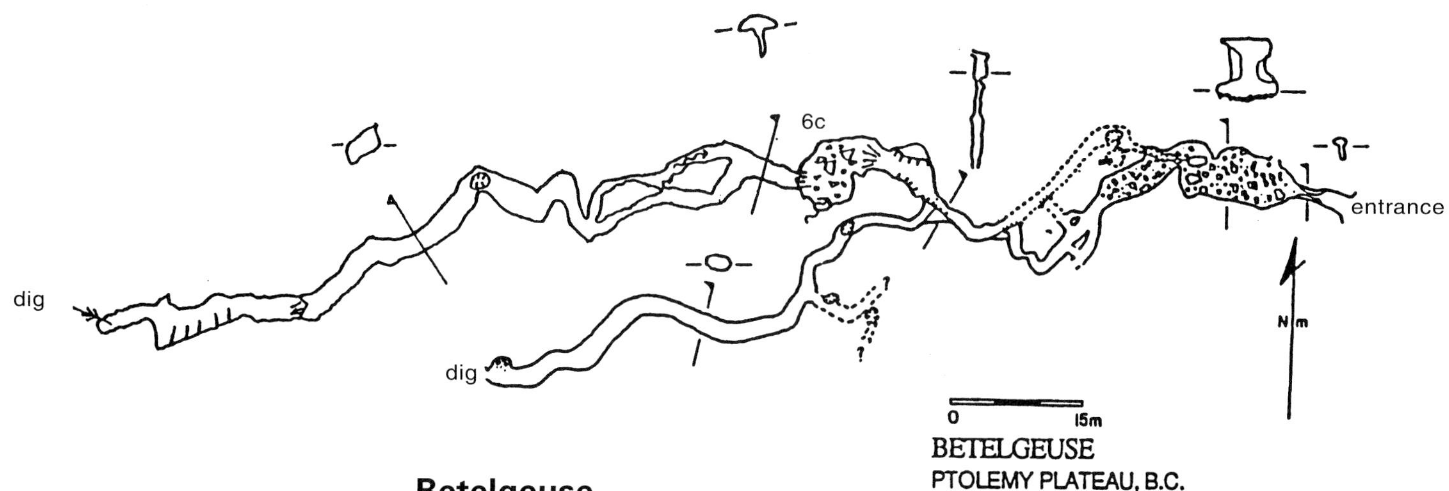

Betelgeuse

Map 82 G/10 732916
Jurisdiction BC Forests (Elk)
Entrance elevation 2348 m
Number of entrances 1
Length 234.4 m
Discovered 1984, by Tom Barton

Location 150 m east of Merlin.

Description Betelgeuse consists of a partially blocked rift passage that drops into a small chamber with a stream flowing in. From here two high level phreatic tubes lead off, both ending in drafting, sediment plugs.

FERNIE AREA

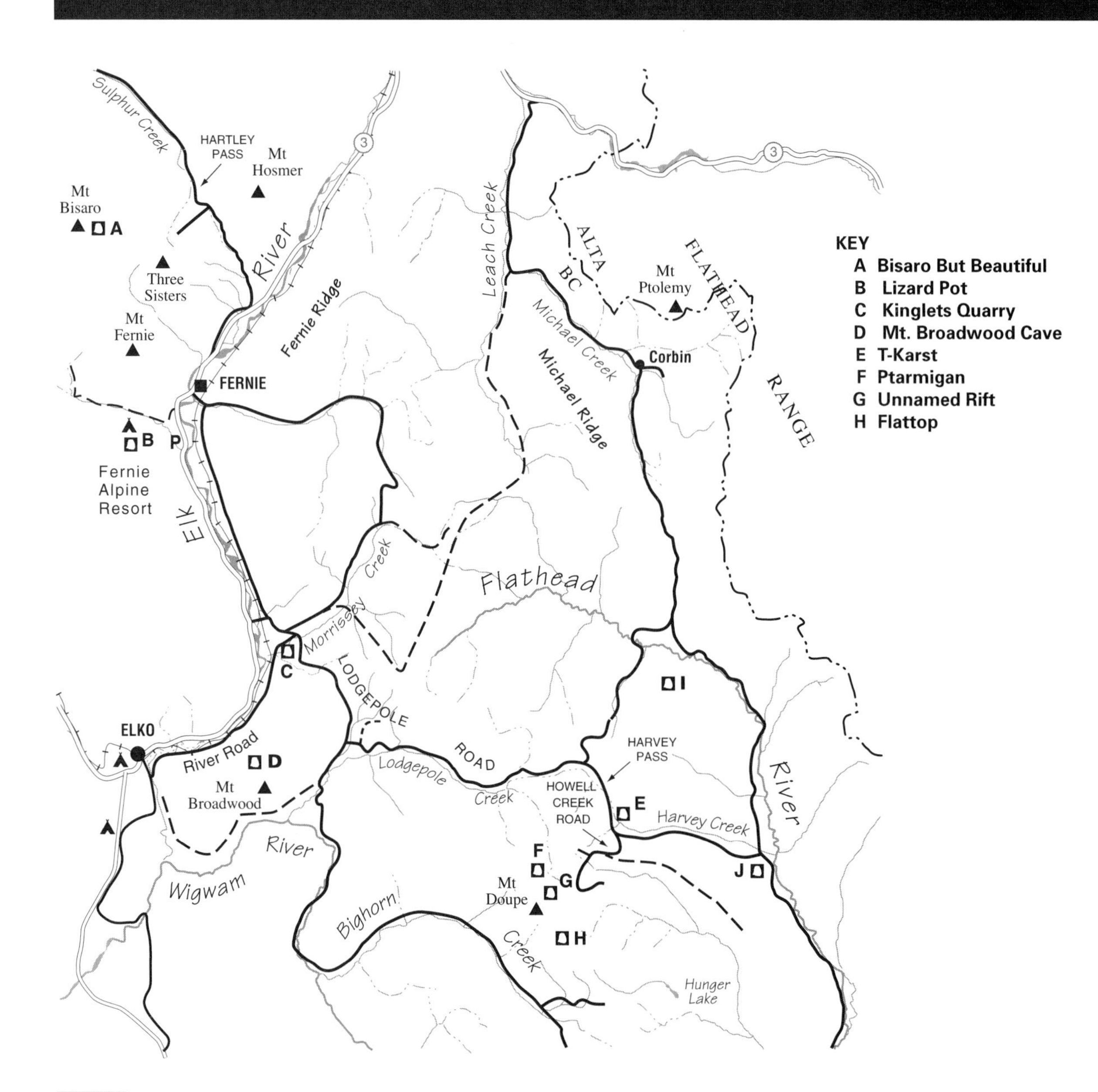

Cavers were first attracted to the Flathead area in southeastern British Columbia by the numerous closed depressions shown on maps. Usually a good indication of karst, closed depressions can also be caused by landslides, impossibly-choked dolines or glaciated talus-filled valleys. A few pits over 10 m deep were found in the 1970s by Peter Thompson, Tich Morris and others, but interest waned in the area when caves were found in the nearby Crowsnest Pass. Hunger Lake was a depression that revived interest in the Flathead area in 1981. It turned out to be a landslide feature. ASS cavers returned to the area northwest of Hunger Lake (the area originally visited by Thompson) and numerous pits were checked on the plateau and up the valley east of the main depression marked on the map. Above this valley, Flattop Cave was discovered.

Interest in the area was rekindled in 1995 when Mark Crapelle, a Vancouver Island caver, moved to Fernie. Mark, together with Stephanie and Stephen Meinke, also from the Island, Dave Ritchie from Manitoba and two local cavers, Tom Foley and Cathy Koot, have explored numerous promising areas of karst.

Exploration to date has focused on the Ptarmigan area that consists of two karst plateaus at around 2000 m separated by a pass of 2200 m. Both plateaus are about two kilometres long and a kilometre wide, with no surface water and a series of huge sinks. Ptarmigan Cave, a classic Yorkshire Pot type cave with multiple pitches, is now one of the ten deepest caves in Canada. Recently, some small caves have been discovered on the north edge of the Watluk Creek drainage (GR 660550), including a 95 m-long narrow rift named Thin Man's Temptation.

The T Karst (named after an inscription left on a cave wall by Tich Morris) has yielded several small caves and the impressive Incredible Cream Hole with passages averaging five to six metres wide.

The Fernie area contains the most recently explored karst in this guide, many caves being discovered during the summer of 2000/2001. Locations and access to some of the caves has not yet been firmly established.

Ptarmigan Plateau

Access for all caves Drive south from Fernie on Hwy.3 for 14 km (measured from the second big bridge next to Rip and Richard's Restaurant leaving town), then turn left (east) onto Morrisey Road. Cross a bridge and railway tracks and follow the main road as it swings right, past junctions on your left and right. At 3 km beyond the bridge turn right onto Lodgepole Road. Follow this without making any turns past junctions for River Road and Ram Creek Road, continuing on to signposted km 47 where you turn right onto the smaller Howell Creek Road (might not be signposted).

Follow this road across the low Kisoo Pass into Howell Creek. Keep left and cross another low pass into Twentynine Mile Creek drainage. Keep right and cross a side stream via a culvert. In 1 km is another junction at Watluk Creek. This junction marks the Watluk Creek access to The Old Slot and River-Gone Dream caves. Keep right and follow this road as it climbs up Twentynine Mile Creek to the bowl above. Watch for some road edge collapses on the Howell Creek Road, that with care can be circumvented. Vehicles with low clearance may need to park 3 km short of the trailhead.

From the trailhead a steep trail (flagged) leads up to south Ptarmigan Plateau, accessing Flattop, The Pellet Factory and Unnamed Rift. With a 4-wheel drive vehicle these caves are a fairly easy day trip. A high pass at GR640542 accesses north Ptarmigan Plateau and Ptarmigan Cave.

Another way to Ptarmigan Cave is to turn right off the Howell Creek Road at GR 682561 and drive to the end of the road (or 1.5 km short of the end if you are in a 2-wheel drive vehicle). A flagged trail leads over the col at GR 645570. On the west side traverse south on scree slopes to the plateau. Whichever way you go, Ptarmigan Cave requires overnight camping.

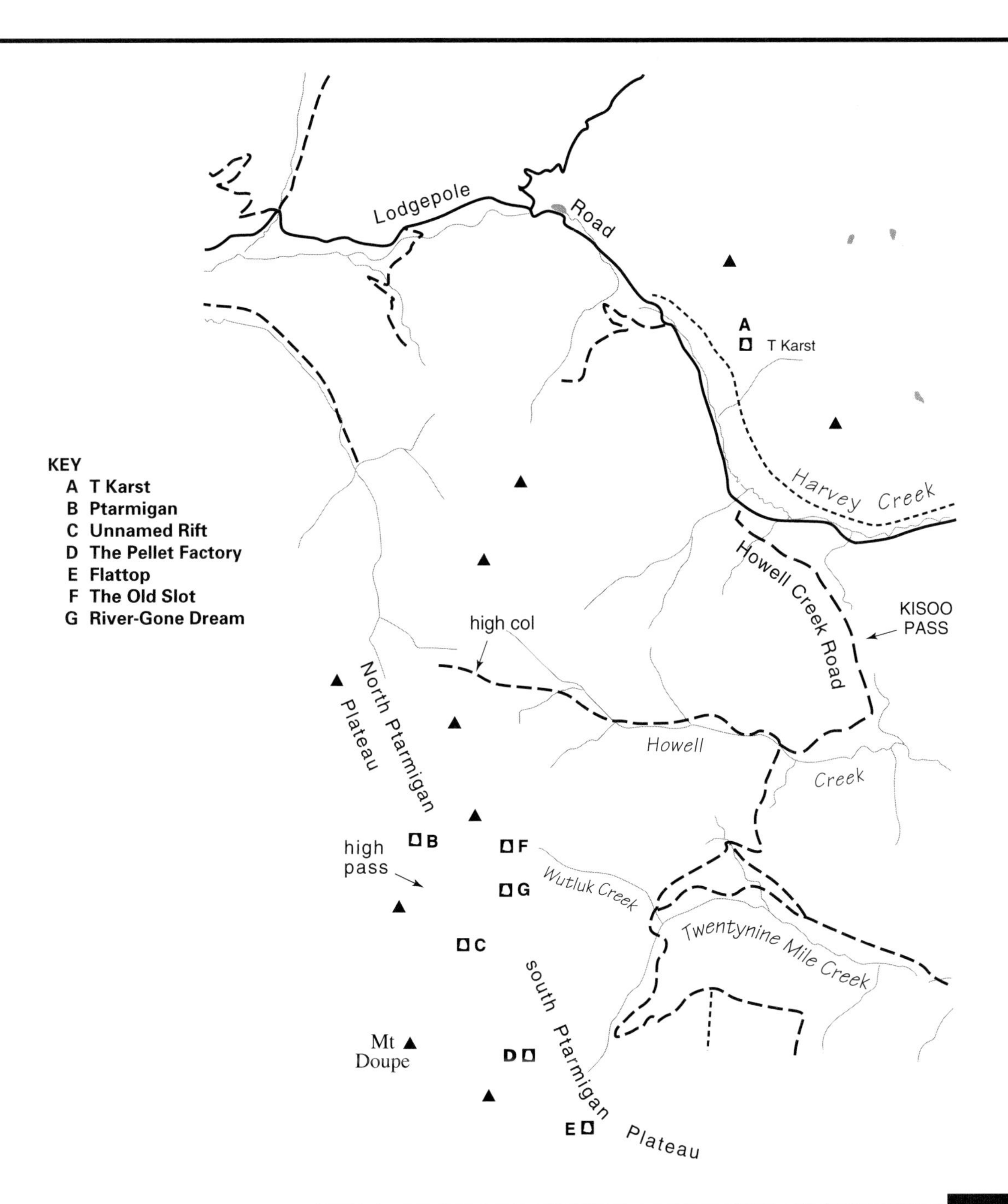

Flattop Cave

Map 82 G/2 662516
Jurisdiction BC Forests (Flathead)
Entrance elevation 2000 m
Number of entrances 1
Length, Depth 156.8 m, 80.52 m
Discovered July, 1982, by Dayle Gilliatt & Randall Spahl

Location On the south side of Ptarmigan Plateau. The lower of two entrances, a circular opening about 1.5 m across.

Description The cave descends at 30° to a boulder choke. From there, a narrow, sinuous passage can be followed to a small junction room. One branch is a phreatic tube which ascends steeply and gets too tight. (This was noted as a potential dig, probably connecting to the upper entrance.)

The lower branch required some digging before a wide stoop passage was accessed. From the passage a couple of tight side leads end in ice, but the main way continues on to the top of a 5 m drop into a large well-decorated room (Pretties Room). The bottom end of the room dips down to a 40 m blind pit. The other lead, past the Tea Room, was not pushed, although passage could be seen continuing past the choke.

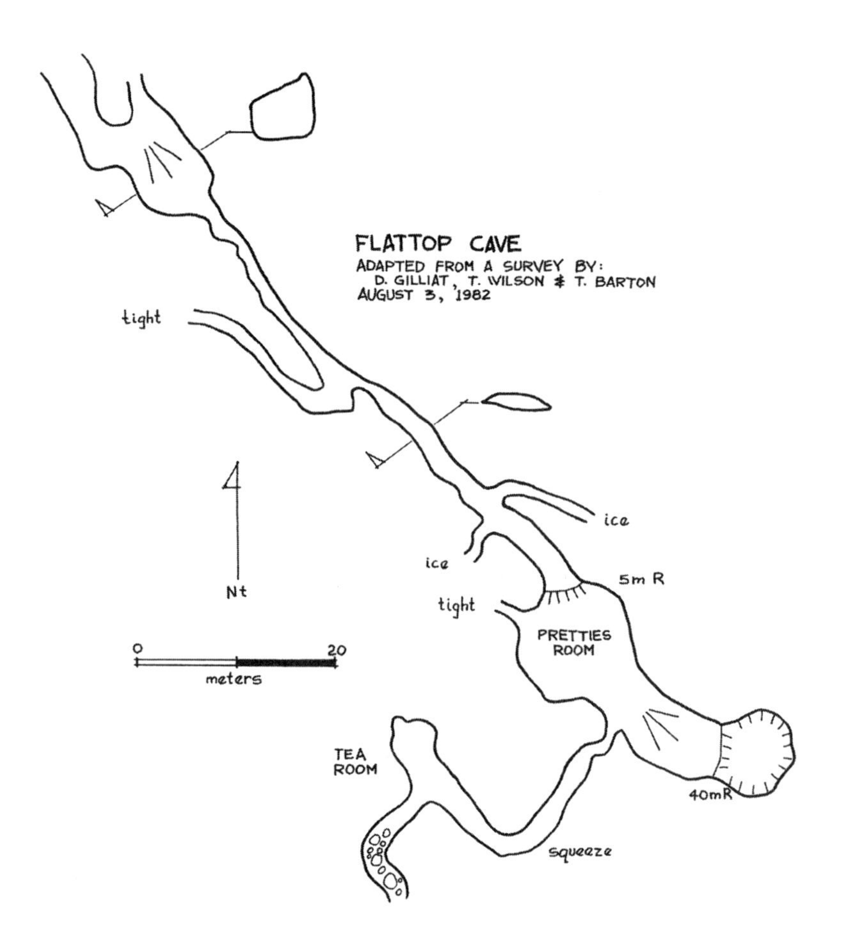

The Pellet Factory

Map 82 G/2 652521
Jurisdiction BC Forests (Flathead)
Entrance elevation 2220 m
Number of entrances 1
Length 442 m
Discovered 2000, by Taco van Ieperen

Location A low, crawl-in, drafting entrance located at the top of a scree slope on south Ptarmigan Plateau.

Description A dry horizontal rift leads quickly to a horizontal keyhole passage and some large chambers. This undemanding cave contains large amounts of pack rat droppings (hence the name) and some spectacular ice formations. Some leads requiring a rope remain to be checked.

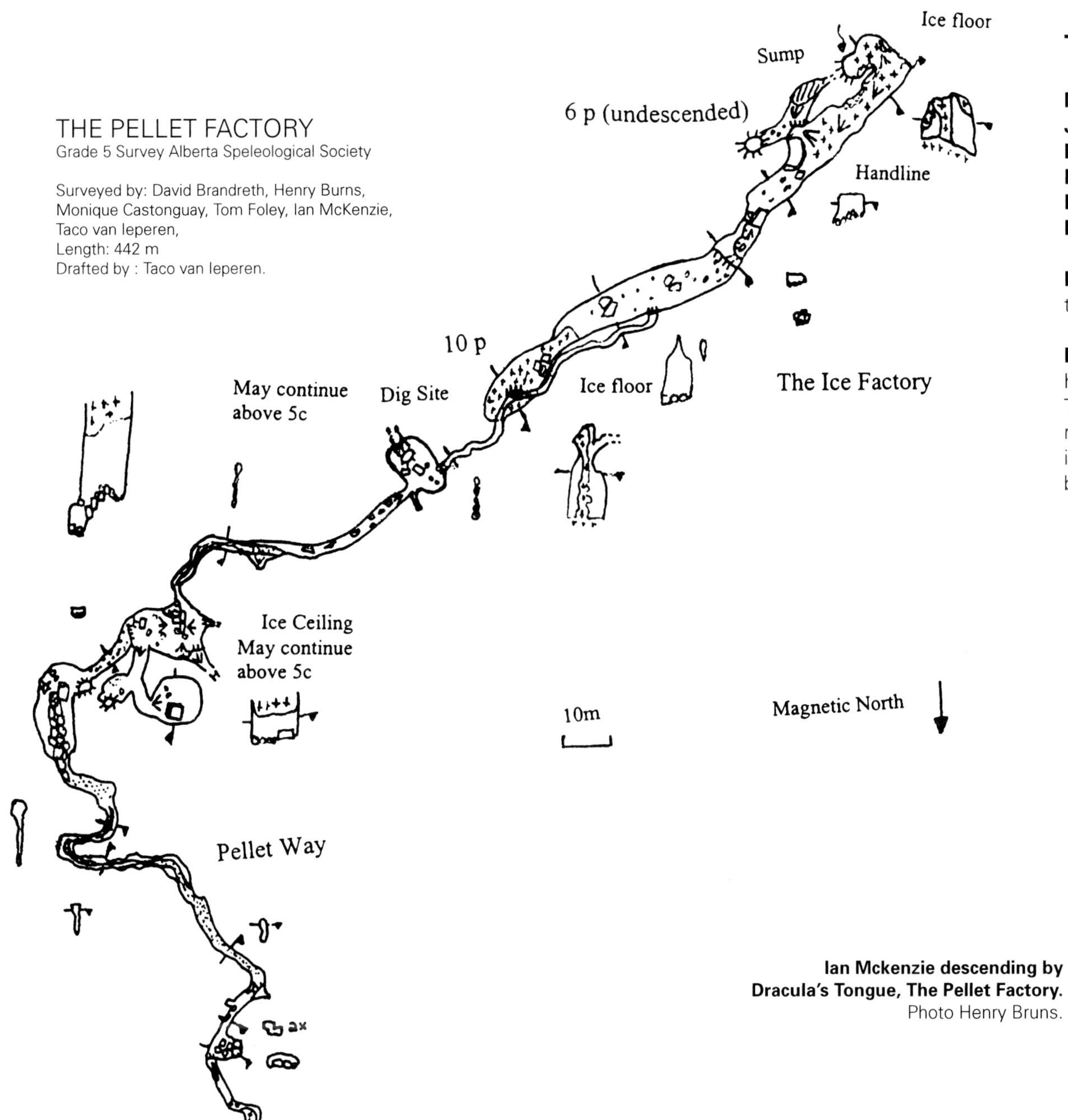

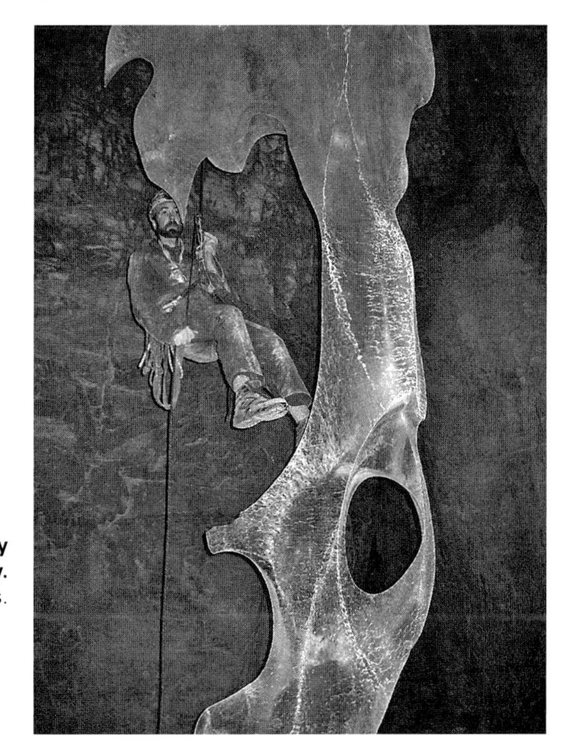

Ian Mckenzie descending by Dracula's Tongue, The Pellet Factory. Photo Henry Bruns.

Entrance to Unnamed Rift. Photo Henry Bruns.

Unnamed Rift

Map 82 G/2 646536
Jurisdiction BC Forests (Flathead)
Entrance elevation 2194 m
Number of entrances 1
Length c. 30 m
Discovered August, 2000, by Ian McKenzie

Location A slot entrance, snow-plugged until late summer.

Description Following the first drop, the rift extends both north and south, with further drops and several leads, none of which were draughting at the time of exploration.

The Old Slot

Map 82 G/2 653546
Jurisdiction BC Forests (Flathead)
Entrance elevation 2000 m
Number of entrances 1
Length Approximately 30 m
Discovered September, 2000, by ASS

Location From the access road bushwhack up Watluk Creek, then hike up scree to the base of the cliffs. Traverse south to an obvious long horizontal recess (frost pocket), then walk another 20 m south to the walk-in entrance to Old Slot.

Exploration & Description Old Slot was explored on the way up to River-Gone Dream Cave, and is contained in the same shallow dipping fracture.

After about 14 m, the walk-in entrance connects with a 4 m diameter phreatic tube. The tube descends to a T junction, both arms of which peter out in low, ice-lined crawls.

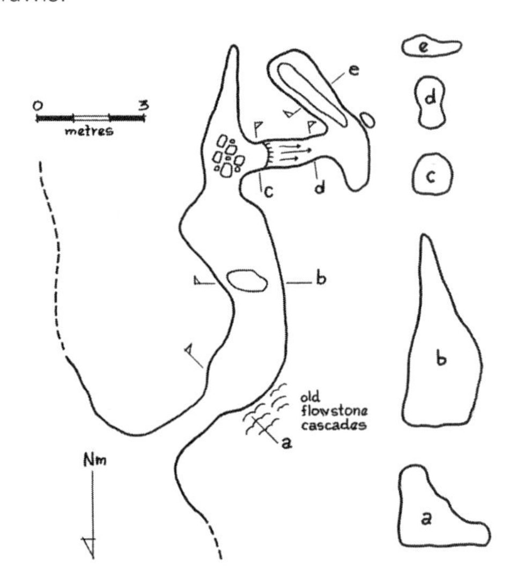

THE OLD SLOT
ADAPTED FROM A CONTROLLED SKETCH BY TOM FOLEY, CATHY KOOT, DAVE McRITCHIE

Entrance to River-Gone Dream Cave. Photo Henry Bruns.

River-Gone Dream Cave

Map 82 G/2 652544
Jurisdiction BC Forests (Flathead)
Entrance elevation 2050 m
Number of Entrances 1
Length 313.19 m
Discovered September, 2000, by ASS

Location Continue south from Old Slot along the base of the Palliser cliffs, then climb up scree to the caverous 10 m + opening visible from a considerable distance away.

Description This horizontal cave contains broad passageways punctuated with breakdown, which eventually lead to a choke that was dug to daylight visible through a crack.

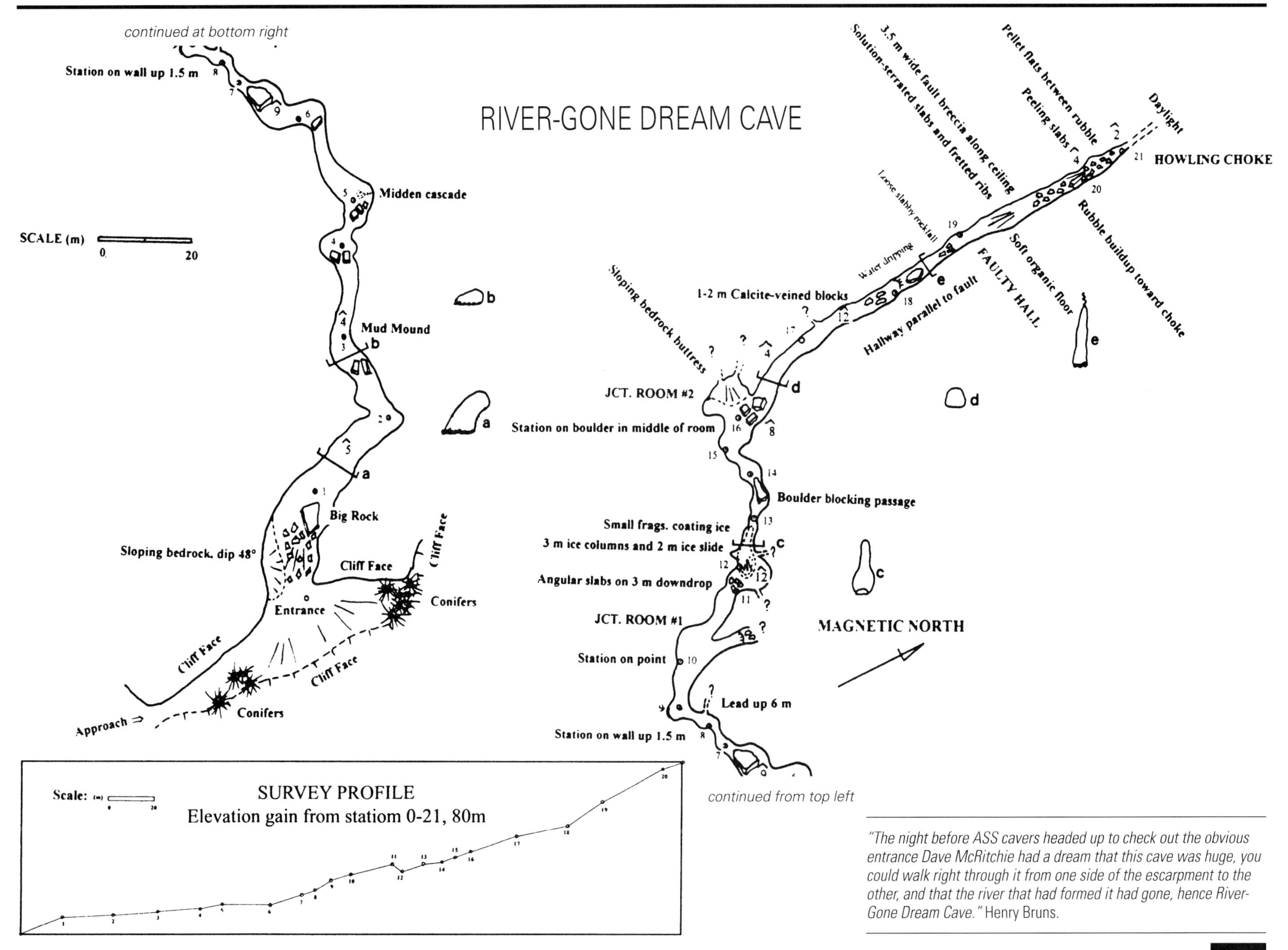

"The night before ASS cavers headed up to check out the obvious entrance Dave McRitchie had a dream that this cave was huge, you could walk right through it from one side of the escarpment to the other, and that the river that had formed it had gone, hence River-Gone Dream Cave." Henry Bruns.

Ptarmigan Cave

Map 82 G/2 638546
Jurisdiction BC Forests (Flathead)
Entrance elevation 2,200 m
Number of entrances 1
Depth 315 m
Discovered 1998, by Mark Crapelle & Chris Lloyd

Location At the bottom of one of the biggest sinks on north Ptarmigan Plateau.

Exploration & Description The major cave of the area is named for a mother ptarmigan and her three offspring seen enjoying the cool breeze blowing out of the entrance.

The entrance contains a 4 m downclimb to 30 m of sloping passage, following which the cave leads in two directions.

Exploration of the cave has been described as the slowest in Canada. Though first entered by Stephanie and Stephan Meinke in 1995, serious exploration did not begin until 1998, when Mark Crapelle and Chris Lloyd surveyed for 60 m in the right-hand Gruntose passage and for 80 m down to the Side Dome in the left-hand passage. In 1999 the cave was pushed to the notorious Satan's Solution, and in the summer of 2000 the cave was finally bottomed at –315 m.

The way on via the left-hand passage is mainly vertical with numerous descents down narrow fissures, much in the same vein as the entrance pitches of Yorkshire Pot. It has a sporting squeeze after the fourth pitch (Satan's Solution) and "ends" at a duck. This duck has been passed using wet suits and reveals a terminal sump. A promising lead, the Plunger, is still going with a good draught.

Warning This cave should only be attempted by serious cavers: there are numerous pitches that are often wet. Satan's Solution can be very awkward to negotiate on the way back up.

Mark Crapelle in Ptarmigan. Photo Henry Bruns.

Ptarmigan Pot entrance. Photo Henry Bruns.

"The Squeeze (Satan's Solution) which had been exciting going down, turned out to be an absolute nightmare coming up." Mark Crapelle.

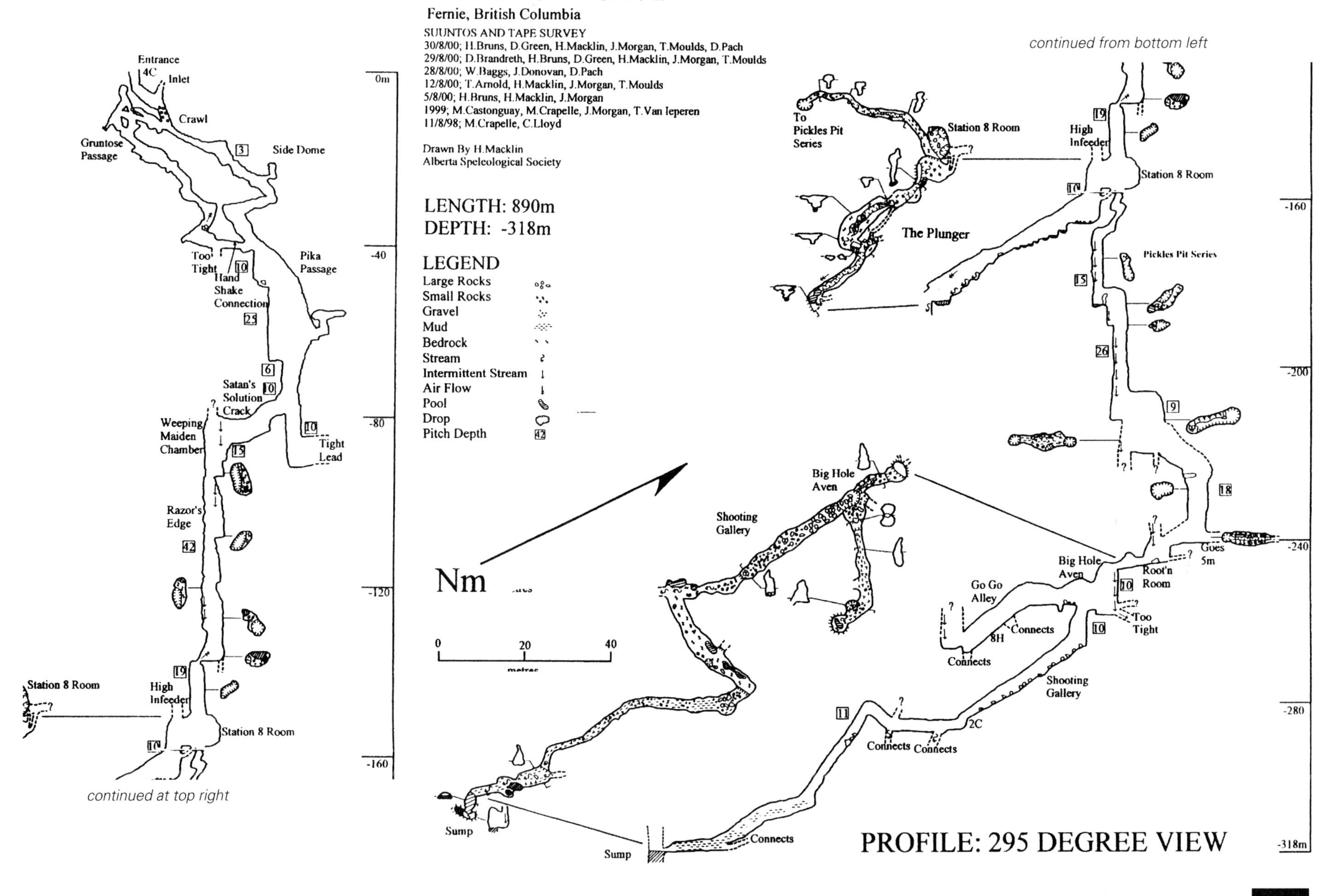

PTARMIGAN CAVE
Fernie, British Columbia
SUUNTOS AND TAPE SURVEY
30/8/00; H.Bruns, D.Green, H.Macklin, J.Morgan, T.Moulds, D.Pach
29/8/00; D.Brandreth, H.Bruns, D.Green, H.Macklin, J.Morgan, T.Moulds
28/8/00; W.Baggs, J.Donovan, D.Pach
12/8/00; T.Arnold, H.Macklin, J.Morgan, T.Moulds
5/8/00; H.Bruns, H.Macklin, J.Morgan
1999; M.Castonguay, M.Crapelle, J.Morgan, T.Van Ieperen
11/8/98; M.Crapelle, C.Lloyd
Drawn By H.Macklin
Alberta Speleological Society
LENGTH: 890m
DEPTH: -318m
LEGEND
Large Rocks
Small Rocks
Gravel
Mud
Bedrock
Stream
Intermittent Stream
Air Flow
Pool
Drop
Pitch Depth
Nm
0
20
40
metres
PROFILE: 295 DEGREE VIEW
Entrance
4C
Inlet
Crawl
Gruntose Passage
Side Dome
Too Tight
Hand Shake Connection
Pika Passage
Satan's Solution Crack
Weeping Maiden Chamber
Tight Lead
Razor's Edge
Station 8 Room
High Infeeder
continued at top right
continued from bottom left
To Pickles Pit Series
The Plunger
Pickles Pit Series
Big Hole Aven
Shooting Gallery
Go Go Alley
Connects
Root'n Room
Goes 5m
Sump
0m
-40
-80
-120
-160
-200
-240
-280
-318m

T Karst

Access for all Caves Drive south from Fernie on Hwy. 3 for 14 km (measured from the second big bridge next to Rip and Richard's Restaurant leaving town). Turn left (east) onto Morrisey Road. Cross a bridge and railway tracks and follow the main roads as it swings right, past junctions on your left and right. At 3 km past the bridge/railway tracks turn right onto Lodgepole Road. Follow the Lodgepole Road without making any turns, past junctions for River Road and Ram Creek Road. Continue on to the T Karst parking area between between signposted km 45 and 46, located just before the Howell Creek junction.

Across the valley to the northeast, at the base of a small red cliff, can be seen Garden Hose Cave resurgence. A hundred metres to the left and higher up are some cliffs with obvious holes. The Incredible Cream Hole (T.I.C.H.) is not visible, being located around the corner in the cliff to the right.

Garden Hose Cave

Map 82 G/7 678604
Jurisdiction BC Forests (Flathead)
Entrance elevation 1,800 m
Number of entrances 1
Length, Depth 56 m, 4.5 m
Discovered 1977, by Tich Morris

Location Hike up the drainage. Garden Hose Cave is the first cave you come to.

Exploration & Description Garden Hose is a low-flow resurgence cave that ends in a terminal sump four metres after a miserable duck. Jason Morgan and Heidi Macklin went all the way to Fernie to buy a hose with which to drain the sump. Unfortunately, after bringing it back to the cave they discovered they had mistakenly purchased a soaker hose...(Golden Garbage Can recipients for 2000).

T 77

Map 82 G/7 678606
Jurisdiction BC Forests (Flathead)
Entrance elevation 1900 m
Number of entrances 1
Length, Depth 20 m 4.5 m
Discovered 1977, by Tich Morris

Location A pocket in the headwall above Garden Hose.

Description A fossilized phreatic passage leads via a dig through organic material to a rubble-blocked active dome-pit.

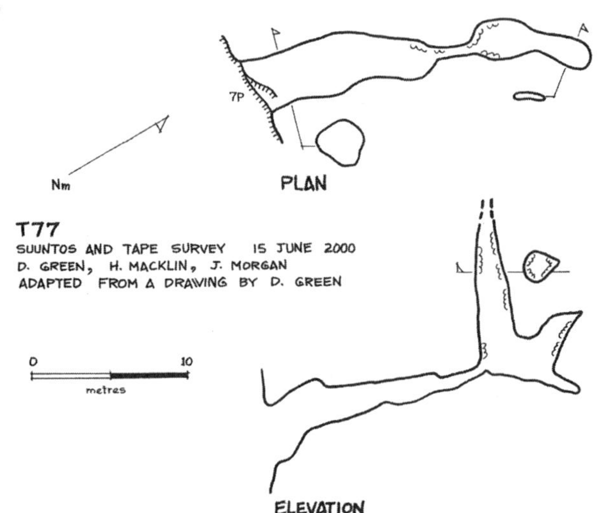

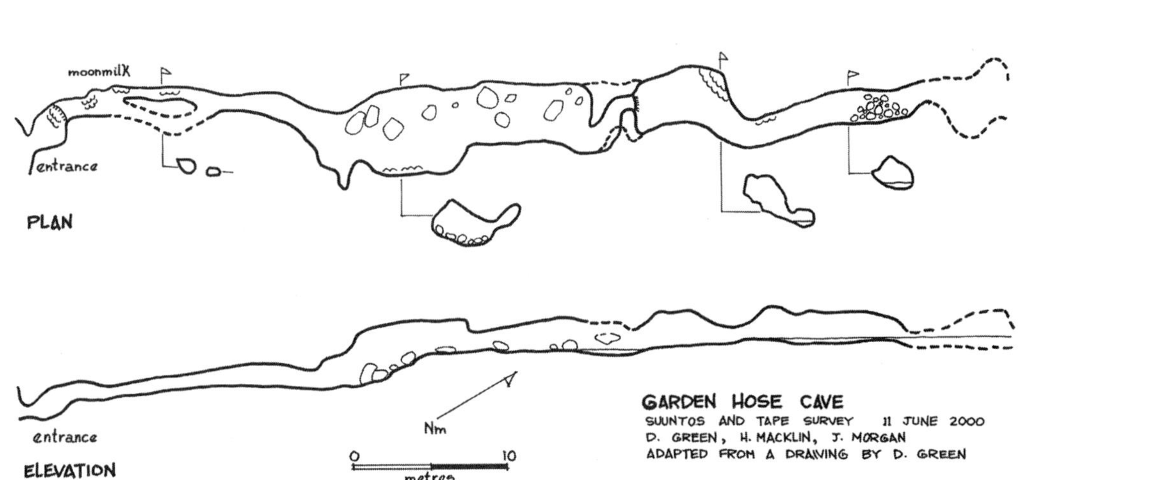

The Incredible Cream Hole

Map 82 G/7 679606
Jurisdiction BC Forests (Flathead)
Entrance elevation 1900 m
Number of entrances 1
Length, Depth 203 m, 14 m
Discovered June 2000, by the ASS

Location A short distance from the bench east of T 77.

Description A large, straight, trunk-passage cave averaging 5–6 m in diameter ends with a promising dig. The awful name originates from Tich Morris's initials left on the wall.

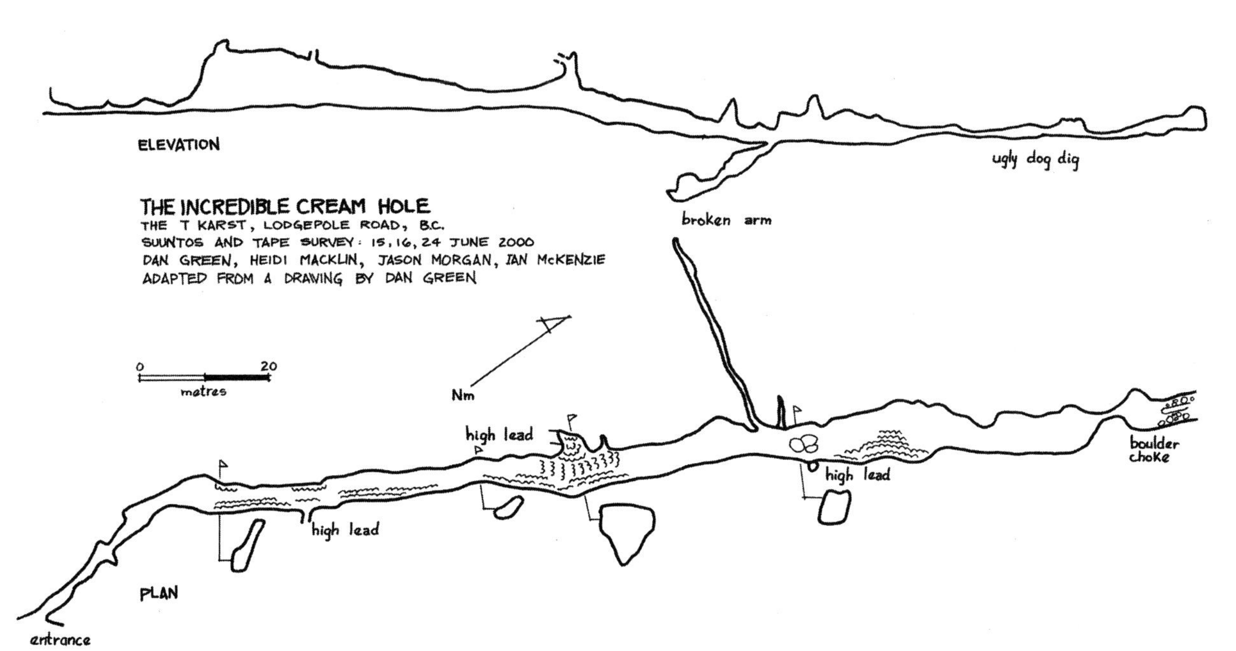

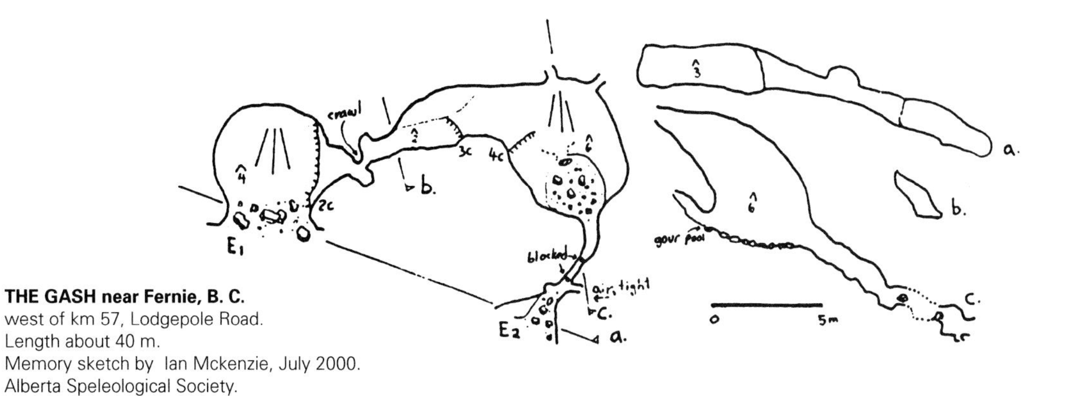

THE GASH near Fernie, B. C.
west of km 57, Lodgepole Road.
Length about 40 m.
Memory sketch by Ian Mckenzie, July 2000.
Alberta Speleological Society.

The Gash

Map 82 G/7 location uncertain
Jurisdiction BC Forests (Flathead)
Entrance elevation unknown
Number of entrances 2
Length 40 m
Discovered 2000 by the ASS

Location The Lodgepole Road at km 57, just before the Flathead junction. Bushwack 1 km west of Lodgepole Road. The large entrance is noticeable as a splash of orange rock on air photos.

Description An impressive entrance to a long-abandoned resurgence leads via three short climbs back to another blocked entrance.

Mount Broadwood Cave (The Hole)

Map 82 G/6 454628
Jurisdiction BC Forests (Elk)
Entrance elevation Approximately 2100 m
Number of entrances 1
Length c. 200 m
Discovered Unknown

Access The great rounded keyhole entrance, located in the northwest-facing Palliser limestone cliffs of Mt. Broadwood, can be seen from Hwy. 3 between Fernie and Elko. Details about access are unclear; those having visited the cave giving conflicting accounts. However, all parties climbed Mt. Broadwood via its easier southwest slopes. One suggestion follows:

From Hwy. 3 south of Fernie turn east into Elko. In 1 km from Hwy. 3 turn left and cross the Elk River onto the River Road. About 2.3 km from the bridge a good trail on the right leads to Silver Springs Lakes. From the vicinity of the first lake head up the west slopes of Mt. Broadwood.

Location The cave is reached by a 30 m rappel. Note carefully the location of the entrance from the highway as it is hard to locate the rappel point from above.

Exploration & Description Despite its relatively inaccessible location, this cave has received at least three visits: by Peter Fuller, Peter Thompson and Mike Goodchild in August of 1968, by Frank Petrella (date unknown) and by Chris and Kim Smallwood in October, 1983. The cave contains some flowstone, soda straws, moon milk, and two possible leads: an aven, and a dig at the end of the cave with water sinking into it.

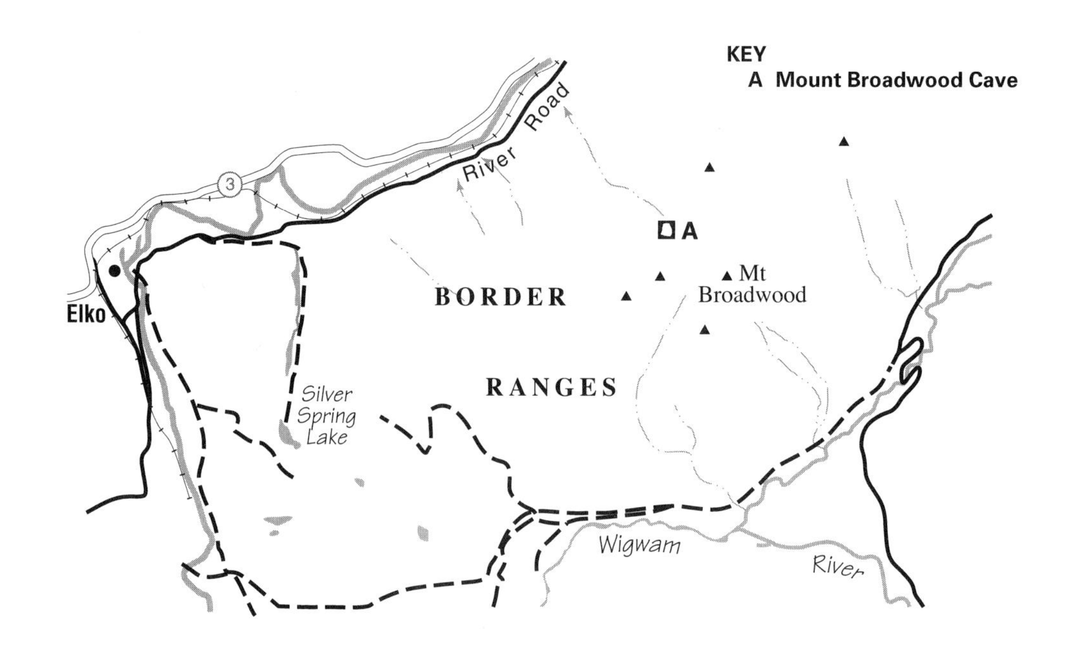

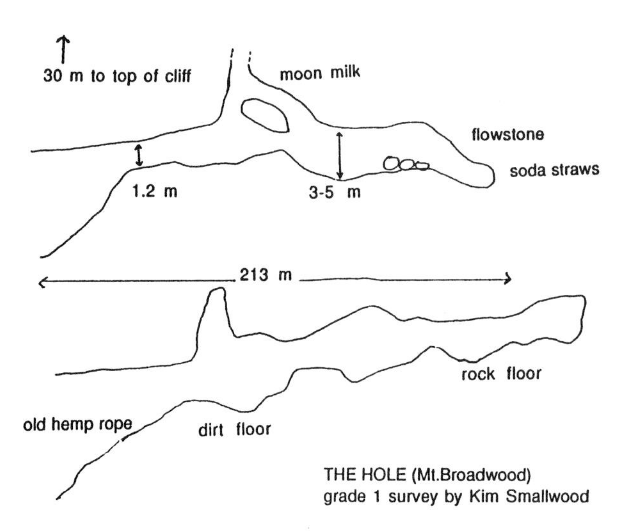

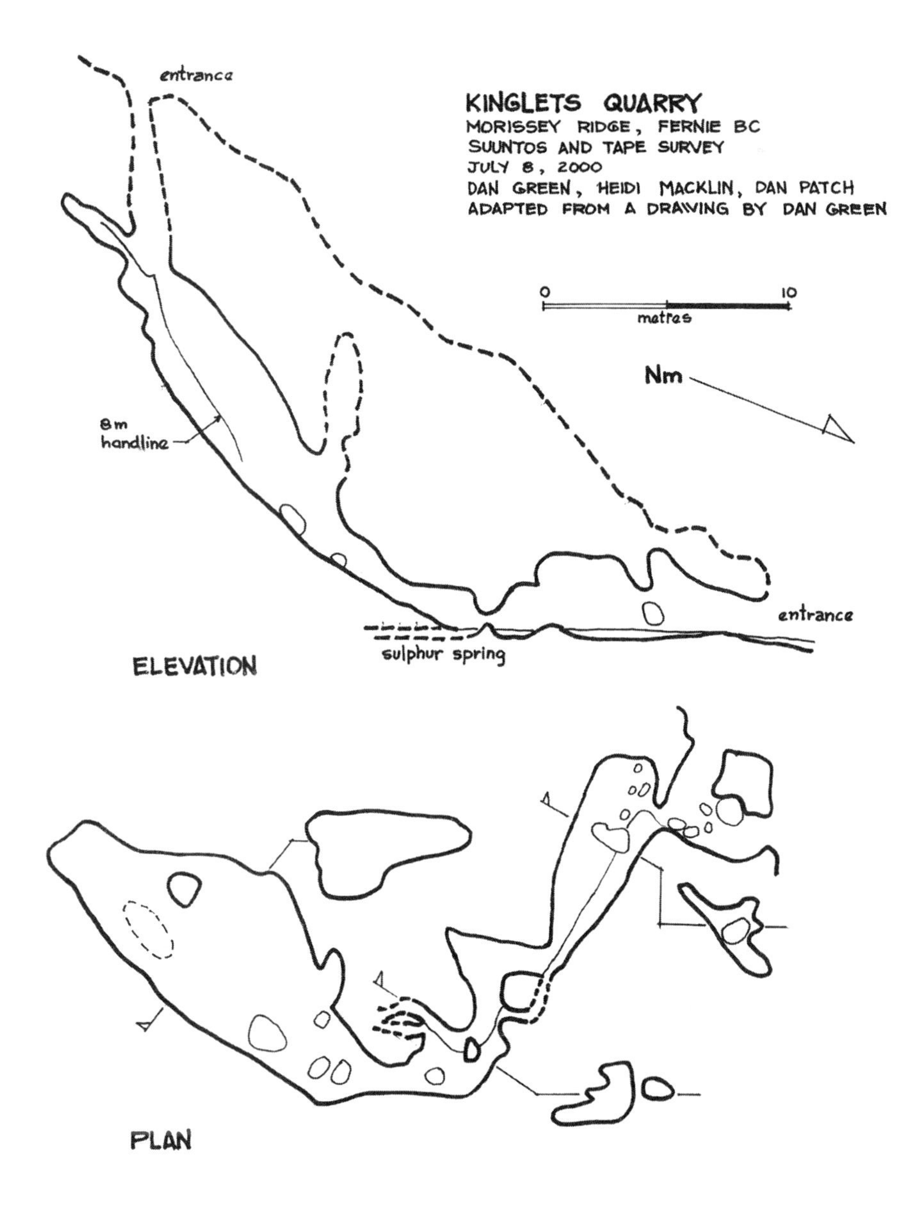

Kinglets Quarry (formerly Stinky Hole)

Map 82 G/6 451681
Jurisdiction Mount Broadwood Conservation Area
Entrance elevation 1000 m
Number of entrances 3
Length, Depth 54m, 21m
Discovered Unknown

Access Drive south from Fernie on Hwy. 3 (measured from the second big bridge next to Rip and Richard's Restaurant leaving town). Then turn left (east) onto Morrisey Road. Cross a bridge and railway tracks and follow the main road as it swings right, passing junctions on your left and right. At 3 km past the bridge/railway tracks take a right fork onto Lodgepole Road, then turn right (south) onto River Road which is followed for 2.7 km.

Location From a pond on the left (east) side of the road follow the inflowing creek to the cave entrance.

Exploration & Description A small sulphurous resurgence cave. During an ASS visit in 2000, the third skylight was excavated and the cave mapped.

Lizard Pot

Map 82 G/6 373805
Jurisdiction Fernie Alpine Resort
Entrance elevation 1700 m
Number of entrances 1
Length, Depth 150 m, 90 m
Discovered Date unknown by Heicho Socher (former ski-hill owner).

Access Hwy. 3, 4.5 km south of Fernie. Follow signs to Fernie Alpine Resort.

Location Located high in Lizard Bowl, Lizard Pot can be reached only in summer by taking the Elk Quad chair to the foot of the bowl. Permission is required from Fernie Alpine Resort at (250) 423-4655. In winter the small entrance is covered over and has a snow fence around it to prevent unwary skiers from falling victim to this unusual hazard. A skier was in fact injured falling into a small hole in a different bowl.

Description A classic pothole, Lizard Pot descends at an angle of 65° for 90 m down a series of steeply dipping shafts where it ends in a narrow drafting rift. There are between five and seven pitches (depending on how it is rigged), loose rock and some tricky route-finding. Despite ease of access, this cave sees little visitation due to the technical nature of its passages.

The same band of limestone extends at similar elevations along the entire northeastern side of the Lizard Range, and although numerous other cave entrances have been checked, all are blocked or become too tight. The deepest goes for about 30 m.

Warning This is a serious cave and should be attemped only by experienced cavers. Careful rope rigging is required to avoid abrasion on the fluted limestone.

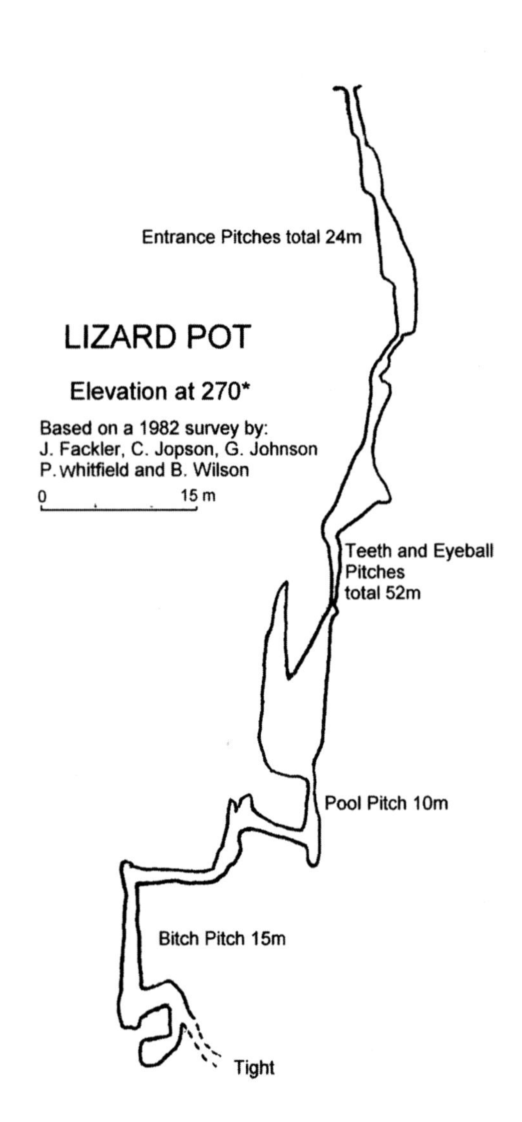

Geology Contained in overturned beds of the grey crystalline limestone of the Rundle Group (Mississippian), the inclined shafts of the cave are formed along enlarged joints. Short sections of essentially horizontal passage that connect the pitches follow the dipping beds. Weak black shale bands, which confine the water horizontally, separate the limestone beds. Small chambers, interlinked by narrow shafts, were developed through enhanced solution at the shale partings.

Above the cave, the back wall of the bowl rises to 2300 m, beyond which is located a parallel high valley developed in shales and the impure limestones of the Exshaw and Banff formations. Next in line is a second band of hard limestone of the Palliser Formation in which several cave entrances were found, but all were choked with debris. This is the same lithological sequence that contains the caves below the Andy Good Plateau in the Crowsnest Pass.

Hydrology Although water sinks over a 25 km^2 area, only one resurgence has been found, at 1250 m in a valley to the southwest. This has a 1 m^3/sec flow in dry weather, with an overflow channel 0.5 km beyond it. If this is the resurgence for Lizard Pot, then the depth potential for the cave is no more than 396 m. (Geology and hydrology adapted from a report by Ben Lyon.)

Bisaro But Beautiful

Map 82 G/11 363968
Jurisdiction BC Forests (Elk)
Entrance elevation Approximately 1500 m
Number of entrances 1
Length, Depth 149.3 m, 21.8 m
Discovered 1991, by Tom Volkers & John Pollack

Access From Fernie head north on Hwy. 3 for 7 km. Here, turn left (west) onto a loop road that heads back towards Fernie. Drive for 500 m, then take the first right and follow Sulphur Creek Road over Hartley Pass (lake). 2 km farther on, before another small lake, turn left (west) and drive the logging road for 2 km to a landing and park.

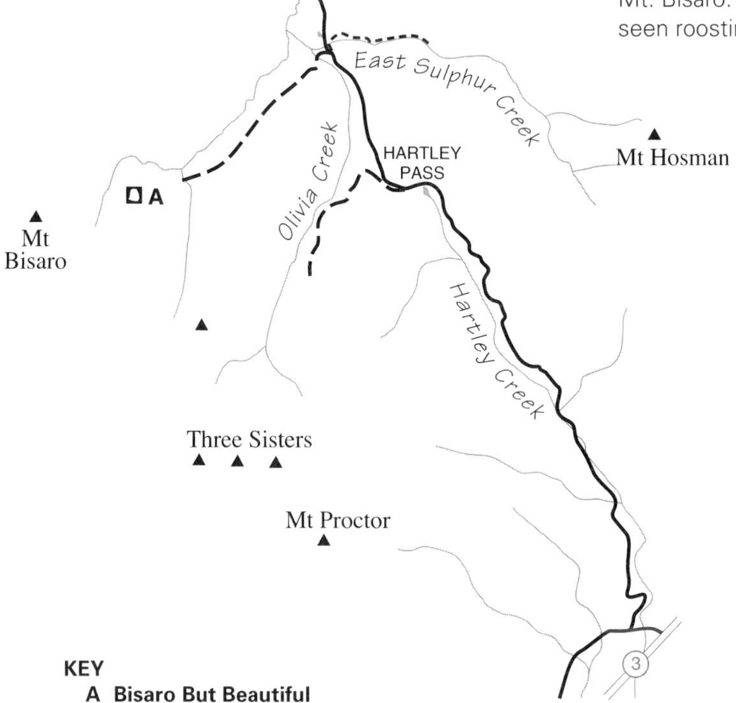

The final approach involves 150 m of height gain over about 1 km and used to be a horrendous thrash up a gully full of slide alder. Fortunately, in 2002 a trail was cut from the landing to the cave, which makes access much easier.

Location 1 km east of the summit of Mt. Bisaro is a resurgence cave with a huge 30–40 m–diameter entrance.

Description Unfortunately, the roaring stream sumps a short distance into the cave. A spring such as this obviously drains a large karst area, probably that above on Mt. Bisaro. Little brown bats have been seen roosting in the cave.

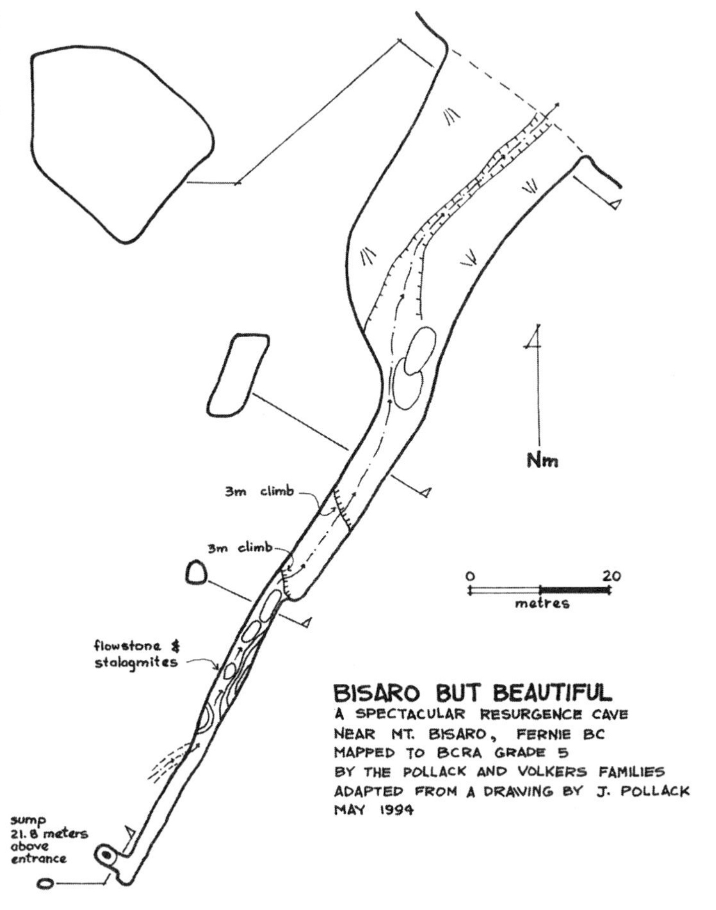

WHITESWAN PROVINCIAL PARK

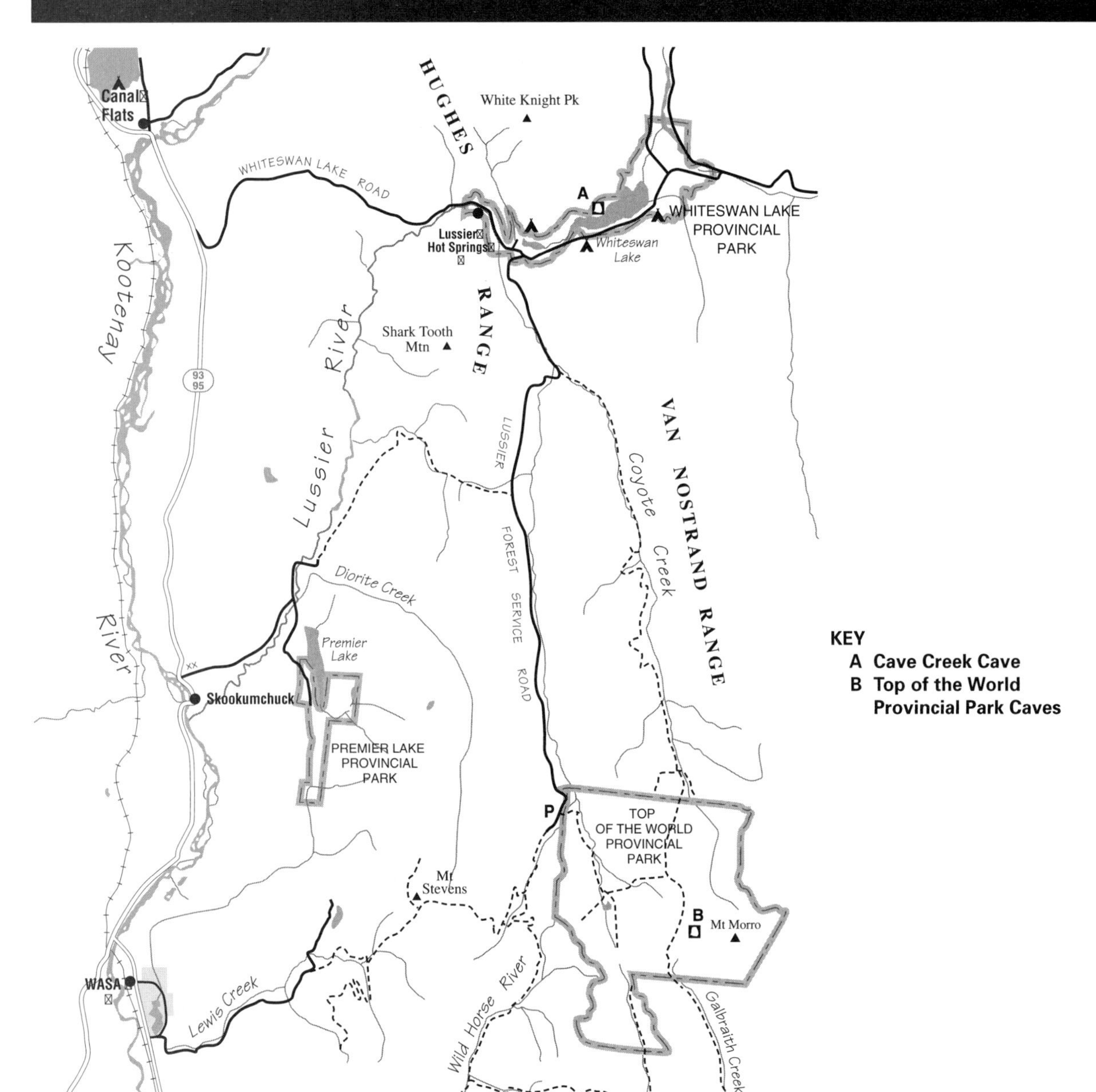

Cave Creek Cave

Map 82 J/3 075553
Jurisdiction Whiteswan Lake Provincial Park
Entrance elevation 1250 m
Number of entrances 1
Length 18.3 m
Discovered October 1991, by John Donovan, Eric Neilsen & Jon Rollins.

Access From Hwy. 93/95 south of Canal Flats take the Whiteswan Lake Road for 25 km. En route check out Lussier Hot Springs near the park entrance at Km 8. Camping is available in the provincial park. Call Kootenay District Office at (250) 422-4200.

Location This is NOT the obvious "main" cave which is visible from the road above the northeast shore of Whiteswan Lake, although the access is the same. From the parking area a well-used trail follows the shoreline heading west and soon crosses a number of creek beds that are dry in summer. The second major drainage, just past where the lake narrows is Cave Creek. Scramble up Cave Creek a short distance, then take the right-hand tributary that brings you to a small, low, crawl-in cave entrance.

Description The cave is evidently a fossil spring, consisting of a short section of crawl-sized phreatic tube with some small pools and popcorn formations. The scallops on the walls indicate the water flow direction towards the entrance.

Commonly visited, the obvious "main" cave goes back only 15 m and contains lots of sheep dung.

TOP OF THE WORLD PROVINCIAL PARK

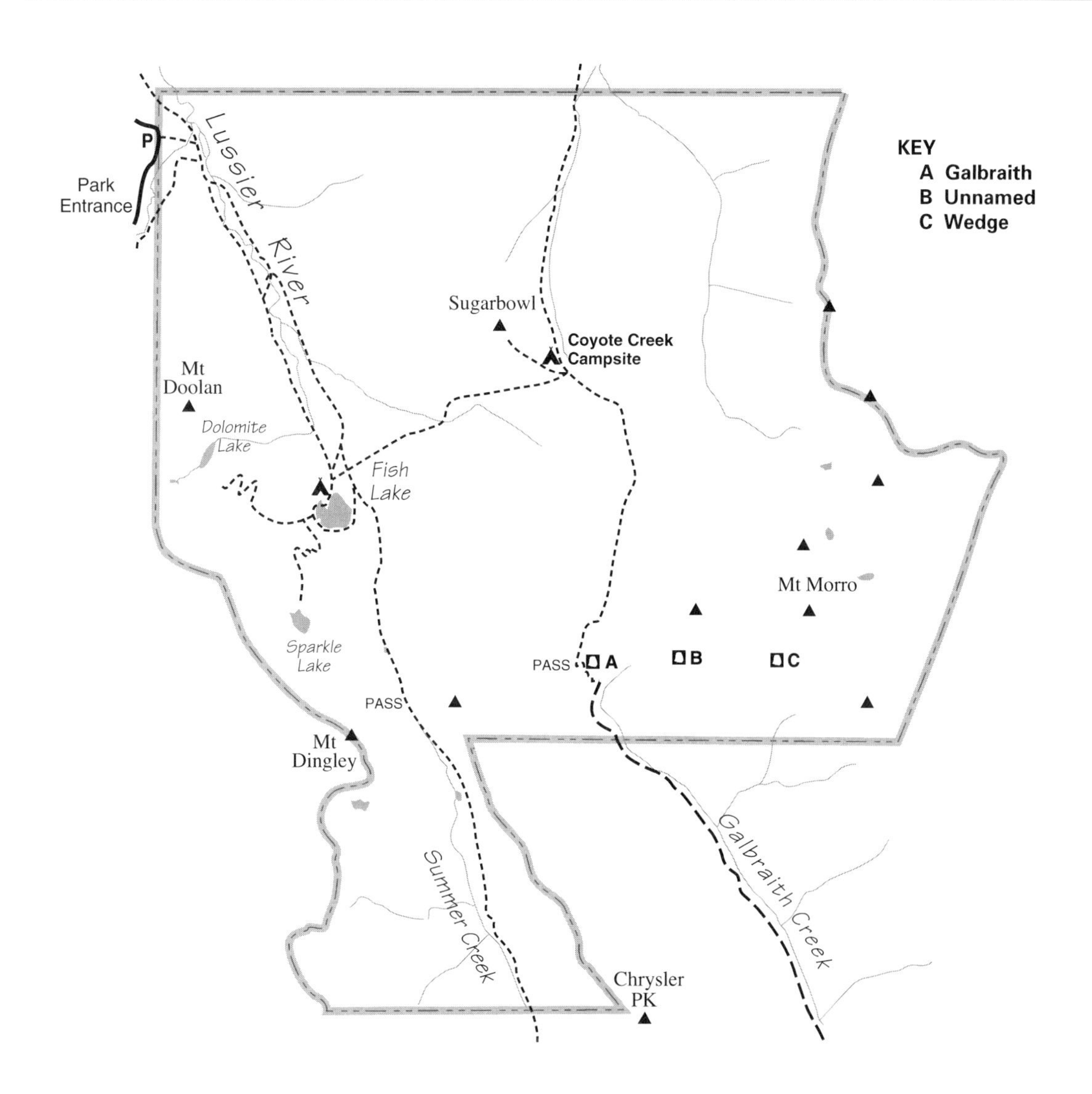

While looking at topographical maps of the Canadian Rockies, Mike Boon of the ASS came across an area near the Bull River, British Columbia, known as Top of the World. The area was extremely flat for a mountainous region, and there were closed depressions indicated on the map. On the first visit to the area in May of 1973, Wedge Cave was discovered, and on a second visit, Galbraith Cave, located just above the resurgence to Galbraith Creek. In July eight cavers visited the area, checking out several pits on the plateau including Unnamed Cave and Kitty's Cave.

The level karst plateau of Top of the World is pocked with shafts. However, long-lasting snow-plugs have hindered exploration (the caves are not accessible until mid-July). The area continues to hold considerable promise for more discoveries and is deserving of further visits.

Access for all Caves See map on page 100. 4.5 km south of Canal Flats, turn east off Hwy. 93/95 onto the Whiteswan Lake Road. At km 29.6 turn right and cross Coyote Creek. Continue straight at km 30.7, staying on the main road until reaching km 52 at the entrance to Top of the World Provincial Park. The trail begins here.

Follow the easy trail to Fish Lake (6.7 km) where there is a campground and a cabin (stove and firewood) that accomodates 20 people overnight. Call the Kootenay District Office at (250) 422-4200.

To reach the caves from Fish Lake follow the Coyote Creek campground trail for 5.6 km, then branch right (south) and climb a similar distance to the plateau below (west of) Mt. Morro.

Wedge Cave

Map 82 G/14 185198
Jurisdiction Top of the World Provincial Park
Entrance elevation 1982 m
Number of entrances 1
Length c. 60 m
Discovered 1973, by Ian Drummond.

Location The first cave discovered at Top of the World is situated 1.5 km east and slightly south of Unnamed. The entrance is a 3.5 m-diameter hole.

Exploration & Description The entrance leads into a wedge-shaped rift that becomes phreatic before ending in breakdown after 60 m. On the left side of the rift passage, a short way into the cave, is a frozen waterfall. A spectacular echo here indicated to the first explorers there was more passage above. Using crampons and ice axes, they reached a higher chamber and climbed another ice slope to an eventual boulder choke. Not surveyed.

Unnamed Cave

Location 82/G 14 172201
Jurisdiction Top of the World Provincial Park
Entrance elevation 2318 m
Number of entrances 1
Depth 60 m Pit
Discovered 1973, by Ian Drummond

Location A shaft located 1 km south of Mt. Morro on a small south-projecting spur.

Exploration & Description The shaft was located in deep snow when first explored in July of 1984 and was descended to a small chamber. Not surveyed.

Kitty's Cave

Map 82 G/14 (GPS location unknown)
Jurisdiction Top of the World Provincial Park
Entrance elevation c. 2200 m
Number of Entrances 1
Depth c. 12 m
Discovered 1973, by Kitty Dunn

Location Uncertain.

Description An attractive oval passage along a joint leads to two different drops, 12 m and 3 m deep respectively. Not surveyed.

Galbraith Cave

Map 82 G/14 155205
Jurisdiction Top of the World Provincial Park
Entrance elevation 2013 m
Number of entrances 1
Depth c. 170 m
Discovered 1973, by Ian Drummond

Location 15 m northwest of Galbraith Creek resurgence.

Description Galbraith Cave is an intermittent overflow with a clean-washed passage 0.6 to 2 m high. 90 m from the entrance, a low crawl from the right wall leads off; a further 60 m leads to an aven, a duck, and 10 m farther on, to a final sump.

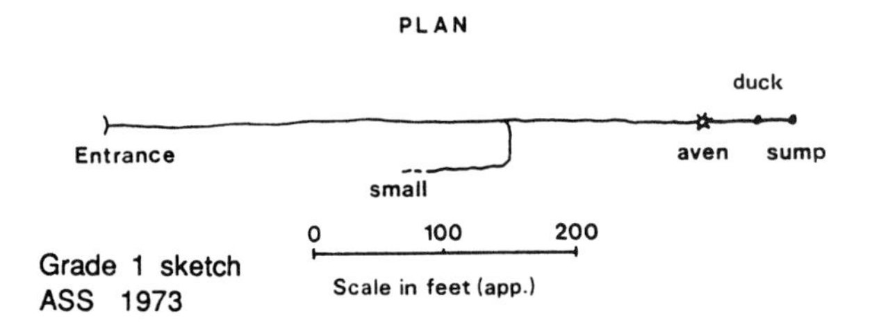

Frozen water seep in Wedge Cave. Photo Ian Drummond.

Forgetmenot Pot

Map 82 J/15 574263
Jurisdiction Alberta Forests (Bow Crow)
Entrance elevation 1980 m
Number of entrances 1
Depth 80 m
Discovered August 1969, by Dave Doze

Access A (on foot) Hwy. 66 (Elbow Falls Trail) at Cobble Flats. From the parking area follow a trail across a wooden bridge, then ford the Elbow River. Use caution at high water. Look upstream for a wooden bridge over a small tributary and an uphill trail connecting to Wildhorse Trail at a signpost. Turn left onto Wildhorse Trail and follow it to where it is intersected by a logging road. Shortly after the trail descends to a large boggy meadow with pond.

Leave the trail and contour around the boggy area, passing through a thin band of trees into clear-cut. Head south across the clear-cut, then through dense trees. After crossing the Howard Creek drainage (ephemeral), thrash steeply up the mountainside (300 m elevation gain) onto blocky scree and finally onto a grassy whaleback ridge. Distance about 16 km.

Entrance to Forgetmenot Pot. Photo Jon Rollins.

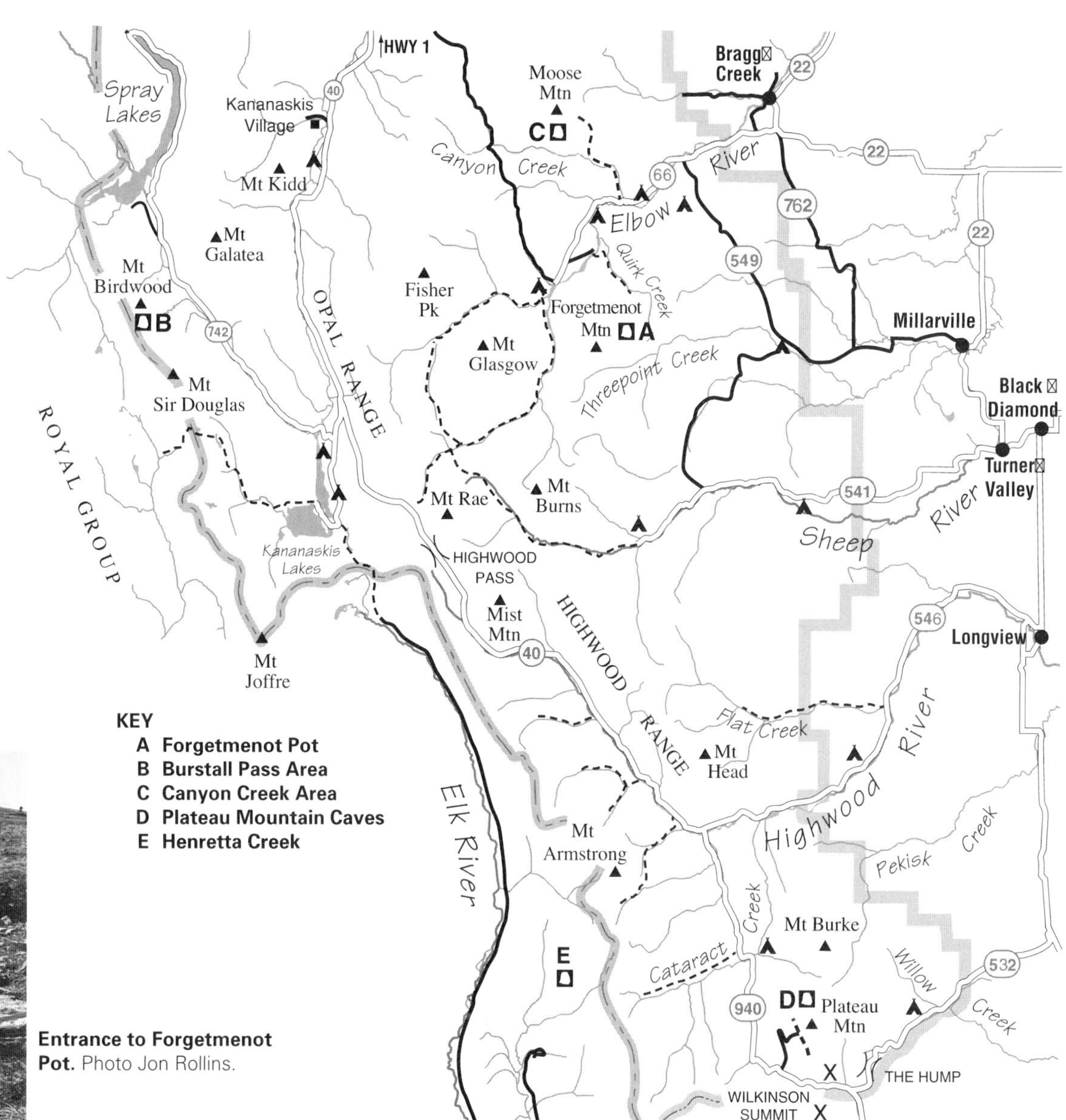

Access B (on foot, mountain bike or via OHV) Hwy. 66 (Elbow Falls Trail) at Cobble Flat. Ford the Elbow River, using caution at high water. Then walk, bike or drive a 4-wheel drive vehicle along the Quirk Creek exploration road to the well site on a side ridge. Descend the other side into the Howard Creek drainage. Turn right onto the Howard Creek logging road. Keep straight twice and enter the clearcut.

On foot head left (south) into dense trees and follow directions for access A.

Access C (via OHV) Hwy. 549 (McLean Creek Trail) at McLean Creek Recreation Area. Using a 4x4 vehicle, drive Elbow River Trail to Silvester Creek (4.5 km). Turn left on Silvester Trail and follow it past Fish Creek Trail to a T-junction with Fisher Trail West (7.2 km) in upper Muskeg Creek. Turn right heading for Quirk Creek valley. Here cross the bridge over Quirk Creek. Keep straight and cross a culvert over Howard Creek into the Howard Creek drainage. Keep straight three times on the Howard Creek logging Road and enter the clear-cut.

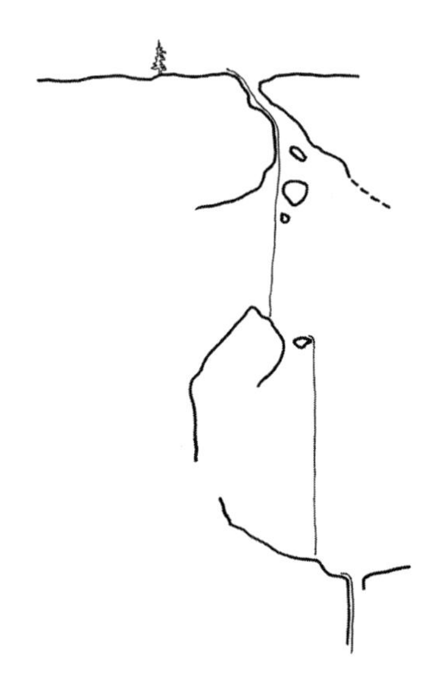

On foot head left (south) into dense trees and follow directions for access A.

Location Clumps of dwarfed Engelmann spruce mark a line of small depressions just where the ridge begins to flatten out, at the end of which is a large sandstone slab partially covering the cave entrance.

Cave Softly The ridge is an environmentally sensitive area with thin soils and delicate vegetation. Do not camp here or light fires.

Description Forgetmenot Pot, once the deepest known cave in Alberta, has an unlikely location on a grassy windswept shoulder of Forgetmenot Mountain.

Below the cap-rock, the cave — probably a former sink-point — extends down over 80 m in three vertical drops. Rarely wider than bridging width, the cave eventually ends in a damp, tight rift that is too narrow to traverse horizontally. There are no formations, but some bat and other unidentified small mammal bones were found at the bottom.

Warning Make sure you are conversant with SRT before attempting this cave.

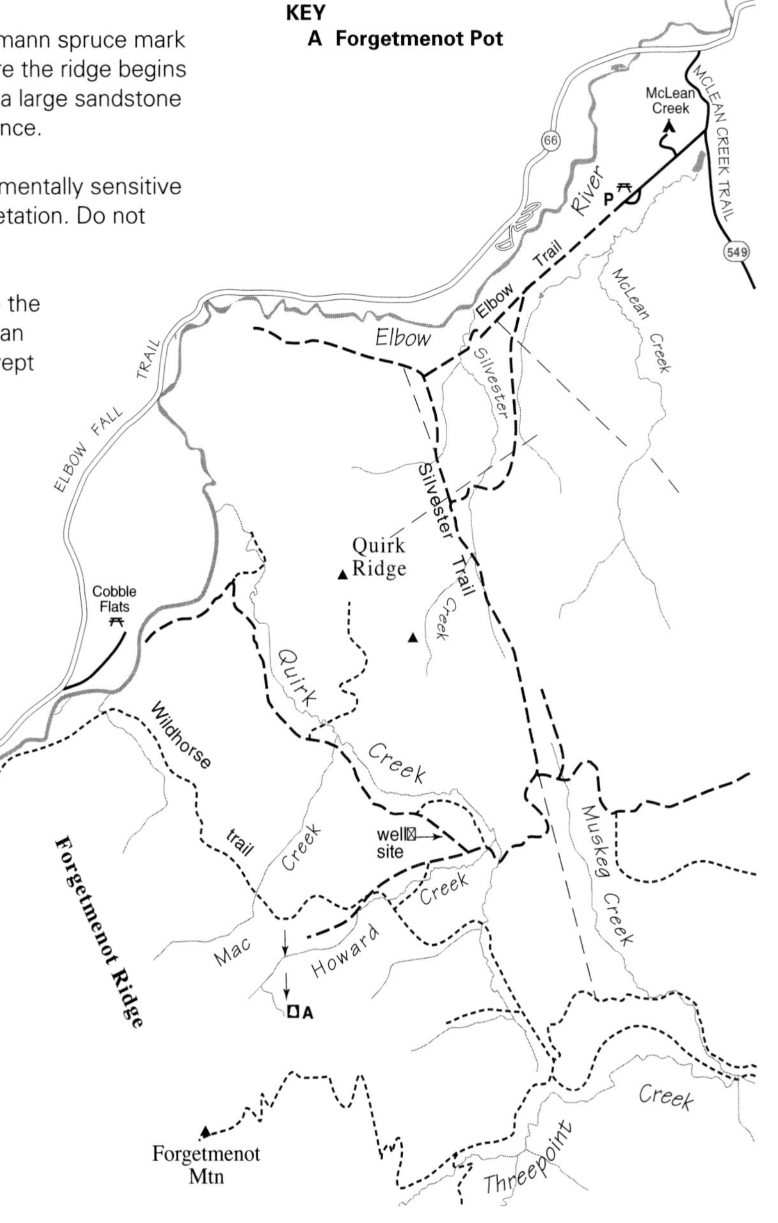

Burstall Pass Area

The Burstall Pass area contains some of the most accessible karst in the Rockies, with large expanses of limestone pavement, numerous closed depressions and sinkholes and some fine elliptical shafts. Two quite large stream risings exist lower down and to the west, towards the Spray River. The large spring feeding Watridge Creek (Karst Spring) is thought to drain the area to the south of Burstall Pass, although to the author's knowledge no water tracing has ever been carried out.

Three main karst areas can be identified. From north to south:

1. The karst at Burstall Pass is mantled with moraine material that provides a fooring for trees, grass and heath. About 1 km to the north is a series of sinks.
2. Towards South Burstall Pass is a more austere karst area consisting of flat limestone pavements. At the summit of the pass itself are blocked shafts. The main area of interest, however, lies just to the northeast below Whistling Rock Ridge and Burstall Slabs.
3. South of South Burstall Pass is a cirque of shattered karst full of depressions and blocked shafts. We are probably one ice age too late to find caves here, but a poke around the numerous collapsed areas is definitely worthwhile.

Shattered karst below Mt. Sir Douglas.
Photo Jon Rollins.

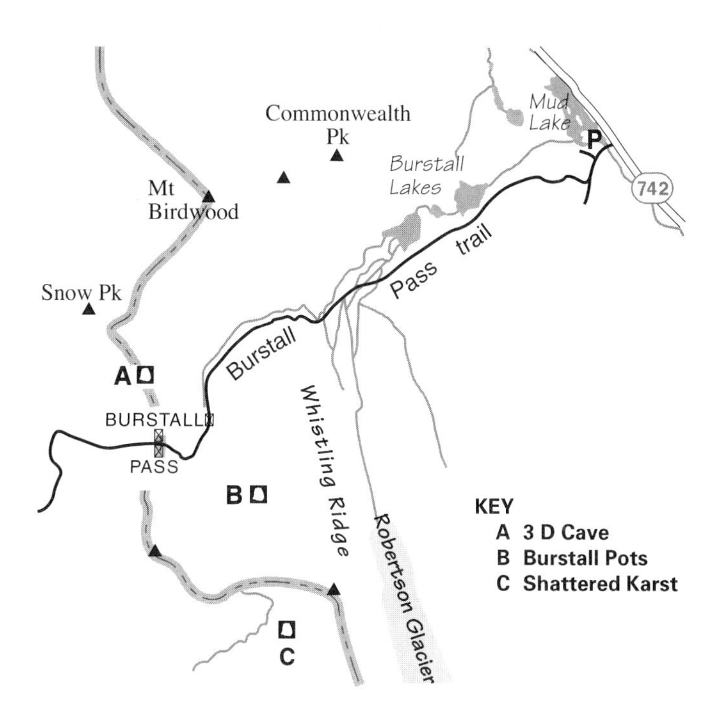

Area Exploration The McMaster cavers visited the area briefly in the 1970s, checking out some of the shafts at South Burstall Pass, and more recently ASS cavers checked out a series of pits just north of Burstall Pass at the foot of Snow Peak. Although some of the pits are fairly deep (up to 50 m), none of the exploration to date has revealed much horizontal passage. However, there could well be some hidden passage to be found as the ice in some of the pits melts back.

Access for all Caves Hwy. 742 (Spray-Smith-Dorrien Highway) at Mud Lake parking lot. Take the Burstall Pass trail (mountain bikes allowed for the first 3 km to the Burstall Creek ford) into the meadows of the pass area. Allow 2 hours (8 km).

For 3 D Caves follow the main trail to Burstall Pass. For Burstall Pots take a left fork towards South Burstall Pass.

Snow lingers well into July. The area is best visited in fall before the first snowfall when the larches are turning colour and most of the pits are dry.

Warning The sharp-edged, ragged limestone of the Burstall karst is hard on ropes and makes rigging tricky. Take lots of gear for tie-backs. Rope protectors are useful.

Cave Softly Please don't place bolts where they can be seen from the surface. This is a heavily visited area.

3 D Caves

Map 82 J/14 145251
Jurisdiction Peter Lougheed Provincial Park
Entrance elevation 2409 m
Number of entrances 3
Depth c. 46 m
Discovered 1989, by Dennis Weeks

Location The caves lie 183 vertical m above Burstall Pass on the ridgeline extending north to Snow Peak. At the base of a small cliff line.

Description The largest pit drops 40° to 50° for 24 m during which a transition from assorted loose boulders to sheer ice occurs and the second entrance connects from above. A slight left twist develops into a further 27 m vertical drop beside huge icicles. A few more metres down the snow slope is a snow plug in a 15 x 6 m elliptical chamber.

The beautifully sculpted easternmost pit is a straightforward free-hang of 30 m in a round shaft ending in a loose rock canyon, too tight in one direction and blocked with debris in the other.

Easternmost entrance to 3D Caves. Photo Jon Rollins.

Looking north from Burstall Pass to Snow Peak. Entrances to 3D Caves can be seen slightly below and right of the centre of the photo. Photo Jon Rollins.

Burstall Pots

Map 82 J/14 155235
Jurisdiction Peter Lougheed Provincial Park
Entrance elevation 2470 m
Number of entrances Many assorted shafts
Depth up to 30 m
Discovered Unknown

Location The most promising pits lie southeast of Burstall Pass towards South Burstall Pass. Once you reach the first pavements head northeast back towards Whistling Ridge and Burstall Slabs where small streams plunge into beautifully-sculpted elliptical shafts of sharply-fluted limestone.

Description One pit drops at 40° to 50° for 24 m during which a transition from assorted loose boulders to sheer ice occurs. A slight left twist develops into a further 27 m vertical drop beside huge icicles. Beyond this, a few more meters down a snow slope, is a snow plug in a 15 m x 6 m elliptical chamber.

The less promising pit is a straightforward free-hang of 30 m in a round shaft ending in a loose rock canyon, too tight in one direction, blocked with debris in the other.

Canyon Creek Area

An access road to sour gas wells winds up Canyon Creek, then up the narrow canyon of Moose Dome Creek, which opens out onto the lower slopes of Moose Mountain. Aside from the well-known Canyon Creek Ice Cave, several other short caves are located in the vicinity, including Canyon Rill Cave, Danger Cave (no information) and The Tube.

Access to all Caves Highway 66 (Elbow Falls Trail) to Canyon Creek. Turn north onto a gas well access road and just before a locked gate turn right into Ings Mine parking lot. Walk or bike a further 5.7 km along the road to a junction. For Canyon Creek Ice Cave and Canyon Rill keep left to the end of the road. For The Tube stay right on the road that winds up Moose Dome Creek.

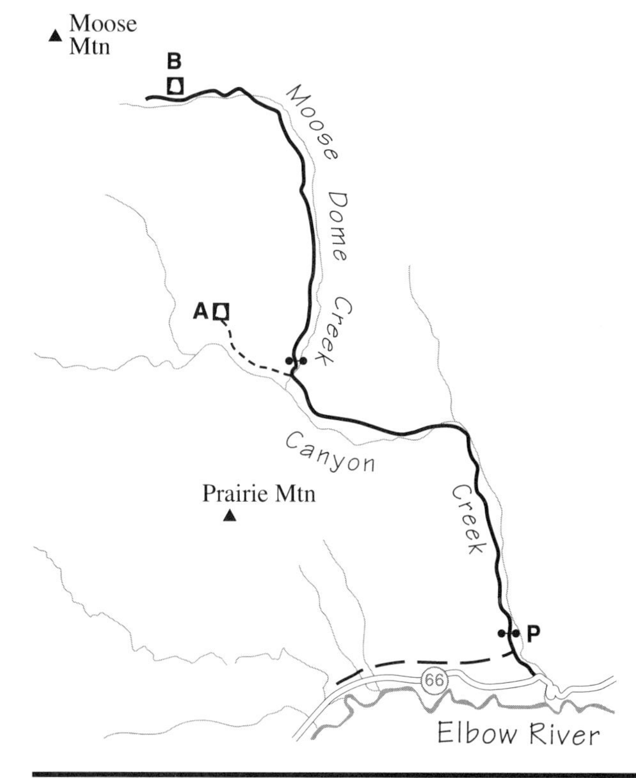

Canyon Creek Ice Cave from Canyon Creek.
Photo Jon Rollins.

KEY
A **Canyon Creek Ice Cave & Canyon Rill**
B **The Tube**

Canyon Creek Ice Cave (Bragg Creek Ice Cave, Moose Mountain Ice Cave)

Map 82 J/15 543418
Jurisdiction Ings Mine Provincial Recreation Area
Entrance elevation 1769 m
Number of entrances 1
Length, Depth 727.12 m +89.4 m
Discovered 1905, by Stan Fullerton

Location From the end of the road a trail traverses up to the cave, its large slot entrance — 15 m high by 6 m wide — is unmistakable. Do not climb up talus slope from the valley immediately below the entrance.

Warning Although no more dangerous than most steep trails, the presence of large numbers of visitors, many not very agile, means that loose rock is a hazard. A German tourist was killed by rock falling from above the cave entrance a number of years ago, and people stranded on steep terrain above the cave entrance have had to be rescued. Take care not to dislodge rocks on people standing below the entrance. Don't climb anything you can't reverse!

Description With the exception of the Cave and Basin in Banff, Canyon Creek Ice Cave is undoubtedly the most visited cave in the Canadian Rockies, its large entrance being clearly visible from the Canyon Creek valley bottom.

The cave consists of a large canyon passage heading into Moose Mountain for some 150 m before breakdown and ice meets the ceiling. This, and a few small climbs and side passages near the entrance constitute the only permanently enterable passages of the cave at the time of publication.

Depending on the amount of ice, a tight squeeze at the end of the Canyon leads to a further 600 m of passage, the Crowbar Section, the Corkscrew Climb and the Weasel Extension. These parts of the cave were first accessed in 1968 through an ice dig, which

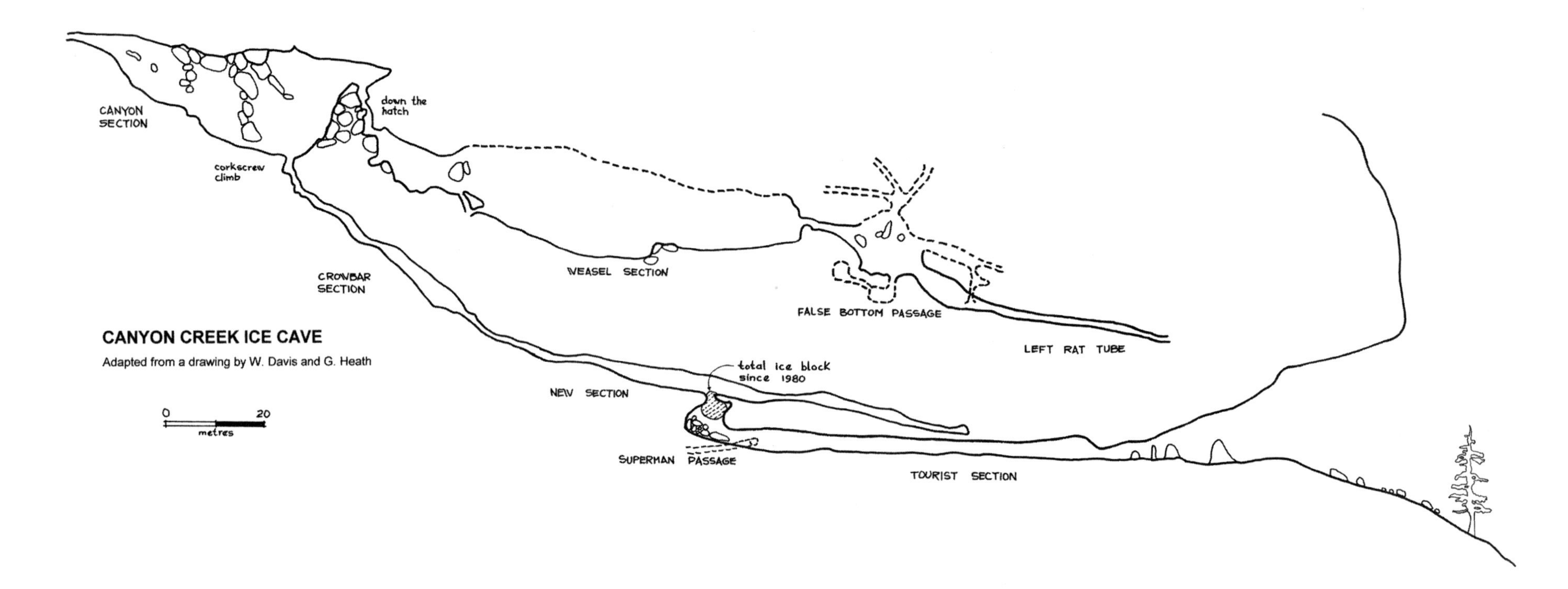

CANYON CREEK ICE CAVE

Adapted from a drawing by W. Davis and G. Heath

required lying flat out in freezing cold water. Once opened, a strong draft started. However, the squeeze had to be repeatedly dug out to prevent it from freezing over. In recent years it has frozen shut again. Recently the ice has retreated from the north wall, opening up a short section of tight rift passage.

The origins and what causes the fluctuations in the large amounts of ice in Canyon Creek Ice Cave are not fully understood. However, the Cold Zone phenomena is probably an important factor. The ice is one of the major attractions of the cave and is quite thick, forming a steeply inclined ramp towards the back of the canyon passage. Sometimes in the winter impressive floor to ceiling ice columns form just inside the entrance. In the summer the surface of the ice melts, forming large pools of water.

Cave Softly Canyon Creek Ice Cave has been known about for a long time; there was even an article on it in the *Calgary Herald* (30 October, 1963). Most weekends in the summer saw fifty to a hundred people make the short steep hike up the talus slope to the entrance. As a result, discarded pop bottles, cans, candy wrappers and batteries can be found amongst the boulders at the entrance, and the walls have been extensively spray-painted. The large number of visitors has caused erosion to the cave entrance and the talus slope leading up to it, damaging vegetation and leaving ugly white scars. The closure of the access road to motorized vehicles and the emergence of a trail from the east will no doubt relieve this problem.

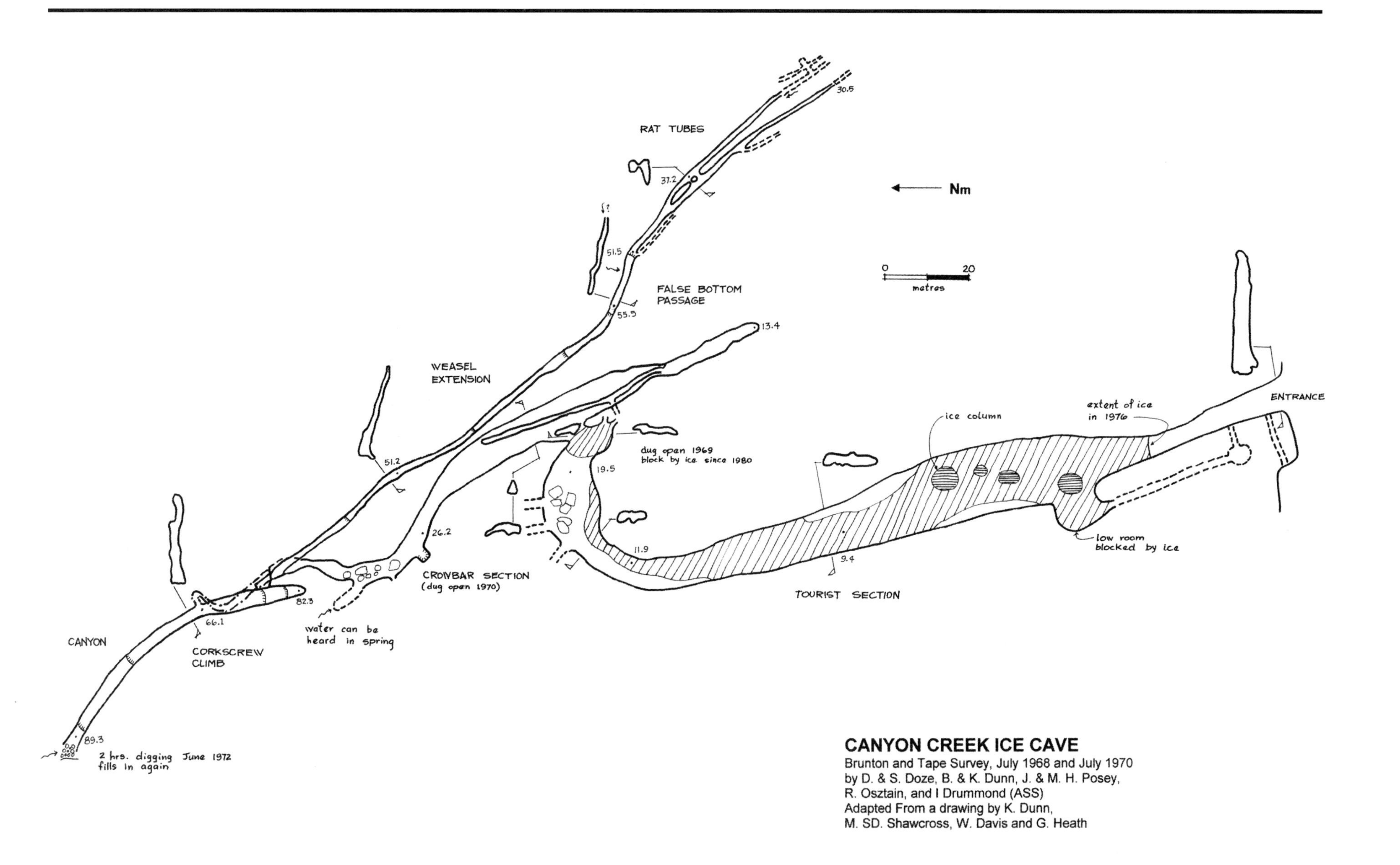

CANYON CREEK ICE CAVE

Brunton and Tape Survey, July 1968 and July 1970
by D. & S. Doze, B. & K. Dunn, J. & M. H. Posey,
R. Osztain, and I Drummond (ASS)
Adapted From a drawing by K. Dunn,
M. SD. Shawcross, W. Davis and G. Heath

The Tube. Photo Jon Rollins.

Entrance to Canyon Creek Ice Cave. Photo Dave Thomson.

Canyon Rill

Map 82 J/15 543418
Jurisdiction Ings Mine Provincial Recreation Area
Entrance elevation 1796 m
Number of entrances 1
Length 25 m
Discovered Unknown

Location A few metres to the west of Canyon Creek Ice Cave.

Description Canyon Rill is a tiny, tight rift, with a small stream on the floor. Squirming your way along its 25 m length provides a workout totally out of proportion to the distance travelled.
The ice that once partially blocked this cave has now all melted.

The Tube

Map 82 J/15 536444
Jurisdiction Alberta Forests (Bow Crow)
Entrance elevation 1850 m
Number of entrances 2
Length 125 m
Discovered March 1977, by Tom Barton

Location Follow the road up Moose Dome Creek for about 3.5 km to where it turns left (west). The caves are in the cliff above the road to the left (south).
Access may be restricted owing to well activities. Contact the Elbow Ranger Station at (403) 949-3754.

Description The cave consists of a series of small phreatic tubes and short rift passages which contain some ice, breakdown and a good draft. The end of the cave is close to and probably connects with the surface.
Two other nearby caves noted by Tom are Danger Cave (short, potential dig), and an ice-blocked cave heading upwards.

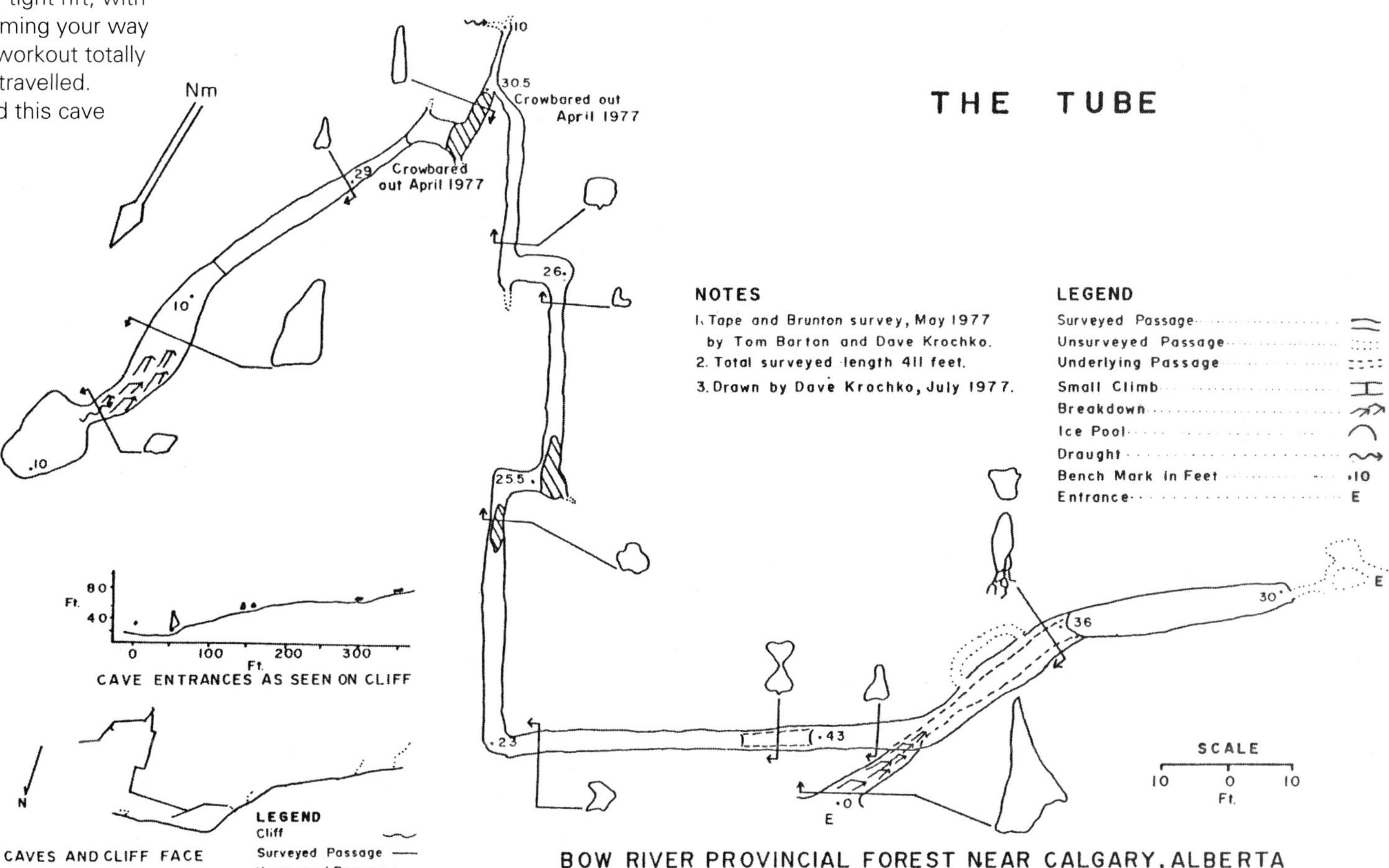

Plateau Mountain Caves

Plateau Mountain is unique in several ways. The extensive flat top lies at 2500 m, and has areas of arctic-like tundra and permafrost. Frost-developed rock features include 'nets' or 'frost polygons,' and shattered 'fell fields' — earth hummocks of possibly pre-glacial origin. The rock polygons are cited as evidence that Plateau Mountain was not glaciated, and are characteristic of freeze thaw cycles in sediments adjacent to glacial ice margins (Jim Burns, 1990). The area also contains rare, disjunct plants and insects.

In the 1950s a road was cut and several sour gas wells were drilled on the top of Plateau Mountain. Two of the wells owned by Husky Oil are good producers of sour gas.

When Husky Oil realized the ecological significance of the Plateau Mountain area, and that it was a candidate for an ecological reserve, the company decided to promote protection of the area, including a commitment to do no new drilling and no deepening of existing wells within the reserve boundaries. The company agreed to immediately begin reclamation of roads and well sites that were not crucial to their operations. As well, Husky hired a wildlife biologist to assess the area's wildlife and their habitats and to develop a management plan for the lands surrounding Plateau Mountain, which are within the Savanna Creek gas field. In exchange, Husky planned to operate the two wells within the ecological reserve and to try some directional drilling from well sites outside the reserve in order to recover more of the gas lying under the reserve. They also requested some amount of royalty holiday for the two Plateau Mountain wells. The final ecological reserve is a compromise. It is smaller than the original proposal for protection, and contains roads and two active sour gas wells which will probably be active for the next 20 years (AWA, 1991).

Area Geology The limestones of Plateau Mountain are outcrops of the Livingstone Formation and are relatively horizontally bedded as a result of their position on the top of an anticline. The bedding appears to have enhanced cave development in the area. The highest caves, Plateau Mountain Ice Cave and Sheep Cave, at an elevation of 2226 m, were formed beneath the water table. They contain pre-Wisconsin flowstones and may have originated in pre-Pleistocene times. January Cave, at 2040 m and 102 m higher than the present valley floor, may be pre-Wisconsin in age (Mike McEachren, 1970).

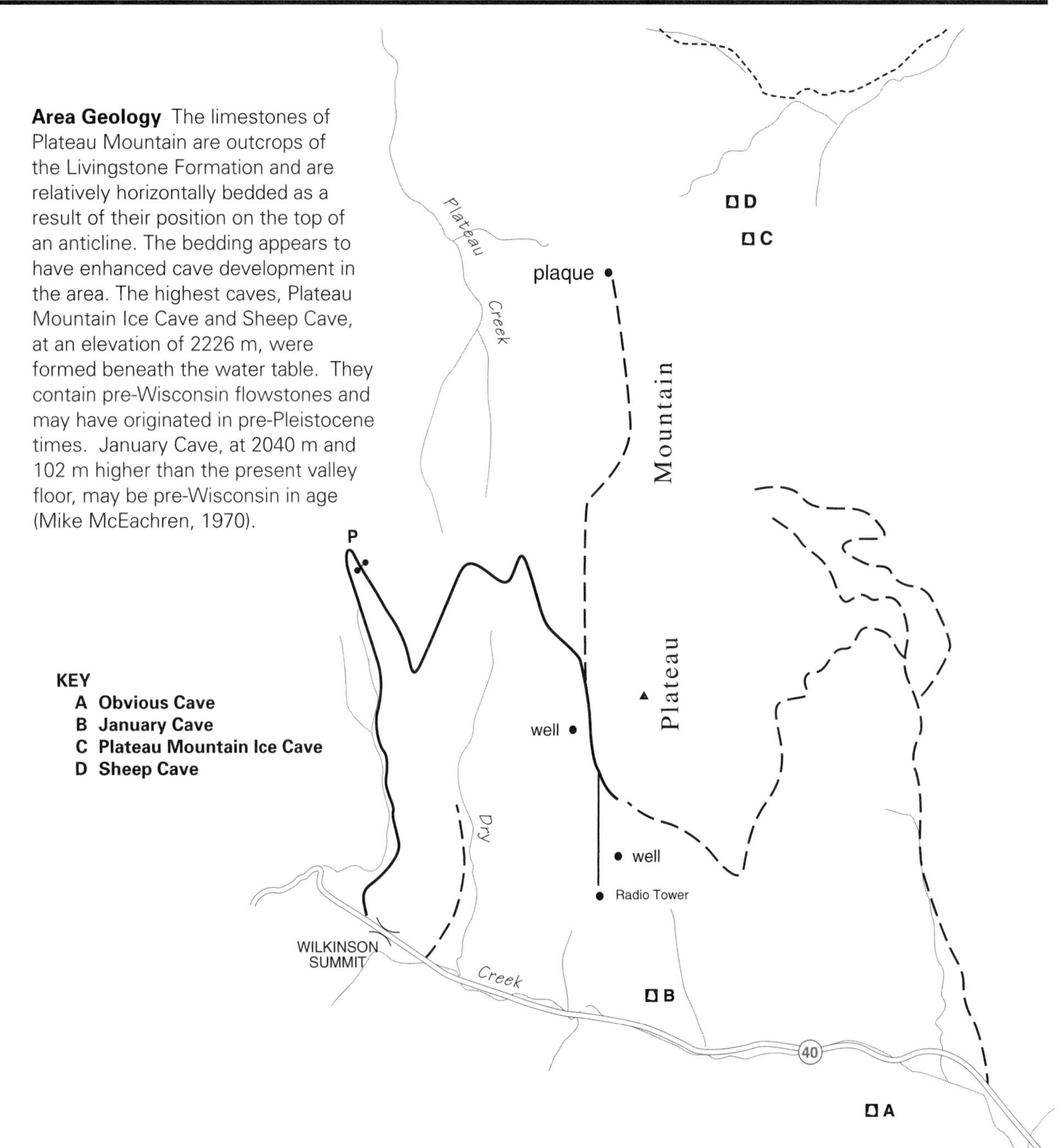

Obvious Cave

Map 82 J/7 798615
Jurisdiction Alberta Forests (Bow Crow)
Entrance elevation 1900 m
Number of entrances 1
Length 20 m
Discovered Unknown

Access Hwy. 940 (Forestry Trunk Road), 1 km west of the junction with Hwy. 532. Hike a short way up the north-facing slope. The large entrance is visible from the highway, located at the base of a long cliff above a scree slope. A large cairn has been constructed at the entrance.

Description Despite the promising looking entrance, the passage only extends back for 20 m. However, a gentle inward draft (summer) indicates that more passage may exist beyond the large slab breakdown.

Obvious Cave.
Photo Jon Rollins.

January Cave

Map 82 J/7 765623
Jurisdiction Plateau Mountain Ecological Reserve
Entrance elevation 2040 m
Number of entrances 1
Length 20 m
Discovered 1963 by Harvey Gardner

Access Hwy. 940 (Forestry Trunk Road) at Wilkinson Summit. About 2 km east of the junction with Plateau Mountain Road hike up a south-facing draw to the large cave entrance visible on the SSW slope of Plateau Mountain above Dry Creek.

Description From the entrance, a short passage leads back to a small chamber 18 m long.

Although short, January Cave contains extensive bone deposits in the sediments, both in the entrance, and in the chamber behind. These deposits have been excavated by Jim Burns of the Alberta Provincial Museum. "The sediments in both portions were rich with the small bones and teeth of 34 species of mammals birds and fish, the largest regularly occurring species being the hoary marmot. The age of the deposits is the subject of some discussion. The consensus is that the bones accumulated in slow but steady fashion over a climatically stable period from about 33,000 to 23,000 years ago. This corresponds to a non-glacial interval during the last, or Wisconsin glaciation" (Burns, 1990).

Plateau Mountain Ice Cave

Map 82 J/7 773700
Jurisdiction Plateau Mountain Ecological Reserve
Entrance elevation 2226 m
Number of entrances 1 (gated)
Length 90 m
Discovered Unknown

Access Hwy. 940 (Forestry Trunk Road) at Wilkinson Summit. Drive north on the Plateau Mountain Road to the gate and park (3.7 km). Continue along the road to a T junction on the plateau. Follow the left-hand road to its end at a cave interpretive sign. Walk northeast along an outlying ridge until about halfway along, then descend the southeast flank (right) for about 60 vertical m. Allow a day for this 18 km-long return hike. A mountain bike can be used to the end of the roads.

Location The cave is located above a narrow rock ledge to the right of a frost pocket and is gated.

Getting Permission Permission (and the key) are very hard to obtain. At the moment, a committee with members from the AWA, ASS and the Alberta Forest Service manages Plateau Mountain Ice Cave. Criteria for gaining permission to enter the cave are not clear, and some sort of access policy, other than complete denial of entry needs to be worked out.

Exploration & Description The road built by Husky Oil enabled people to drive almost to the entrance of Plateau Mountain Ice Cave. As a result, summer weekends saw dozens of tourists driving to the top of the mountain. Some came to take in the scenery, but many came just to visit the cave. "On Sunday afternoon, we counted approximately 60 people either going to or coming out of the cave. One group contained 20 people, all of them loaded with ropes, hammers, lights, etc" (ASS newsletter August, 1971). In 1972 the cave entrance was gated. The small rise in air temperature caused by people in the cave was having a harmful effect on the delicate ice formations.

Currently, Stuart Harris of the University of Calgary is using the cave as a laboratory to study permafrost and has installed thermistors to measure small variations in temperature. A rise of 2°C has been recorded in the permafrost surrounding the cave and the ice formations are showing signs of melting (Personal communication Bill MacDonald, 1992).

The entrance consists of a low passage that leads after six metres to a large breakdown chamber with a fracture line running through the center of the ceiling. From here two passages lead off to a series of chambers. The whole cave is decorated with ice formations, the most spectacular being large individual ice plates up to 15 cm in diameter (in 1978).

"What industry has been able to achieve in a couple of months, environmentalists have been working on for 20 years." AWA on Husky Oil being instrumental in Plateau Mountain becoming an ecological reserve.

Sheep Cave

Map 82 J/7 772703
Jurisdiction Plateau Mountain Ecological Reserve
Entrance elevation 2200 m
Number of entrances 1
Length, Depth 50 m
Discovered Unknown

Location As for Plateau Mountain Ice Cave to the northeast outlying ridge, but on the other (left) side of the ridge. Descend the northwest flank (left) for 150 vertical m.

Description Sheep Cave is short but impressive with some nice avens, vugs and scallops. It makes one wonder if there isn't more cave passage to be found in Plateau Mountain. Permafrost lies only 22 cm below the sediment surface in the cave.

Top: Ice crystals in Plateau Mountain Ice Cave. Photo Ian Drummond.

Left: Entrance to Plateau Mountain Ice Cave. Photo Gillean Daffern.

Entrance to Sheep Cave. Photo Jon Rollins.

Henretta Creek

Amos Cave

Map 82 J/7 580722
Jurisdiction BC Forests (Elk)
Entrance elevation 2250 m
Number of entrances 1
Length, Depth 250 m, 35 m
Discovered A local trapper named Amos

Access Because of access problems with the Fording Mine route, the usual access is now from Kananaskis Country. Start from Hwy. 940 (Forestry Trunk Road) 6 km south of Etherington campground access road and 1.5 km north of the bridge over Cataract Creek. Bike or hike the logging road up Cataract Creek to the point where a cutline (snowmobile trail Cataract Loop, red markers) turns off to the right in a huge cutblock. Higher up, intersect the Great Divide Trail and continue west on cutline to its end (15 km). Then scramble over the 2887 m-high pass between Mt. Etherington and Mt. Scrimger (8 km). In winter be aware of avalanche hazard. Traverse north to the cirque at the headwaters of Henretta Creek (8 km).

Location The cave entrance is a fossil spring at the axis of a syncline. It lies at treeline at the foot of a four square kilometre drainage basin delineated by Courcelette Peak, Mount Cornwell and Baril Peak.

Area Exploration The cirque above the cave is characterized by beautiful open sub-alpine meadow with occasional karstic bedrock exposures. Several sink-points were noted, but none were large enough to be enterable. A line of active sinks is located at the foot of a lateral moraine ridge below Courcelette Peak. A short distance below the cave entrance is a streambed, which leads down into a beautiful bedrock canyon containing spectacular waterfalls and several springs which emerge from the walls through bedrock fissures. The area now appears to be drained by several active springs 30 m lower and some distance to the south.

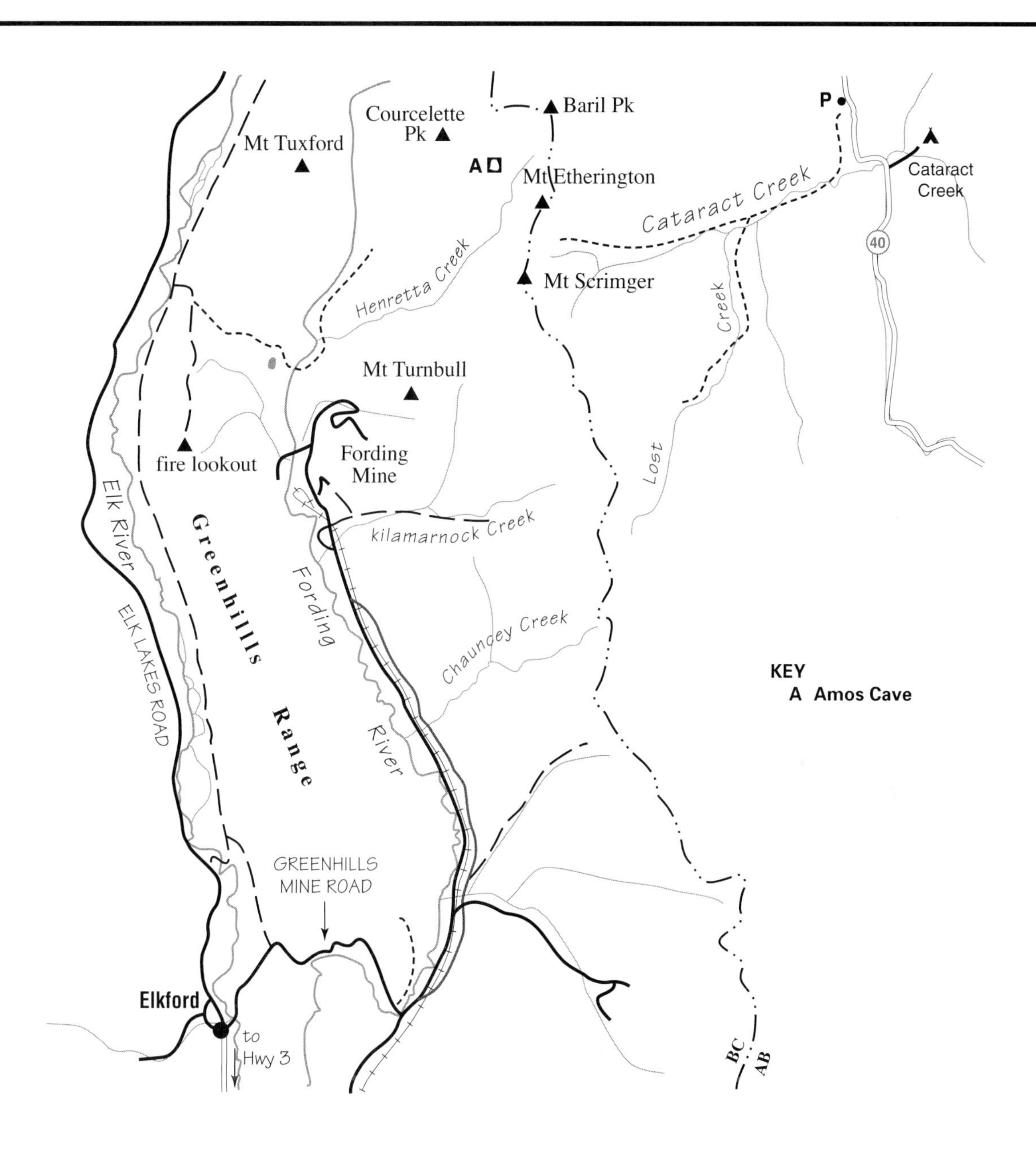

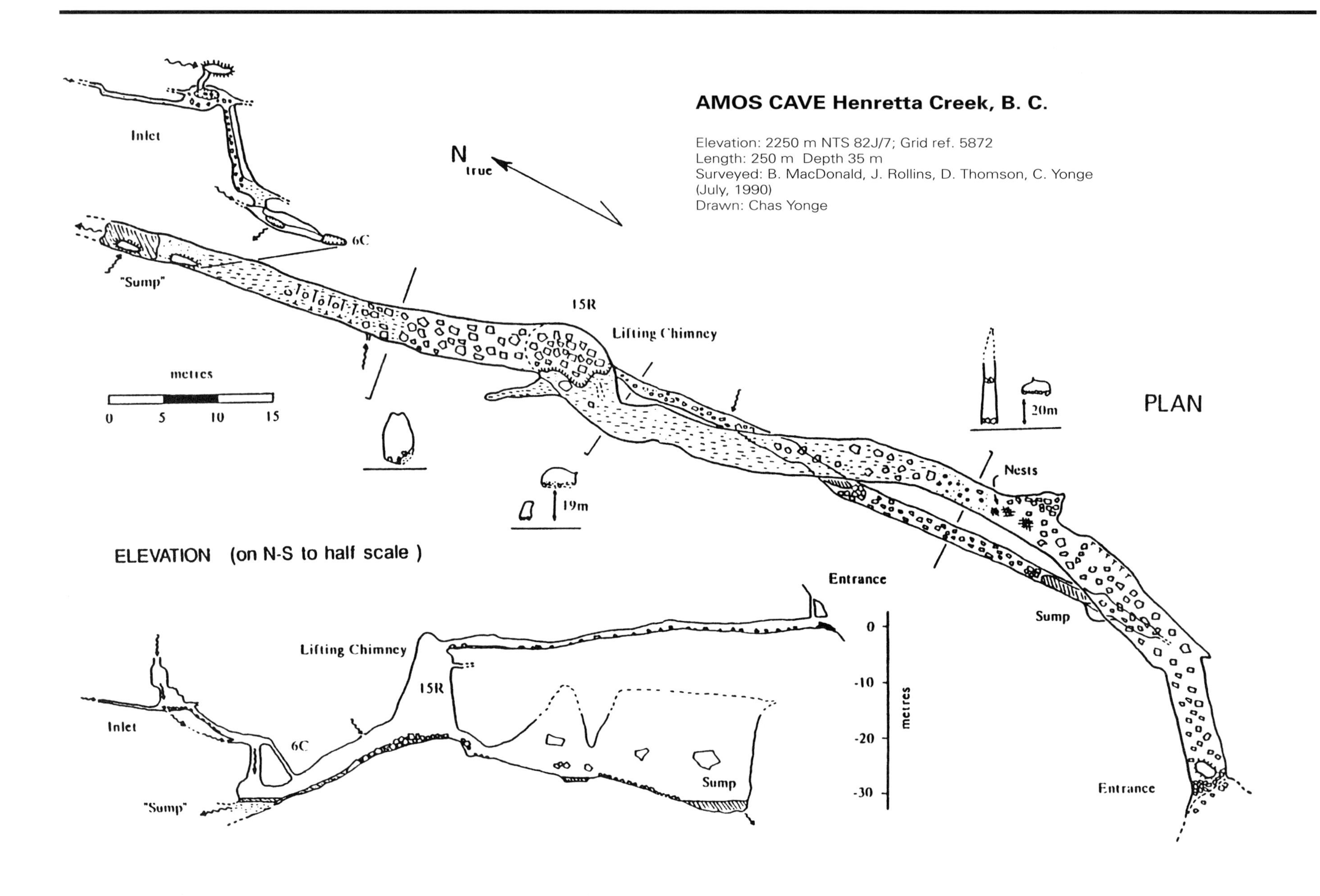
AMOS CAVE Henretta Creek, B. C.
Elevation: 2250 m NTS 82J/7; Grid ref. 5872
Length: 250 m Depth 35 m
Surveyed: B. MacDonald, J. Rollins, D. Thomson, C. Yonge
(July, 1990)
Drawn: Chas Yonge
N true
Inlet
6C
"Sump"
15R
Lifting Chimney
metres
0 5 10 15
19m
20m
PLAN
Nests
Sump
Entrance
ELEVATION (on N-S to half scale)
Entrance
Lifting Chimney
15R
Inlet
6C
"Sump"
Sump
0
-10
-20
-30
metres

Exploration & Description Much wildlife has been observed; it is obviously an important summer grazing area for elk and moose. Mountain goats were seen in the winter. The area is popular with hunters, so it is not surprising that a local trapper named Amos first discovered the cave.

The first attempt to locate the cave by ASS members was hampered by having to a make a lengthy diversion around Fording Mine, which effectively blocks access to the whole of the Fording River Valley. Henretta Creek was first reached by crossing the Elk River on a cable way and by following a steep logging road over the intervening ridge. Subsequent trips were made from the Alberta side via Cataract Creek and a 2287 m pass below Mount Etherington.

The cave entrance is a 100 m phreatic tube of stooping height that accesses the large active cave passage via a 15 m slightly overhanging pitch. This would have been a lifting tube up which the water was forced before emerging from the entrance. The large lower passage sumps in both directions, the downstream section being characterized by a 20 m-high vadose rift. Upstream water enters through a small tube in the ceiling above the sump pool. This was pushed through a series of miserable, tight, wet climbs in the hope that it would provide a way over the sump. It was obviously, however, heading for an inlet from the surface. Attention then turned to the upstream sump that was checked thoroughly during a quick swim. On a later visit to the cave in September the sump had drained and all that remained was a small trickle of water running through sand under the wall. Plans for a winter trip were made in the hope that it would have completely dried up, and that the removal of some of the sand would allow access to continuing passage. An ASS group skied in from the mine but was unable to locate the cave entrance due to deep snow. Avalanches below Mount Etherington thwarted another winter trip from the Alberta side. A more recent winter trip skiing in from the mine located the cave entrance after excavating a 10 m tunnel through snow. The inlet stream was still running and the cave, alas, was still sumped. The absence of other cave entrances, surface dolines or shafts indicates that Amos cave could continue uninterrupted for some distance under the cirque. Maybe by diving the sump, a longer cave system will be found in the future. Amos Cave obviously takes in large amounts of water during spring melt.

Right top: Bill MacDonald checks the downstream sump in Amos Cave.
Photo Dave Thomson.

Right bottom; Amos Cave entrance.
Photo Chas Yonge.

MOUNT ASSINIBOINE PROVINCIAL PARK

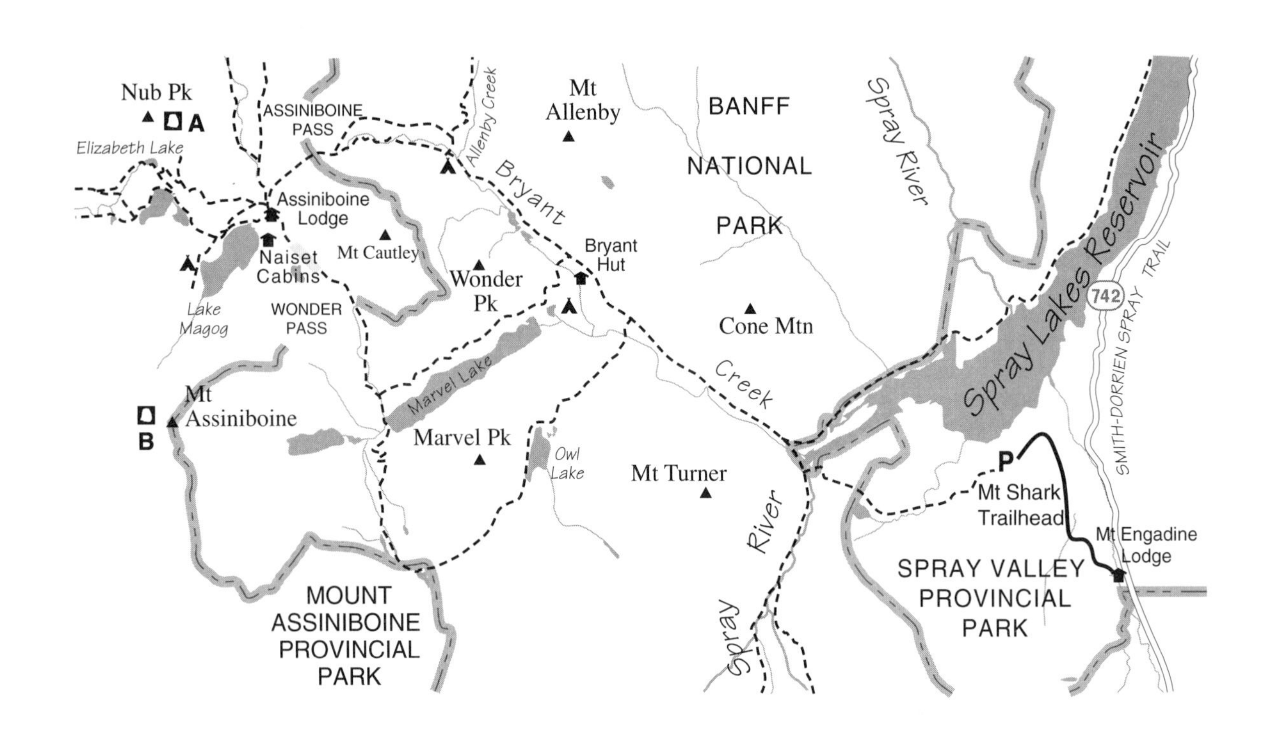

KEY
A **Nub Pot**
B **Assiniboine Sink**

Access to Assiniboine Via Hwy. 762 (Spray Smith-Dorrien Trail). About 38 km south of Canmore, turn west and drive the Watridge logging road to Mount Shark trailhead, en route passing Mount Engadine Lodge, and then the helipad on the left.

Walk the Watridge Lake trail to Bryant Creek trail junction, then follow the Bryant Creek trail over Assiniboine Pass to Lake Magog, a total distance of about 25 km. At the time of writing, the Bryant Creek trail is closed for mountain biking.

Alternately, helicopter in to Lake Magog, or have your gear flown in and walk. For schedule information contact Alpine Helicopters in Canmore at (403) 678-4802.

Accomodation is available near Lake Magog at Assiniboine Lodge (403) 678-2883, and at the Naiset Cabins (much cheaper, first come, first served), operated by B.C. Parks. A campground is located near the lake's west shore. The Hind Hut (operated by B.C. Parks, first come, first served) is located above Lake Magog (GR 946373) at the base of the north face of Mt. Assiniboine.

Assiniboine Sink

Map 82 J/13 945372
Jurisdiction Mount Assiniboine Provincial Park
Entrance elevation 2775 m
Number of entrances 3
Depth c. 30 m shaft
Discovered Tim Auger in the 1990s

Access From the west end of Lake Magog, take the climber's access "trail" to the Hind Hut that climbs the headwall to the right of the gully to the snowfield above. The hut is located on a buttress about 500 m back from the headwall.

Location A short distance above the Hind Hut. The sink is usually hidden by snow, and in the summer is taking a stream that makes it difficult to explore.

Exploration & Description Late summer through fall is probably the best time to visit. When open, the entrance consists of three holes taking water. In 1991 Dave Chase descended about 30 m to a chamber from where the stream continued down through a slot.

Warning The approach requires mountaineering skills. Good SRT skills, probably in a waterfall, are needed for the sink.

Nub Pot

Map 82 J/13 950420
Jurisdiction Mount Assiniboine Provincial Park
Entrance elevation 2379 m
Number of entrances 1
Depth 61 m shaft
Discovered 1975, by park wardens & Erling Strom

Access From Lake Magog campground take the Assiniboine Pass trail north for 1.5 km to a junction. Turn onto the Nub Peak trail. In another 400 m continue ahead. In 0.5 km keep right and follow the Nub Peak trail, keeping right, as it climbs and emerges above treeline onto the Nublet. Leave the ridge leading to Nub Peak, and traverse about 1 km north to reach the barren cirque between Nub Peak and Jones Hill.

Location The pot's entrance is narrow and ledged.

Exploration & Description It took three trips by assorted VICEG and ASS members before this shaft was finally bottomed in 1976.

From the entrance the shaft descends at an angle of 80°, with daylight penetrating 12 to 20 m. At 20 m it narrows to a one-metre hole before the final drop to the bottom where a narrow rift leads off, soon becoming too tight. The shaft was not surveyed.

Warning Nub Pot requires good SRT skills.

Peter Thompson at the entrance to Nub Pot.
Photo Chris Smart.

Nub Pk
Jones Hill
A
Elizabeth Lake
Nublet
to Bryant Creek
Cerulan Lake
Sunburst Lake
Ranger Cabin
Assiniboine Lodge
Sunburst Pk
Lake Magog
Naiset Cabins
Wedgwood Pk
climbers route
Mt Strom
B
ACC Hut
Mt Magog
Mt Assiniboine

KEY
A **Nub Pot**
B **Assiniboine Sink**

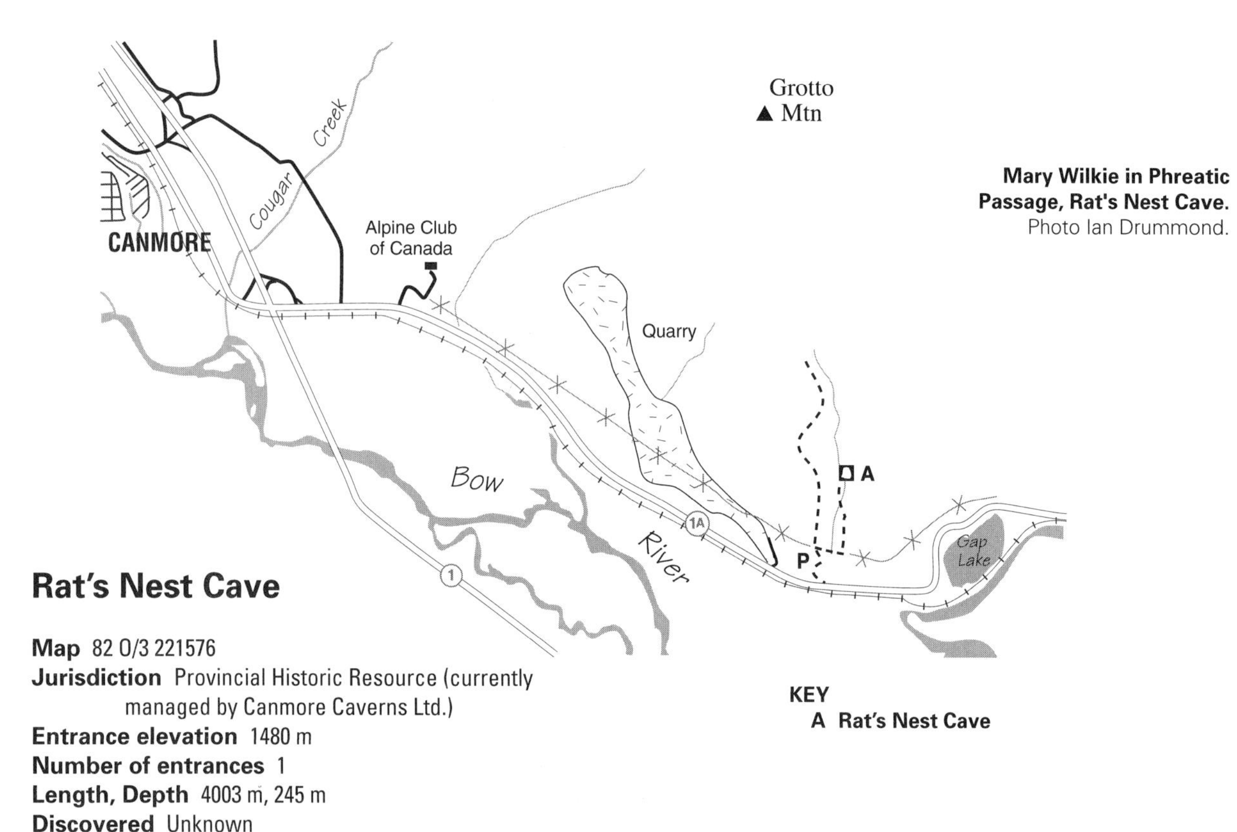

Mary Wilkie in Phreatic Passage, Rat's Nest Cave. Photo Ian Drummond.

Rat's Nest Cave

Map 82 0/3 221576
Jurisdiction Provincial Historic Resource (currently managed by Canmore Caverns Ltd.)
Entrance elevation 1480 m
Number of entrances 1
Length, Depth 4003 m, 245 m
Discovered Unknown

Access Hwy. 1A, 7.5 km east of Canmore or 7.5 km west of Exshaw. The entrance is situated near the southern tip of Grotto Mountain (2706 m), about 300 m above the floor of the Bow Valley. A rough trail starts from a gravel pit parking area adjacent to Gap railway siding (sign re cave access), and climbs to a powerline right-of-way on a glacial terrace. Turn right and cross a gully (shallow at this point), then head left on the trail that winds up through trees before dropping into the gully shortly before the cave mouth.

Look for a slot on the right (east) side of the gully.

An exception among Rockies caves, Rat's Nest Cave is located below treeline and a mere 20-minute hike above Highway 1A. Not only is Rat's Nest easily accessible, it is also, at just over four kilometres long, Canada's eleventh longest cave.

In 1987 a square mile around the cave entrance was designated a Provincial Historic Site. The principle reason for this was Bone Bed Pit, one of the most important paleontological sites in Alberta. The cave entrance is located on a quarry lease held by Greymont Ltd., who have gated the cave and restricted access to guided trips.

Guided Trips "Canadian Rockies Cave Guiding" of Canmore (Jon Rollins) caters to the University of Calgary Outdoor Pursuits programme and the City of Calgary Parks and Recreation Department. The company also runs private trips for both novices and experienced cavers. Call 1-888-450-2283, www.caveguiding.com. "Canmore Caverns" of Canmore (Chas Yonge) offers half-day and full-day adventure tours for the public. Call 1-877-317-1178, www.canadianrockies.net/WildCaveTours.

Exploration The first visitors to the cave were probably aboriginal people. With a year-round temperature of 4.5°C (the warmest cave in the Rockies), the entrance would have provided a pleasant location in which to shelter from winter storms. A Pelican Lake style point (arrowhead-shaped) was found in the Bone Bed Pit at the entrance and dated at about 3,000 years old. In the early 1960s, following rumors from climbers in the Banff and Canmore area, ASS members visited the entrance and dropped the Bone Bed Pit, but finding it blind did not look farther. During a later visit, ASS members dug through pack rat material and found their way into the cave via the chamber to the right of the pit.

By the mid 1970s the cave was being visited regularly by the ASS and other local cavers. All the "tourist" part of the cave — the Grand Gallery down to the Grotto, and the Dip Tube and Hose-pipe Passage as far as the Duck — was explored to the base of Bone Bed Pit, and then back north along Wedding Cake Passage (so named after a beautiful tiered stalagmite) to a low point where the ceiling dropped to meet a cracked clay floor. Anxious to protect the beautiful mineral formations in such an accessible location, the existence of the cave was not broadcast, no surveys were published and Bone Bed Pit was not excavated.

In 1979 and 1980 divers, including Paul Hadfield, Tom Barton, Randy Spahl, Dayle Gilliat and David Sawatsky dived the sump in the Grotto through to three other sumps.

In 1979, excavations at the end of Wedding Cake Passage finally broke through the sediment-plugged tube (the Birth Canal), the final digging sessions being orchestrated by Norman Flux who was an expert cave digger from England. Exploration carried out by ASS members Chas Yonge, Dave Thomson, Bill MacDonald, Randy Spahl and the author led to the Terminus Room, where the cave seemed to end. However, a search revealed a way on past beautifully decorated passages to a 40 m drop and the complexity of passages beyond. Ranger Way, a passage leading off to the left just before the Wedding Cake, was explored by a group of Bow Valley Park rangers led by Frank Gee as far as the boulder choke. The author and Dave Thomson later broke through to reach the lowest point in the cave.

Visitation to the cave increased dramatically in the early 1980's, and many large and small formations started disappearing from the passages leading to the sump, and in the sump chamber itself. Removal of some of the larger formations must have required extensive work, presumably with a hammer and chisel. Concerns regarding damage to mineral formations and disturbance to the stratigraphy of the bone bed led to the cave being gated, the first of a series of gates being placed in 1987.

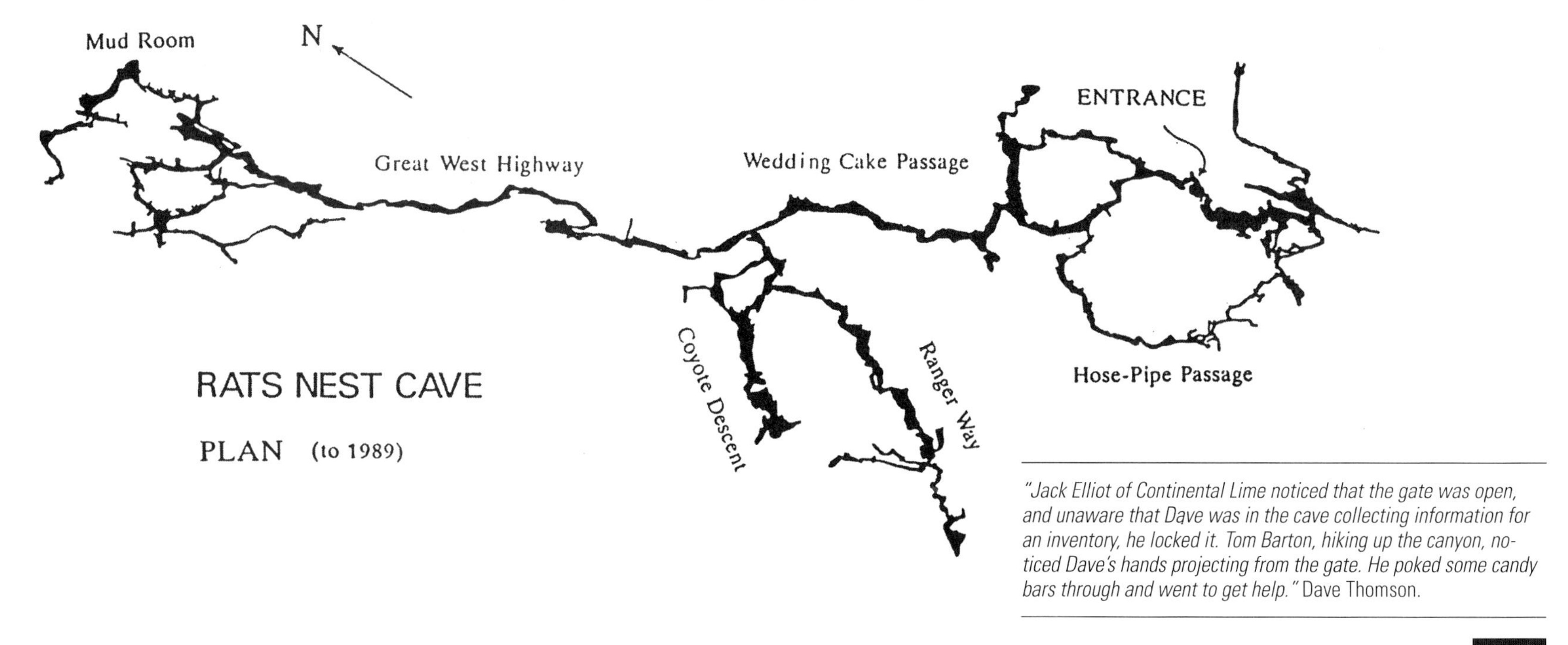

"Jack Elliot of Continental Lime noticed that the gate was open, and unaware that Dave was in the cave collecting information for an inventory, he locked it. Tom Barton, hiking up the canyon, noticed Dave's hands projecting from the gate. He poked some candy bars through and went to get help." Dave Thomson.

Top: Digging in the Birth Canal, Rat's Nest Cave. Photo Dave Thomson.

Bottom: Diving in Rat's Nest Cave. Photo Dave Thomson.

Description Just inside the five metre-wide entrance alcove is a 15 m pit that leads to the Bone Bed, and if the squeeze beyond is open, to most of the cave.

Guided trips, however, go through a small opening at the right side of the entrance alcove. A slide down a slab, followed by a 4 m down-climb (hand-line) leads to a small chamber with a beautiful water-sculpted ceiling. A 2 m climb at the end of this chamber accesses a short section of slanting rift passage, opening into the top of an 18m shaft. Prior to 1999, descending this pitch using ladders and ropes was the only way on. Nowadays, a hole excavated in the floor of the first chamber shored up with wood (the Box), enables the pitch to be bypassed. At the bottom of the 18 m pitch is a large chamber with several holes in the ceiling that can be accessed from a passage that heads off from a ledge halfway down the pitch.

A brief stoop under the lowering roof at the bottom of the large chamber accesses the Five-way junction. A passage at ceiling level leads via a short section of bedding-plane passage to the Grand Gallery — the largest chamber so far discovered — with a ceiling rising to a height of 30 m. The long straight crack that runs the length of the ceiling is believed to be a fault that passes through Grotto Mountain.

More intrepid parties take a lower passage from Five-way junction to the Laundry Chute, a small phreatic tube with a right-angle bend that requires tall people to perform some entertaining manoevers. The Laundry Chute descends even more steeply (hand-line), and opens into the Dip Tube, a wide phreatic passage that eventually ends in a sediment plug.

Near the top of the Dip Tube, a cool breeze can often be felt blowing from a hole high in the right wall. This is the start of the Hose-pipe passage, a tight, wet, arduous phreatic tube that, following a duck through a pool of water, eventually connects with the beginning of Wedding Cake Passage and Bone Bed Pit. Before Bone Bed Pit was excavated, negotiating the Hose-pipe was the only way to access the rest of the cave.

Most guided parties ignore the Hose-pipe and climb back up the Dip Tube, past where they emerged from the Laundry Chute, and slide on down the Treacherous Slab. From here, two small chambers joined by short crawls decorated with stalactites and stalagmites lead to the Grand Gallery.

A short spiralling climb below the Grand Gallery accesses a beautifully decorated sump chamber. Large amounts of flowstone cascade down the walls and many long, slender soda straws hang delicately from the ceiling. A clear pool with a line attached to the ceiling marks the start of the four successive sumps explored by cave divers. The passage in the first two sumps is very contricted, but past sump 3 opens up into a large descending rift passage with a 15 m pitch in a waterfall. (Sump 3 is actually a duck, having just enough air apace to keep your nose out.) The passage divides at sump 4, both continuing passages becoming too tight to continue.

Guided trips end at the sump chamber. Clients head back through the Grand Gallery, bearing right at an obvious junction, climbing over large breakdown blocks to enter the bedding plane passage that leads back to the Five-way Chamber.

Going Farther The passages beyond Bone Bed Pit are quite different in character to those experienced by the guided parties, consisting mainly of stooping and crawling passages sandwiched in a low but often wide bedding plane.

Ranger Way leads off to the left before the Wedding Cake in Wedding Cake Passage. Negotiating this beautifully decorated passage requires special care as it descends steeply to a boulder choke. Beyond the choke is a clear sump pool at the lowest point in the cave, 165 m below the entrance.

Beyond the Birth Canal (which no longer deserves its reputation as a tight squeeze), a rising passage leads to the Terminus Room, followed by a low crawl to the base of the Slimy Climb. Beyond is easy caving and some impressively decorated passages where cavers are required to move carefuly to avoid soda straws, helictites and flowstone formations. After the "high point," the cave descends 40 m in two 20 m pitches. Beyond the 20/20 a rope swing before the base of the shaft accesses ongoing passages. (20/20 can be bypassed by way of a low crawl over pools — the Rabbit Warren).

To the left, the Pearly Way (several nests of cave pearls) leads to a clear flowing stream-way and sump. To the right, a a low crawl in water (Hose-pipe Pool) leads to some steeply descending, muddy down-climbs, a 15 m and 10 m pitch, and finally, a mud choke — the ignominous end.

Passages beyond the 40 m drop are rarely travelled. The difficulty of the passages and the long round-trip time from the cave entrance — approaching 10 hours — deters most cavers.

Geology Rat's Nest Cave is contained in limestone of the Livingstone Formation, Mississippian in age, underlain by the Exshaw and Banff Formations and overlain by the Mount Head Formation (*Studies at Rat's Nest Cave: Potential for an Underground Laboratory in the Canadian Rocky Mountains* by C. J. Yonge, Cave Science, vol.18, no. 3, December 1991). Although the cave appears to be developed on a low angle thrust fault, evidence is hard to find because solutional activity has removed slickenside polished surfaces.

Speleogenesis Passage morphology indicates extensive phreatic development of a fault-guided bedding plane fissure. Water probably entered the system in the vicinity of Cougar Creek and flowed towards an outlet low on the southeastern flank of Grotto Mountain. The recharge head may have been provided by the glacial blockage of the Bow Valley, causing drainage to flow through the Fairholme Range towards Lake Minnewanka and Devils Gap. Following drainage of the cave system, there is little evidence of the shafts and trenches normally associated with vadose development, the vertical features in the cave being associated with the phreatic exploitation of prominent joints. Glacial retreat probably removed the source of recharge, and with little subsequent water input, vadose development did not occur. Analysis of sediments and speleothem growth in the cave provides evidence of four glacial advances in the Bow Valley (*Studies at Rat's Nest Cave: Potential for an Underground Laboratory in the Canadian Rocky Mountains* by C. J. Yonge, Cave Science, vol.18, no. 3, December 1991).

Likely the present entrance was created by a later breaching of a passage loop by the down-cutting surface gully. After the breach occurred, the entrance may have become a spring. A good-sized spring currently exists below the highway, just to the east of the Gap railway siding.

Cave pearls, Rat's Nest Cave.
Photo Jon Rollins.

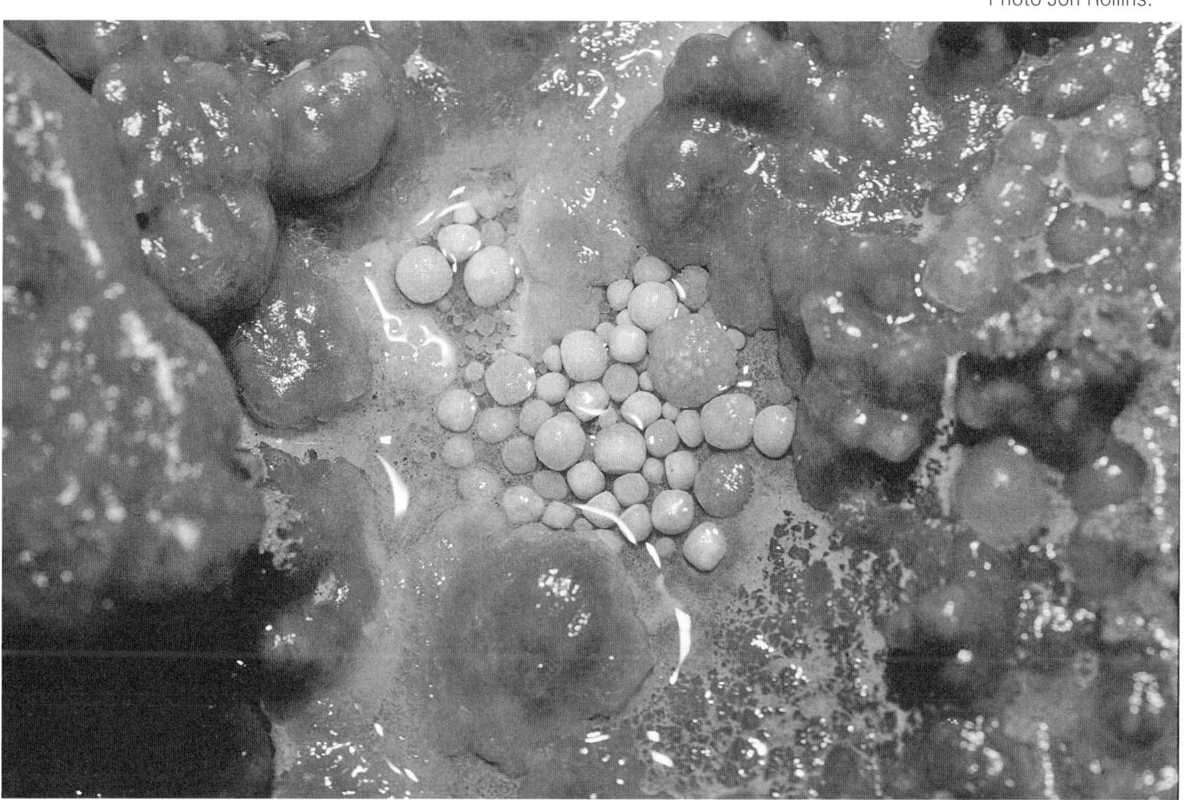

Cave Fauna Large packrat nests can be found on ledges in the entrance alcove. Originally they blocked the fissure entrance to the cave, but following installation of the large alcove gate the packrats moved their nests a short way into the cave. Little brown bats have been seen on several occasions throughout the cave, suggesting a roost exists, possibly near an unknown fissure entrance. Mosquitoes, harvestmen 'spiders' and occasional crickets winter in the chambers near the entrance.

Paleontology Bone Bed Pit, a natural trap, is a 20 m shaft containing large deposits of bones collected from animals that have either fallen down the shaft or been dragged in by predators. The site has been partially excavated by Dr. Jim Burns of the Provincial Museum and been found to contain the remains of over 30 different animal species, one locally extinct (swift fox), and an extinct bird (passenger pigeon). The oldest bones (using carbon-14 dating technique) are believed to be about 7,000 years old.

Mineral Formations Rats Nest Cave contains some of the best decorated cave passages in the Canadian Rockies, and includes soda-straws, stalagmites and stalactites, columns, flowstone, draperies, rimstone dams, helictites, cave pearls and moonmilk. Distributions of speleothems range from single isolated formations to spectacularly decorated grottoes. Due to the low bedding plane structure of much of the passage, formations are vulnerable to accidental damage from inexperienced visitors.

For further information on Rats Nest Cave see *Under Grotto Mountain Rat's Nest Cave* by Charles J. Yonge, Rocky Mountain Books, 2001.

The Wedding Cake in Wedding Cake Passage. Photo Dave Thomson.

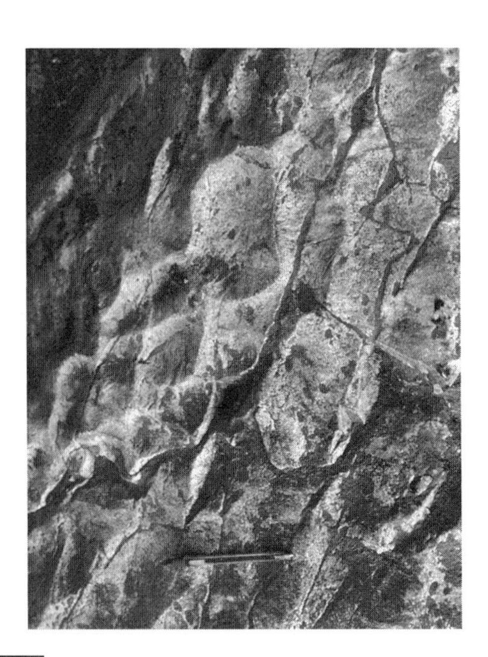

Far left: Scallops in the entrance of Rat's Nest Cave. Photo Jon Rollins.

Left: Helictites in Rat's Nest Cave. Photo Jon Rollins.

RAT'S NEST CAVE

GROTTO MOUNTAIN, ALBERTA

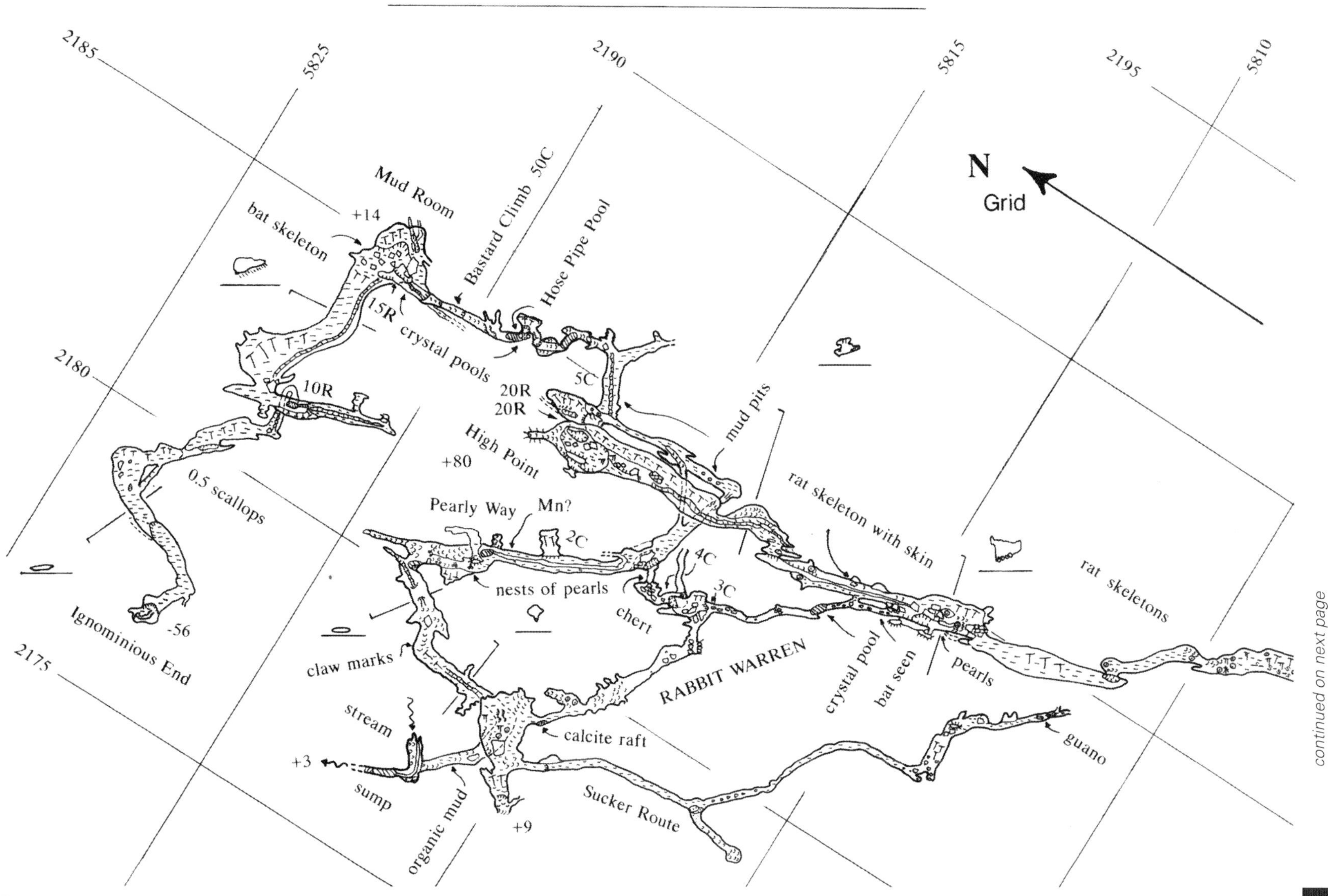

continued on next page

continued from previous page

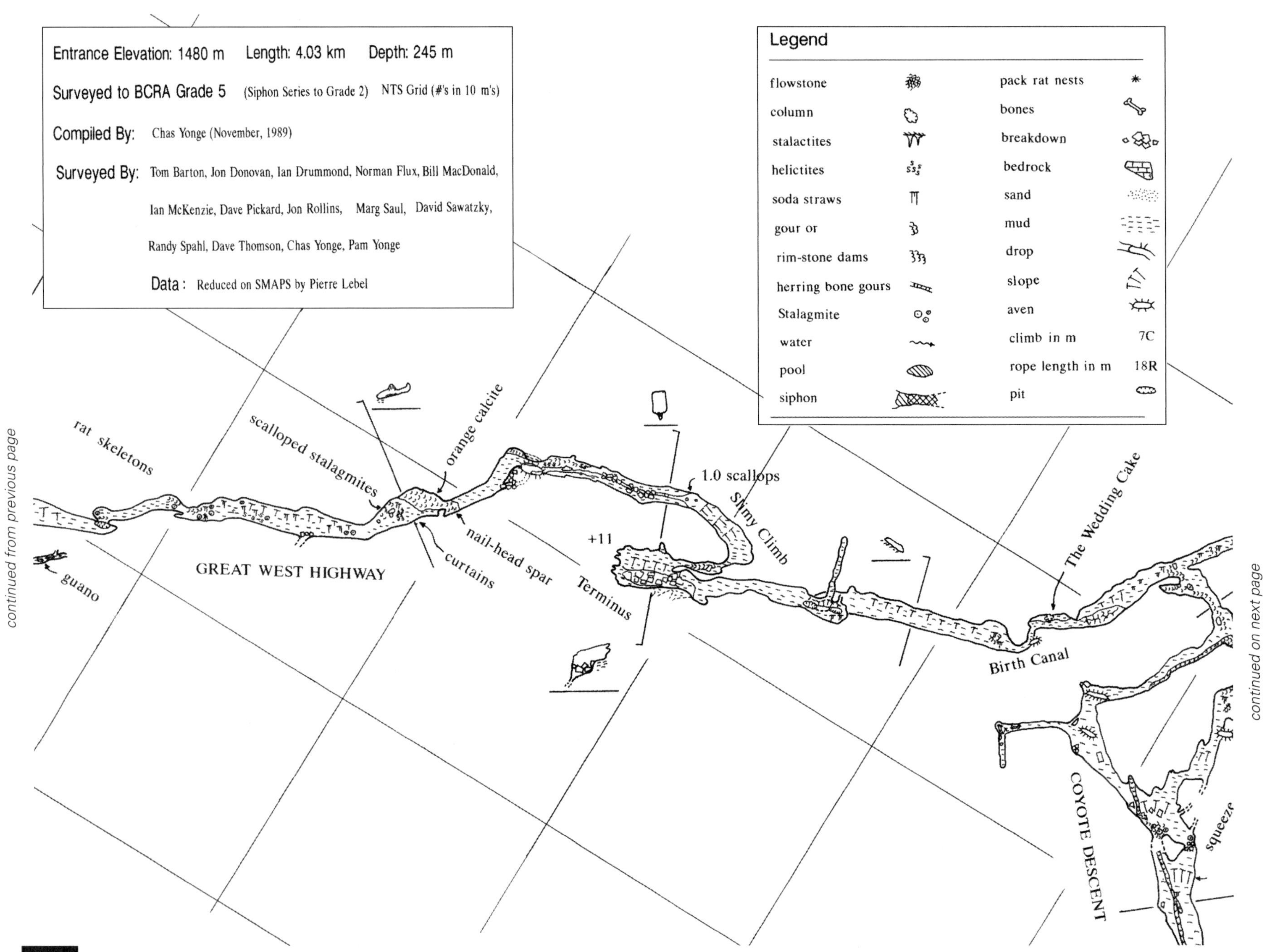

continued on next page

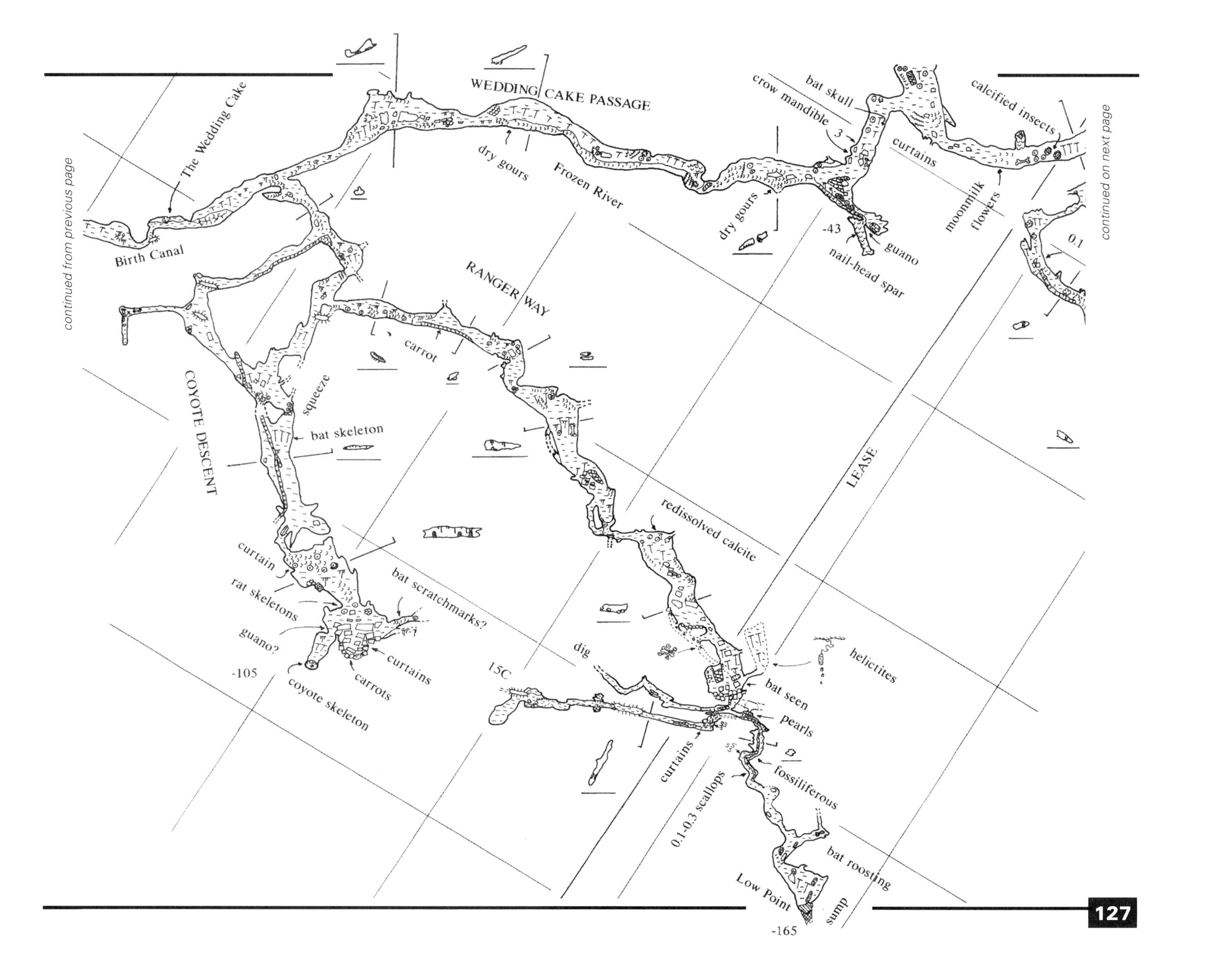

continued on next page

continued from previous page

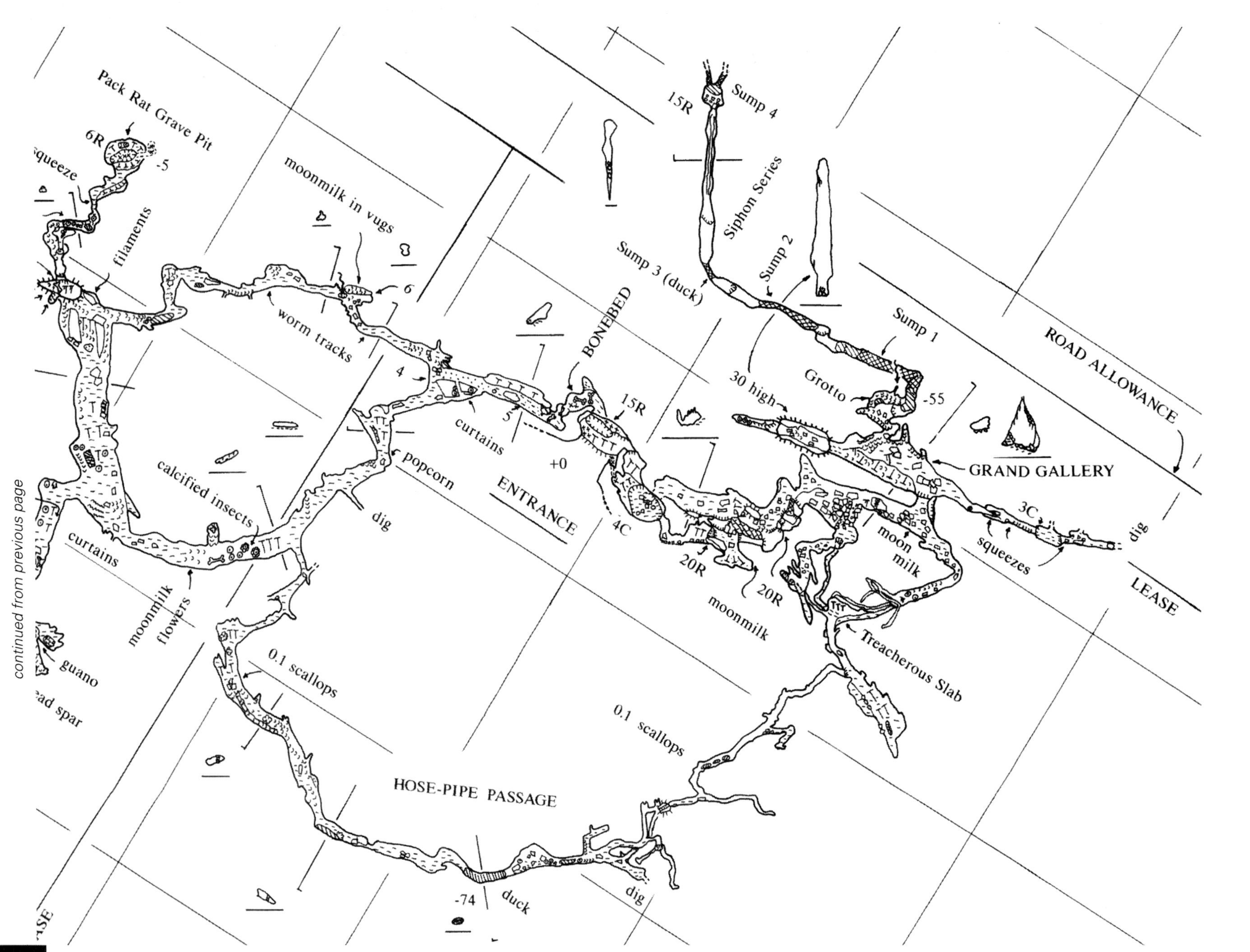

Ya Ha Tinda Steaming Pits

Map 82-0/13 (location uncertain)
Jurisdiction Alberta Forests (Clearwater- Ram FLU zone)
Entrance elevation Unknown
Number of entrances 2
Depth c. 45 m
Discovered 1968, by Larry Gingras

Access Hwy. 940 (Forestry Trunk Road) at the Red Deer River crossing. On the north side of the river turn west onto the road that leads up the Red Deer River valley. Just past Bighorn Falls, the road turns up Scalp Creek to Ya Ha Tinda Ranch, located at 980344.

Location Reputedly in a valley to the north of the ranch.

Exploration & Description Larry Gringras was out hunting in the fall of 1968 near Ya Ha Tinda Ranch when he noticed steam rising from a hole in the hillside. He returned with ASS members Roslyn Osztian, Dave Doze and Maryhelen Posey, and the 'Steaming Pit' was subsequently explored to a depth of approximately 30 m. The passageways they chimneyed down were formed in cretaceous sandstone, the cave being essentially a fracture line running across the hillside northwest to southeast.

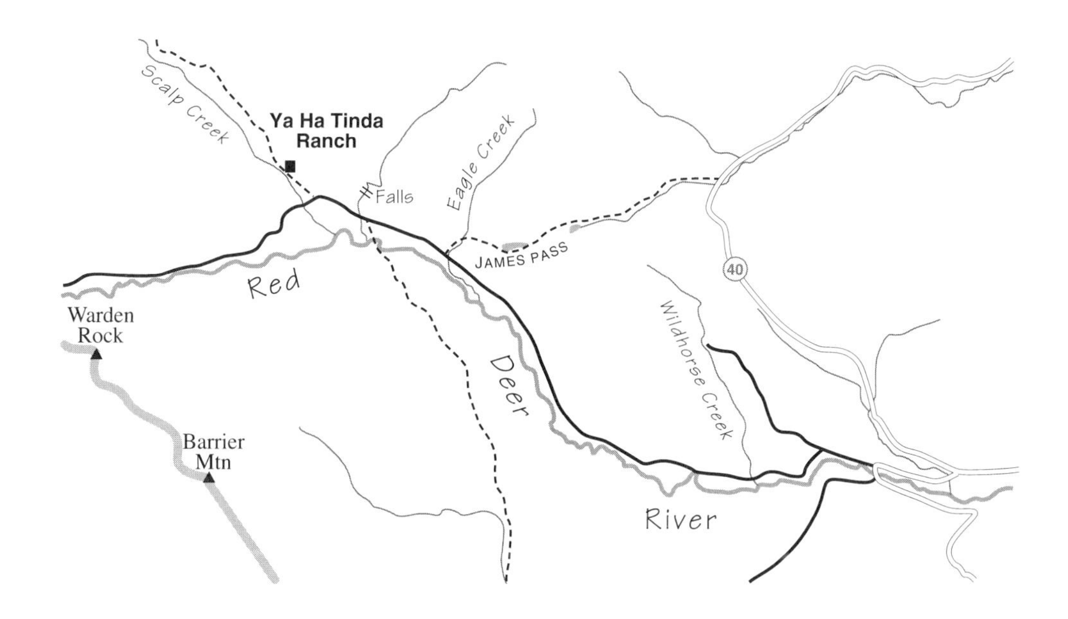

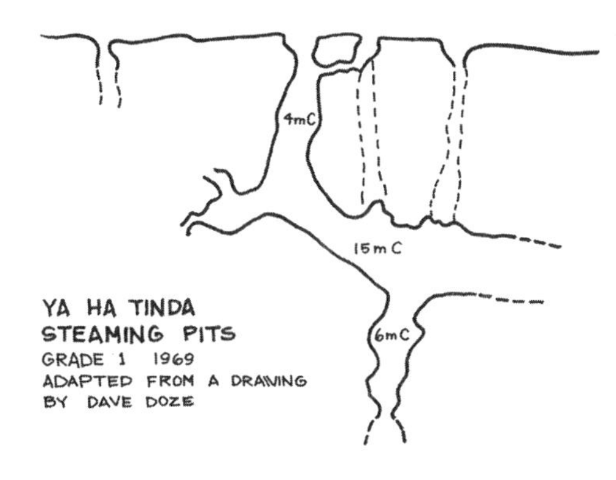

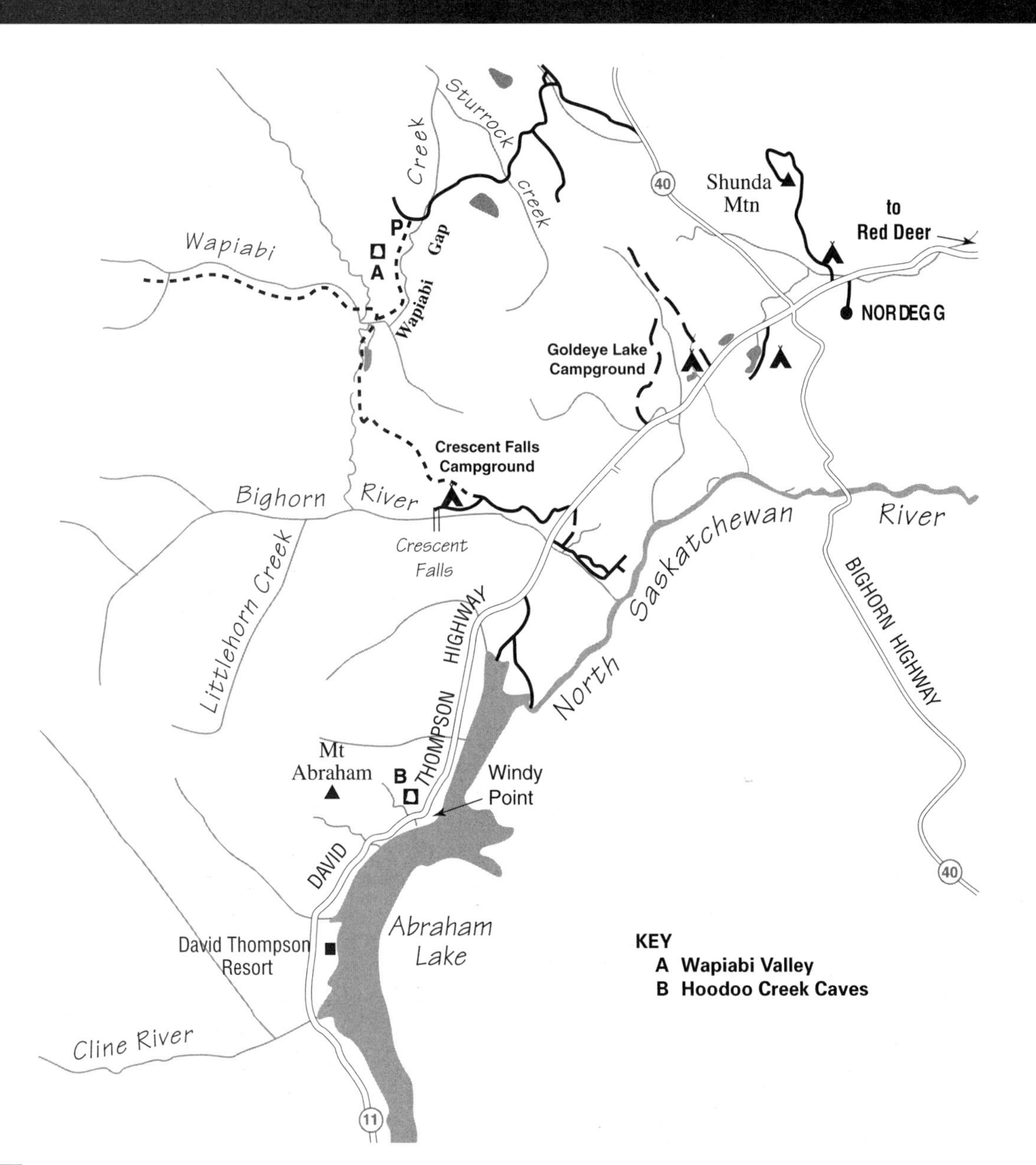

Wapiabi Valley

While the Wapiabi Valley contains numerous springs and other karst features, Wapiabi and Little Wapiabi are the only caves discovered to date. Developed on a bedding plane near the top of a ridge, Wapiabi Cave is located in the same Mississippian limestones that are found at Crowsnest Pass.

Wapiabi (Chungo) Cave

Map 83 C/8 388147
Jurisdiction Alberta Forests (Wapiabi-Blackstone FLU zone)
Entrance elevation 1940 m
Number of entrances 1
Length, Depth 539.8 m, 151.5 m
Discovered 1910, by Russ McFall

Access From Nordegg take Hwy. 940 (Forestry Trunk Road) north for about 20 km to where a rough road, signed Chungo Road to Blackstone Campground, heads southwest for 20 km towards Wapiabi Gap. In summer the road is passable in a regular vehicle, except following rain or snow when you need a 4-wheel drive. At about 7 km watch for a sharp hairpin bend where a road heads off to a gas plant. Continue across a bridge over Wapiabi Creek to a parking area. From here a trail leads alongside the river to Wapiabi Gap (4.5 km). Starting east of the cave, a steep trail climbs 600 m up to the cave entrance.

Location A square-shaped opening in a rock face just below the top of the ridge.

Guided Trips Wapiabi Cave is currently being used by a cave guiding business. Contact Reg Banks of Alpine Ventures (Nordegg) at (403) 721-2171.

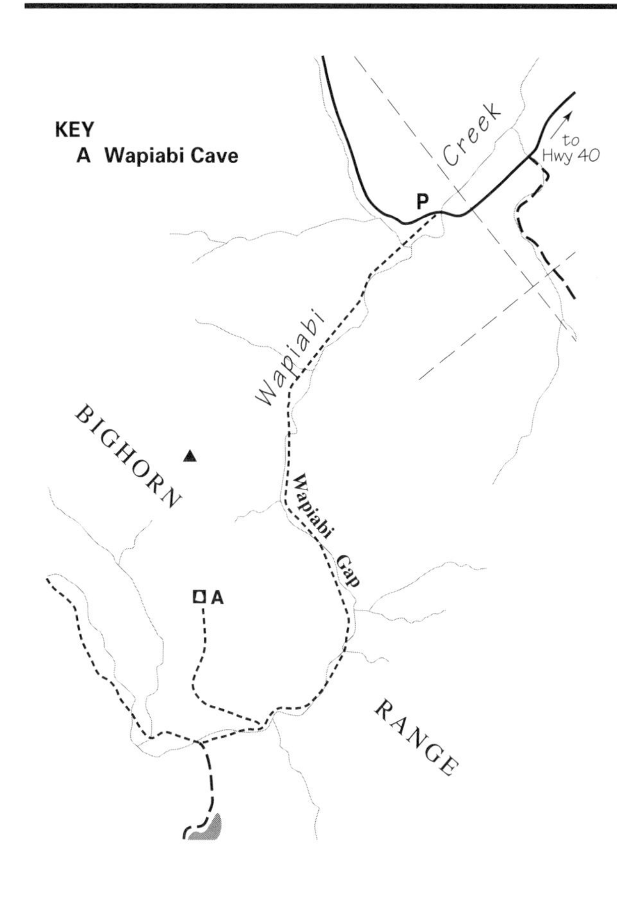

"On an autumn evening back in 1910 Russ McFall of Red Deer was returning to his camp on the Kootenay Plains, ninety miles west of Rocky Mountain House. He was on a big game hunt, but hadn't fired a shot that day and now he had to hurry to reach camp before dark. Russ was high up on a mountain shoulder, so he looked around for a good game trail to lead him down to the valley. And just then, he spotted a squared opening in the face of a nearby mountain wall."
Kerry Wood, 1955.

Exploration & Description Wapiabi Cave was explored by Calgary mountaineering instructor Gordon Huchenson in 1976. He told the ASS about the cave and in the course of three visits in 1971, most of the cave was surveyed. In 1976, a passage marked on the old survey as being "too tight," just beyond The Tube, was pushed to the head of a 6 m pitch. Following a series of trips to the cave plagued by vehicle problems and deep snow, cavers eventually bottomed this new section of cave known as the "French Extension" (owing to the involvement of Quebec cavers), thus making the cave just over 150 m deep.

Wapiabi Cave contains some interesting sections of phreatic passage such as the impressive Journey to the Center of the Earth. This voluminous passage, which descends steeply at 35° to its abrupt end in a sediment choke, contains some nice soda-straws, stalactites and stalagmites. In contrast, the French Extension is a series of connected vertical pitches, some of them quite loose, containing large amounts of breakdown material. There is doubtless more passage beyond the sediment and breakdown plugs of the Journey to the Center of the Earth and the French Extension passages.

Cave Fauna Little brown bats have been seen in the cave on numerous occasions, and packrats make themselves known if you stay in the entrance for long.

Cave Softly Avoid visiting the cave in the winter (October to April) when the bats are hibernating. Please remove any garbage you find and report spray paint or other damage to the AFS.

Phreatic Tube in Wapiabi Cave. Photo Ian McKenzie.

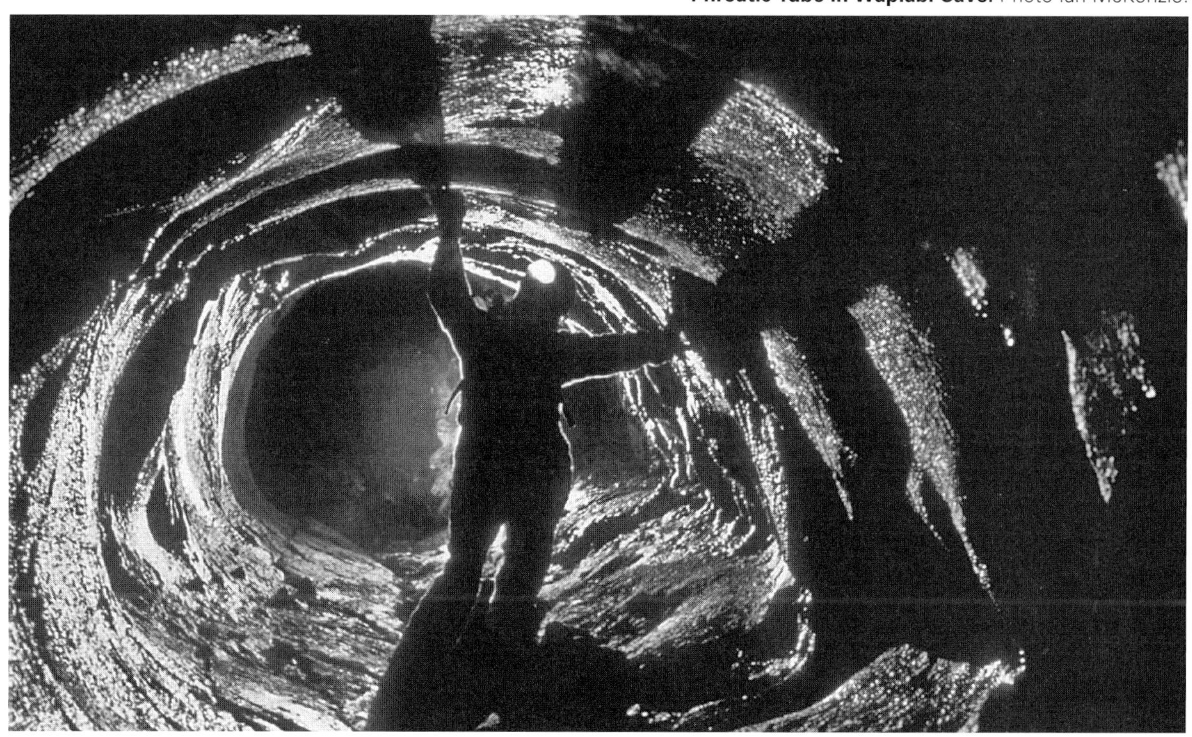

"A few more pieces of the cave puzzle scattered across the Rockies after the box was dropped by clumsy old glaciation."
Ian McKenzie, 1996.

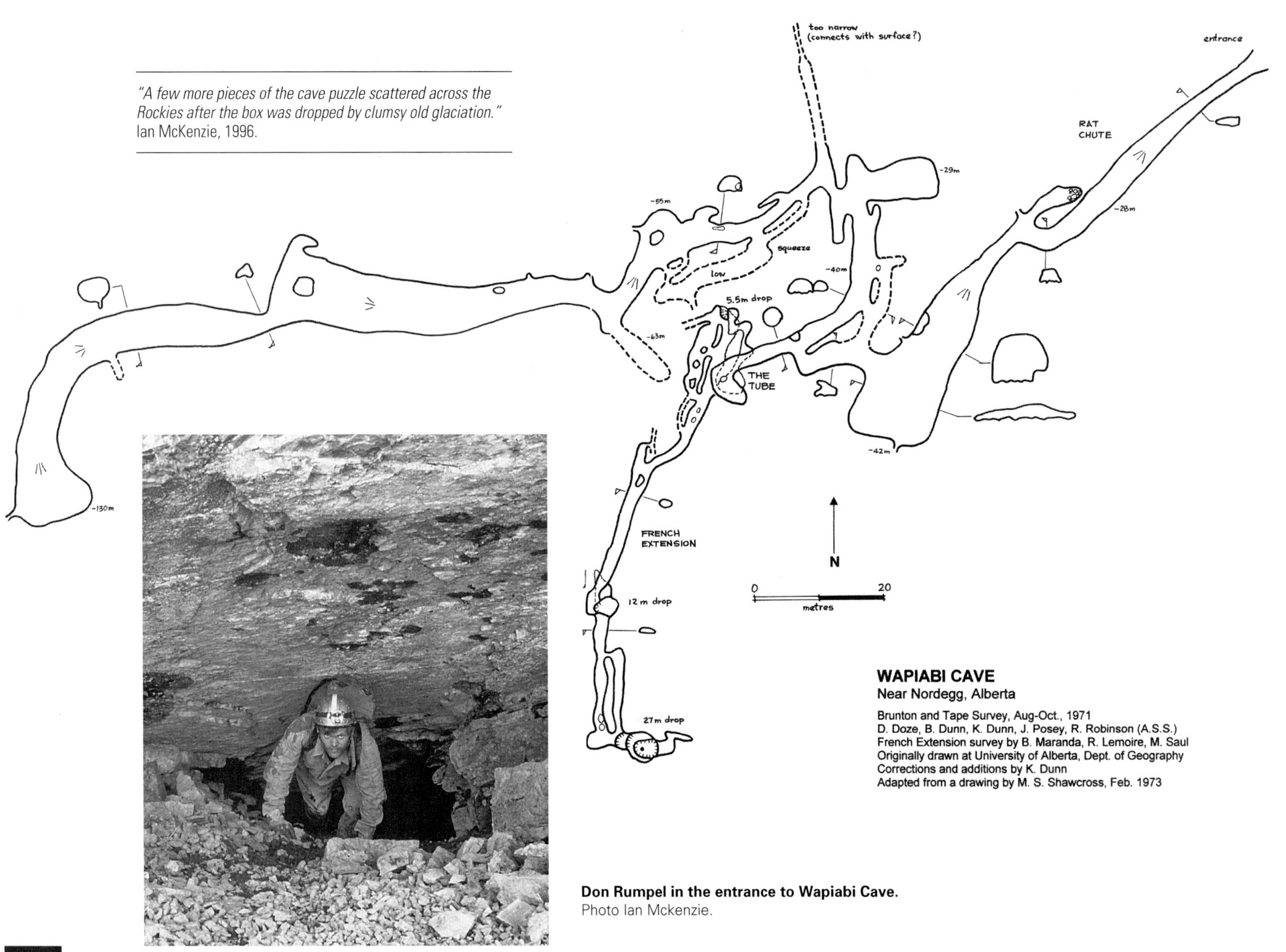

Don Rumpel in the entrance to Wapiabi Cave.
Photo Ian Mckenzie.

Little Wapiabi

Map 83 C/8
Jurisdiction Alberta Forests (Wapiabi-Blackstone FLU zone)
Entrance elevation 1700 m
Number of entrances 1
Length, Depth c. 60 m, 15 m
Discovered 1983, by Ian McKenzie. It may have been known previously.

Location A tight hole in the streambed below Wapiabi Cave.

Description The small entrance leads to a large, elliptical, steeply descending passage with four tight leads at the bottom. One climbs into a 3 m-high passage with formations, then ends in a dirt choke. Another leads to a boulder choke above a rift which delivers some water to a small pool about 0.3 m deep. The other two leads quickly end in chokes.

Hoodoo Creek Caves

Map 83 C/1, 83 C/8
Jurisdiction Alberta Forests (Rockies-Clearwater)
Entrance elevation unknown
Number of entrances 2 caves
Lengths 50 m & 7 m
Discovered Unknown

Access Hwy. 11 (David Thompson Highway), 38 km west of Nordegg or 45 km east of the Banff National Park boundary. Park at Hoodoo Creek and follow the creekbed and trails on one side or the other for 2.4 km, en route passing two hoodoos.

Location The spacious entrance of the larger cave is located up the righthand slope, and is accessed by a short, steep trail.

Description A steep climb and a squeeze past a calcite column is followed by a rift that ends in both directions.

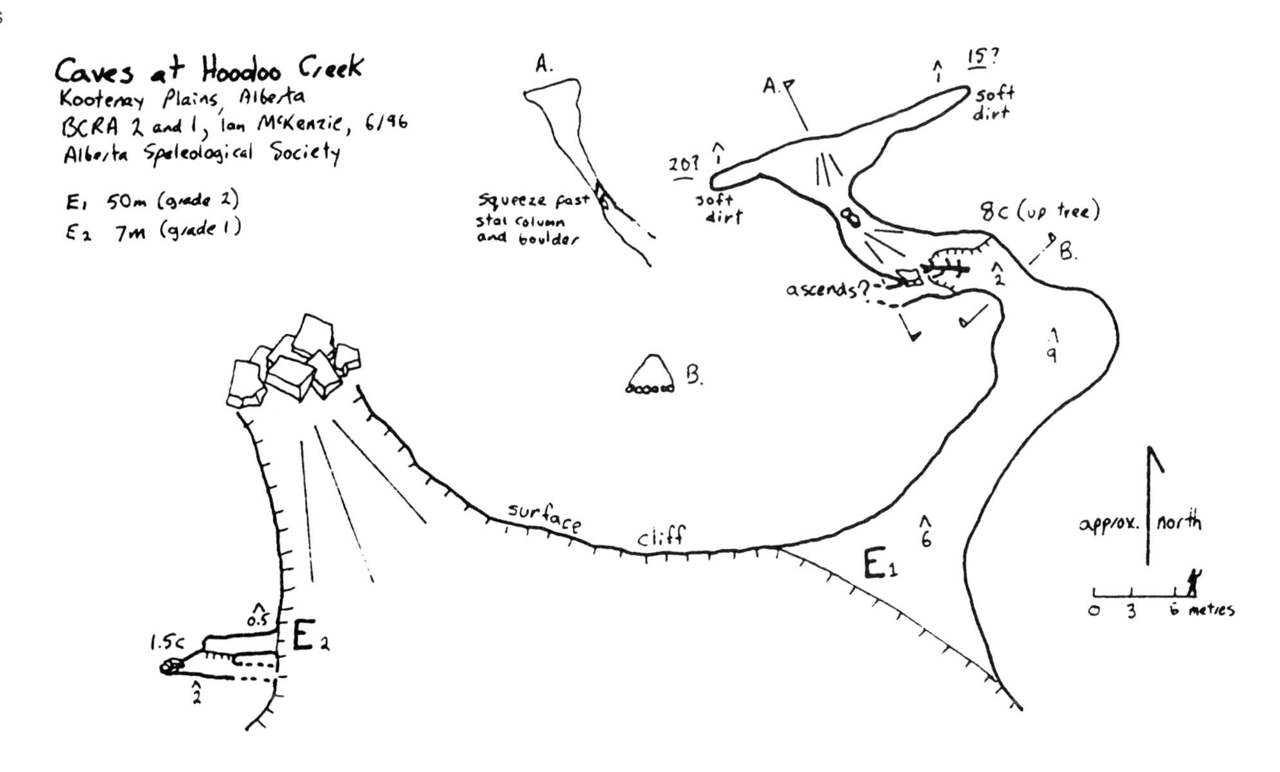

CADOMIN AREA

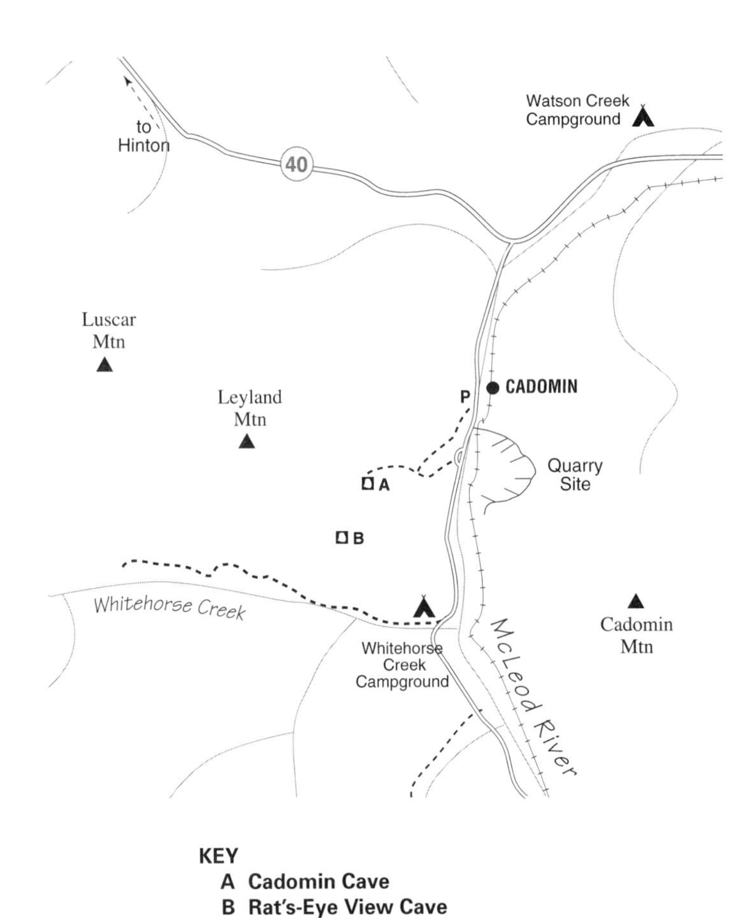

KEY
A Cadomin Cave
B Rat's-Eye View Cave

Since 1997 the ASS have been stewards of Cadomin Cave for Alberta Environment. Thanks to hard work by caver Terry Wachniak, in 1999 the ASS was awarded a plaque honoring their environmental efforts."
Canadian Caver, 1999.

Large block breakdown with graffiti in East Gallery. Photo Jon Rollins.

Top: Sump in Lower Gallery. Photo D. Weeks.

Cadomin Cave

Map 83 F/3 759739
Jurisdiction Whitehorse Wildland Park
Entrance elevation 1891 m
Number of entrances 1
Length, Depth 2791 m, 220 m
Discovered Unknown

Access The hamlet of Cadomin (acronym for Canadian Dominium Mining), 40 minutes drive south of Hinton on Hwy. 40. Drive through Cadomin, heading south on Hwy. 40 alongside the McLeod River. About 2 km south of the hamlet, turn right into a gravelled parking area with washrooms, picnic tables and a sign-board giving information on cave access.

From the southwest corner of the parking area a trail heads over a wooded ridge to connect with the old access trail. Turn right at the junction. Follow a well-used trail up and adjacent to a streambed for 1.5 km to a large boulder, then climb steeply up the north-facing slope of Leyland Mountain. The entrance is reached shortly before the ridge line.

Location The large entrance is hidden because it initially descends steeply into the mountainside. However, the deeply eroded approach trail makes it easy to locate.

Guided Trips Contact Jerry at Inroads Mountain Sports in Stony Plain at (780) 817-1512 or at the field office in Hinton at (780) 817-1512. Or email jerry@inroadsmountainsports.ab.ca.

Warning In spring and fall there may be snow on the access trail. The final approach to the cave can be quite icy, and possibly avalanche prone depending on conditions. Take an ice axe if necessary. If you plan to descend below the Mess Hall, take proper caving gear and make sure you are familiar with SRT.

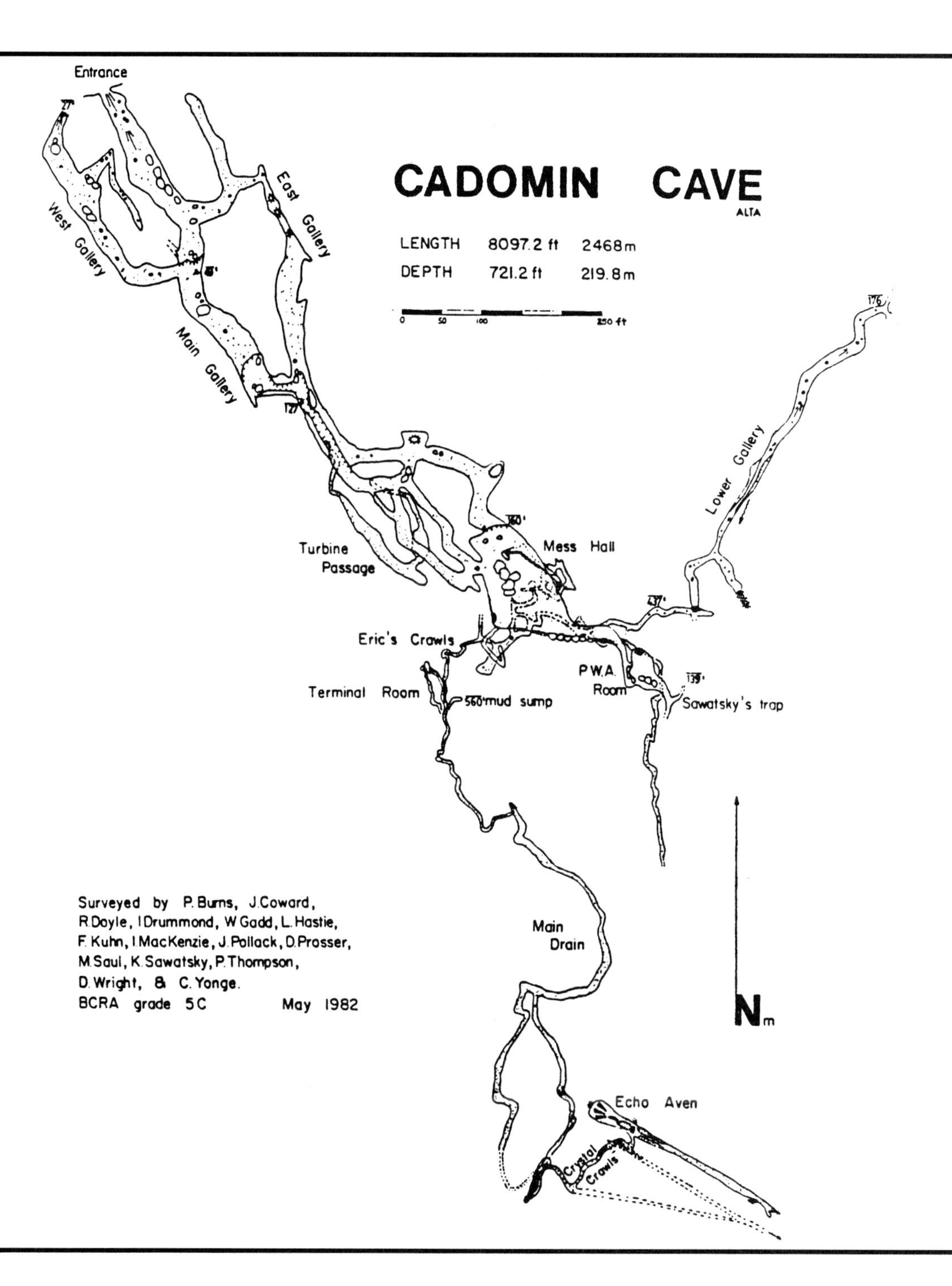

Traditionally, Cadomin is one of the most popular caves in the Canadian Rockies. Currently the cave is closed October to May to protect hibernating bats. As we go to press Alberta Community Development is initiating plans to control access. All persons should call ahead to Alberta Community Development in Hinton (780) 865-8264 for the latest information. If going beyond the Mess Hall you may be required by to show proof of caving experience and ability.

Exploration Probably the earliest known cave of any extent in Alberta, Cadomin Cave is believed to have been discovered by prospectors early in the 20th century (Prosser 1972). As with many well-known caves, colourful rumors circulate about the length of the passageways and the existence of other entrances. It is said that in their free time coal miners used to travel 32 km through the mountains to Miette Hot Springs for a soak — a round trip of impressive length. Not content with this accomplishment, it is said the miners found a back entrance to Cadomin Cave and were thus able to make the trip underground. Cavers, while eager to believe this because it would produce a cave of staggering length, have yet to find the kilometres of passage.

In 1959 W.L. Bigg and R.S. Taylor first surveyed the cave. A later survey was produced in 1972, showing the upper part of the cave much as we know it today.

A remark made in 1972 by Don Prosser, an ASS member and former provincial geologist, is interesting in retrospect. "There are rumours that a Cadomin miner claimed to have found twice again the length of passage shown in the survey. Supposedly, the entrance to it was through a squeeze in the Mess Hall, and after one has wedged under many dozen blocks to find only mud and wet marmot nests (sic), one's enthusiasm rapidly declines." Coincidentally, it was in this spot in 1977, by squeezing under some blocks in the Mess Hall, that caver Keith Sawatsky descended the pitches, and the cave was eventually doubled in length.

Description From the entrance a scree slope descends for about 20 m to a relatively flat area where you can pause and allow eyes to adjust to the darkness. In May and often into June, when snow spills into this entrance, an ice axe is useful to break your descent.

About 45 m in from the entrance the passage divides.The obvious way on is to the right where a short slippery down-climb accesses the junction of the West and Main Galleries. Less obvious, over boulders and via a low crawl up to the left is the East Gallery.

The West Gallery slowly climbs past some chiselled stal. remnants and large boulders until it ends in breakdown at a location close to the cave entrance.

Returning to the Main Gallery, two further slippery down-climbs lead beneath a large boulder and then traverse a shelf to the junction with the far end of the East Gallery. At this point a short climb down into a canyon is necessary to continue towards the Mess Hall. As the canyon trends to the left, two interconnected passages lead off to the right. The first ends at a dig, the second passes through a winding constriction to a climb up through a hole into the impressive Mess Hall.

A muddy slope at the south end of the Mess Hall leads up some awkward climbs and squeezes to access the PWA Room and Sawatsky's Trap.

A steep climb at the north end of the Mess Hall (watch for loose rock) brings you to an impressive aven and further passage connecting with the Main and East galleries. The East Gallery is a wide low passage punctuated by three high avens. The climb up from the Main Gallery as you complete the circuit on your way out of the cave is a bit of a work out, with some route-finding necessary before you glimpse daylight and drop back into the entrance passage.

The Turbine Passage is a vadose rift accessed from the base of the canyon passage south of the Main Gallery. Watch for loose rock in the ceiling towards the end of the passage.

Going Further The lower passages of Cadomin Cave (do not attempt unless you are equipped for and experienced in SRT) start beneath boulders in the centre of the Mess Hall where a search reveals a solid phreatic tube 1 m in diameter. Below the first (18 m) pitch a spiral passage gains the Gravel Pit. A second (10.6 m) pitch accesses an impressive phreatic tube 4 m in diameter named the Lower Gallery. The down-stream end of the Lower Gallery ends in a sump. Numerous efforts that have included the use of 76 m of garden hose and a pump have been made to drain the sump, but all were unsuccessful. In 1978, Paul Hadfield attempted to dive the sump, but was stopped by poor visibility, although he located a tight continuation. In 1983 Dayle Gilliat dived again, but could not find a way on.

The upstream end of the Lower Gallery, an aven, was climbed for 33 m leading to a further 35 m of large passage that ended in a choke of huge stream-rounded boulders. In 1984, using the cave radio, it was determined that the upper end of the Lower Gallery was very close to a band of cliffs overlooking the McLeod River.

A passage from the Gravel Pit leads to Eric's Crawls and the Terminal Room. Digging beyond the Terminal Room has produced a further 600 m of tortuous passage ending at Echo Aven. An attempt to climb Echo Aven in 1982 ended after a 21 m climb when the passage choked.

The complexity of passages below the Mess Hall contains several drafting leads. Digs and squeezes in these sediment-plugged passages will doubtless reveal more passage in the future.

This lower section of the cave contains some interesting mud and mineral formations including stalagmites and stalactites, helictites, cave pearls and soda-straws.

Fauna Little brown bats occur throughout the cave with a total hibernating population estimated at about 2,000. The occasional long-legged, long-eared and northern long-eared bats have also been seen. Bat counts range from 400 in the year 1895 to 809 in 2000 (Dave Hobson, Alberta Environmental Protection). These annual bat counts are always limited to the upper sections of the cave, although a similar number of bats are thought to be roosting in the lower sections.

Other than bats, the ubiquitous pack rats have a nest just inside the entrance, and a preliminary survey of invertebrate fauna found sone crane flies, mites and ice insects.

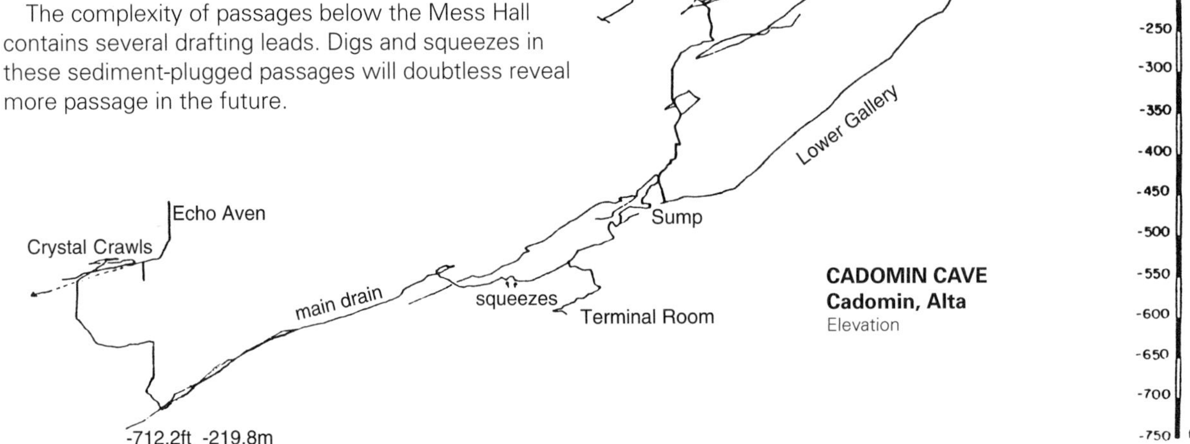

CADOMIN CAVE
Cadomin, Alta
Elevation

Cave Softly Don't disturb the bats in any way, including using flash photography.

Because the large easy passages of the cave's upper section have been known to tourists for many years, they now contain little in the way of formations, those that once existed having been broken off and taken. The walls have been extensively spray painted, but due to frequent clean up efforts by ASS members, the cave is currently in good shape. The spray paint, when it has been applied over mud, is easy to remove, but even the use of wire brushes and hydrochloric acid have little impact on bare rock that has been painted. In recent years spray-painted arrows have been found as far in as Eric's Crawls, together with a motley collection of knotted ropes and rope ladders. Use low impact caving techniques and leave no sign of your passing.

Geology & Hydrology The cave is contained in the Palliser Formation, the base of which is exposed in the McLeod River valley floor. The large chambers of the West Gallery are, in fact, entrenched phreatic passages of the classic keyhole type. The entrenchment has produced the slippery climbs and narrow passages, sometimes roofed by breakdown, low in the West Gallery. The Mess Hall is a classic breakdown chamber with a stoped ceiling and large breakdown boulders littering the floor.

Large amounts of stratified sediments indicate the cave must have undergone a period of quiescence or slow water flow. Spectacular avens in the East Gallery and towards the Mess Hall indicate probable inputs of water from above the cave. A spring located just off the road 3 km south and down-dip, may be the resurgence.

A significant karst exists above the cave with many bedrock karren features. However, no enterable caves have been found.

Rat's-Eye View Caves

Map 83 F/3
Jurisdiction Whitehorse Wildland Park
Entrance elevation Uncertain
Number of entrances 2 caves
Length 15 m & 106 m
Discovered 1984, by Chas Yonge & Jon Rollins

Access As for Cadomin Cave. Above and slightly north of Cadomin are two caves, reached by rappelling from the cliff top above.

Location One entrance (easily seen from the highway) is an alluring solutional hole. Spectacular views back out to the valley are seen from the entrances.

Description The first cave goes down dip for about 15 m to a huge pack rat's nest and blockage.

The second cave, after a crawl and stoop, goes in an up-dip strike direction before joining a fine dip tube with some formations. To the right is a tight crawl on flowstone that soon becomes very tight. To the left, down-dip, the tube gets lower and ends in a pool over which a variable strong draft is blowing, presumably leading to the surface. The cave ends close to the east wall of the amphitheatre. Both caves contain pack rat nests.

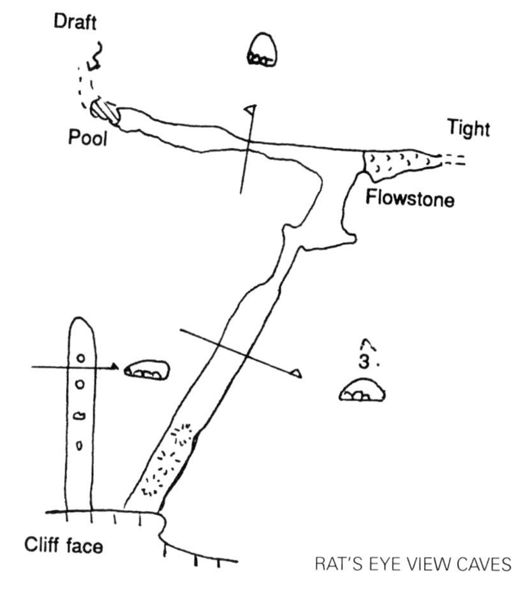

RAT'S EYE VIEW CAVES

Rollins & Yonge
Feb. 1984
Grade 1 sketch
Total length 137 m

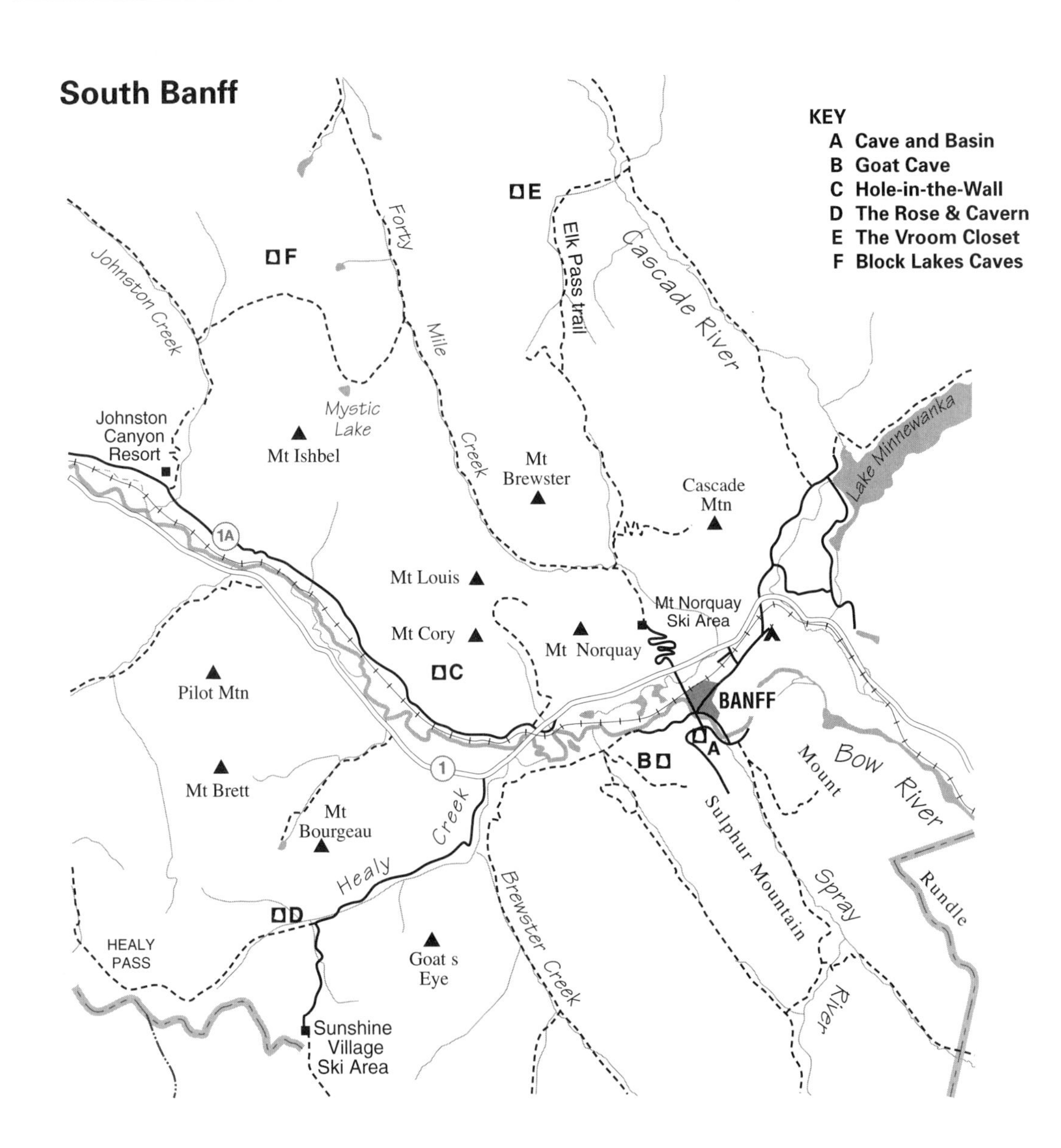

Cave and Basin

Map 82 O/4
Jurisdiction Banff National Park
Entrance elevation 1403 m
Number of entrances 2
Length, Depth Various
Discovered Aboriginal Peoples

Access In Banff, turn right after crossing the Bow River bridge on Banff Avenue and follow Cave Avenue (signs) to the Cave and Basin parking lot.

There are six main hot springs on Sulphur Mountain with several small caves formed in the travertine deposited by the mineral-rich waters. The largest and best-known cave is the one associated with the Cave and Basin complex. It is not managed as a show cave, but rather as a historic site, with the emphasis on the anthropological perspective as the birthplace of Canada's national park system.

Discovery & Exploration The hot springs of Sulphur Mountain were known by aboriginal peoples long before the more recent immigrants came on the scene with their colourful accounts of "discovery." Excavations near Vermilion Lakes indicate that pit dwellings there had been occupied almost continuously between 2800 and 440 years ago, with an earliest date for occupation dating back to 10,700 BP (Gwyn Langemann, *A Description and Evaluation of Eight Housepit Sites in Banff National Park, Alberta*. Canadian Archeological Society, May 1998, p.2). In his book *These Mountains are our Sacred Places*, Chief John Snow describes how they were often used by medicine men and women to cure illnesses. He also referred to the makutibi, or "little human beings" that the Stoneys believed to live under the ground.

The first recorded mention of the springs was by James Hector of the Palliser Expedition in 1859: "To the right of the trail I observed some warm mineral springs which deposited iron and sulphur, and seemed to escape from beds of limestone." In 1874 propector Joe Healy (after whom Healy Creek is named) reported discovering a small stream of hot water flowing into the Spray River. "He followed the stream to its source and thereby discovered the upper hot springs. He claimed to have also visited the cave and basin springs the following year, or perhaps the one after, and although he reported his discoveries to his brother John and a few others, never took any action to have them recognized." (E. J. Hart, *The Place of Bows*, p. 95.)

In 1883, CPR workers Billy McCardell and Frank McCabe claimed discovery. The following account is from the Peace Enquiry, held in 1886 to decide compensation following the Canadian government's decision to declare the site a reserve. "We came across the River, we ran up with a hand car (on the railroad) and we laid off here, and we spotted Sulphur Mountain several times, and we though we would like to go to the bottom of it, and we made a ferry and crossed the river and we discovered a big pool by the Cave..." McCardell then went on to describe how he built a cabin at the spring: "I was two or three days building the shack, and then I returned to Padmore, and the following winter I visited the cave occasionally." (National Archives of Canada, R.G.15, Item 137193.)

This description reads very dull when compared with his famous description made 50 years later, which inspired the mural at the current Cave and Basin interpretive centre. "Soon we were climbing up the strange formation of which the bank was composed (tufa). Once on the top of the first bench we rushed around to try and locate the place where the warm water came from. This did not take long. Out of a beautiful circular basin, jammed full of logs all the way in size from four to eight inches in diameter, some ten inches, seeped this warm water. Our joyousness and the invigorating thrill we experienced in thus locating this strange phenomenon, and hidden secret of the wilderness, knew no bounds. So we moved on up to another bench and to our utter astonishment we came upon a great cave 35 to 40 feet in depth. This was the greatest climax to a major discovery that we had ever seen. Frank McCabe, Tom McCardell and myself just stood in silence looking at this mysterious grotto where warm clear water bubbled from its depths."

McCardell then went on to describe the famous scene where a tree was used to enter the cave. "I shouted that I had reached the bottom alright, and that as soon as I had accustomed my sight to the darkness of the cavern, I would call to them at the top. This was the grandest sensation I had ever experienced in all my life. Taking the candle and matches from my pocket I struck a light and again the flame affected my sight. But leaning up against the tree it was not very long until I could pretty well see, but could not see very far.

"I scrambled around the edge of the water and I began now to witness one of the grandest spectacles, which I believed very few men had ever beheld. Beautiful stalactites hung in great clusters from the roof of the great amphitheater-like cavern. When I shouted up to Tom and Frank, it seemed as though I was speaking out of an immense drum, the sound rumbling and echoing in an erratic way. But the beautiful glistening stalactites that decorated this silent cave were like some fantastic dream from a tale of the 'Arabian Nights.' Glittering crystals and stalactites bedecked that dome-like cavern." (McCardell's memoir *Reminiscences of a Western Pioneer*, Whyte Museum of the Canadian Rockies, p. 250.)

Other claimants to discovery included Nova Scotia MP, D.B. Woodworth, who tricked McCabe and McCardell into signing over their interests in the cave, and went so far as to purchase a hotel at Silver City (Castle Mountain), intending to move it to the site. Eventually, Prime Minister Sir John A. MacDonald became involved and urged that the springs and cave be placed in a reserve to protect them from private interests. In 1885 an Order in Council reserved ten square miles around the springs — thus creating the oft touted "birthplace" of Canada's national parks. McCabe and McCardell were later paid $675, and Woodworth $1,000 in compensation.

David Drummond Galletly (1838-1916) who served as caretaker and tour guide for the Cave and Basin from 1897 to 1910 mirrors McCardell's discovery description: ".....they felled a couple of long trees, trimmed the branches, improvised a ladder out of these, lowered it down through the hole, and came down that way. Then the sight that met their gaze when they came in here must have been a glorious one. For the whole of the roof was decorated with beautiful stalactites, six, seven, eight, nine and ten feet long, and beautiful stalagmites, standing round the base, like so many sentinels at their post. In fact the whole of the interior of this cave was literally covered with beautiful crystals. Now, wouldn't you have thought, and wouldn't you have expected also, that these men, surveyors, educated, intelligent men, — would have to be to follow that vocation — stumbling accidentally, as they did, on such a lovely cave as this, that nature had so lavishly and gorgeously adorned with crystalline beauty, that they would have been so pleased, and so proud of being discoverers that they would naturally have refrained from vandalism, themselves, and they would have protected it from others. But alas, this was not the case. These are the men that robbed this cave of all its crystal adornment. This would have been one of the prettiest caves on this continent, if they had left it alone, — left it as they had found it and as nature made it. The poet says 'a thing of beauty is a joy forever.' Well! that depends upon the thing of beauty being in its proper place, and in this cave, at any rate, the proper place of these things of beauty, these crystals, these stalactites, and stalagmites, was right here where the creator placed them, and not as they are now kicking around the houses of these cave robbers, and I have not the slightest doubt but the good wives every time they go house-cleaning wish to goodness they were back where they came from. They are of no intrinsic value. You can't make anything of them they are so brittle and they deteriorate in color from being confined and I feel assured that they have by this time lost all interest in them.

"Probably they have them confined in boxes, stowed away in some cellar, attic or outhouse, whereas if they had left them here what a glorious sight it would have been to the thousands and thousands of visitors who come here year after year, to have seen this cave in its natural beauty. Now we are deprived of that pleasure and privilege, just from the selfishness, covetousness, cussedness, wickedness of these surveyors going to work and despoiling this beautiful 'temple not made by hands' and robbing it of all its crystal adornment."

Originally descending through the vent in the ceiling was the only way in, then in 1887 a tunnel was driven into the cave allowing easy access. For those opposed to structural alterations being made to caves, it is heart-warming to know that even at the turn of the century there was opposition to this project, and a heated debate occurred, as recorded in the *Calgary Herald*: "We were astonished to read in the columns of the Herald that a petition was to be forwarded to Ottawa and that a telegram had been sent ahead, to protest against the construction of the tunnel into the cave. Everyone I have interviewed on this subject declare that they neither saw nor heard anything of the kind, so in my opinion the whole thing is canard. However as regards the tunnel, several seem to think it will destroy the romantic beauty of the cave by doing away with the use of the curious natural entrance through the hole in the top. That may be very well for those who go down only for the romance of the thing, but as this place is essentially a health resort, I think it will be a great boon to invalids and rheumatic persons who may wish to try the curative properties of the water, to have an easier mode of access to it, and if Mr. Stewart carries out his design of constructing a pretty rustic cottage at the entrance, with ladies and gentlemen's dressing rooms and refreshment rooms attached, I think it will make it one of the most favorite places of resort in the Park." (*Calgary Herald*, 18 February, 1887.)

In 1909, workmen painting the Cave and Basin buildings reported another cave just above the existing one. A lamp was found, indicating it had been entered previously. "Several parties entered the new cave on Sunday morning, having previously learned, however, that its existence was known to several old timers. The mouth of the cave is small and drops almost straight down for several feet, where it becomes too narrow for a stout person to get through. The heat and humidity in the cave entrance are such that a bathing suit and a pair of boots form sufficient costume for the exploit. Once through the narrow hole the visitor stands in a large chamber, with several side rooms opening in different directions from it. Immediately in front is a hanging screen of rock and stalactites, and passing through an aperture in the rock, entrance is gained to another chamber. The floor of the cave slopes sharply down, and sulphur water runs over it to a pool about 10 inches deep and four or five feet in diameter. One stream finds its way out on to the hillside a little way below the entrance to the cave and the rest runs through lower crevices probably finding its way to the stream of water feeding the lower cave. Mr. Galletly has been careful to prevent as far as possible any damage being done by visitors and those wishing to see the new cave must first get his permission." (*Calgary Herald*, May 15 1909.)

D.D. Galletly beside the pool at the Cave and Basin. Photo courtesy Whyte Museum of the Canadian Rockies.

In its current situation, with artificial lighting, the cave carries little of the appeal it had in the days of Drummond Galletly's tours. The following is a description of the main cave as reported in 1912 in *Nor-West Farmer*, a Winnipeg newspaper. "The visitor enters a long, dark, gruesome tunnel and is conducted by a Scottish guide through a length of darkness and along its winding course whose walls are the bold rocks worn away in ages past by the running water, until the cave is reached. But the whole story of the cave is best told by our friendly guide, he of the Highland accent.'Ma friends' he says, as he casts the light of his oil lamp now here, now there, on the glistening walls, 'ye'll observe ye are now in the very heart o'the geyser. This whole chamber was scooped out by the awful force of water and gas. See how the giant fretted his prison walls, tearing off chunks of rock and grinding them to mud. The gas kept the water in a constant turmoil. Friction from the water and condensed steam, created heat, and when the heat got to a certain intensity, sky-ward it flew out at the orifice o'its own makin' sendin' mud and water many feet into the air with a terrible roar, then the cold air rushed in and filled the vacuum

Banff (Sulphur Mountain) Hot Springs		
Name/Location	Temp.°C	Flow l/min
Upper Springs The Aquacourt high on Sulphur Mountain above the Banff Springs Hotel.	47.3	545
Kidney Springs Beside the road 200 m before the Upper Springs.	39.2	91
Middle Springs In a housing development above the Parks Building and 400 m west of the road to the Upper Springs.	34.8	225
The Cave East end of the Cave and Basin	32.8	500
The Basin West end of the Cave and Basin (into an artificial pool)	34.5	680
The Pool Reached by the trail that climbs above the hole in the top of the cave of the Cave and Basin	32.0	550

From *Handbook of the Canadian Rockies* by Ben Gadd

"There are at least three things which every tourist visiting the Canadian National Park at Banff should see: the Cave, the Basin, and the good-natured old Scotchman in charge." Early tourist brochure.

and the war of Titans went on as before; but when he wounded himself in the side, when the pressure tore that breach in the wall by which we entered, then the gas and water escaped with a roar and it spouted 'nevermore.' He just gave himself a 'solar plexus' and knocked himself out o' business; he e'en smote himself under the fifth rib and gave up the ghost. Like Samson when he got his locks shorn, his strength has left him. The cave underwent an operation for appendicitis, the operation was successful, but the patient died and its been smelling like rotten eggs ever since.'"

Galletly went on to describe formations, such as: "Britain's eminent statesman, Hon, J. Chamberlain, Nero, Emperor of Rome, a mountain sheep, King of the Cannibal Islands, and a replica of the Venus de Milo."

Cave Softly The Banff Springs snail *(Physella johnsoni)* has evolved owing to the unique chemical, geological and biological forces found only in the springs on Sulphur Mountain. It was first identified in 1926 and subsequently found in nine locations around the springs. In 1996 it was found to be extirpated from four of the original nine sites and has been listed as "endangered."

The microscopic snail, with a shell length of only 5 mm, lays its eggs on floating algae mats and the movement of rocks and logs, together with construction and pollutants from the encroaching human communities, have been blamed for the extirpation. Parks Canada has issued closures of some springs to suppress further damage.

"These springs are worth a million dollars." William Van Horne.

Goat Cave

Map 82 O/4
Jurisdiction Banff National Park
Entrance elevation c. 1850 m
Number of entrances 1
Length 20 m
Discovered Unknown

Access Goat Cave is located on the northwest slope of Sulphur Mountain (Samson Peak 2270 m), above Cave Avenue in Banff and is visible from the Trans-Canada Hwy. across Vermilion Lakes. Formerly accessed by a trail directly up from the Cave and Basin, the recent designation of a wildlife corridor necessitates a slightly longer approach.

In Banff, turn right after crossing the Bow River bridge on Banff Avenue and follow Cave Avenue (signs) to the Cave and Basin parking lot.

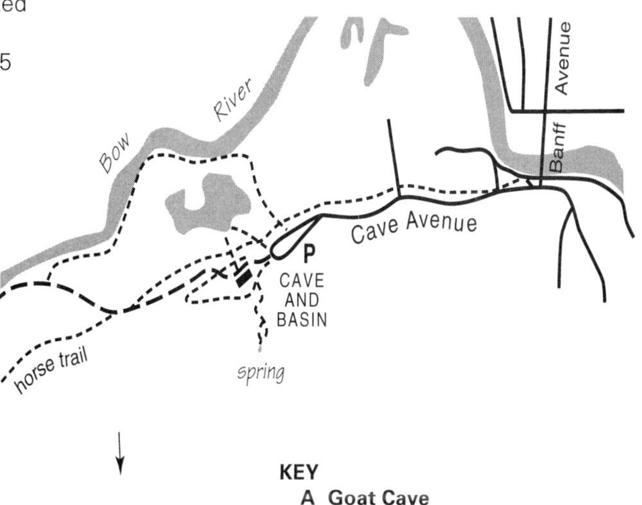

A

Follow Sundance Trail (paved) to a small spring with a pipe coming out of it. Take the horse trail left for a few minutes until you reach the base of a ridge (beyond, the horse tail drops to cross a gully). A faint trail, sometimes marked with cairns and flagging, leads to the left up through lodgepole pines with occasional smooth bedrock exposures. Higher up, keeping to the right (west) side of the ridge, you break out onto open slopes dotted with impressive old Douglas fir. Where a steeply bedded cliff intersects the ridge, turn left (watch for cairns) onto a narrow trail that soon climbs into an impressive embayment backed by steep cliffs. The cave is at the east end of the embayment.

Allow at least 40 minutes for the 300+ m elevation gain approach. Depending on winter snowfall, the final section of trail below the cliffs may be snow-covered well into May.

Description Goat Cave is named after the sheep and goats that use it for shelter during stormy weather. Although not of much interest to cavers — it is very short and doesn't go beyond daylight — the pleasant hike and fine views over Banff have made it a popular picnic destination for hikers since the turn of the century.

Hole-in-the-Wall

Map 82 O/4
Jurisdiction Banff National Park
Entrance elevation 2013 m
Number of Entrances 1
Length 30 m
Discovered Unknown

Access Take Hwy. 1A (Bow Valley Parkway) west of Banff to the Muleshoe picnic area. From here the cave can be seen high on the south face of Mt. Cory. The distinctive zigzag overhang on the steep cliffs above the entrance is a useful landmark. For the most direct access, park at a small pull off (not signed) 1 km east of Muleshoe picnic area on the south side of the road.

From here hike a short distance east along the road to a sign indicating "Divided Highway" ahead. Then head north beneath power lines and through burned Douglas fir and up a low ridge, keeping the boulder outwash creek below you to the right. At distinctive clean-washed slabs where the gully ends at the lower-gradient slopes, a well-worn trail climbs steeply up the ridge. Follow it to a break in slope beneath steeper wooded bluffs where the trail contours left to a cairn. Climb the bluffs (steep and exposed in places) until a small flat area finally allows a good view of the cave.

Drop a short distance into a subsidiary gully on your right and climb back up to a small tree with orange flagging. From here an easy traverse right accesses the main gully (that leads to the cave) just above a vertical drop. Two short water-washed steps up the gully access the unconsolidated mud and rocks that form a tongue down from the cave. Avoid this by traversing up the treed slope to the right, stepping back left onto the steep grassy slope that extends into the cave entrance. Allow two hours.

Entrance to Hole-in-the-Wall seen from the approach ridge.
Photo Jon Rollins.

Exploration & Description The Hole-in-the-Wall is probably the most famous cave entrance in the Canadian Rockies. It appears on maps, and on table-mats in Banff restaurants, and despite the long hard approach has provided a popular destination for hikers since the early 1900s and was probably visited by aboriginal people before that. It was certainly visited by prospectors Billy McCardell and Frank McCabe in 1883 prior to their "discovery" of the Cave and Basin. A photo taken by Byron Harmon in the 1920s shows a party of ten enjoying the view of Muleshoe Lake from the entrance. Another of the same period shows two men, one leaning on a shovel. They may have been trying to "push" the cave by digging, but more likely used the shovel to negotiate the steep slopes below the cave entrance.

Although having a large entrance, the cave doesn't go out of daylight. It was visited by Derek Ford in 1966 and judged a possible candidate for extension by digging.

Warning Only tackle the steep and in places exposed approach during good weather when no snow is present. Beware of rockfall if people or sheep are above you.

The Rose & Cavern System

Map 82 O/4
Jurisdiction Banff National Park
Entrance elevation 2135 m
Number of entrances The system is composed of the Rose & Cavern, and the Upper and Lower caves, with numerous entrances leading off the main canyon.
Length, Depth Total length of passages 309 m, Vertical range 37 m
Discovered Pete Charkiw (date unknown)

Access Hwy. 1 west of Banff. Turn onto the road signed Sunshine ski area and drive to Borgeau parking lot. Hike about 200 m west along the north side of Healy Creek until an avalanche slope can be climbed steeply to the base of the large south-facing cliffs of Mt. Bourgeau. Contour the base of the cliffs for about 500 m as they bend north to the gorge that contains the caves.

Location The main cave is located across a waterfall high in the left (west) wall of the gorge. Access to the cave entrance often requires a rope swing.

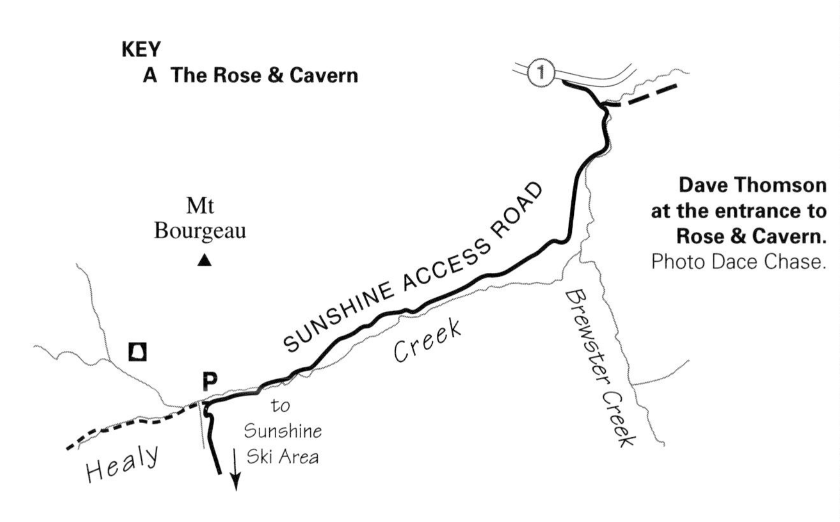

Just southwest of the Borgeau parking lot for Sunshine ski area, a spectacular canyon breaks through the end of large prominent cliffs. In the walls of this canyon are a number of small caves and one longer one: the Rose & Cavern. Although not one continuous cave, the numerous sections of passage cleaved by the recent intrusion of the canyon are all included under the same name, and shown on the same survey. Sporting rather than dangerous, this cave is best approached outside periods of high snowmelt.

Exploration & Description The longest section of cave was discovered by Pete Charkiw, a local climbing guide, who crawled along it for 65 m to where a large cobble was jammed across the passage. He passed this information on to ASS members and subsequently several visits were made to try and remove the obstruction. The cobble, which must have been thrust into the cave with incredible hydraulic force, did not succumb easily, but eventually fell to the persistent efforts of Dave Thompson and Dave Chase, who named the cave after the local pub in Canmore and a rose bush growing tenaciously at the cave entrance. The lucky couple next got to crawl through a water-filled tube (Bar Slops Crawl) before a tight squeeze ejected them into a walking-sized canyon passage 8 m high. Alas, after only 75 m this passage ended just around the corner with a small stream disappearing through a hole in the floor.

The Rose & Cavern (longest cave segment) provides a caving experience out of all proportion to its length. The thundering 30 m waterfall adjacent to the entrance can be felt for some distance reverberating through the bedrock phreatic entrance tube. A thorough shower at the Tap Room is merely a precursor to the flooded crawl that follows. A small stream runs over the lip of the squeeze into the walking passage beyond. When going back through the squeeze, your body dams the path of the water, which gradually builds up in your face.

Further downstream, a short cave remnant, the Upper Cave, passes through the side of the canyon adjacent to a spectacular 30 m waterfall. Water gets diverted through this during periods of high flow. By climbing up a short rift below the Upper Cave one can access the Lower Cave, now abandoned and dry. The adjacent plunge pool contains a sink which can be entered at low water for 20 m to a gravel choke. This may possibly connect with the spring down on the west side of Healy Creek about a kilometre away.

Dave Thomson at the entrance to Rose & Cavern. Photo Dace Chase.

Warning Don't attempt this cave during periods of high water (i.e. during spring melt or after heavy rain).

Speleogenesis The different sections of passage composing the Rose & Cavern system were doubtless once connected, but have since been invaded by the down-cutting gorge. The vadose canyon passage indicates that a more extensive cave system exists, but is now plugged with the glacial till which can be observed forming the floor of the infamous squeeze. The removal of some of this material by aggressive water action has left sections of suspended calcite "false floors" attached to the walls of the canyon passage. Part of the Rose & Cavern cave system is still active, carrying water from the floor of the plunge pool below the 30 m waterfall.

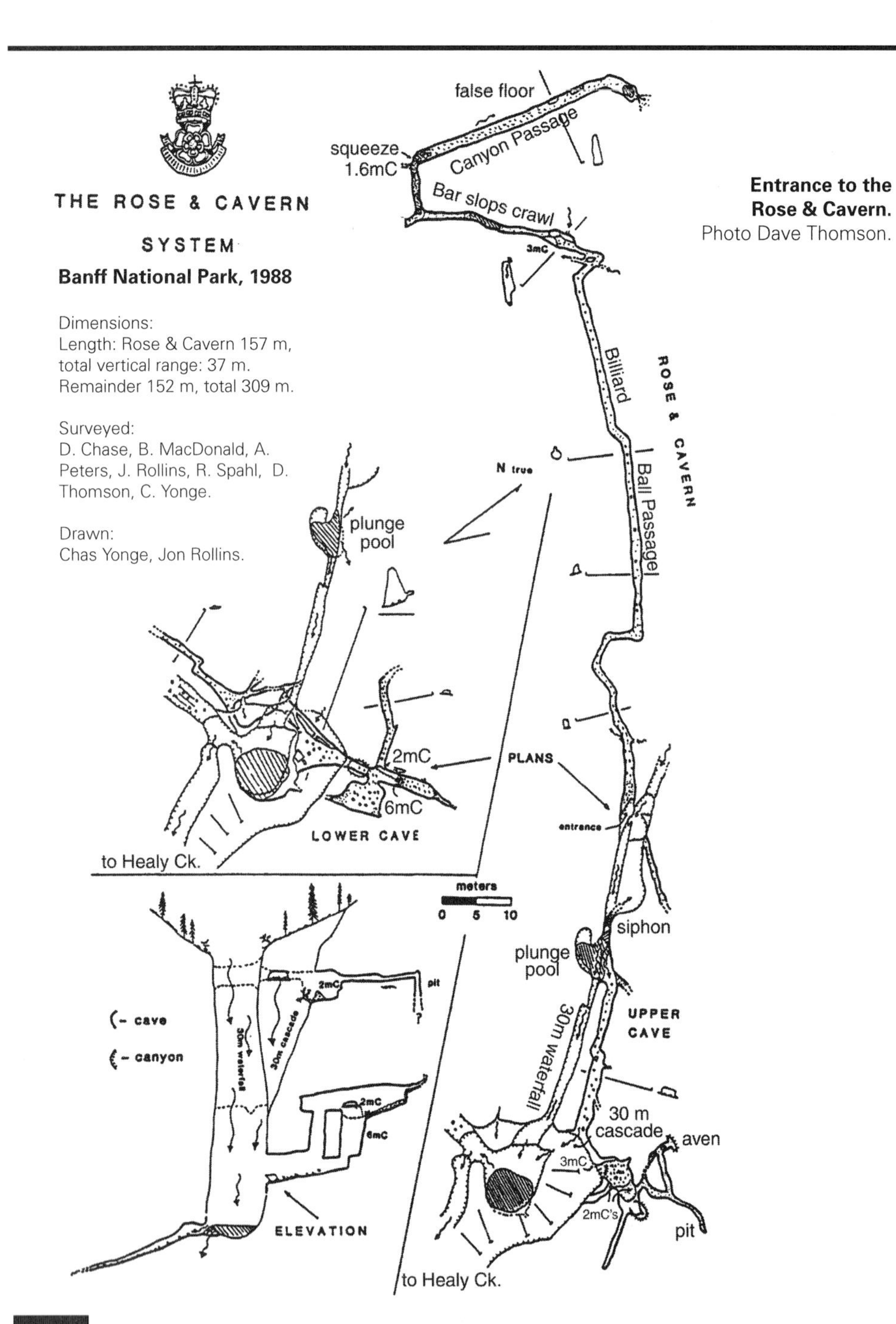

Entrance to the Rose & Cavern.
Photo Dave Thomson.

The Vroom Closet

Map 82 05/E
Jurisdiction Banff National Park
Entrance elevation 2257 m
Number of entrances 1
Length, Depth 106 m, 60 m
Discovered 1975, by Bill Vroom

Access From the Trans-Canada Hwy. east of Banff drive the access road towards Lake Minnewanka. Park at the Upper Bankhead picnic area.

Hike or mountain bike up the Cascade fire road for 15 km to the junction with Elk Pass trail. Turn left onto Elk Pass trail and cross a bridge over the Cascade River. Follow the trail through a gorge for about 2 km to a bridge where the trail turns south. Leave the trail and follow the creek in a northwesterly direction (no trail) through boggy willow flats for another 2 km to the second drainage emanating from the Vermilion Range to the west. Follow this drainage southwest for 3 km, keeping right at a fork and climbing steep treed slopes to gain a rugged basin.

At the foot of a small ridge shaped like a rough fin is a depression in the scree containing the 2.4 m-wide opening.

By using a bike on the Cascade fire road, Vroom Closet can be done in one long day.

Exploration & Description The Vroom Closet was found by Banff National Park warden Bill Vroom while searching for grizzly denning sites. Only one other visit is known to have taken place since its discovery.

A 9 m inclined tube is followed by a 30 m climb down a rubble strewn slope into a chamber. Two passages lead off the chamber, but quickly pinch out. The only possible lead appears to be an aven that would require bolts to reach.

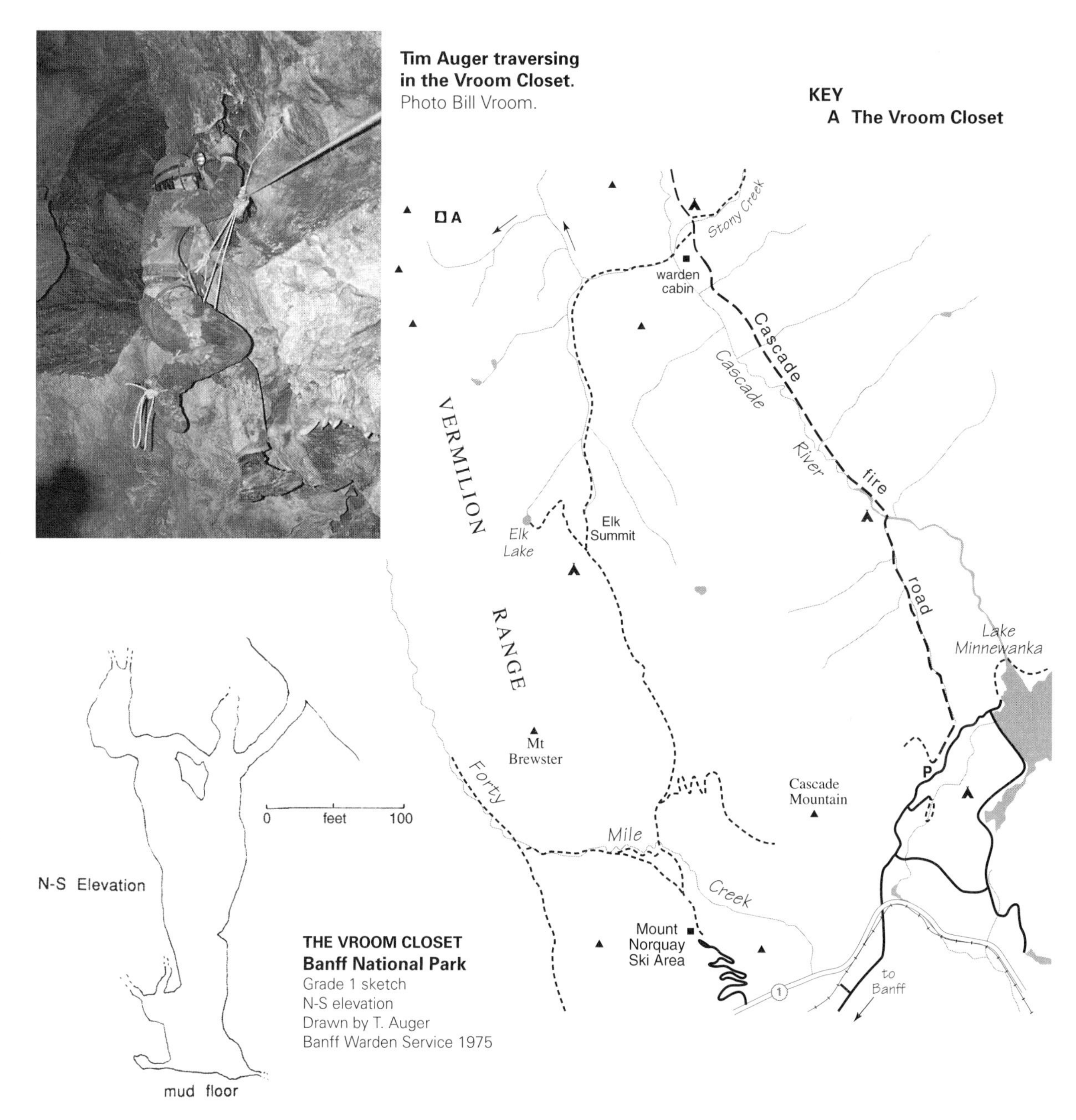

Tim Auger traversing in the Vroom Closet. Photo Bill Vroom.

Block Lakes Cave

Map 82 O/5
Jurisdiction Banff National Park
Entrance elevation 2300 m
Number of Entrances 2
Length, Depth 227 m, 39 m
Discovered 1979, by Ian Drummond

Access Hwy. 1A at Johnston Canyon Resort. Hike the trail to the Ink Pots. Then follow the Mystic Lake Trail up Johnson Creek. 1 km after the Mystic Lake Trail bears right, cut left (north) up a drainage that eventually leads over a high pass to Block Lakes. 2 km north of the Mystic Lake Trail, the cave entrance is visible high on the east side of the valley at a bend. Scramble up right for about 400 m. The large entrance lies just above tree line at the bottom of some slabs.

While the cave can be done in one long day, the use of the campground just beyond the bridge over Johnson Creek is recommended in order to avoid hiking out in the dark.

Exploration & Description During July of 1979 three ASS members hiked up the access trails to check it out. So certain were they it was only a frost pocket they carried only one carbide light between them! After a 400 m scramble, they reached the entrance, a large opening that gets smaller after a few metres. "This distinctly smelt like a cave and a discernable draught was blowing out of it. Dave quickly checked to the limit of daylight and reported that it still went. I took the skeptical look off my face and lit our carbide light. Beyond the limit of daylight, the cave opened up to 5 m wide by 4-6 m high and headed steeply uphill. After about 80 m the passage appeared to be blocked, but a steep, gravel-floored crawl took us past this blockage and into a chamber with daylight streaming in from a small upper entrance. This chamber had an accumulation of aeons of rat shit, as did most of the rest of the cave" (Peter Thompson, 1979).

This cave was surveyed in 1997. A previously unexplored passage ended in an ice plug.

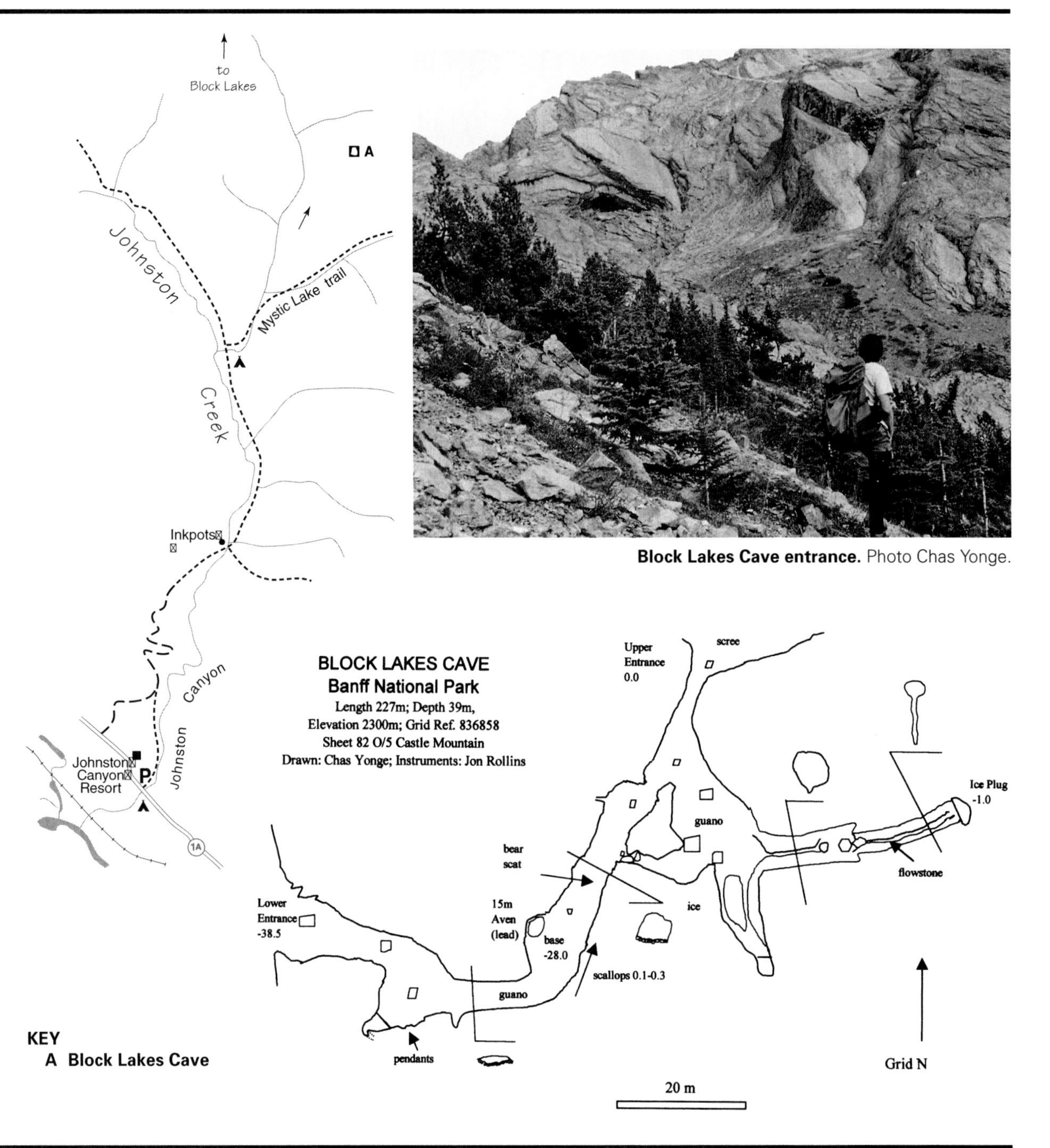

Block Lakes Cave entrance. Photo Chas Yonge.

North Banff

KEY
- A Jaw Bone Cave
- B Pinto Lake Cave
- C Castleguard Cave

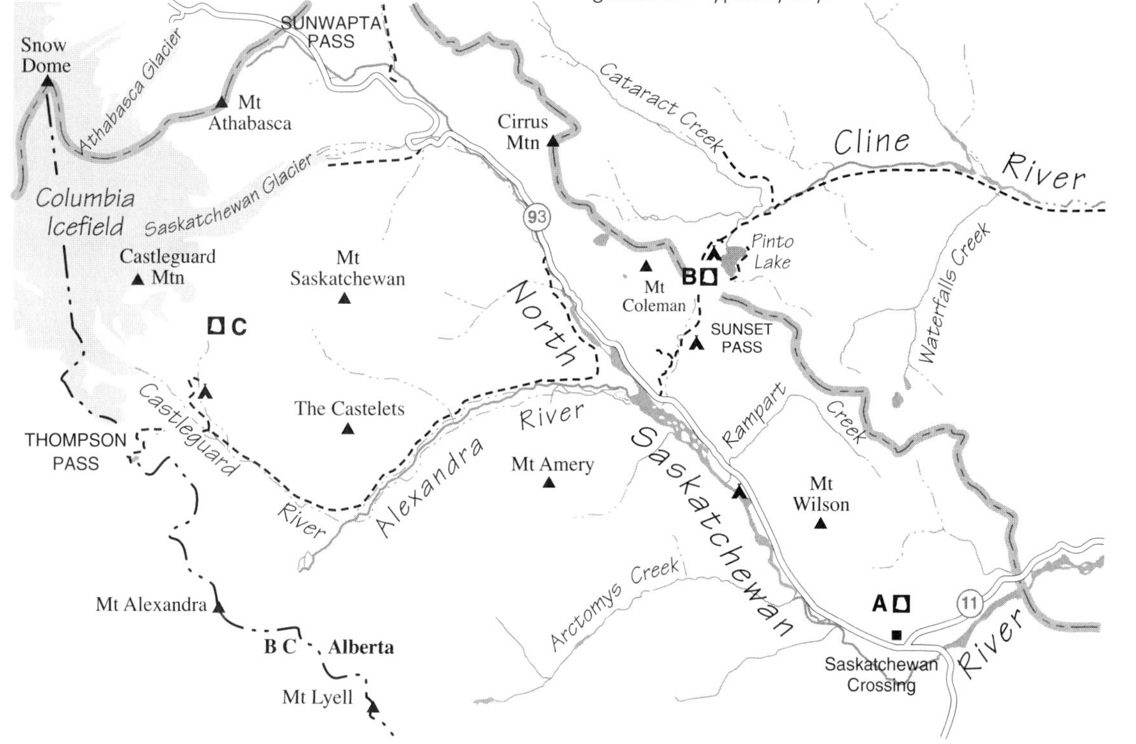

Jaw Bone Cave

Map 83 C/2
Jurisdiction Banff National Park
Entrance elevation 2013 m
Number of entrances 1
Length 9.76 m
Discovered Unknown

Access Hwy. 93 (Icefields Parkway) just north of Saskatchewan River crossing. Take Hwy. 11 east for approximately 200 m to the first gravel turnoff to the left (north), then turn immediately right (east) along a rough road to an old gravel pit. Park here and view your objective — a spring issuing into a gully on the south face of Mt. Wilson. The good-sized resurgence stream can be seen flowing down the gully when adjacent gullies are typically dry.

Hike towards the base of the gully, using the occasional game trail through the dense bush (1 hour). En route pass a short deep gully in white boulder-clay, the resurgence of the sping water. Dense willow and head-high cow parsnip mark the base of the gully — watch for bears. A narrow sometimes indistinct game trail wanders up the right side of the gully, leading to steep grassy slopes. Where an obvious subsidiary gully joins from the left, cross the stream and climb up the nose of trees dividing the two gullies. Keep the stream in earshot. After you've hiked about 40 minutes up the nose, the sound of the stream fades and you traverse back right and negotiate the steep slopes leading down to the spring. Allow 2 hours from the road.

Location The spring consists of a rapidly flowing 1.2 m x 2 m deep stream, the bedding plane opening of which is covered by a large collapsed block. Two diminutive caves, each no longer than 6 m, are located above.

Exploration & Description The spring on Mt. Wilson is well known, and was noted in the Geological Survey of Canada (Baird, 1967). But don't expect to be doing much caving. Possibly dry in winter, it may be enterable, but is located in an obvious avalanche gully.

A second smaller resurgence is situated 15 m higher up the gully, and was dug out for 3 m to a low room 2 m in diameter and 1 m high. From this room a narrow flooded rift leads off.

A third dry entrance located another 15 m above this was explored and a short section of cave passage found containing the jaw and horn of a mountain goat. A stream can be heard through a small fissure.
Sinks probably feed the springs that originate from the glacier on the summit of Mt. Wilson. Caver Peter Marshall noted that the well-jointed limestone dipping at an angle of 30° offers good possibilities for cave formation.

Warning The gully below the spring is steep and slippery. Move cautiously. Only attempt this hike when no snow is visible — the gully is an avalanche chute.

ELEVATION

horn

jaw

jaw

rat nest

PLAN

N

JAWBONE CAVE

Mount Wilson, Banff Park, Alberta
Adapted from a drawing by P. Marshall, August, 1972

0 2
metres

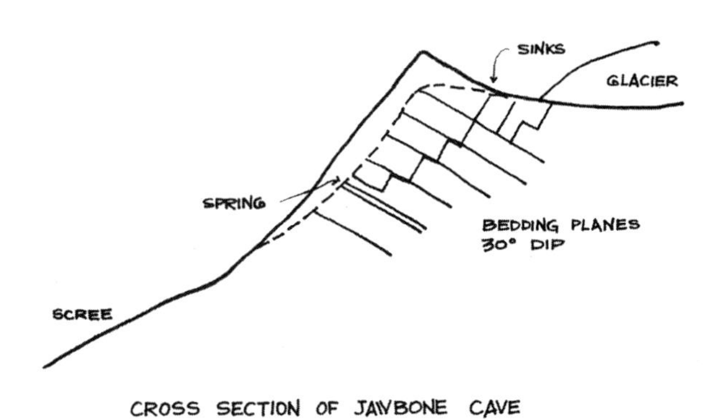

CROSS SECTION OF JAWBONE CAVE

Pinto Lake Cave

Map 83 C/2 091736
Jurisdiction Alberta Forests (Clearwater)
Entrance elevation 2013 m
Number of entrances 2
Length, Depth 610 m, 96 m
Discovered Unknown

Access Hwy. 93 (Icefields Parkway). 3 km north of Rampart Creek Hostel, take the Sunset Pass trail. This trail gains about 400 m in the first kilometre, but is then fairly flat over Sunset Pass until the steep decent to Pinto Lake, located just beyond the Banff National Park boundary. A camping area by the lake at 13.7 km from the trailhead has a toilet, firepit and picnic table.

Location In the cliffs southwest of Pinto Lake, a streambed runs down to the lake from the resurgence. Although the resurgence is reasonably accessible, reaching the upper entrance requires some very steep scrambling. A middle "entrance" is a resurgence and not enterable.

Although the lake margins are heavily wooded and floored with feather moss, the area around the cave is very steep and devoid of vegetation. The drainage catchment area probably runs over into Jasper National Park, where areas of boggy terrain, linked by small streams, contain some small sinkholes.

The lake is a popular fishing spot for people flying in by helicopter, and as a result its shores are becoming badly impacted with numerous campfire sites and piles of rusty tin cans. Please use the designated camping area.

Exploration Information about the cave came from ASS member Don Prosser, whose uncle noticed the resurgence when fishing at the lake. In June, 1979, ASS cavers visited the resurgence and found two dry, higher entrances farther up the cliff. The cave was finally explored and surveyed in July of 1979. An attempt to push the cave on a winter trip in 1986 failed when the party had to camp in a blizzard above the headwall.

Description The resurgence cave entrance is impressive, being 12 m wide and 18 m high, with a small lake 20 m in. A tricky traverse or cold wade accesses a high rift which has been climbed for about 15 m until it becomes too tight. Probably it connects with the cave above. A waterfall usually comes down this rift and into the lake, which then overflows down a surface stream channel to Pinto Lake.

Pinto Lake Cave resurgence entrance.
Photo Jon Rollins.

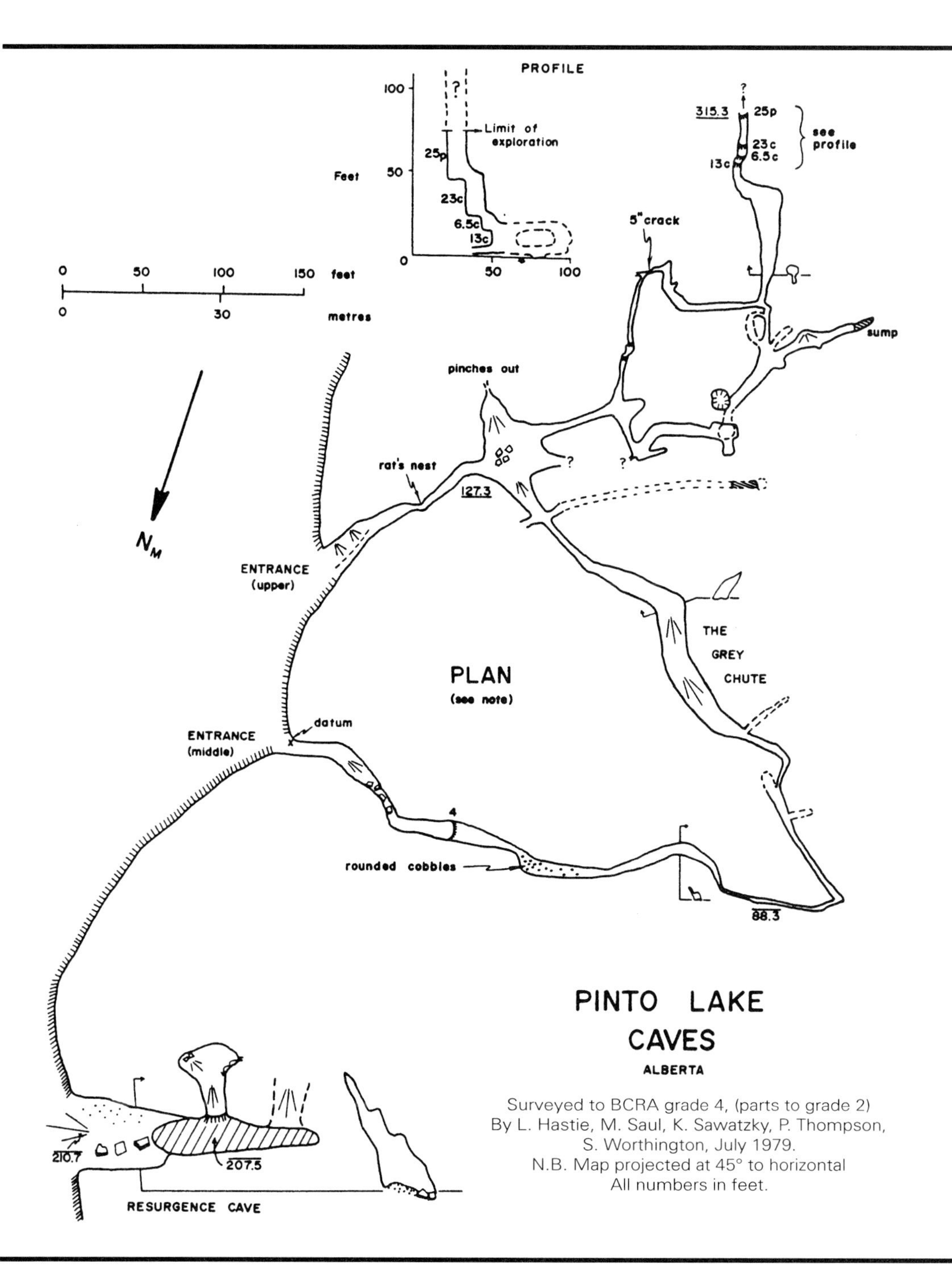

Surveyed to BCRA grade 4, (parts to grade 2)
By L. Hastie, M. Saul, K. Sawatzky, P. Thompson, S. Worthington, July 1979.
N.B. Map projected at 45° to horizontal
All numbers in feet.

The upper entrance, originally stuffed to the ceiling with pack rat material, leads to a loop passage connecting with the middle entrance. The passage consists of a series of water-washed chutes, the coarse grey limestone providing good climbing. Attractive clusters of cave coral and some small stalactites decorate the ceiling where steps in the passage cause water to drip. The highly polished cobbles in the passage just below the middle entrance indicate that in times of flood this section of the cave becomes a hydraulic lift similar to the 8 m pot in the entrance of Castleguard Cave.

The passage leading upwards from the upper entrance has short phreatic tubes with some fine sediment containing small animal bones. Attempts to climb a steep passage via three climbs ended when the party ran out of bolts. This high passage suggests the presence of another loop from an even higher fourth entrance. A thorough search of the area on the cliff top above the cave has not turned up any sinks. However, a possible entrance has been seen not far below the cliff top and appears accessible by a short rappel.

Geology "The caves are formed predominantly in a steeply-inclined thrust fault which forms a prominent feature in the near-vertical limestone cliffs bordering the lake. The fault runs diagonally across the cliff and into the mountain. Water at one time apparently flowed down the Grey Chute and up the ramp to the now dry entrance above the present resurgence. It now appears to sink through the gravel to the resurgence cave. There is evidence of ponding in the lower part of the Grey Chute, probably during the spring and early summer melt" (Keith Sawatsky, 1979).

Notes for survey The cave is two-dimensional in shape: when viewed along the line of the fault the cave map would degenerate almost into a straight line. The best view of the cave, as used in the plan, has been obtained by looking perpendicularly at the fault plane. This almost cancels out the depth component and creates a true plan view.

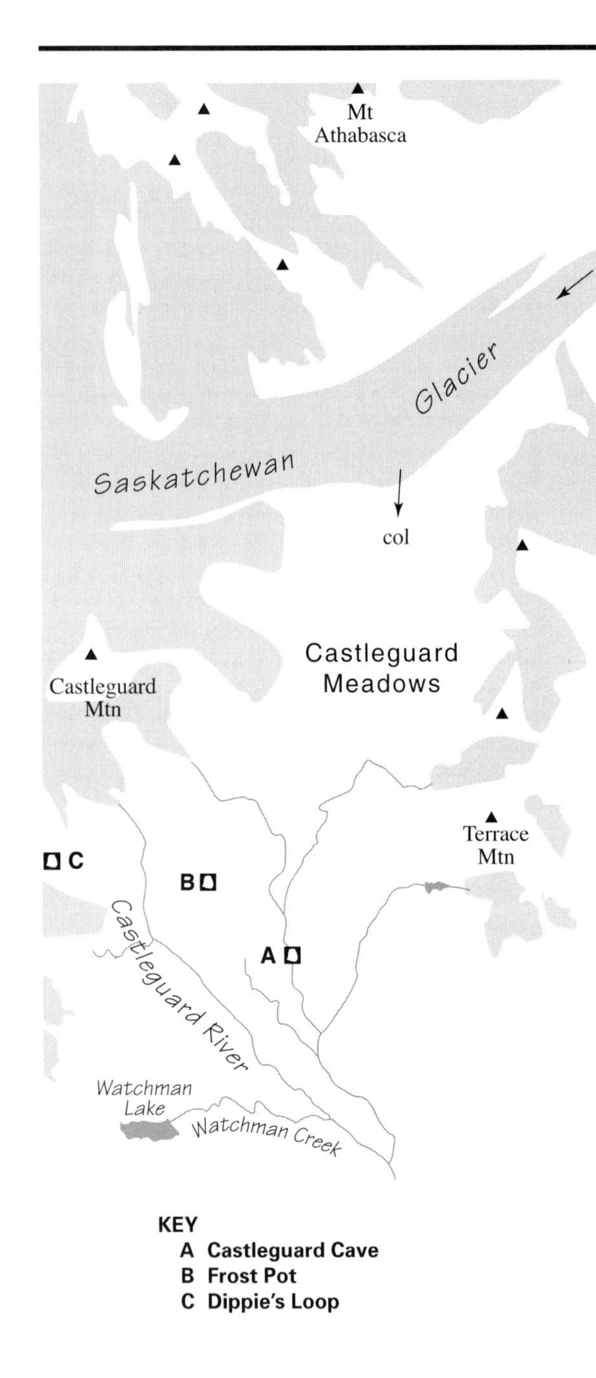

Castleguard Meadows

Castleguard Meadows provides one of the most spectacular locations for caves in the Canadian Rockies, being located high above the Saskatchewan Glacier in the midst of the Columbia Icefield. Castleguard Cave itself is fittingly impressive, being the longest cave in Canada and the fourth deepest.

Because it floods in the summer, the entrance to Castleguard Cave can only be accessed in winter conditions. Trips usually take place around Easter when longer daylight hours provide some margin for reaching the cave. Only Frost Pot is accessed in the summer when extensive karst pavements, including numerous shafts, close to the Saskatchewan/Castleguard Col are exposed at snow melt.

Winter Access Hwy. 93 (Icefields Parkway) at the "Big Bend," 12 km south of Sunwapta Pass. Ski south a short distance, crossing the small creek to an old road in the trees on the south bank. Follow the road over a hill (avalanche hazard) to the gravel flats leading to the Saskatchewan Glacier. Ski up the easy-angled glacier until it is possible to head south up a steep moraine ridge (avalanche danger) to the Saskatchewan/Castleguard col. Ski south down Castleguard Meadows toward the impressive Watchman Peak.

With the inevitable heavy caver's pack and assuming good weather, it takes about 6 hours to ski in. Allow at least an hour of daylight to locate the cave entrance.

Summer Access The route is the same. However, the Saskachewan River must be forded and getting onto the Saskachewan Glacier is not as easy.

Warning Glacier travel has its dangers. In summer take crampons and ice axe. In winter be aware of avalanche danger.

Area Geomorphology The meadows contain the largest area of surface karst in the four mountain parks: 10–15 km^2 (Ben Gadd, 1995). However, due to recent glacial activity, few large surface karst features remain. Ice has retreated from 0.5 to 2 km in the last 50 years. In 1925, maps showed the Saskatchewan Glacier to hold an ice-dammed lake on the upper meadows that no longer exists, the ice now being at least 50 m lower. The 100 or so shafts discovered towards the col separating Castleguard Valley from the Saskatchewan Glacier were formed by sub-glacial melt water, or by melt water at ice margins during previous glacial advances (Chris Smart, 1983). Castleguard Cave is the largest visible karst feature remaining today. Fresh moraine covers much of the area, deposited during the glacial retreat following the end of the Little Ice Age 700 years ago (Derek Ford, 1985). Surface karst features include various forms of karren, shafts, limestone pavements, felsenmeer, sinks and springs. Micro features include sub-glacial precipitates and sub-glacially-formed mineral formations.

Area Hydrology Besides the documented caves and the numerous shafts near the Saskatchewan/Castleguard Col, there are numerous sinks and springs worthy of note.

During his fieldwork on the hydrology of the Castleguard karst in 1983, Chris Smart divided the springs into those at the base of the Eldon Formation in Castleguard Meadows, those near the cave entrance and those in Castleguard River valley. The major cave-vicinity springs are in the Cathedral Formation, and situated 300 m above the floor of the Castleguard River valley. They are: Castleguard Cave entrance, Forest Spring, and Red Spring.

Forest Spring and Red Spring are situated 30 m below Castleguard Cave entrance. Forest Spring flows intermittently. Normally a drafting boulder/scree slope, it has a flow rate approaching 1 m^3/s when in flood.

Groundwater Hydrology for Castleguard Meadows.
Adapted from diagram by Chris Smart, 1983.

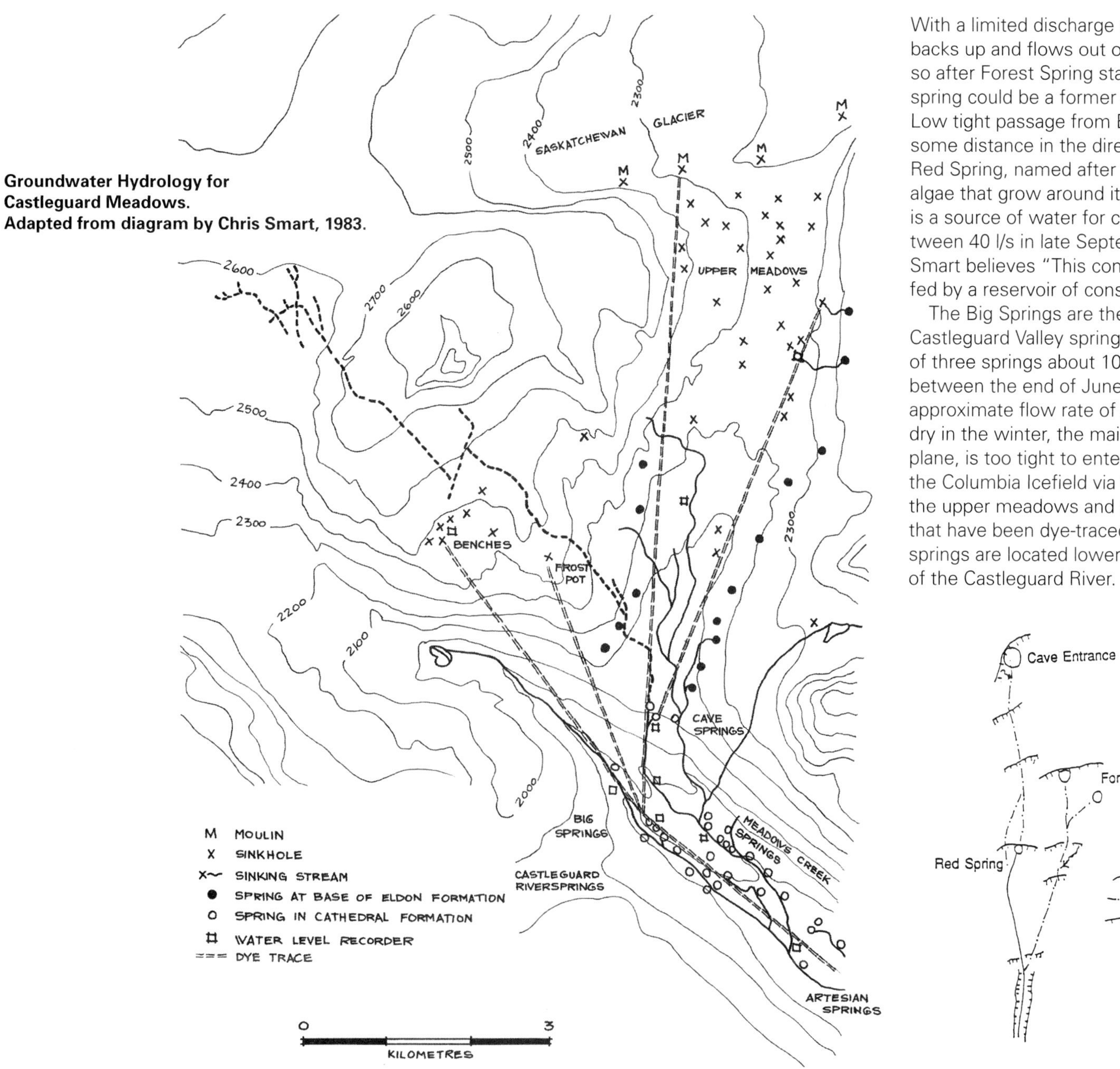

With a limited discharge capacity, the water quickly backs up and flows out of Castleguard Cave an hour or so after Forest Spring starts flowing. It is possible this spring could be a former entrance into Castleguard II. Low tight passage from Boon's Sump has been pushed some distance in the direction of Forest Spring. The Red Spring, named after the colours of the mosses and algae that grow around it, flows through the winter and is a source of water for cavers. Flow rates vary between 40 l/s in late September and 1 l/s in April. Chris Smart believes "This continued flow suggests that it is fed by a reservoir of considerable size."

The Big Springs are the most impressive of the Castleguard Valley springs, and consist of a group of three springs about 100 m apart that usually flow between the end of June and mid-September, with an approximate flow rate of nine to11 m^3/sec. Although dry in the winter, the main point of seepage, a bedding plane, is too tight to enter. They are probably fed by the Columbia Icefield via Castleguard II, and by sinks in the upper meadows and on the Saskatchewan Glacier that have been dye-traced to this spring. Many more springs are located lower down in the banks and gravel of the Castleguard River.

THE CAVE SPRINGS

Castleguard Cave

Map 83 C/3
Jurisdiction Banff National Park
Entrance elevation 2016 m
Number of entrances 1
Length, Depth 20122 m, 390 m (-18, +372)
Discovered 1921, by Cecil Smith

Location The cave entrance is located below treeline, about 500 m beyond the lip where Castleguard Meadows drops down to the Alexandra River. Look for it slightly west of the major surface gully and drainage. Capped by a large overhang, the cave entrance is difficult to see from above.

At over 20 km, Castleguard is the longest cave in Canada — over 8 km longer than its nearest rival, Yorkshire Pot — and the fourth deepest. The cave's position under the Columbia Icefield and its great length, combined with harsh winter surface conditions, make Castleguard a unique caving experience. It is surrounded by impressive mountains including the 3077 m-high Castleguard Mountain (under which the cave passes), and across the valley of the Castleguard River, the 3009 m-high Watchman Peak which is framed in the cave entrance. Emerging into such fine scenery after three days underground must rate as one of the highlights of any caving expedition in the Canadian Rockies.

The current trend is for small lightweight groups (as few as three persons) who ski in and out from the cave. Without the use of helicopters, Castleguard is an isolated cave, requiring the carrying of a large pack (food and caving gear for five days minimum) on the long ski in from the Icefields Parkway 17 km away. Going alpine style is now the ethic for long caves, and the ability to reach the cave with a large pack is part of the selection process.

With the discovery of a climbable by-pass to the 24 m Pot, the Ice Plug can now be reached without rope, except for the 8 m drop at the entrance. An ascender is needed to climb the fixed rope at Boon's Aven.

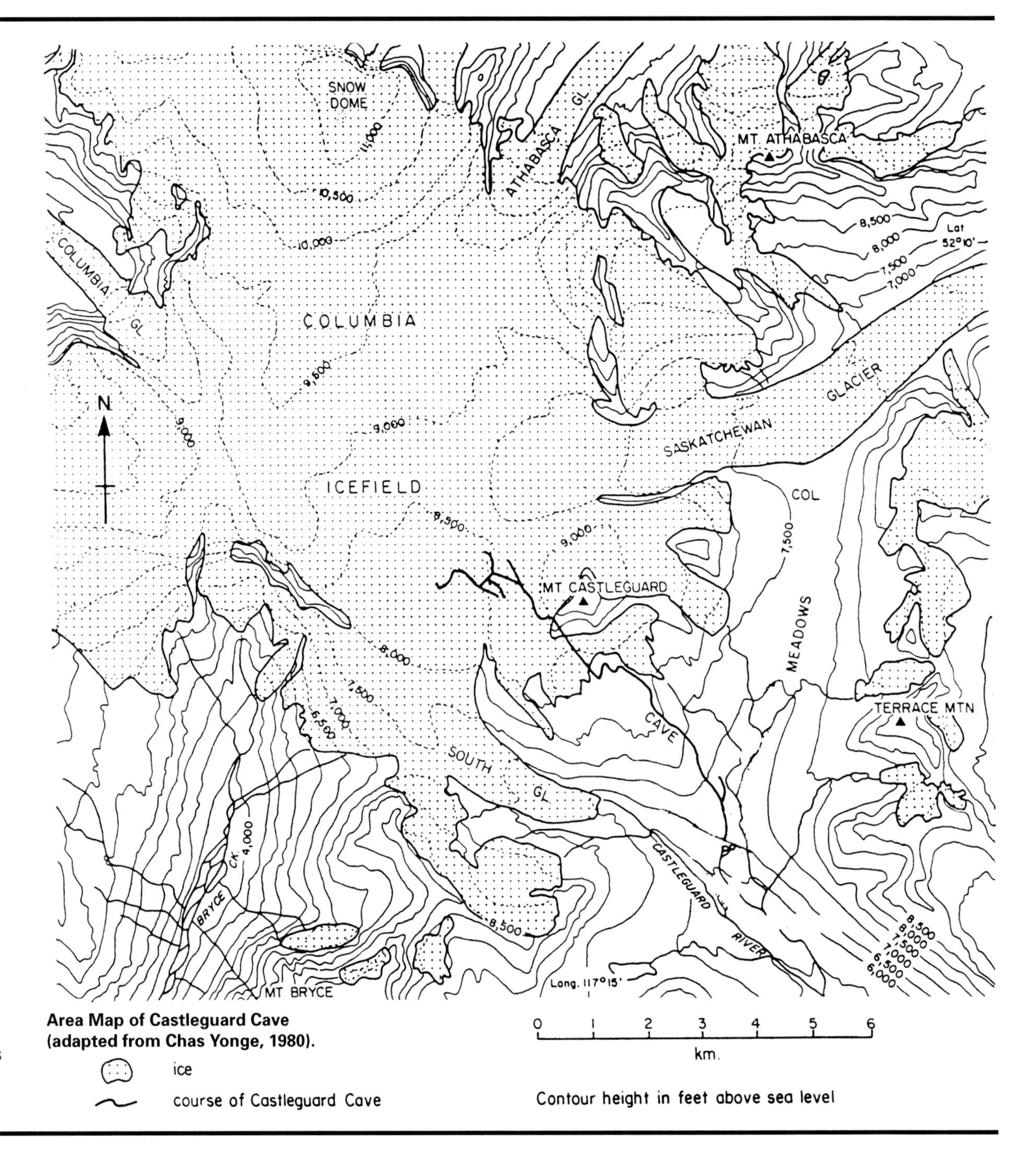

Area Map of Castleguard Cave (adapted from Chas Yonge, 1980).

Top: Castleguard Meadows. Photo Chas Yonge.

Bottom: Castleguard Cave in flood. Photo Chas Yonge.

Warning The entrance series flood in summer when the cave entrance often becomes a spring.

An immobilizing accident beyond the first fissure could easily become a fatality, the logistics of rescue being incredibly complex. Stabilizing the victim with a tent, sleeping bag and hot food/drinks until medical attention could arrive would probably be the best course, most of the cave being dry with plenty of flat spots. However, don't have an accident in this cave; it will take all the cavers in Canada just to get your body out.

Cave Softly In order to visit the Ice Plug most cavers take about three days underground, camping at Camp One just before the Holes-in-the-Floor. This introduces the problem of human waste disposal underground. Urine, being sterile, presents less of a problem, providing it is kept away from water sources. Formerly, Bog Alley was used as a toilet area until it was cleaned up in 1983. Since then cavers have been encouraged to carry out their own feces in zip-lock bags filled with deodorizing kitty litter.

All survey marks and lettering smoked on the walls during early exploration, and most of the telephone line left in from the filming trip has been removed. In recent years the ASS has organized numerous trips to laboriously haul garbage from the underground camps. A few random items remain at The Emergency Depot and at Camps One and Two, but all caving trips are now expected to be self-sufficient. Don't leave ANYTHING in the cave. Respect and replace marker tape around pools and easily damaged formations in the main passageways. Take a sponge to clean muddied flowstone close to water sources.

Outside, please don't cut firewood or have fires in the cave entrance. Parks have made Castleguard Meadows a prime protection area. To keep warm on the surface take a tent and stove.

Discovery On July 5, 1980, Ron Seale taped an interview with discoverer Cecil Smith (available at the Whyte Museum of the Canadian Rockies in Banff). By way of an introduction Seale describes how "Cecil Smith guided people up to Nakimu Caves and worked for Brewster Transport from 1917 to 1919 with horses. In the spring of 1921, Cecil guided Mr. and Mrs. Walcott through the mountains. They were both over 70 years old. Mr. Walcott was head of the Geological Survey for the U.S. government and president of the Smithsonian Institute. Mrs. Walcott took photos of wild flowers, making paintings from them on her return home. Mr. and Mrs. Walcott persuaded Cecil to take them to the Columbia Icefields although he had not been there before. They had a packer named Billy Lewis and a coloured cook named Arthur who had cooked for Mr. Walcott for 25 years on his expeditions. They travelled over the Pipestone, down the Siffleur over to the ford on the North Saskatchewan, then up the Alexander River and into Castleguard Valley. It took six days. There were trails through the Siffleur but they had to bush-whack up the Alexander."

Cecil Smith takes up the story. "One night we had gone to bed when I heard the horses come dingle-dingle, dangle-dangle down the little swayle below. I said to Billy Lewis — 'I've sure got to get up early to get those horses or they'll be in Lake Louise'— so I got up early the next morning and I followed the horse tracks down and the first thing I knew I came round the corner and here was this big cave in front of me and the horses had forded that, turned round and gone back up into the meadow. I didn't stay long, just a glance more or less at the cave. I told Mr. and Mrs. Walcott about the cave at breakfast, they were very interested. I took them both down to the cave. The Walcotts spent some time in the cave.

"Now I'm sure if they'd been any kind of lettering around that cave we would have noticed it. Mr. Walcott said, 'Well Cecil, now I think there's a first for you.' I'm certain in my own mind no one had been there. We spent two weeks camped on the meadows from the beginning of August and kept going back to the cave. We saw it in flood. I was taking lunch with Mrs.

Walcott, all of a sudden there was a big roar that came out of the cave. Mrs. Walcott said 'That sounds like a waterfall,' then out gushed the water and it flowed down the 'stairway' [the stepped pavement below the cave entrance]. That was four in the afternoon on a hot day, we saw it everyday at that time of the day and it slowed down in the night and by morning it was dry [the cave only resurges during hot weather]. We went in as far as the sump-hole [the 8 m drop], we only had matches.

"During my time in Banff after the discovery of the cave I fished everybody out; George Harrison, Bill Potts, Jimmy Simpson and all those old timers. I tried to find out if anyone had been up in there and there wasn't one of them. I felt the cave was very interesting, I was going to get a gold pan and see if there was any gold in those caves."

At the end of the interview Ron Seale asked Cecil if as original discoverer of Castleguard he would like to rename it. "Yes, I would, one man who I don't think has been appreciated enough, he has given more work to people than any other man in Banff: Jim Brewster [brother of Pat Brewster]."

Exploration Probably the second visit to the cave entrance took place in 1924 during a major expedition involving 21 people. Those present included J. Monroe Thorington, Conrad Kain, Jimmy Simpson, Dr. W. S. Ladd and Ulysse La Casse. The event was recorded in photographs taken by Byron Harmon. The team also located and photographed the Big Springs and theorized, even back then, that due to its glacial colour that it must drain from beneath the icefields.

Opposite left: Crossing La Grande Gueule, Castleguard Cave. Photo Chas Yonge.

Opposite far left: Second Fissure, Castleguard Cave. Photo Ian Drummond.

From the late 1920s onwards pack-train/tour operators guided groups to the meadows and visited the cave in the late afternoon in the hope of seeing floodwater pour from the entrance. The National Geographic carried a story on Castleguard Cave in 1925.

Although the entrance of Castleguard Cave has long proved an attraction for summer hikers and a good site for sheltering from the elements on winter ski trips, no recorded exploration beyond the 8 m drop occurred until the arrival of the McMaster cavers in July, 1967. Following a tip from Red Creighton who ran the Maligne Lake boat concession, Mike Goodchild visited the cave entrance, and the rest, as the saying goes, is history.

Since 1967 there have been at least 20 recorded expeditions to the cave and several small unrecorded trips. The annual Castleguard expeditions have acted as a focus of attention for cavers internationally.

Description Beyond the initial 8 m drop, the first 600 m of Castleguard is relatively straightforward, consisting of a low, wide bedding plane. Initially the caving is fun, sliding along on a hard ice floor. However, farther in the cave temperature begins to assert itself and angular breakdown separated by pools of water makes progress unpleasant. 200 m in, a particularly low and often flooded section of passage known as The Duck is reached. Just beyond this the passage divides.

The right-hand passage, Boon's Blunder, leads via 180 m of braided passages to a sump, from which floodwaters emanate in the summer. Many an unwary caver has taken this passage by mistake.

The main left-hand passage, following an abrupt turn to the left (south), begins to enlarge and at last it is possible to walk for short distances. This is fortunate because at least two deep pools block the passage and must be waded. The passage eventually loses its bedding plane appearance and becomes more rift-like, with soda straws decorating the ceiling. Pass a mud flat on the left side, site of the old Emergency Depot which was put in place following the incident when Peter Thompson and Mike Boon were trapped by rising water. Shortly after, the head of the 80' Pot is reached at 1.6 km into the cave

Although usually rigged and descended by rappell, this pitch can be bypassed via a rifty down-climb accessed off to the right of the main passage shortly before the pitch head. The lower section of this bypass joins the main pitch about 5 m from the bottom, where a hand-line is usually in place.

Discovery of the next section of cave by Peter Thompson must rate as one of the high points of Canadian cave exploration. "The bottom of the boulder-strewn pit at first appeared blind, but a crawl out of which issued a strong breeze seemed to be the way on. The muddy crawl dropped into a narrow trench which gradually grew bigger... and bigger...and bigger! The trench opened out into a nearly circular 3 m tunnel (The Subway), a perfect example of the classic phreatic shape. Pete galloped along this passage for a while to make sure it did not close down around the corner — but there was no corner! The passage continued almost dead straight for 300 m and was still continuing when he turned around" *(Cave Exploration in Canada).*

After the 450 m-long Subway is the start of the First Fissure, 2 km of ascending rift passage punctuated halfway by the Waterfall Room, where a large cascade enters from the ceiling. A ledge up to the left provides a welcome respite from a kilometre of bridging, and a a much needed drink of water. Beyond the Waterfall Room, awkward bridging between mud-caked ledges eventually leads to a hand-line up into a low bedding-plane passage marking the start of The Grottoes. This series of low bedding plane passages contains some superb calcite formations. Where The Grottoes takes an abrupt turn to the south is The Next Scene, a low, sandy, parallel crawlway that soon gets too small. Camp One is located at the end of the Grottoes, where the nature of the passage changes once again to a large keyhole shape, with some tricky traverses along sloping mud ledges between narrow clay bridges (The Holes-in-the-Floor). Towards the end of The Holes-in-the-Floor passage, the vadose floor trench swings abruptly off to the left, accessing the spectacular 200' Aven (63 m), measured using a hydrogen balloon on a length of string.

The main way on is the Second Fissure, much like the First Fissure but with some even more tricky climbs, some rigged with hand-lines. Approximately 2 km through Second Fissure is the Crutch, where the original explorers headed right into Thompson's Terror, named after some loose climbs that caused them to turn back. This is hardly surprising, because by this point Mike Boon and Peter Thompson had been caving for 9 km in one push from the entrance! It was on their return that they discovered the cave had flooded at The Duck, and out of carbide and energy, they huddled in a shallow, sandy depression to await rescue.

Turning left at the Crutch, you follow the passage for 370 m to an aven, first climbed by Mike Boon during his remarkable nine-day solo trip in November of 1970. "As soon as I saw the aven I was delighted because I knew at once it would go. The actual climb itself was to one side of a cylindrical ascending shaft. Leading into the cylinder was a rift. By putting in a piton with two etriers I was able to climb onto a ledge which ran across the face of the shaft. I was now halfway up. I hammered a bolt into the face to use as a running belay should I fall, and then free-climbed a strenuous blade of rock to the head of the shaft. Here at last was my chance to find out where Castleguard Cave went. Tying off the line for my return, I galloped up a wide, low tunnel to a large chamber with an ominous black shaft to the left. Beyond, the passage became lower and more filled with mud until I could finally go no further without digging. There was no draft, and although one goes caving to find caves rather than to prove that they do not exist, I was delighted to have reached an end. But the cave was not finished. On the way back, before the boulder chamber, I noticed a low passage to the left. Crawling in, I found an attractive rising tunnel fringed with stalactites. Again I explored for about 240 m where I was stopped by a brilliant blast of reflected light. Intensely curious as to what it was, I moved forward to find ice. The passage was completely plugged with a glittering wall of ice, presumably pushed down from the glacier above."

With the discovery of the Ice Plug, the general feeling was that the cave was more or less finished. However, this was to prove far from true, as much new passage has been found in the headward reaches, emphasizing the cave's dendritic nature. The 1979 to 1987 expeditions resulted in a 50% addition to the total length of the cave, making it just over 20 km long.

In 1979 the ominous black shaft noticed by Mike Boon (24 m deep) was descended by Quebec cavers. It led to 1 km of new passage, the Boulevard du Quebec, but of major interest were a number of pitches in this section (see survey). If these could penetrate to the lower system called Castleguard II that feeds the Big Springs located 300 m below the cave entrance, then a major extension to the cave might be possible. The deepest of the pitches, known as La Grande Gueule, eventually sumped at 139 m, possibly halfway to Catleguard II. In 1980 a tricky traverse across the top of La Grande Gueule by Steve Worthington accessed the 500 m-long Third Fissure, and along the way Ooley Gooley Pit — a blind 75 m single drop. There are still drafting leads to be explored in this section of the cave.

The other major lead was Thompson's Terror, a large continuing fissure passage that because of difficult and loose traverses was initially abandoned. Ironically, exploration has terminated at the last climb, and in 1983 a kilometre of very pleasant passage was mapped to a second ice plug, the Nice Plug, around which were several smaller passages also culminating in ice. Like the main ice plug, the ice was coarsely crystalline with the same composition as ice samples taken high up on the Columbia Icefield. There is little doubt the ice is from the base of the icefield and has been intruded into the cave passages. According to data on the distance ice will intrude into rock cavities, it seems likely there is not much more that 30 to 40 m of cave passage beyond the ice front (Derek Ford, 1985). In 1984 a cave radio was taken underground in an effort to locate the position and depth of the cave under the Columbia Icefield. Based on a tentative cave "height" of 300 m and an entrance elevation of 2016 m, the Ice Plug is estimated to be some 200 m below the surface. Owing to poor surface conditions (a full scale blizzard) there were rendezvous problems and the results were not conclusive. Nevertheless, tone contact was made in the same general area as shown by the survey, indicating that there were no grave errors in cave mapping. However, from the strength of the signal, the distance the signal could be heard over and the lack of nulls, it seems that either the end of the cave is deeper than the supposed 200 m, or the ice adversely affects radio transmission.

New passage has also been found closer to the entrance. A lead near Holes-in-the-Floor went for 0.5 km before pinching out. Boon's Blunder also yielded some low passage, but no way round the sump was found. This complexity of passages comprises a braided drainage emanating from Boon's Sump which was dived in 1987 to a large conduit passage with a steady downstream current (see hydrology).

In recent years exploration in Castleguard cave has become increasingly difficult as points of exploration have been pushed further from the entrance. In the headward complex, a far greater system of passages exists than the cave survey might suggest, because not every bypass and crawlway has been surveyed. A new survey for Castleguard is long overdue. The data exists, but translation into a survey showing passage detail would require a lot of work, and at a comprehensive scale would probably approach 4 m in length!

Opposite right: Boulevard Du Quebec. Photo Tom Miller.

Opposite left top: The Subway. Photo Derek Ford.

Opposite left bottom: The Ice Crawls. Photo Dave Thomson.

Opposite right: Holes-in-the-Floor. Photo Ian Drummond.

Opposite far right: The 80′ Pot, Castleguard Cave. Photo Derek Ford.

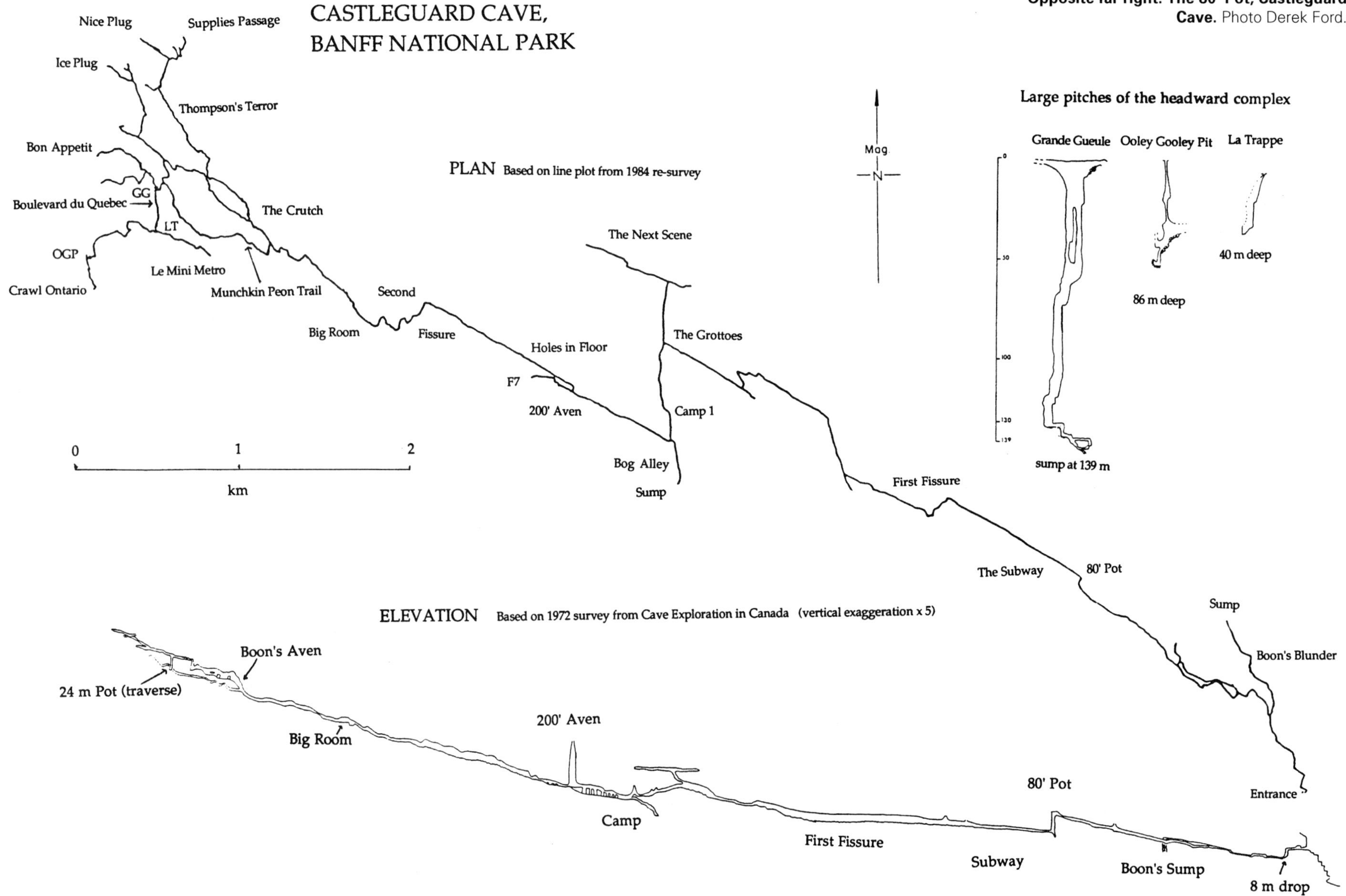

continued at bottom right

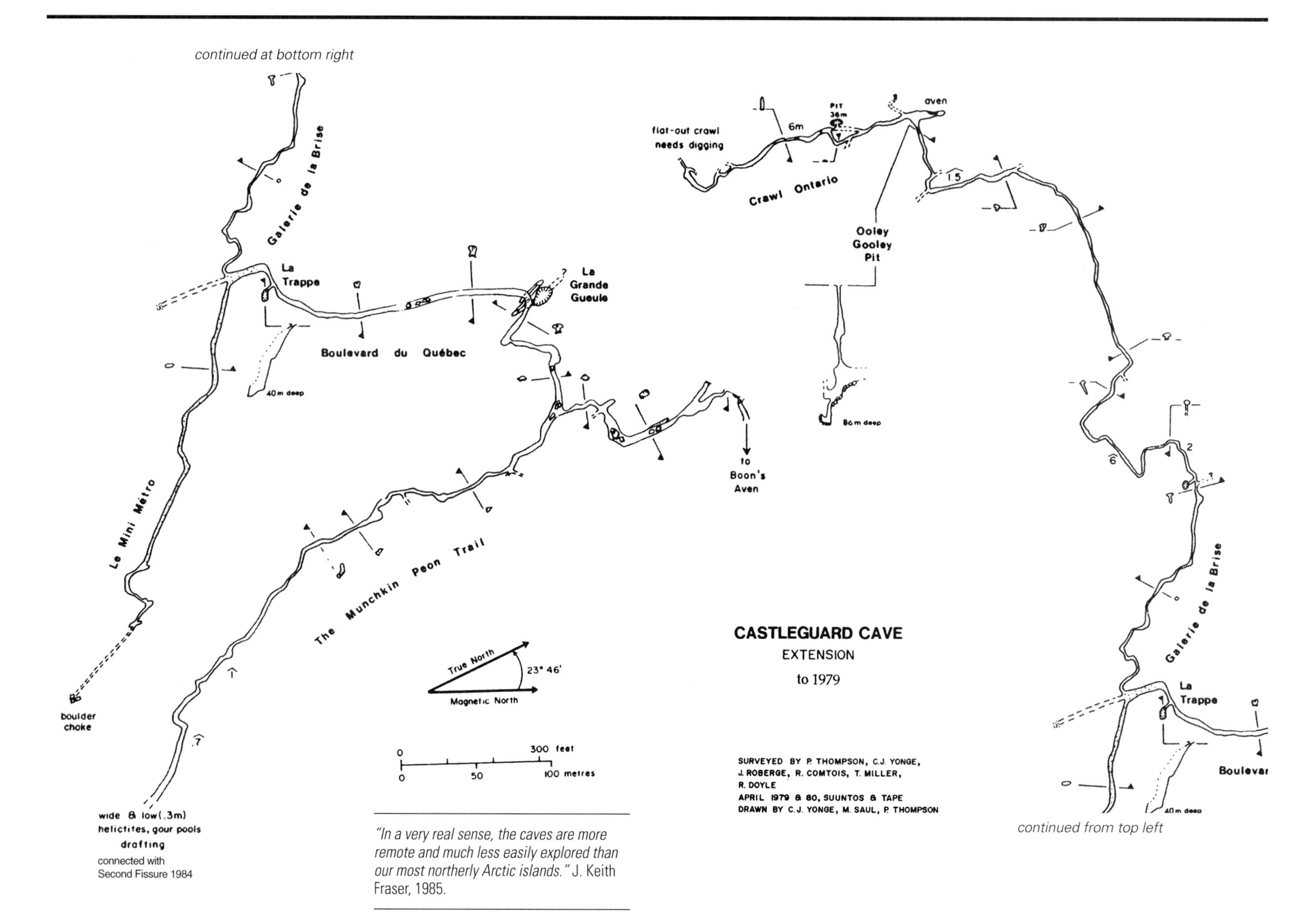

"In a very real sense, the caves are more remote and much less easily explored than our most northerly Arctic islands." J. Keith Fraser, 1985.

continued from top left

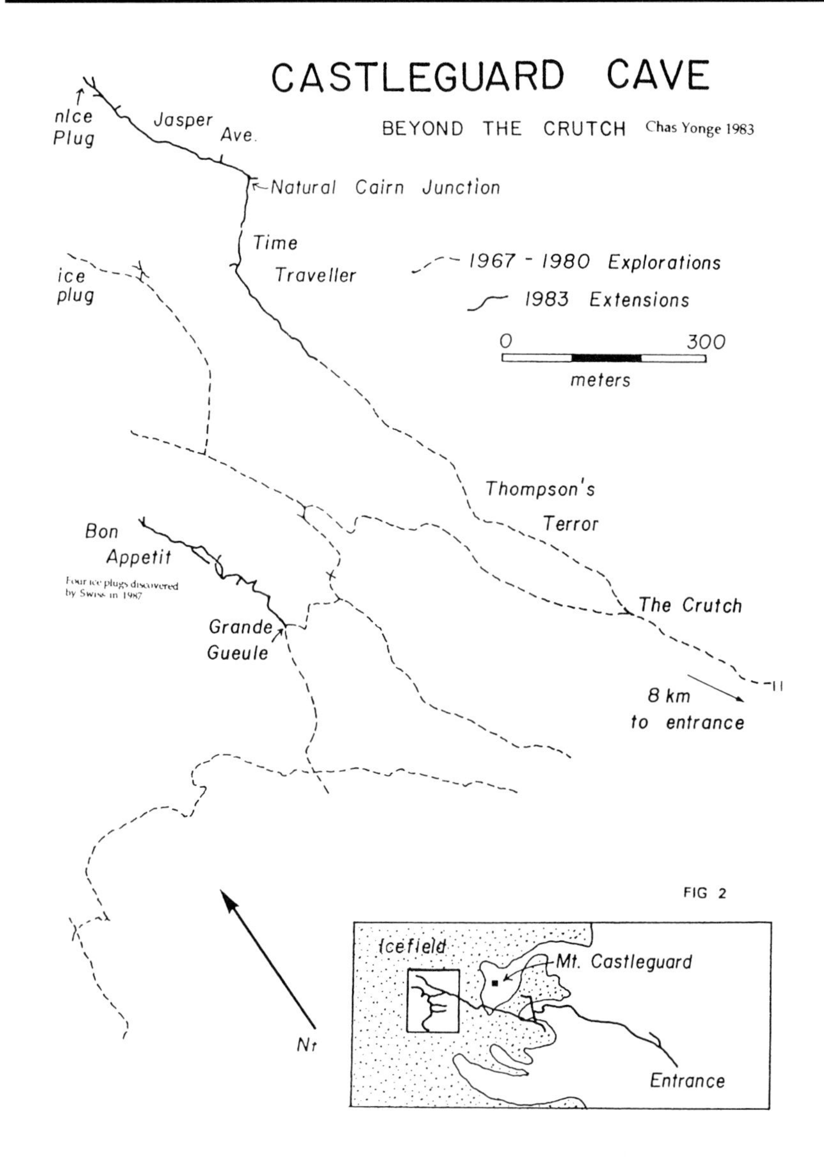

The Ice Plug, Castleguard Cave. Photo Dave Thomson.

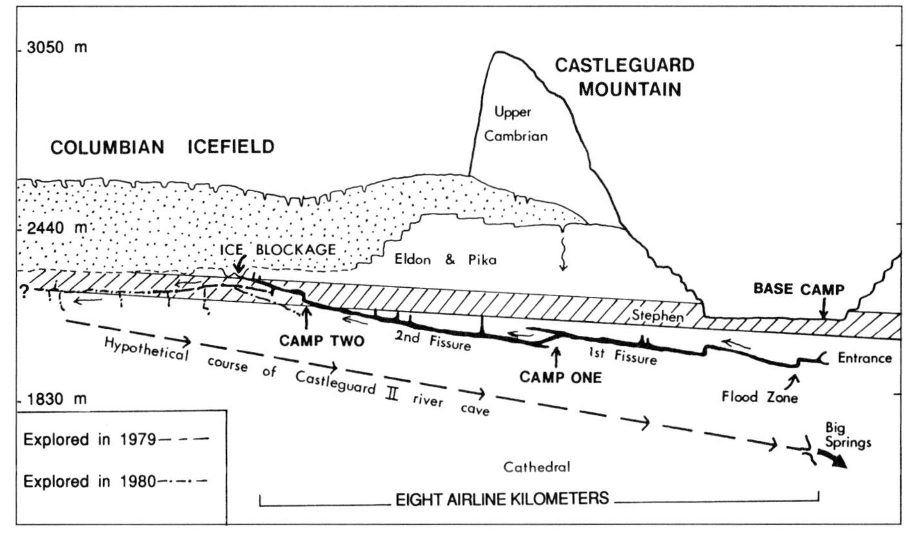

Northwest/Southeast section through Castleguard Mountain (adapted from Chas Yonge, 1980).

Geology Castleguard Cave is contained in Middle Cambrian limestones of the Cathedral Formation, which can be divided up into the Upper and Main Members. The explored passages are contained in the Upper Member, which is more varied in texture and porosity than the massive Main Member containing the flooded Castleguard II system feeding the Big Springs. Above the Cathedral Formation is the Stephen Formation, which consists of a mixture of shales, limestones, and dolomite that forms the ceiling and breakdown material in the headward sections of the cave beyond the Crutch. The Eldon and Pika Formations, above the cave where it passes under Castleguard Mountain, are relatively impermeable. This explains the lack of inlets and thus side-passages in the straight, middle section of the cave. Castleguard has formed generally in the direction of the bedding planes, down dip, southeast at an angle of five to six degrees.

Because of the massive nature of the Cathedral Formation, the few joints that guide the cave passages tend to go a long way. Smaller geological features that intersect, and have guided the formation of the cave passage, are narrow intrusions or dikes 20 to 40 m thick (Derek Ford, 1980). Formerly called Neptunian, in reference to their apparently sedimentary origins, it is now believed that these dikes are in fact igneous intrusions (Ben Gadd, 1986). Because the dikes are insoluble, they cause passage diversions and resultant ponding and lifting such as what occurred at The Grottoes. An eventual breech of the dike in Bog Alley provided a lower flow route. A dike also makes its presence felt at Boon's Aven where a nick-point has developed, perching the headward complex above the Second Fissure.

Speleogenesis Chris Smart postulates the existence of four separate cave systems:

1. Castleguard 0 A pre-glacial feature, flowed north south under Castleguard Meadows. The upstream section was truncated and blocked with debris when the Saskatchewan Glacier breached the meadows valley. The exit to this cave may be a moraine-covered lifting shaft, or could be Forest Spring.

2. Castleguard I The existing cave, the sink point of which would have been located in a closed depression seismically interpreted as lying beneath the Columbia Icefield northwest of Castleguard Mountain, now the Ice Plug. At this point, the Cathedral Limestone would have outcropped beneath the Stephen Shale, and would have conducted drainage to the current entrance owing to the hydrological focus provided by Castleguard 0.

Castleguard I is now apparently an abandoned cave, no longer conducting water except for the entrance complex, which still becomes active on hot summer days when Castleguard II overflows up Boon's Blunder and out the entrance. Paleomagnetic dating has indicated that the Grottoes drained through Bog Alley at least 700,000 years ago suggesting that Castleguard Cave as we now know it had formed by this date (Alf Latham, 1972).

3. Castleguard II The current water-filled system underlying Castleguard I and feeding the Big Springs developed due to the deepening of Castleguard Valley to the south, and the subsequent increase in hydraulic gradient.

4. Castleguard III occupies the route of the former Castleguard 0 but is a rejuvenated and recent feature referred to as the Meadows System."Through dye tracing in the upper meadows, Chris Smart has determined that 20% of the Big Springs water comes from the meadows. This leads us to postulate the existence of Castleguard III, a cave dropping 732 metres in elevation with miles of passage" (Derek Ford, 1983).

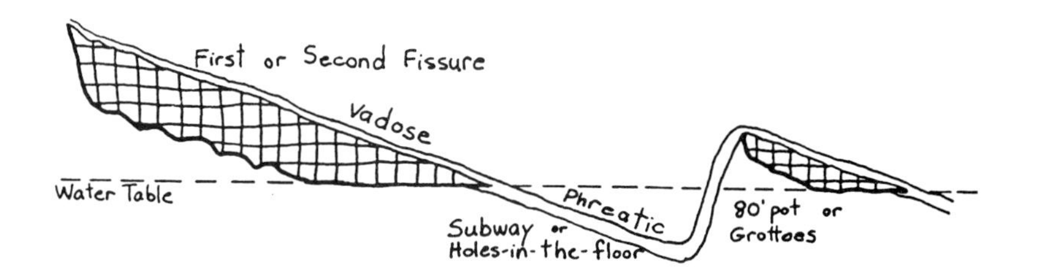

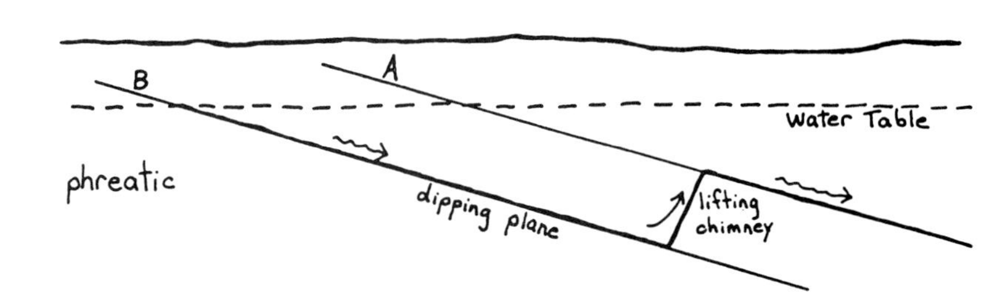

Vertical Developement of Castleguard Cave. (P. Burns 1980.)

Rick Blak in Castleguard Cave.
Photo Ian McKenzie.

Speleothems Calcite formations exist the length of Castleguard Cave and, characteristic of Canadian Rockies caves, are remarkable for their brilliant whiteness. More unusual and exotic formations include round and cubic cave pearls (cubic cave pearls have only been found in a few other locations in the world), wind-blown erratic straws and straws with flags (sometimes adjacent straws have flags pointing in opposite directions), helictites and crystal flowers.

Minerals occurring other than calcite includes aragonite and gypsum. Aragonite flowers exist in the Central Grottoes, and rocks on the floor of Supplies Passage (The Time Traveller) are coated with gypsum, giving them a shiny appearance as if water had just left the passage. Areas of the cave with particularly splendid collections of formations include the Holes-in-the-Floor, The Central Grottoes, The Next Scene, Supplies Passage and The Boulevard Du Quebec. In many areas, passage ceilings are well decorated with straws, but are high up and so go unnoticed.

Flagged stalactites, Castleguard Cave.
Photo Chas Yonge.

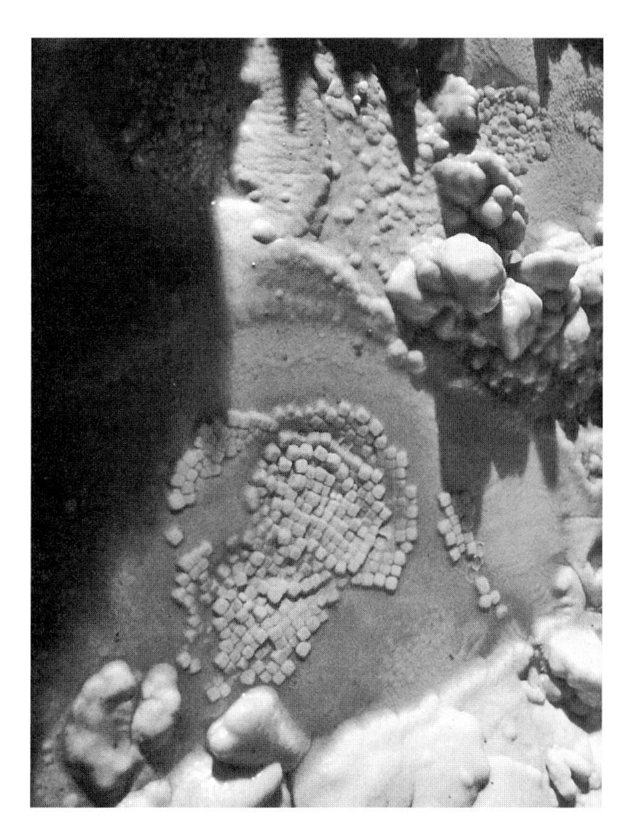

Top right: Cubic cave pearls, Castleguard Cave. Photo Jon Rollins.

Bottom right: Ice detail in the Ice Crawls, Castleguard Cave. Photo Jon Rollins.

Sediments Sediments occur throughout Castleguard Cave except near the entrance where the resurgence of Boon's Blunder flushes out all loose material. The sediments can be divided by size into gravel, sand and silt-clay laminates. The schematic below indicates the distribution of these sediment types through the cave. The larger material is generally found near the entrance of the cave, especially at the base of the 24 m and 8 m lifting chimneys where they have become rounded due to agitation. The pebbles at the base of the 8 m shaft have become polished through hydraulic agitation. Finer sediments exist throughout much of the cave in the form of "varved clays," the banding in which indicates three distinct periods of deposition (Derek Ford, 1983). In places, laminated clays can be found close to the ceiling, indicating the passage may at one time have been totally blocked by sedimentation and was later flushed out by aggressive rather than depositional waters. Some passages in the headward complex are still completely blocked by laminates (Chris Smart, 1984).

Right: Varved sediments in Castleguard Cave. Photo Chas Yonge.

Far right: Detail of ice crystals, The Nice Plug, Castleguard Cave. Photo Jon Rollins.

Ice Formations During winter months, floor to ceiling ice-columns form in areas of local seepage in the entrance series of the cave. Pools of water left over from summer flooding freeze and sometimes make the Ice Crawls impassable, as occurred in 1990. The resultant ice is dense, clear, and very hard, with striking arrangements of streamlined, elliptical air bubbles trapped within. A fine dusting of glacial-flour often covers the ice. An ice-heave or "pingo" is usually located just before the initial 8 m drop. At the end of the cave are the ice plugs. Currently, eight have been discovered, and all are thought to consist of glacial ice injected a short distance into the cave. The Nice Plug has ice flowers and beautiful pine tree-shaped ice crystals several centimeters high on its surface.

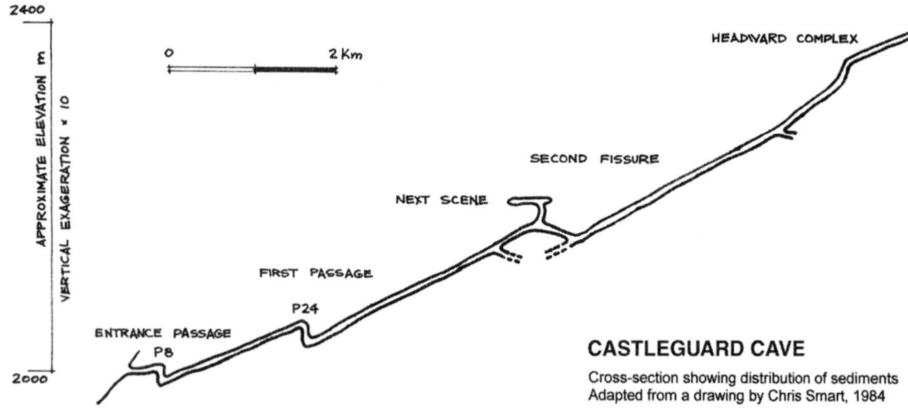

CASTLEGUARD CAVE

Cross-section showing distribution of sediments
Adapted from a drawing by Chris Smart, 1984

Cave Fauna Ralph Ewers made the first observations of animal life in the cave during the filming of the Castleguard movie in 1974. Starting in 1977, biologists John Mort and Anneliese Recklies made several visits to the cave as part of the team. The following is an account of their findings:

"Generally the fauna in Castleguard Cave is very sparse, presumably due to the extremely limited nutrient supply as well as the low temperatures in this environment. One of the most exciting discoveries during our visits was that of a previously undescribed species of an eyeless, unpigmented amphipod. Specimens were found only in a series of pools at the end of The Subway, and their population seemed to be rather small. These animals showed quite a remarkable agility that made the collection of specimens somewhat of a challenge. The amphipods were identified by John Holsinger as *Stygobromus canadensis* Holsinger, and extend the northern record for troglobitic amphipods.

"The major species found in the cave belongs to the order isopoda. They were first observed in the pools following the Ice Crawls and thereafter in almost every pool encountered all the way to the end of the Second Fissure. Specimens were identified by T. E. Bowman as *Salmasellus steganothrix* Bowman. This is an eyeless, unpigmented isopod previously described near various springs in different areas of Alberta. The population of the cave is quite large, often a dozen or more specimens were counted in a small pool. Undisturbed sediments at the bottom of the pools often show an abundant pattern of their tracks. The finding of this species in the Second Fissure is of particular interest as this area of the cave is known to be fed with the melt water from the Columbia Icefield and further in some passages are blocked by ice. These creatures are thus living successfully in a presently glaciated region, or more precisely, beneath it."

In 1983, a colony of cave-adapted mites were found living only a few metres from the Ice Plug. These were later identified by I. M. Smith of Agriculture Canada as a new species of *Rhagidiidae foveacheles*.

Hydrology Castleguard is usually dry between October and April; the ponded water in the entrance series freezes creating the Ice Crawls that characterize the first kilometre of travel through the cave. In the summer, the cave entrance acts as an intermittent spring, disgorging substantial floods on hot summer days. Usually starting between 4:00 pm and 7:00 pm, the flood pulses may last until midnight, or during very hot spells, for several days. The floods rise from a sump at the end of Boon's Blunder, which was dived in 1987. "When we arrived at the sump we found a rift about 2 m wide, 6 m long with the water about 2 m below the lip. I dropped to 17 m and then followed the floor off to the left to a depth of 21 m. I had broken out into huge passage with 10 m plus visibility! There was a noticeable current from the right to the left and the passage was about 3 m high by 7 to 10 m wide. The sump was a rift that broke into a major drain" (Keith Sawatsky, 1987).

This dive would seem to confirm the belief than the cave resurgence is an overflow mechanism to the major active conduit (Castleguard II) that underlies Castleguard Cave and feeds the Big Spring.

Mike Boon. Photo Eric Neilson.

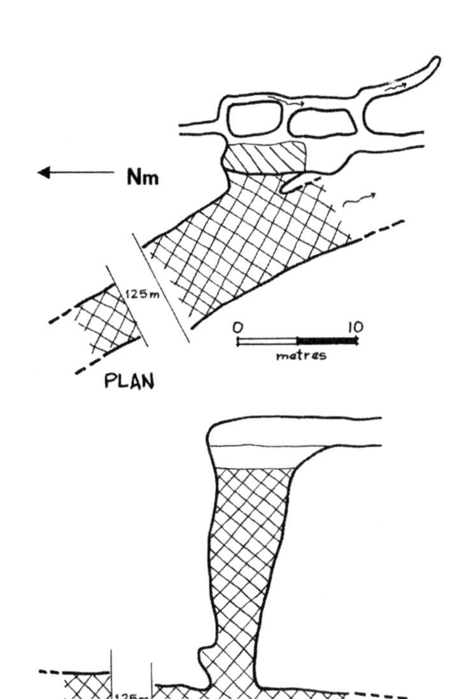

BOON'S BLUNDER

Adapted from a Grade 1 Survey by Keith Sawatsky, 1987

The Cave Environment The surface mean annual temperature is a chilly zero Celsius. In the central regions of the cave, where it is well insulated by bedrock beneath Castleguard Mountain, the cave temperature is around 3-4°C, getting lower towards the entrance and in passages close to the ice plugs.

Castleguard drafts strongly inward in the winter (from about October until April) and outward in the summer. Barely perceptible in voluminous passages, the draft becomes a wind strong enough to blow out a carbide light in constricted areas, as at the bottom of the 24 m Pot. On a daily basis, the winds are strongest when the temperature differential between the interior of the cave and the surface are the greatest. Thus, one can expect the strongest draft at night when the outside temperature drops. For cavers spending several days underground, the draft is strong enough to penetrate uncovered sleeping bags. When not sleeping, cavers must move constantly in order to keep warm.

Humidity levels in the cave vary. The extreme dryness of the entrance series in the winter is caused by the cold dry surface air blowing in. Further in, the cave temperature rises until just past the junction with Boon's Blunder where there are three successive ponded sections of passage through which one is forced to wade waist deep. Further into the cave, water is encountered where inlets occur in the passage ceiling, notably in the Waterfall Room. In the summer it's a whole different ball game, and Peter Thompson's descriptions of long wading passages, large waterfalls and being trapped by floods make exciting reading. Cavers exiting the cave in the early spring, during warm conditions, get a sample of summer conditions. The Ice Crawls start melting, a trickle of water starts emerging from Boon's Blunder and a low area near the junction, known as The Duck, becomes a small lake with lumps of ice floating around. Minimal air space and a large rucksack to maneuver can make this an interesting ending to a week spent underground.

Frost Pot

Map 83 C/3
Jurisdiction Banff National Park
Entrance elevation 2500 m
Number of entrances 1
Depth 35 m shaft
Discovered 1968, by Mike Boon

Location Frost Pot is located in a relatively flat area 2.5 km southeast of the summit of Castleguard Mountain and about 2 km northwest of Castleguard Cave. Look for a cairn. The entrance is only exposed in the summer when the benches are free of snow.

Description Frost Pot is a sink-point situated on the benches below Castleguard Mountain in the Eldon-Pika formation. Surrounded by a sea of felsenmeer, the shaft in summer receives a small stream of water from nearby snowmelt. It has been shown through dye tracing to connect with the Big Springs 4 km away and 750 m lower down, the flow-through time being only three and a half hours.

Warning Make sure you can use SRT in a waterfall!

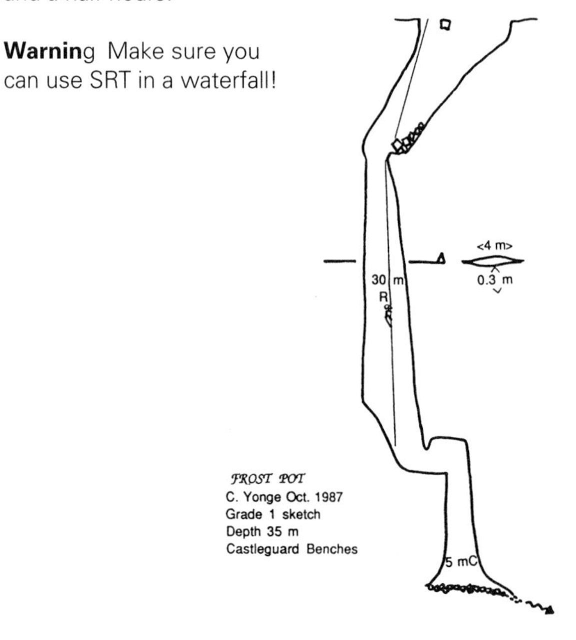

Dippie's Loop (Dippy's Loop)

Map 83 C/3
Jurisdiction Banff National Park
Entrance elevation Uncertain
Number of entrances 1
Length 270 m
Discovered 1968, by Peter Thompson

Location Uncertain and probably best left in the fading memories of the McMaster cavers!

Exploration & Description During the fruitless search for a back door to Castleguard Cave, the McMaster cavers scoured high-level benches to the south of Castleguard Mountain.

Charlie Brown and Peter Thompson pushed the most promising cave for 270 m of flat-out crawling until it looped back to daylight, seen through an impenetrable slot.

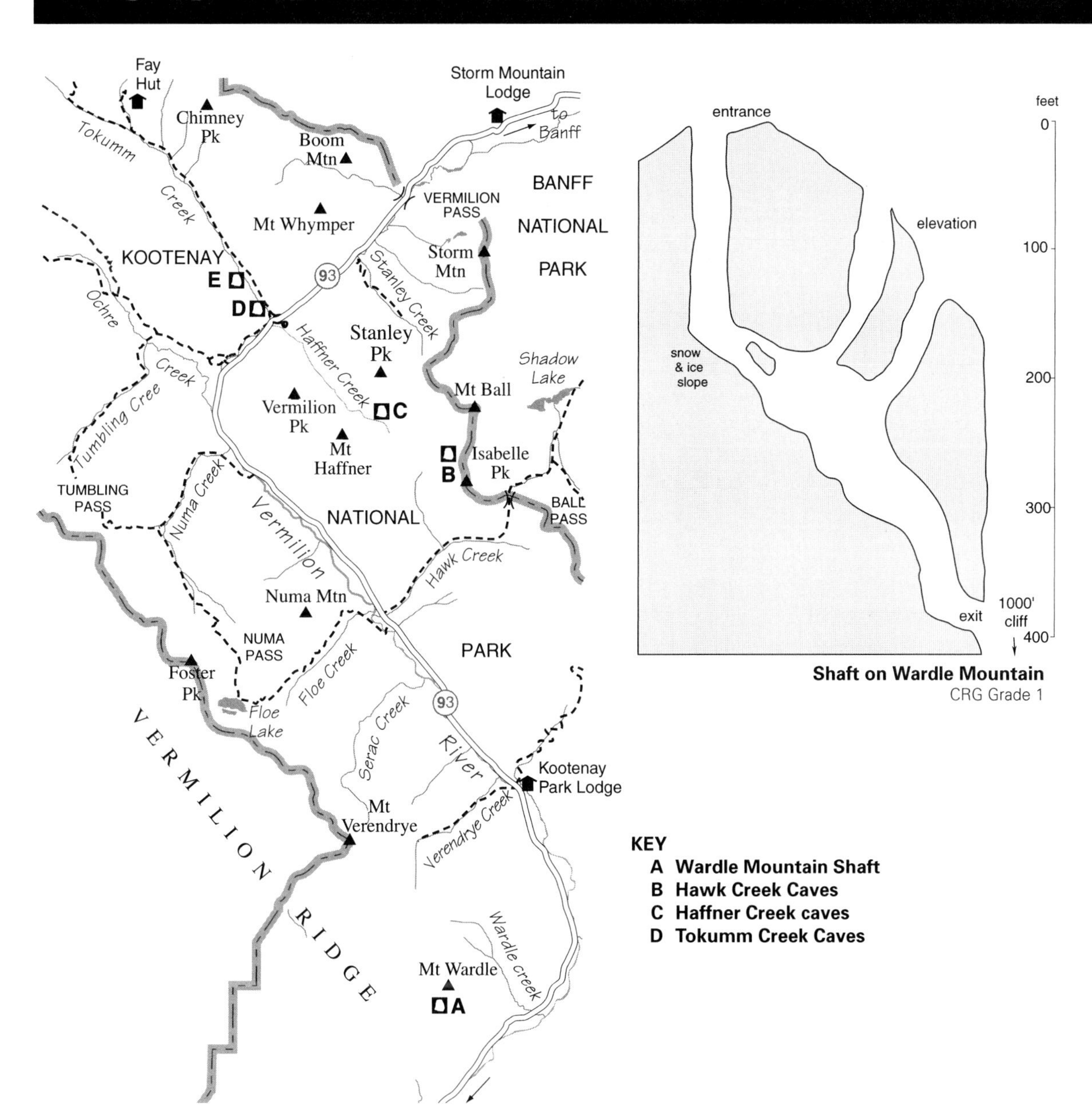

Shaft on Wardle Mountain
CRG Grade 1

KEY
A **Wardle Mountain Shaft**
B **Hawk Creek Caves**
C **Haffner Creek caves**
D **Tokumm Creek Caves**

Wardle Mountain Shaft

Map 82 K/16
Jurisdiction Kootenay National Park
Entrance elevation 2790 m
Number of Entrances 4
Depth c. 122 m shaft
Discovered 1977, by Irv Graham

Access The cliff-face entrance can be seen with binoculars from the Mt. Wardle picnic area on Hwy. 93 (Banff-Radium Hwy.). Start from the Mt. Wardle signpost. Climb Mt. Wardle (2850 m), following the obvious line of the central rock rib to the right of the first rock face, then grass slopes and talus to the first summit. From here traverse first summit cirque and scree and a short distance of exposed ridge (rope required) to the summit.

The entrance is located only 60 m below the summit in a cliff face, and provides spectacular views over the Vermilion River valley.

Description Descend the 6 x 9 m shaft for 60 m to a snow slope. From here the cave continues down a series of steps for another 60 m, the passage width varying between 1 and 6 m, before reaching a lower entrance in the cliff face. Two other entrances bring daylight into the main shaft just below the snow slope.

"Mount Wardle is the only area in the Four Mountain Parks where mountain goats winter at montane elevations." CPS, 1988.

Hawk Creek Caves

Access for all Caves Hwy. 93 (Banff-Radium Hwy.) at the Floe Lake/Hawk Creek parking lot on the west side of the highway, 22½ km south of Vermilion Pass.

Hike the Hawk Creek trail (starting from the east side of the highway) for 4.2 kilometers to a bridge. Cross the bridge and then follow a series of indistinct animal trails that wind northwest up and across the steeply wooded base of Mt. Isabelle. After traversing several avalanche slopes, the rough trail emerges into the larch-covered lower slopes of a hanging valley. Above the larch, wide expanses of karst pavement angle gently (6°) up to a cirque and terminal moraine below a glacier.

The valley is divided in two by a small glacially-shattered ridge. The northern half contains Canyon Remnant Cave, Silver Sands Sink and Resurgence. The southern half of the valley is riddled with shafts (over 30 have been checked), and contains Larch Valley Cave, Double Pot and Provenance Cave.

Allow 3 to 4 hours.

Area Exploration Ken Duckworth, a geologist with the University of Calgary, first hiked over the Hawk Creek karst after some field work at a lead-zinc showing lower in the valley. He was struck by its resemblance to karst areas in his native Yorkshire, England. He passed this information on to Jon Jones, who in turn told Chas Yonge. In February of 1987, Tom Barton and Chas Yonge of the ASS skied up the valley and checked out a spring (Hawk Creek Spring). In September, Chas, Randy Spahl and Jon Rollins hiked up to the karst above.

Area Geomorpholgy In the southern half of the valley, melt water from the glacier runs down over the terminal moraine, which "armours" (renders impermeable) the karst. On contact with the limestone, the water quickly sinks down two potholes that are too small to enter. The drainage in the northern half of the valley is complicated by bands of sandstone and shale. A large sink point at the toe of the glacier is plugged with moraine, and the small sinks lower down are quickly inundated, causing surface streams to flow all the way down the valley during summer afternoons, forming chains of melt water pools. Silver Sands Sink is fed by a stream flowing down cliffs on the north side of the valley.

Hawk Creek provides one of the best examples of glaciated limestone pavements in the Canadian Rockies, and would be an excellent area for further field studies. Large slabs of limestone, devoid of vegetation, have been swept free of weathered debris by recent glaciation and have been divided by solutionally-opened joints or grikes. Solution pits and channels known collectively as karren are present, both rinnen karren (runnels) and the larger kluftkarren (clefts). A few massive sandstone erratics, which were rafted in on the ice and then dumped as it retreated, lie scattered on the pavement. Further down the valley the karst is covered by a thin mantle of soil with encroaching vegetation, solifluction lobes having formed where the gradient steepens towards the valley rim.

Head of Hawk Creek valley. Photo Jon Rollins.

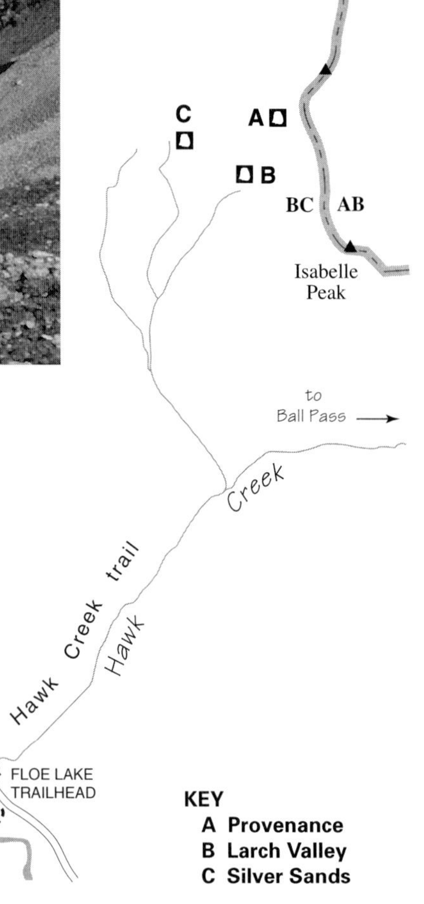

Provenance Cave

Map 82 N/1
Jurisdiction Kootenay National Park
Entrance elevation 2460 m
Number of entrances 1
Length, Depth 256 m 102 m
Discovered 1987, by Chas Yonge, Dave Thomson & Jon Rollins

Location Towards the head of the southern valley, streams emerge briefly from talus on the north-facing slopes of Isabelle Peak before sinking into the karst pavement. The largest cave opening, an east west aligned slot marked by a cairn, is Provenance Cave.

Description Provenance Cave is an exciting series of interconnected shafts, active during snowmelt. Most of the water enters through an extension of the entrance fissure on the bench above the cave and flows down the second 15 m pitch before disappearing through juvenile fissures in the floor.

The first drop in Provenance Cave, Tracery Shaft, is one of the most beautiful pitches in the Canadian Rockies. The initial steeply-angled entrance chute descends for 10 m, before windowing onto an impressive shaft that provides a 30 m free hang. Like the nave of a Gothic cathedral, fluted buttresses of rock arch up to the ceiling and are decorated with frost crystals that sparkle as you descend. Towards the bottom, a snow slope leads off down to the head of another 15 m pitch which takes a roaring waterfall in the summer. Some re-dissolved formations, like yellow shark teeth, decorate a section of passage that continues across the head of the second pitch, shown on the survey to connect with the Attapulgite Room. However, this connection has yet to be established.

From below the second pitch, a rift passage split by a false conglomerate floor leads to the head of a further 25 m pitch. Sediments have obviously been flushed out leaving a surprisingly solid layer of calcite-cemented pebbles bridging the passage. The 25 m pitch bottoms in a tight juvenile stream passage.

A traverse around the head of the pitch leads to the Attapulgite Room and Gay Abandon Passage — the only horizontal passage of any length — which disappointingly ends in a drafting choke. The plug of glacial till is reminiscent of that found at the bottom of Double Pot Cave. In fact, the two caves may be very close at this point. The only remaining lead appears to lie at the bottom of a 25 m pitch off Gay Abandon Passage, where another drop can be seen through a tight slot.

Speleogenesis The Attapulgite Room was so named after curious tissue-paper-like fronds hanging from fractures in the walls. X-ray analysis at the University of Calgary identified the mineral as attapulgite in the palygorskit group, an aluminum silicate with an asbestiform habit. Sometimes called Mountain Leather, the mineral is described in *Cave Minerals Of The World* as "White to dirty gray in color resembling crumpled dirty rags or soggy chewed up cardboard. The grimy cloth-like appearance creates interest amongst the miners, one of whom told me that he had made a pair of underwear out of the stuff!" It is associated with either hydrothermal activity or with montmorillinite clay exposed to dry atmospheres. The showings in the Hawk Creek area and some mineralization seen in joints on the karst pavement may explain the presence of the mineral underground. Provenance Cave represents the first known sighting of attapulgite in a Canadian cave.

Provenance Cave. Photo Chas Yonge.

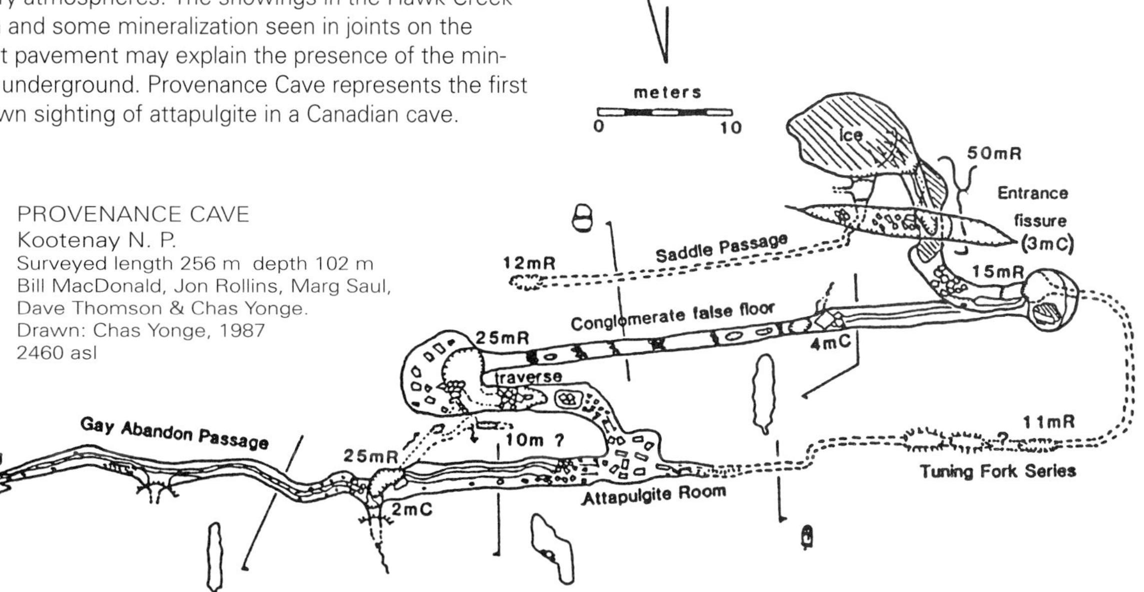

Double Pot

Map 82 N/1
Jurisdiction Kootenay National Park
Entrance elevation 2460 m
Number of entrances 2
Depth 64 m shaft
Discovered 1987, by Chas Yonge, Dave Thomson & Jon Rollins

Location Close to and slightly northwest of Provenance Cave.

Exploration & Description Double Pot was descended to –64 m before pinching out. The shaft lies close to the end of Gay Abandon Passage in Provenance Cave, and may be responsible for breakdown blocking that passage.

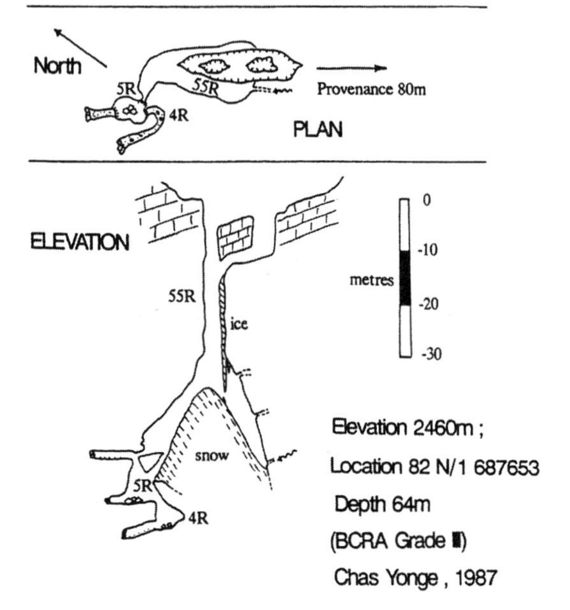

Bill MacDonald exiting Double Pot. Photo Dave Thomson.

Larch Valley Cave

Map 82 N/1
Jurisdiction Kootenay National Park
Entrance elevation 2440 m
Number of entrances 1
Length, Depth 611 m, 39 m
Discovered 1989, by Chas Yonge, Dave Thomson & Jon Rollins

Location One of the first obvious slot entrances you arrive at in the southern valley. If it has a natural thread belay at its lip it is Larch Valley Cave.

Exploration & Description "It was late in the day... and our thoughts turned to the hike out and the possibility of making the Rose and Crown in Banff before last orders. It had been a pleasant but somewhat frustrating weekend; we had descended numerous shafts, all either plugged with ice or blocked by glacial till at around 60 m depth.

"This superb karst area perched on a shoulder of Mt. Ball had so far yielded two beautiful and contrasting caves, Silver Sands and Provenance, but we had yet to find the major cave system which the surface topography so tantalizingly promised. With two melt water streams sinking from a glacier snout at the head of the valley and a good resurgence a thousand feet below and a kilometre away, there just had to be something big down there. It was just a question of checking the 30 or so shafts that lay in between. One of them had to punch through to the main subterranean drainage.

"We ambled over to what we promised ourselves would be the last drop of the day, a steeply descending fissure heading off into the blackness. A convenient natural thread at the lip made rigging the rope easy and all that was left was to argue over who got the pleasure of checking it out. I eventually conceded that it was my turn and with considerable lack of enthusiasm rappelled down into the cold miserable slot. After 20 m the walls belled out leaving me hanging free, slowly spinning in space, my light catching water droplets on the walls and turning them into sparkling beads. Another 10 m and I landed on the tip of an ice plug. A familiar occurrence this; most of the shafts we had dropped at Hawk Creek ended this way. A scan around the chamber revealed the only way on, a small triangular hole mostly blocked with verglassed rocks" (Jon Rollins, *Canadian Caver*, 1989).

Larch Valley Cave proved to be the longest cave discovered in the Hawk Creek karst, but at 611 m it is no mega-system. It is, however, a very interesting cave, and testament to the notion that caves should be appreciated for their intrinsic beauty and character as well as their length and depth.

The main passage runs due north along the axis of the valley toward the glacier and ends in an aven. A higher phreatic passage accessed in the ceiling swings around to the east, heading towards a sink on the benches. A similar sink on the eastern benches feeds the Oscillator, an impressive vadose canyon that joins the main passage at Window Junction. Water in the cave sinks into juvenile fissures that underlie the main cave but are too small to enter.

Jon Rollins in Larch Valley Cave.
Photo Dave Thomson.

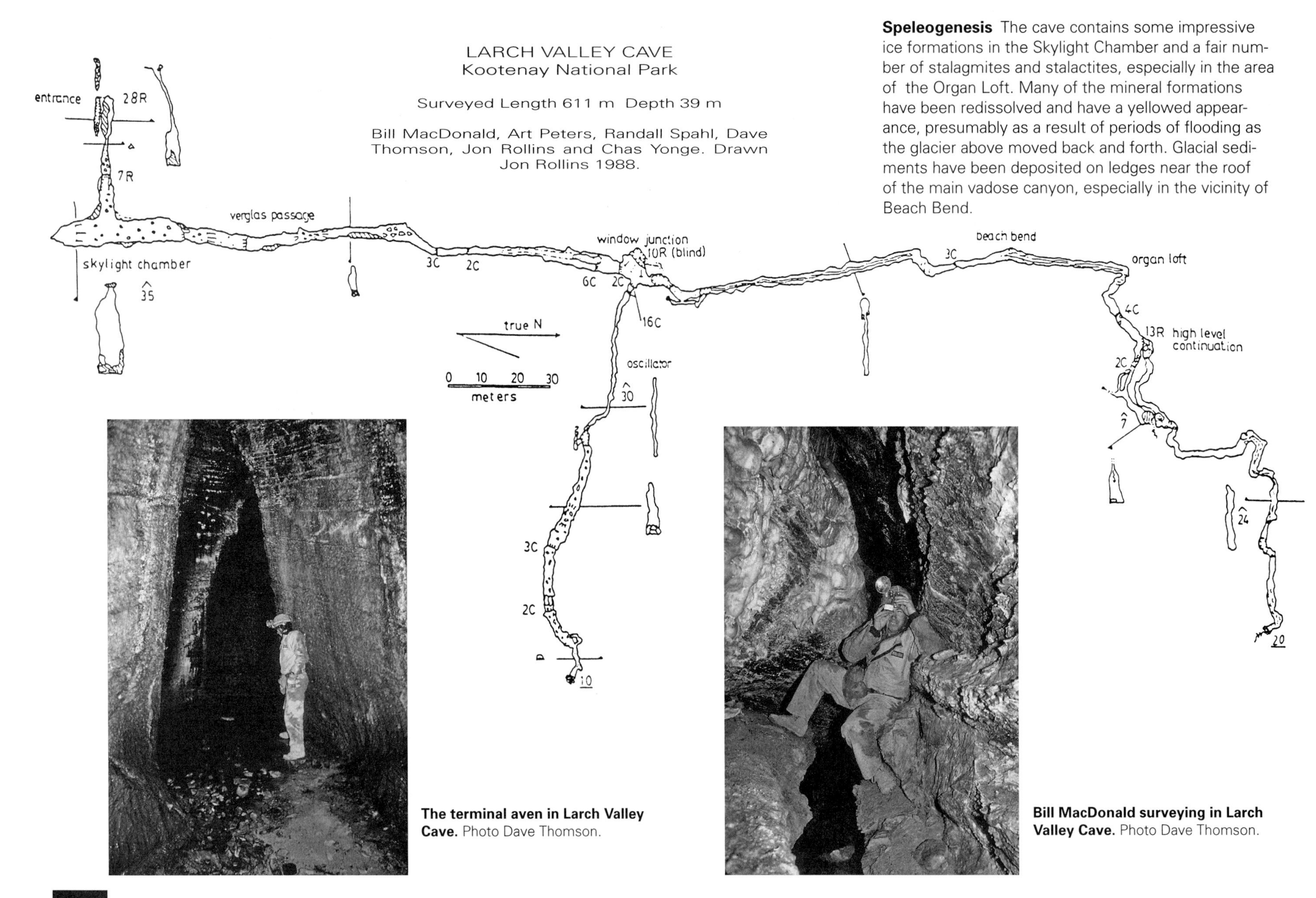

Speleogenesis The cave contains some impressive ice formations in the Skylight Chamber and a fair number of stalagmites and stalactites, especially in the area of the Organ Loft. Many of the mineral formations have been redissolved and have a yellowed appearance, presumably as a result of periods of flooding as the glacier above moved back and forth. Glacial sediments have been deposited on ledges near the roof of the main vadose canyon, especially in the vicinity of Beach Bend.

The terminal aven in Larch Valley Cave. Photo Dave Thomson.

Bill MacDonald surveying in Larch Valley Cave. Photo Dave Thomson.

Silver Sands Sink

Map 82 N/1
Jurisdiction Kootenay National Park
Entrance elevation 2400 m
Number of entrances 2
Length, Depth 254 m, 14 m
Discovered August, 1987, by Jon Rollins & Mary Day

Location A waterfall pouring into a narrow slot to the east of Silver Sands resurgence in the northern valley.

Exploration & Description Towards the end of a summer day, numerous streams flow from the glacier at the head of the valley, spilling over the moraine, and in a series of waterfalls pass through the sandstone and shale bands before flowing across and sinking in the limestone pavements below. One such waterfall marks the entrance of Silver Sands Sink. Not obvious from below, the narrow slot in which the water sinks appears to be barely higher than the rising, Silver Sands Resurgence, which it reaches via 300 m of delightful winding stream passage. The connection was made in 1993, the low wet passage emerging behind a scree cone just inside the entrance alcove of the rising.

In the late afternoon a trip into this cave can be tremendously exciting: the canyon floor contains a swiftly-flowing stream and the sound and spray of water fills the air. A little way in, a small tributary enters from the right wall, carrying a jet of water. Luckily for cavers the canyon walls have razor sharp brittle projections (cherty flags) that enable you to stay out of the water. At head height the sides of the passage form benches bearing glacial sediments, and small tubes lead off, by-passing the meanders of the main passage. Occasional straws, some with protrusions (flags) indicate the direction of air movement into the cave. The passage gets smaller, eventually becoming a flat-out crawl in water. The through trip can only be made when the water is low.

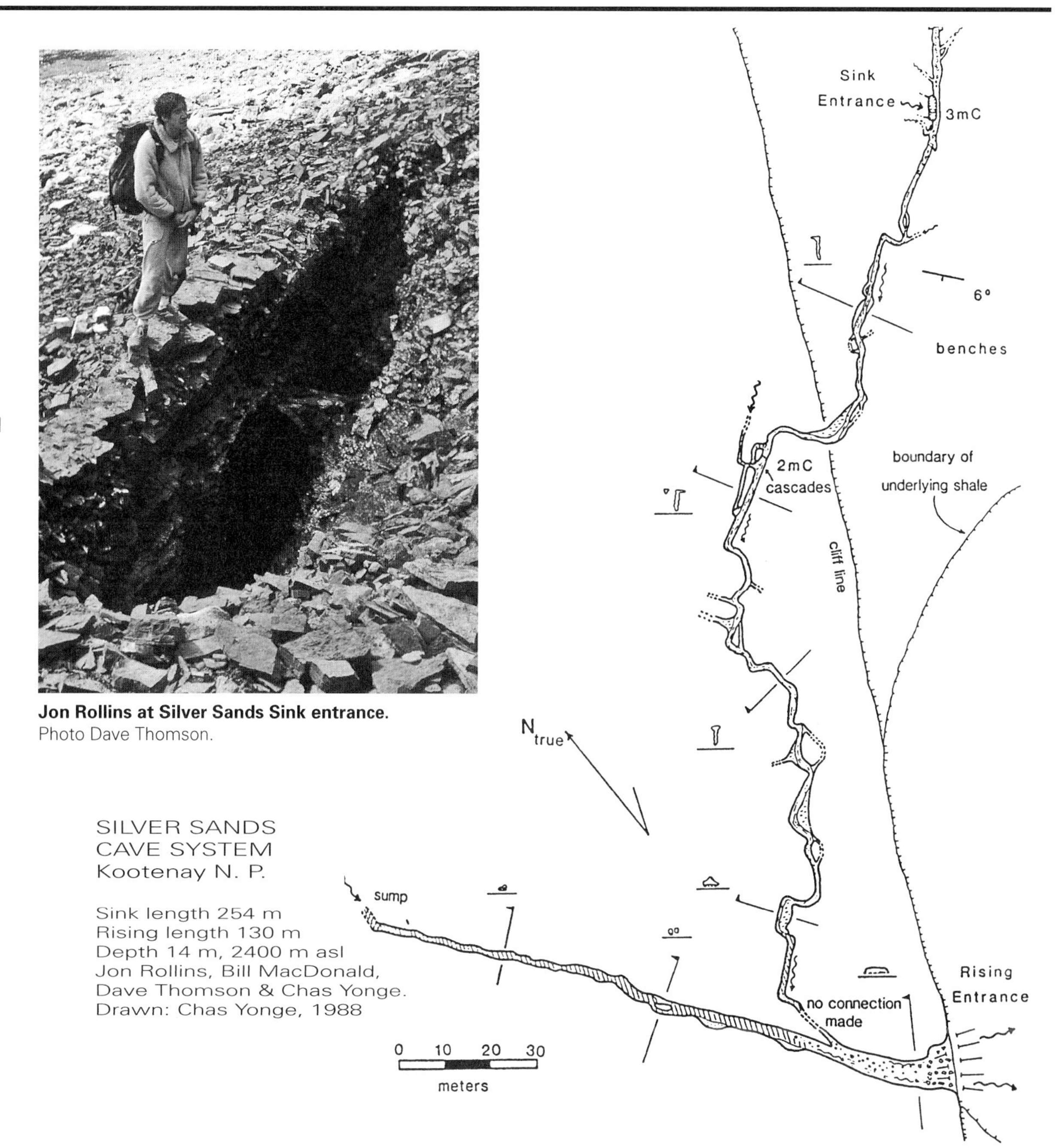

Jon Rollins at Silver Sands Sink entrance.
Photo Dave Thomson.

Silver Sands Resurgence

Map 82 N/1
Jurisdiction Kootenay National Park
Entrance elevation 2400 m
Number of entrances 1
Length 130 m
Discovered 1988, by Jon Rollins & Chas Yonge

Location This rising, partially hidden behind a large talus slope, has the appearance of a classic frost pocket when viewed from across the southern valley. It flows year round, a well-defined path contouring along the foot of the cliff face to the entrance attesting to its importance as a watering hole for sheep and goats.

Exploration & Description Although Silver Sands Sink contributes some of the water, it is probable the majority is coming from beneath the glaciers some distance away on Mt. Ball. There are no signs of sinks or other caves in the valley to the north, or higher in the same ridge, except for a likely looking hole above the cave entrance that is a little too high to climb into and remains to be checked. In the summer water appears briefly below the entrance before sinking again into the rubble. It seems probable that this is the same water that can be heard through a fissure in the side of a fossil sink point lower down the valley.

The cave entrance is a pleasant spot on a summer's evening, the running water reflecting the sun and dappling the cave ceiling with golden light. A short way in, a small passage floored with cobbles leads off from the right wall to a connection (usually flooded) with Silver Sands Sink. The main cave, formed on a shale band, quickly narrows from its wide bedding plane entrance into a narrow water-filled rift which sumps under the right wall after about 130 m. Although the water is very cold, the sandy floor feels quite pleasant in bare feet after a hard day's hiking. Wet suits were used to explore to the sump as there is a brief duck a short way in. It is possible the sump is only short and that a dive will reveal more passage beyond.

Canyon Remnant

Map 82 N/1
Jurisdiction Kootenay National Park
Entrance elevation 2400 m
Number of entrances 1
Length 35 m
Discovered 1987, by Jon Rollins & Chas Yonge

Location About 150 m southwest of Silver Sands Resurgence, in the same cliff face is a small unlikely looking cave-entrance in fissile (plate-like) limestone.

Description A short crawl round two tight bends drops you into a large canyon passage that ends shortly at a packrat midden. A small drafting gap in the left wall leads back to the surface, and high on the right, a tight tube heads back into the ridge.

The large size of this section of passage indicates that it was once part of a larger cave system removed by glacial widening of the valley. A narrow fissure in the roof of the canyon passage contains some very pleasing cave coral.

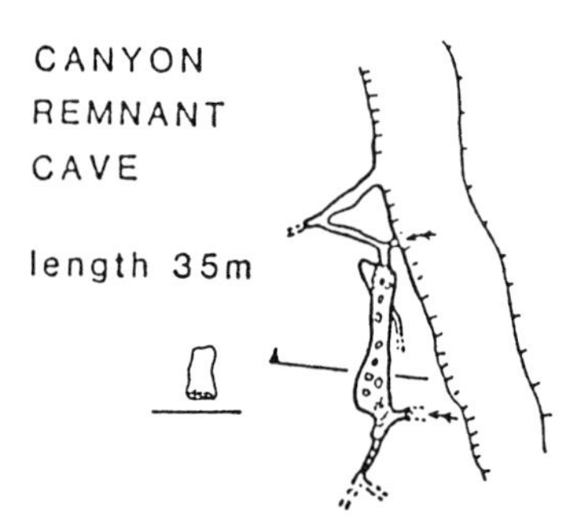

Top: Silver Sands Resurgence. Photo Dave Thomson.

Bottom: Lesley DeMarsh in Canyon Remnant entrance. Photo Jon Rollins.

Hawk Creek Spring

Map 82 N/1
Jurisdiction Kootenay National Park
Entrance elevation Uncertain
Number of entrances 1
Length, Depth 15 m to a sump
Discovered February 1987, by Tom Barton & Chas Yonge

Location About 260 m directly below the lip of the hanging karst valley, a small stream trickles from the base of a short wall at the head of a rocky gully and runs off down a series of mossy steps.

Exploration & Description On the first exploration, Yonge stripped from the waist down and crawled into the cave. He followed a low wet passage for 10 m, beyond which he could see 5 m to a sump.

Haffner Creek Caves

Access to all caves Hwy. 93 (Banff-Radium Hwy.). Park at the Marble Canyon parking lot 7 km west of Vermilion Pass.

From the warden station on the east side of the highway opposite the parking area, follow a rough trail up the north side of Haffner Creek, crossing several avalanche slopes. Close to the end of the valley, climb north up steep slopes to the base of the first rock band where the caves are located.

Allow 4 hours.

Area Exploration Haffner Creek was first recognized as having good potential for caves by Dave Chase. Descending the valley after an attempt on Mt. Ball he noticed a massive resurgence issuing from a cliff face on the east side of the valley. Jon Rollins noticed this same resurgence several years later from a pass adjacent to Hawk Creek.

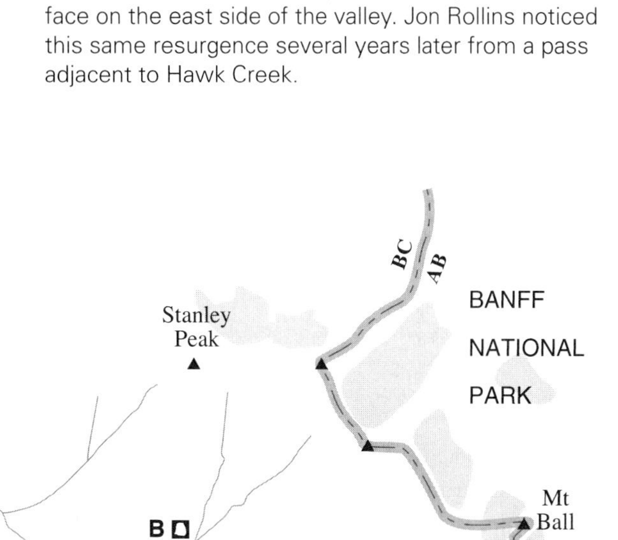

Dave Chase and Dave Thomson made the first caving trip into the valley in February of 1989. They intended to ice climb to the resurgence, but on reaching the cliff they found it bare. They then attempted to rock climb to the resurgence, but were stopped five metres short, needing a bolt. They did, however, find some short caves in the canyon close to the highway.

The next attempt to reach the resurgence was made by Dave Chase and several members of the Edmonton section of the Alpine Club of Canada. They were unable to locate it, finding instead a small resurgence cave they called Tin Can. Dave Chase crawled into it, getting quite wet before building a small cairn and turning back. After this they followed the base of the cliff-band and discovered the impressive entrance of Her Majesty's Cave, and then another small cave, Heinz 57 (no information). They then made their way to the top of the cliffs and found some flat areas of good karst.

During a third trip, Dave Chase and Dave Thomson explored Her Majesty's Cave to a boulder choke, which they pushed to access an aven.

Ironically, no one has yet made it to the large resurgence which originally drew attention to this area. It is hard to locate from below when it is not spouting water.

Her Majesty's Cave

Map 82 N/1
Jurisdiction Kootenay National Park
Entrance elevation 2135 m
Number of entrances 1
Length 240 m
Discovered 1989 by Dave Chase (ASS) & members of the Alpine Club of Canada Edmonton section

Location This cave lies high on the north side of Haffner Creek below a large cliff band. In winter a snow slope partially hides the bedding plane entrance.

Exploration & Description In winter the large entrance chamber is decorated with impressive floor to ceiling ice columns. The chamber ends abruptly 40 m in, but a walking passage leads off from the right wall, eventually becoming a narrow canyon with a small stream in the bottom. Breakdown material creates some easy scrambling and there is a tight squeeze past a large ice boss. Shortly after this, another squeeze through some large boulders leads into a high aven with water entering from the ceiling. The initial passage appears to continue under the aven but is plugged with breakdown material. The back wall of the aven was climbed for 20 m but no obvious continuation could be found. The only way on appears to be a tight phreatic tube in the floor of the aven that has a strong draft blowing out of it. Despite several attempts this has so far proved too tight.

Because the cave heads directly beneath Mt. Ball, there is a good possibility of more passage and a higher entrance to this cave. Apart from the splendid ice formations in the entrance chamber, the cave contains some pretty cave coral and short stalactites. A number of very lively packrats live close to the entrance, and several large middens attest to their longtime occupation of this cave.

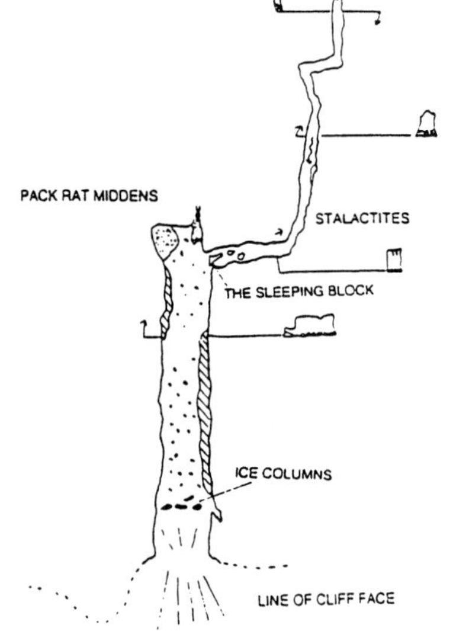

Her Majesty's Cave entrance.
Photo Dave Thomson.

Ice formations in Her Majesty's Cave.
Photo Jon Rollins.

Tin Can Cave

Map 82 N/1
Jurisdiction Kootenay National Park
Entrance elevation 2135 m
Number of entrances 1
Length 60 m
Discovered Unknown

Location Tin Can Cave is located approximately 1 km west of Her Majesty's Cave, along the same cliff-band. The entrance is an idyllic spot: a little pool fed by a small stepped waterfall surrounded by green moss.

Description A low passage heads around a couple of bends before the cave sumps. The cave was named after a tin can found lying in the entrance.

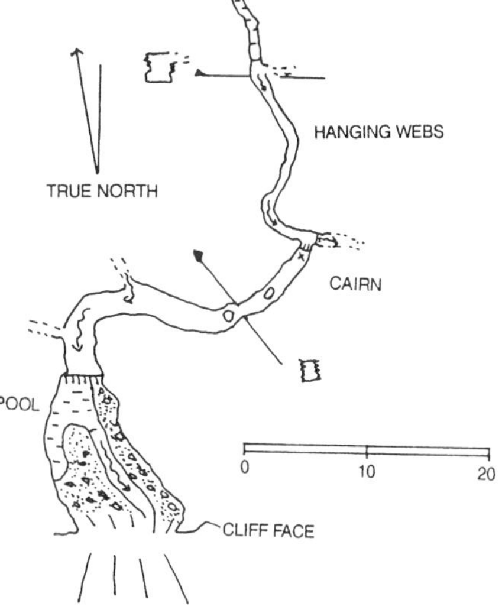

Dave Chase in the entrance to Tin Can Cave.
Photo Dave Thomson.

Tokumm Creek Caves

Access for all Caves Hwy. 93 (Banff-Radium Hwy.) at Marble Canyon parking lot, 7 km west of Vermilion Pass.

Area Exploration Aside from the two named caves low down in the valley, there are possible caves in areas to the north of Tokumm Creek below Mts. Little and Quadra. Located just below the moraine at the glacial margins is some karstic grey limestone sandwiched between less karstic crystalline beds. When hiking down from the Colgan Hut to the Fay Hut, Jon Rollins noticed that the first of these areas had some limestone pavements, which lower down was dotted with larches and very reminiscent of the Hawk Creek area. A poke around in some sinks just below the moraine indicated some promising holes.

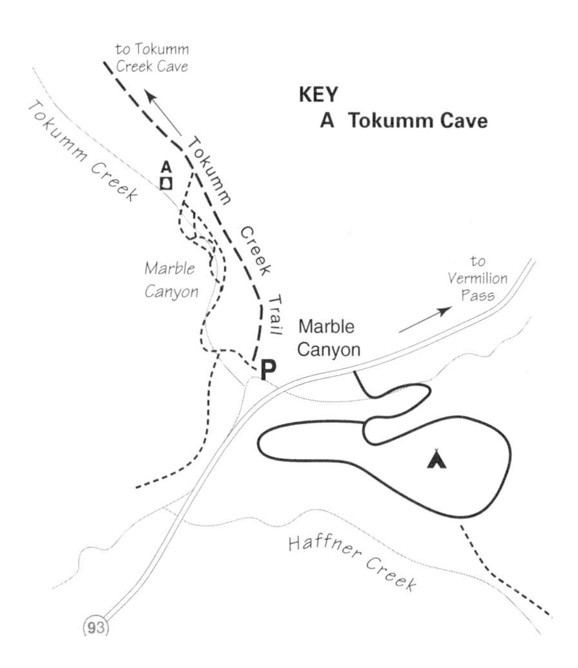

Tokumm Creek Cave

Map 82 N/1
Jurisdiction Kootenay National Park
Entrance elevation 1799 m
Number of entrances 1
Length 60 m
Discovered Unknown

Location Walk approximately 8 km up the Tokumm Creek trail. A 12 m x 12 m entrance can be seen 120 m above the creek at the top of the scree on the south-western wall of the canyon.

Description The entrance leads to 60 m of spacious 9 m x 5 m passage terminated by a bedrock wall.

Tokkum Cave borehole. Photo Jon Rollins.

Tokumm Cave

Map 82 N/1
Jurisdiction Kootenay National Park
Entrance elevation c. 1480 m
Number of entrances 1
Length 436 m
Discovered March 1997, by Dave Thomson

Location Walk up Marble Canyon Trail. A 30 m rappel into the head of the canyon is followed by a short wade or swim to the cave entrance which is located on the north side (left as you face downstream) just across a pool. The cave can only be reached in the winter when water levels are low.

Exploration & Description Every day Dave Thomson was ice climbing in Marble Canyon he noticed water gushing from a hole just above the canyon floor. Wading a pool, he gained the entrance and explored the main rift passage for approximately 100 m to where it ended at an impenetrable bedding plane. Next, he made a short foray into a side passage before establishing that further exploration would require the help of the ASS.

The following weekend a team of six made it out to snowy Kootenay Park. "Exhausted after a Christmas of overindulgence and inactivity, there was much complaining during the ten minute approach hike, but the beauty of the ice-laced canyon revealed during a 40 m rappel made it all worthwhile. A short waist-deep wade through a plunge pool and we were in. The first part of the cave is an elegant, sinuous keyhole passage, high enough to bridge along comfortably. A small stream rushes along the floor and some nice flowstone and soda straws decorate the wall and ceiling at a bend. Unfortunately, it ends all too soon at a waterfall emerging from a bedding plane slot in the roof.

"Next, we checked the two tubes that lead off from the main passage at ceiling height. The Worm Hole was fairly tight and shut down after 30 m, but the Warm Hole, although only slightly larger, went on and on and on. At a few points the passage is high enough to allow crawling on all fours, but for the most part squirming is the method of progress. For the first section the sound of running water provides a pleasing background rumbling sound. Whether this comes from an adjacent stream passage, or from a sound connection with Marble Canyon itself, is unclear. After turning abruptly to the left (north) the passage becomes silent and switchbacks gently up and down, eventually ending in a small sump pool which can't be far short of the parking lot.

"A large resurgence cave, easily visible from Highway 93 heading north as you approach Tokumm Creek, is located in the mountainside approximately 300 m above and to the north of Marble Canyon. Shortly after emerging from this resurgence the water sinks again in a scree slope. Because this occurs pretty much above the cave, it is possible that this is the source of the cave water. A dye trace might be a worthwhile project." (Jon Rollins, Canadian Caver 1998).

Warning Falling ice during the rappel, and collapsing ice over water are very real dangers when approaching this cave. Make sure you know how to use your SRT gear otherwise you will be faced with a long cold swim down the canyon after your caving trip! A wet suit (with gloves) is strongly recommended.

Geology The survey indicates that Marble Canyon, a postglacial feature, has captured the cave and caused down cutting, thus creating the rift floor in the keyhole passage. The wavy black sheets noticed in the keyhole passage are probably not, as first thought, chert, but are more likely dolomite mottling typical of Dolomitic limestone. The cave is formed on an obvious bedding plane weakness, on which small seeps can be seen both upstream and downstream of the cave entrance. It is possible that inter-bedded shale plays a part in the cave development, though none was noticed during exploration. The situation of Tokumm Cave bears a marked similarity to The Rose & Cavern in Banff National Park. This certainly suggests that the floors of other Rockies canyons might be worth checking in winter for resurgence caves.

The 30 m rappel into Marble Canyon. Photo Jon Rollins.

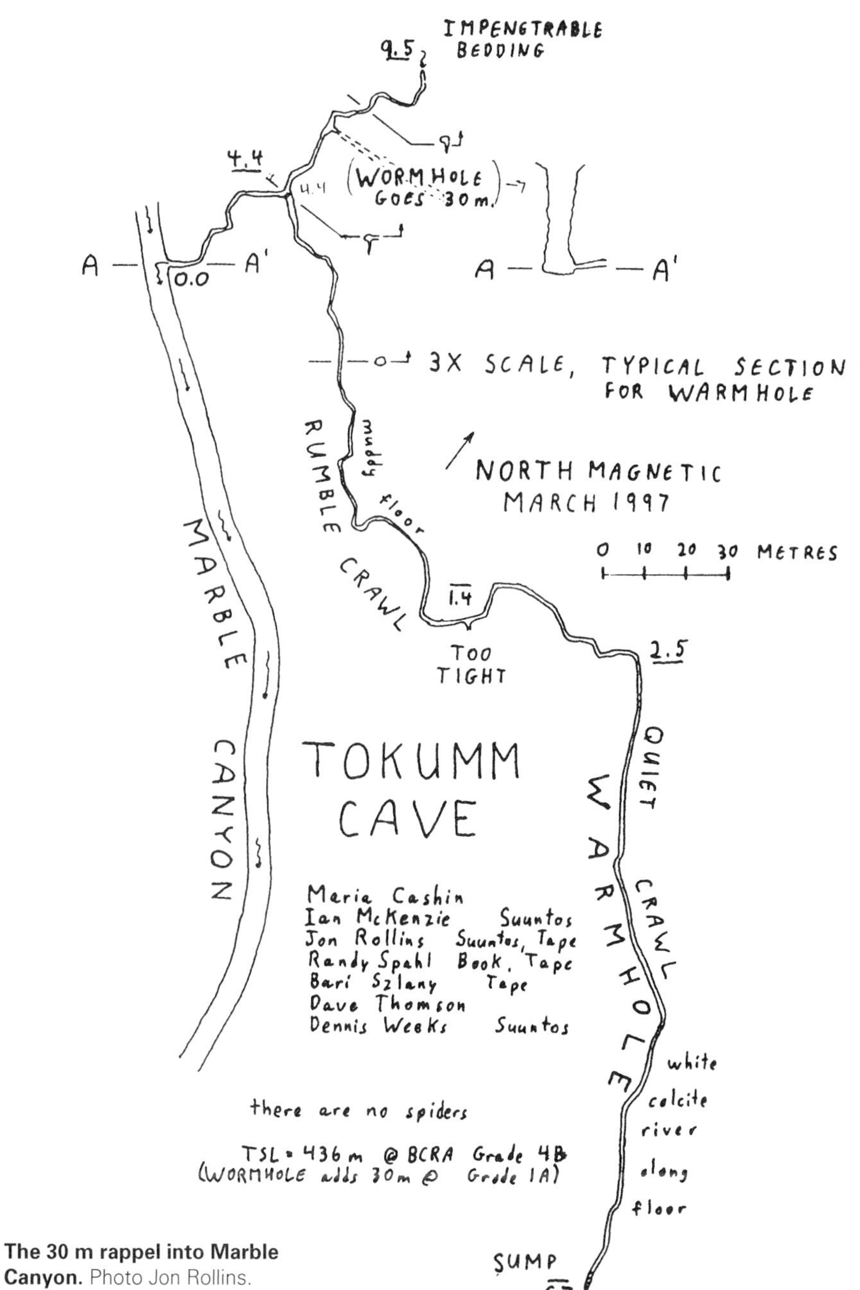

YOHO NATIONAL PARK

Only a few short caves have been found in Yoho National Park.

The numerous holes evident in the cliffs north of Field are mines excavated in the search for zinc, lead and silver deposits, the last closing down in the 1950s. They contain many features of historical interest, including extensive railway systems, cases of abandoned drill bits, shovels, picks and wheelbarrows. Wooden artifacts such as ladders and walkways have survived remarkably well underground, protected from the rigours of surface weathering. Extensive calcite formations have formed in some sections of passage, including stalagmites, stalactites and cave pearls (*Canadian Caver*, vol.15, no. 2). Do not attempt to enter these mines. Unlike caves they are prone to frequent collapse and contain many shafts.

"Will I ever forget our last evening here? Sitting with you at the Crystal Caves and the sunset. It is wonderful now to be able to associate things around here with you, and I think I'll go up to the caves again." Jon Whyte, Letter in the Whyte Museum of the Canadian Rockies, 1980.

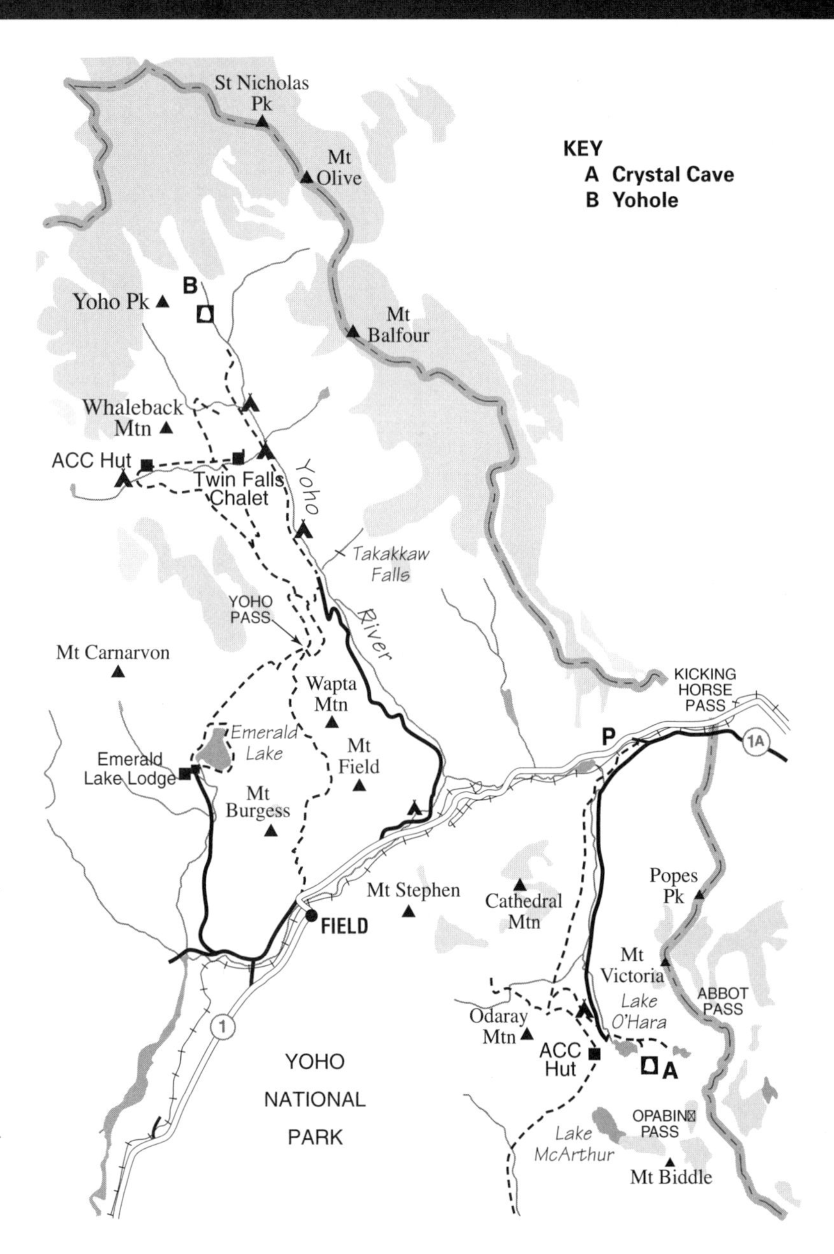

Crystal Cave

Map 82 N/8
Jurisdiction Yoho National Park
Entrance elevation 2074 m
Number of entrances 1 (bricked up)
Length c. 10 m
Discovered Unknown

Access From Lake Louise townsite, drive towards Lake Louise on Hwy. 1A. At the junction with the access road to the lake keep right on Hwy. 1A. Shortly before the highway crosses railway tracks to Hwy. 1, keep straight into the Lake O'Hara parking lot. Alternately, from Hwy. 1, 2.7 km west of Kicking Horse Pass, turn south onto Hwy. 1A. After crossing the railroad tracks, turn right into the Lake O' Hara parking lot.

Walk up the fire road to Lake O' Hara Lodge (11.2 km). Alternately, take the bus from the parking lot. The bus runs from late June to late September and requires prior booking with the Canadian Parks Service (604) 343-6433.

Facilities Campground, 0.8 km before Lake O' Hara Lodge. Elizabeth Parker Hut, run by the Alpine Club of Canada, is located in a meadow west of the Le Relais day-use shelter bus stop (0.7 km by trail).

Location The south shore of Lake O' Hara shoreline trail. Between East and West Opabin trails a side trail leads up to the cave. When viewed from beneath the quartzite cliffs on the east side of the lake, the eroded trail up to the cave stands out as an ugly white scar.

Description Not really a cave, this frost-shattered joint has attracted many visitors over the years owing to the fine quartz crystals that have formed in the rock weakness. A particularly fine example sits on a desk in the archives of the Whyte Museum in Banff.

Present parks policy discourages visits to the site: the trail is not signed or maintained, and most of the cave entrance has been bricked over. "These measures are the result of damage to the site in the past, and the illegal removal of the interesting but valueless quartz crystals. In 1979 and 1980 twenty people were charged for removing rock samples from the area. The present penalty is a fine of $150 plus confiscation of the rocks. Not much can be seen of the cave itself, although a small opening at the top of the brick wall permits anyone with a flashlight to view the cave wall. Below the cave you can find poor specimens of quartz among the loose rock" (Don Beers, *The Magic of Lake O' Hara*,1981).

Yohole

Map 82 N/10
Jurisdiction Yoho National Park
Entrance elevation 1982 m
Number of entrances 2 (entrances not connected)
Length Lower cave c. 122 m
Discovered 1975, by Tim Auger

Access From Hwy. 1 (Trans-Canada Hwy.) just east of Field follow the Yoho Valley Road to its end at Takakkaw Falls.

From the trailhead follow the Parks Canada "super highway" Yoho Valley trail for 9.2 km (bet you wish you could use your bike) to the outwash flats below the Yoho Glacier. As you finally emerge from the trees, the cave is briefly visible to the northwest below a rounded cliff with distinctive orange and white bedding.

Yohole. Photo Peter Thompson.

Continue to follow the less distinct but cairned trail as it traverses below a cliff and unstable gravel slope to some beautifully sculptured, recently glaciated bedrock slabs. Follow these upward past more cairns, then climb a ramp of white limestone with darker grey above to the bottom of the orange and white banded cliff. The entrance becomes visible as you round the corner.

Location The lower entrance can be reached by a tricky traverse (a rope is recommended), or by using a rope from above.

Reaching the upper entrance requires a rope and at least one piton.

Warning These caves are not enterable in the spring or early summer when large volumes of water pour out of them.

Exploration & Description Tim Auger first spotted the lower cave during a rescue on the Yoho Glacier. He returned later and explored about 30 m of very wet passage. A visit by ASS members was made early in the summer of 1976. "A ladder was dropped down to the entrance from above to avoid the tricky climb from below. Within sight of daylight was a nasty crawl in water that they thought might have been the limit of the first exploration. Beyond this the cave continued, quite tight and fairly intricate, with many short upward climbs and low chutes and crawls" (Mike Boon, 1976). Wearing wetsuits the cavers penetrated some 122 m into the cave, which eventually ended in an impossibly narrow bedding plane with water welling out.

By following the prominent band of limestone uphill 366 m from Yohole, Tim Auger discovered the upper cave that he explored to a wet rift. This cave has probably not been revisited.

Other caves and resurgences have been noted in the vicinity, including a 76 m-long inclined tube named Helicopter Hole (location unknown).

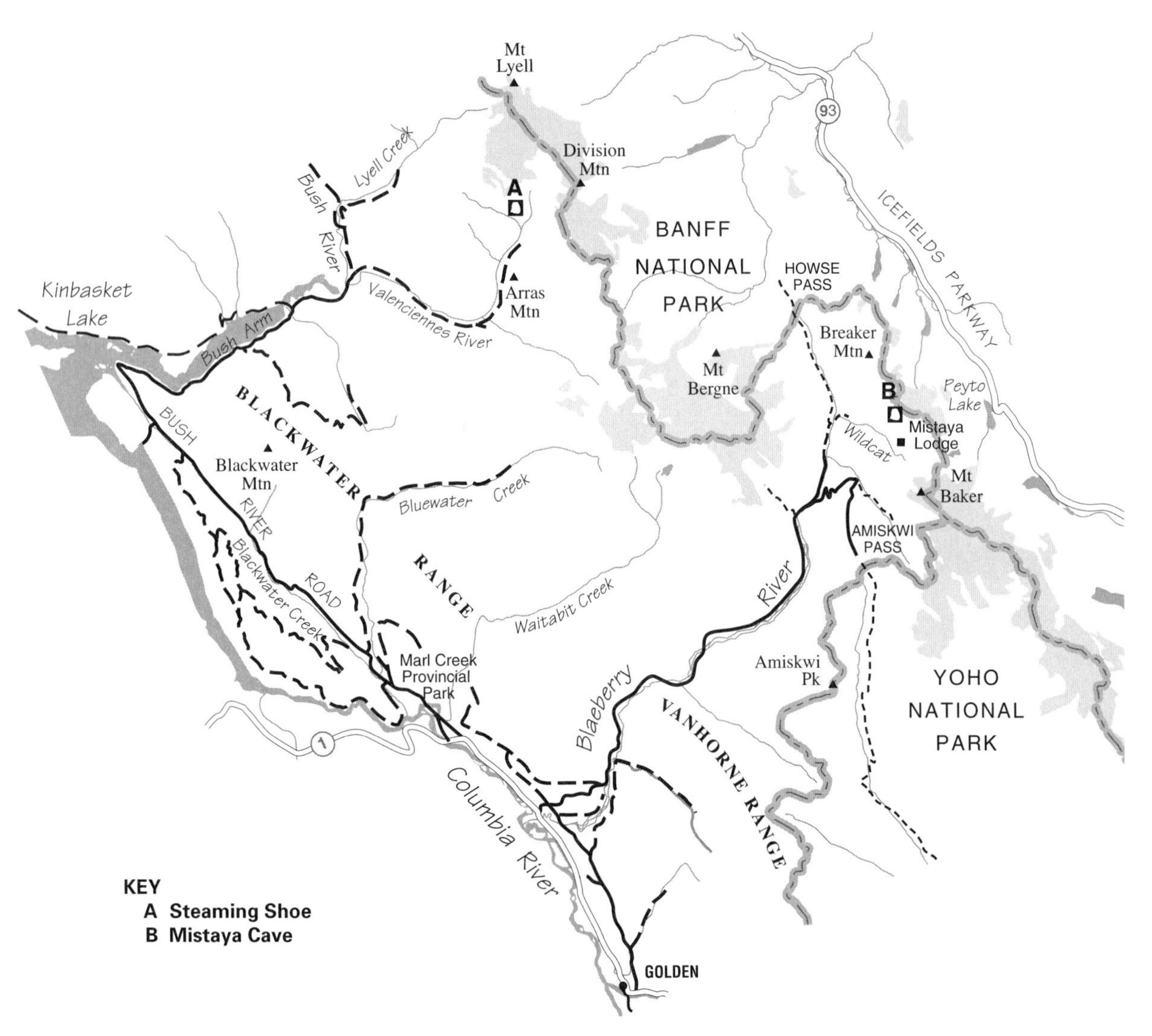

Mistaya Cave

Map 82 N/10 272285
Jurisdiction BC Forests (Columbia)
Entrance elevation 2400 m
Number of entrances 1
Length, Depth 63 m, 23 m
Discovered Unknown

Access Mistaya Lodge, located in the headwaters of the Blaeberry River above Wildcat Creek, is usually accessed by a twenty-minute helicopter flight from Golden. Contact Mistaya Lodge at (250) 344-6689, www.mistayalodge.com.

Alternately, hike in over Baker Col from the Wapta Icefields (2 days). Take Hwy. 93 (Icefields Parkway) to Bow Summit. From the Peyto Lake Viewpoint follow the trail that drops to the creek (bridge), then follow a narrow moraine to the toe of the Peyto Glacier. Go up the centre of the glacier, then make a long arc to the right (northwest) to avoid crevasses, coming back left to reach the Peyto Hut located at 314237 below Mt. Thompson. To overnight contact the Alpine Club of Canada at (403) 678-3200. Allow 5 hours to the hut. The second day cross the Wapta Icefield and climb over Baker Col between Mt. Baker and a small unnamed peak to the northwest. Descend the Baker Glacier (watch for crevasses) until the lower slopes of Trapper Peak can be traversed. Head north for about 3 km to a line of three lakes below Mistaya Mountain.

Location The cave is located above the three lakes. Look for a stream swallowed by a large sink. From Mistaya Lodge the entrance lies to the northeast and approximately 340 m higher up, just to the west of Mistaya Mountain. In winter the entrance is probably buried under considerable snowfall.

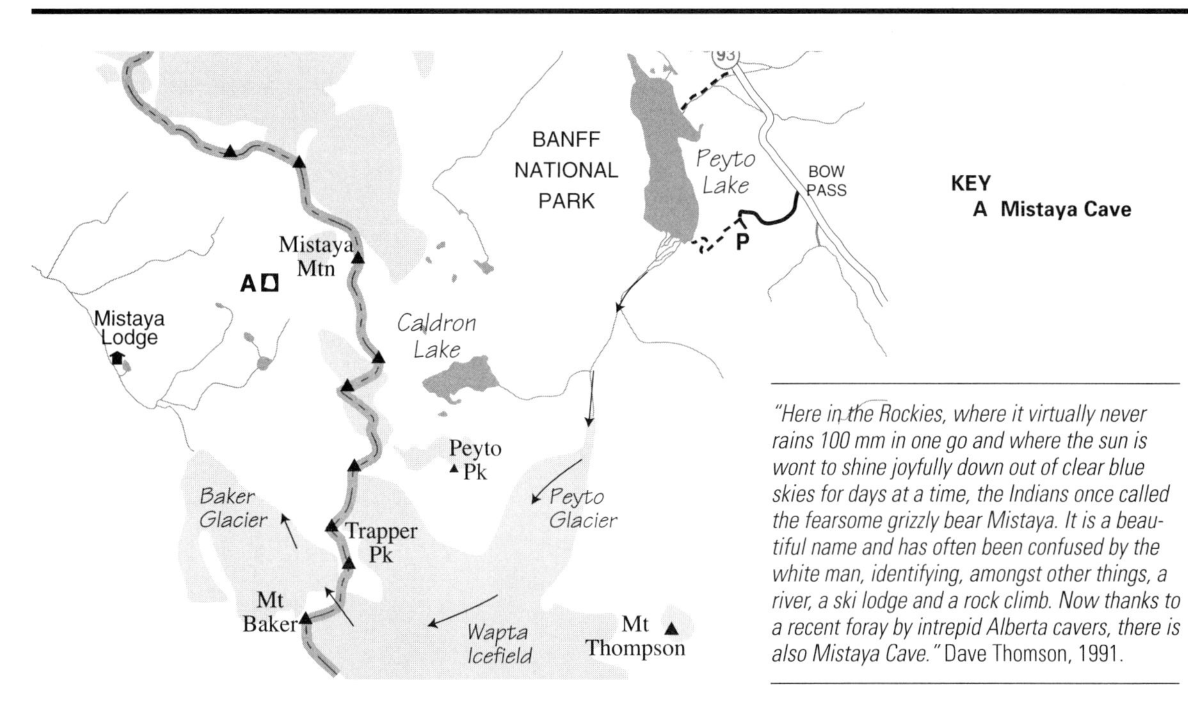

"Here in the Rockies, where it virtually never rains 100 mm in one go and where the sun is wont to shine joyfully down out of clear blue skies for days at a time, the Indians once called the fearsome grizzly bear Mistaya. It is a beautiful name and has often been confused by the white man, identifying, amongst other things, a river, a ski lodge and a rock climb. Now thanks to a recent foray by intrepid Alberta cavers, there is also Mistaya Cave." Dave Thomson, 1991.

Entrance to Mistaya Cave.
Photo Dave Thomson.

Warning The Wapta Icefield approach requires glacier travel with its inherent dangers of crevasses and whiteout conditions.

To avoid drowning, only attempt this cave during periods of low water flow. A wet suit may be useful.

Exploration & Description Mistaya Cave has seen little serious exploration. To the author's knowledge there has been only one foray and the cave remains unsurveyed. The Mistaya Lodge operator mentioned a sinking stream to Peter Thompson who returned with Dave Thomson and Marg Saul in September of 1991. The following is Dave's description of his solo foray.

"The initial drop appeared to be about eight metres. It was somewhat murky and uninviting. From the rubble floor another slot-like hole led down into a metre-wide, almost vertical fissure formed along a joint in the rock (in fact the whole entrance series is an irregular fissure). Rocks dislodged from the walls, if they ever hit bottom at all, failed to produce any sound that could be heard above the roaring waterfall. Mist and spray obscured everything beyond a few metres. This second drop was about 15 m and led to a roomy vadose passage with the stream rushing away along the floor. With visions of another Nakimu, I raced off along it, splashing through rapids, wading thigh deep in narrow pools. After about 40 m of rift, the stream ran into a pond and disappeared (a sump). Dry passage was nice, but away from the flowing water the air was dead and soon another pool appearing in the floor began to take up more and more of the available space. I didn't push the thing to its end, so I can't say it doesn't go. But.."

The resurgence for Mistaya Cave is at a series of sloping slabs about 1 km and 100 vertical m below the entrance.

Icefall Brook

Icefall Brook lies just outside the northwestern boundary of Banff National Park to the west of the Lyell Icefield. The only cave so far discovered is Steaming Shoe, named after the pile of steaming shoes clustered round the camp fire after repeated fording of the creek in order to get to the cave. Attempts to reach lower holes at the head of the gorge have been repulsed by falling rocks which turn this area into a shooting gallery in the summer. Two return trips in the winter have produced some good ice climbs, but as yet no caves of any length.

The head of the gorge is a high-energy environment, and an awe-inspiring place to be, especially after a summer rainstorm when the creek becomes massively swollen and changes within a few minutes to a dark brown colour. This is one of the most impressive gorges in the Canadian Rockies and deserves some form of protection, although it is probably too late to save the few remaining stands of large cedar that grow in this area. Unusual igneous intrusions of diabase basalt occur along Icefall Brook (Ben Gadd, 1986).

Area Exploration Cavers first heard of this area from a helicopter pilot flying people to Castleguard Cave. After the making of the Castleguard movie in April of 1974, Tom Wigly and Charlie Brown (McMaster University) were treated to a visit to the canyon. The visit nearly ended in disaster when an avalanche swept down the 450 m walls of the gorge to come to rest against the skids of the helicopter. Two return visits were made to the lip of the gorge on the southwest Lyell Glacier, which extends to the brink of the headwall. In the massive cliffs several openings could be seen, some having water pouring out of them.

Several years later, the University of Calgary Caving Club (long since defunct) made a visit to the bottom of the gorge. Climbing up a dry streambed on the west side of the gorge, a kilometre before the headwall, they found Steaming Shoe Cave. Helmut Geldsetzer (father of Torsten, member of the U of C Caving Club) found the entrance.

Access Hwy. 1 (Trans-Canada Hwy.) at Donald Station north of Golden. Follow logging roads up the Bush Arm of the Columbia Reach of Kinbasket Lake. From the east end of the Bush Arm a logging road heads east up Valenciennes River, and then up Icefall Brook itself to within a kilometre of Steaming Show Cave. Depending on the state of the logging roads, reaching the area can take a full day from Hwy. 1.

Icefall Brook.
Photo Ian Drummond.

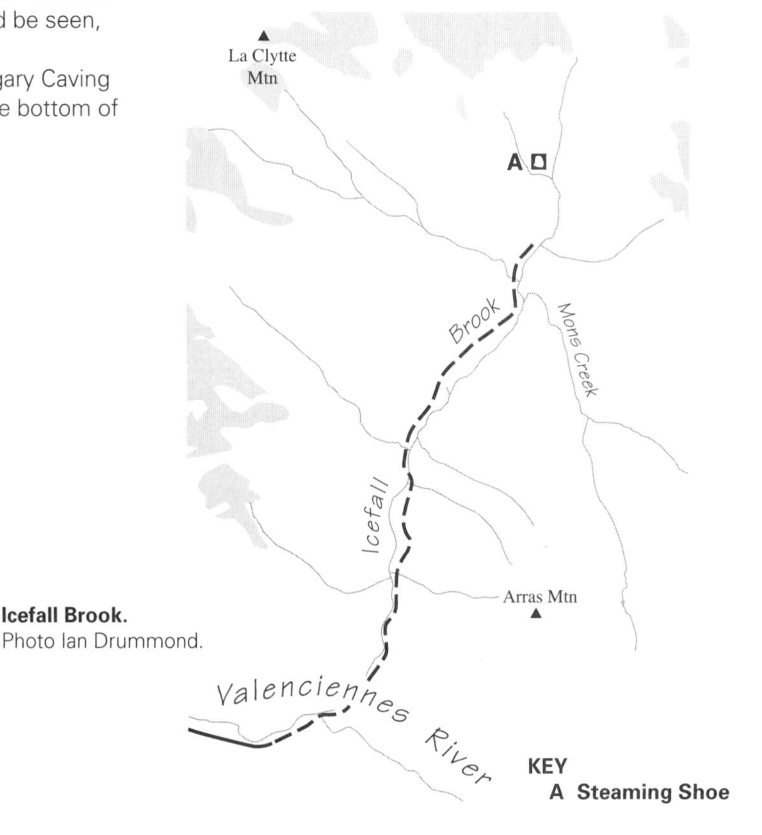

Steaming Shoe

Map 82 N/14 945473
Jurisdiction BC Forests (Columbia)
Entrance elevation 1311 m
Number of entrances 1
Length, Depth 210 m, 7 m
Discovered 1985, by Helmut Geldsetzer

Location 1 km after the end of the logging road and about 2 km before the huge cliffs marking the terminus of Icefall Brook, follow an ephemeral bedrock stream channel west up the cliffs. The obvious cave entrance is to the right of the channel, at the base of the cliffs.

Warning Icefall Brook is only accessible during the summer months. In winter there is considerable danger from avalanches and seracs calving from the icefield above. In summer watch for rockfall. Be aware that during rain storms water levels in the creek rise quickly. Additionally, the cave entrance probably re-surges on occasion.

Exploration & Description The cave was explored and surveyed by Tom Barton and members of the University of Calgary Caving Club.

It consists of a loop, with two entrances in the base of a cliff, one totally blocked with breakdown. The two passages branching into the cliff both end in sumps. The main phreatic passage is decorated with some stalactites and soda straws.

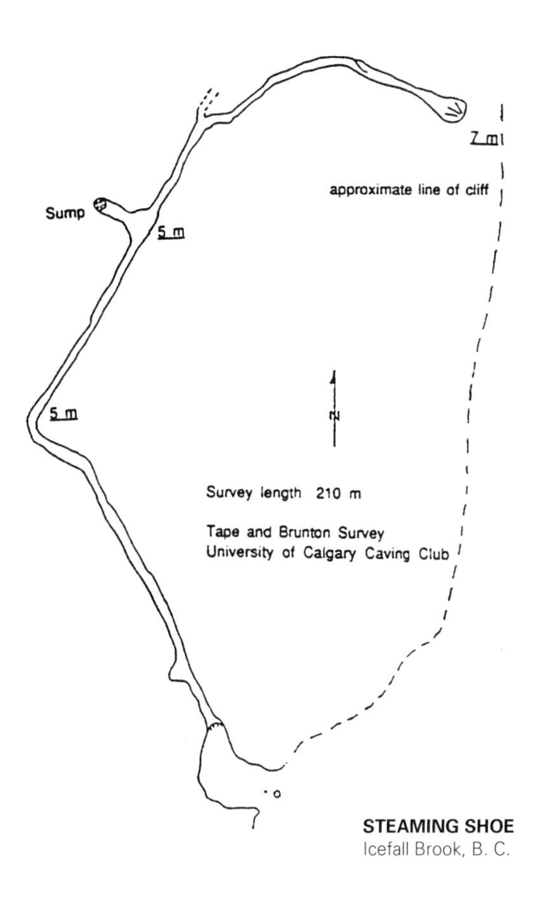

STEAMING SHOE
Icefall Brook, B. C.

"In the brilliant sunshine the spray of the waterfalls caused a kaleidoscope of colors in the void beneath us. We discussed the impossibility of access to some of the more obvious entrances. A 120 m rappel down a 450 m overhanging cliff, with a pendulum to the entrance in the spray of a small stream straight from the glacier. As the ultimate deterrent, we watched a serac collapse, and the green ice go crashing into the depths."
Ian Drummond, 1977.

Lyell Glacier above Icefall Brook.
Photo Ian Drummond.

HAMBER PROVINCIAL PARK

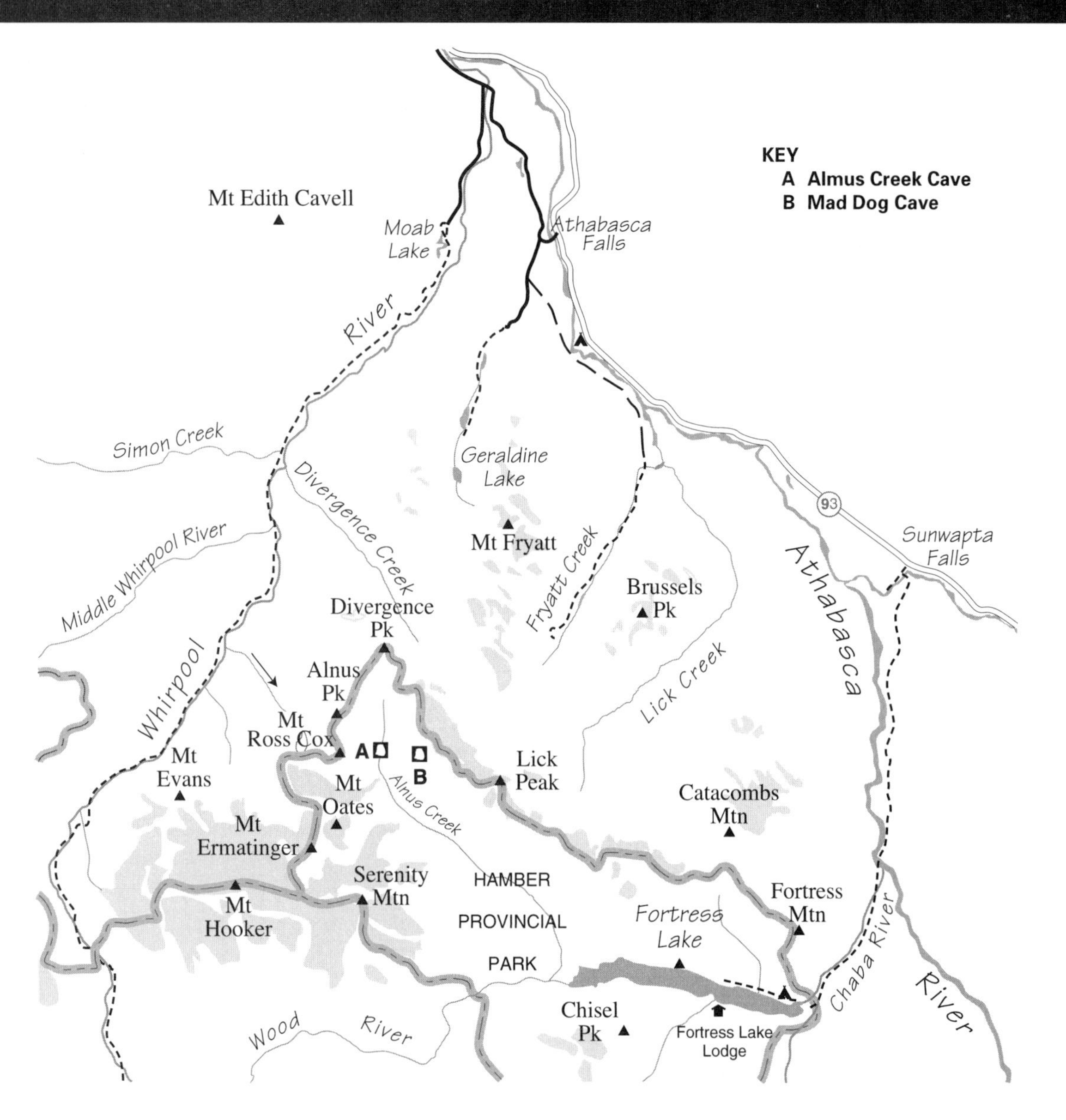

The area at the head of Alnus Creek is a true wilderness and very remote. Above the trees, the high altitude bench containing the Alnus Creek Sinks has stunning views across the valley to the Hooker, Serenity, and North and South Alnus glaciers. At 2300 m, the boundary between the steeply dipping limestone and the underlying quartzite/shale units is revealed in a series of small cirques that even in summer contain snow accumulations. The time between when the snow melts, allowing entry to karst fissures and shafts, and the first snows fall may be a window of only two months between late July and September.

While the beautiful views and grizzly sightings make the Alnus Creek area a memorable location, the 40-km round trip and lack of any dry sizeable cave passage make return trips to the area unlikely in the near future.

Winter Access (40 km) From Hwy. 93 (Icefields Parkway). From Sunwapta Falls follow the east bank of the Athabasca River to Fortress Lake (20 km). Ski up Fortress Lake, then climb northeast up Alnus Creek (400 m elevation gain) to reach the cave (20 km).

Summer Access (40 km) From the junction of Hwy. 93 (Icefields Parkway) and Hwy. 93A at Athabasca Falls, drive north on 93A for 9 km to the Whirlpool Fire Road. Follow the fire road to the Moab Lake parking area (7 km). The first 16 km of trail follows a fire road to a warden cabin. A kilometre past the cabin, ford the Whirlpool River using the sandbars and gravel flats that exist between braided channels. Use caution. On the far side, a strenuous bushwhack up a steep forested side valley leads to an extensive talus field. Traverse a large lake (not shown on the map) on its southwest side, then climb steeply to a snowy col at 2470 m between the avalanche prone upper slopes of Alnus Peak (2976 m) and Mt. Ross Cox (2999 m).

From the col a descent is made to the meadows above, and to the north of Alnus Creek, directly opposite the Alnus Glaciers flowing off the Hooker Icefield. Allow at least a day for this 40 km approach that has only been used in August, when group tactics were adopted to ford the Whirlpool River.

The pass to the northwest of Lick Peak, provides for an alternative route back to the Icefields Parkway via Fryatt Creek. Take gear for glacier travel.

Area Exploration This area was first visited by Dave Thomson when hiking up from Fortress Lake. He came upon the river resurgence and higher sink before finding the sinking streams running off the western flank of Lick Peak. All were too wet to enter, and so a return trip was planned for the winter. Four ASS members skied from Sunwapta Falls to Fortress Lake, then bushwhacked up Alnus Creek to the river cave, a total distance of 40 km. Alnus was explored for 100 m to a flat-out crawl in water. On the way back down Fortress Lake, the party encountered strong headwinds, and the temperature, which at the start of the trip was -35° C and supposed to rise, but actually fell to -42° C. Apart from minor frostbite there were no injuries, but one trip member was repeatedly charged by an angry bull moose on the Chaba River flats.

In August, 1986, the area was visited by the Imperial College Caving Club from London, England. This time the area was accessed via the Whirlpool River and a high pass between Alnus Peak and Mt. Ross Cox. The seven-day trip was plagued by bad weather, lack of food — boiled stinging nettles were a welcome addition to the meager menu — and by repeated encounters with grizzlies. A concerted effort was made to push the sinks on Lick Peak, one of which went for 272 m. Nine sinks in all were investigated, but snow prevented progress in most. Sink number six finally produced the rather wet Mad Dog Cave. On the return to the Icefields Parkway, the party split up, all three groups seeing bear tracks in the snow over three different passes. On one very precipitous pass at 3050 m crampons would have been useful for those without the advantage of claws.

Fording the Whirlpool River en route to Alnus Creek. Photo Dave Thomson.

"When we reached the lake our worst fears were realized. The wind had not changed and was still blasting westwards down the ice creating snow devils in its path. We put our heads down and headed up the mighty white strip of Fortress Lake, guarded by massive quartzite peaks such as the wedge shaped Chisel Peak and Serenity Mountain. Himalayan-style snow plumes whirled from their summit ridges. We were frozen by the time the cabin came up and rested briefly. The next section was endless and it was here that we became frostbitten: noses, faces and fingers. Never was there a more appropriate phrase than 'God, it's hell in the karst'!" Chas Yonge, 1986.

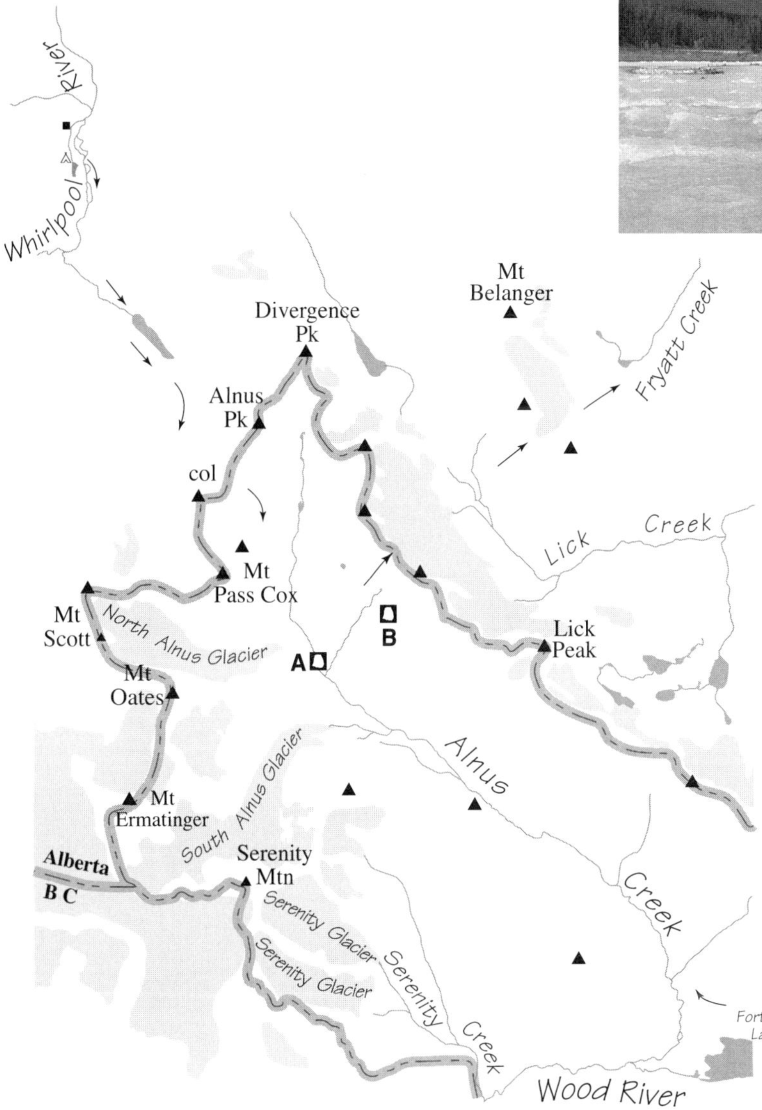

KEY
A Alnus Creek Cave
B Mad Dog Cave

Alnus Creek Cave

Map 83 C/5 329107
Jurisdiction Hamber Provincial Park
Entrance elevation 1830 m
Number of entrances 1
Length, Depth 100 m 15 m
Discovered 1985, by Dave Thomson

Location The entrance — a shaft — is easy to find, being located where Alnus Creek passes through a limestone ridge.

Warning Best visited in the winter when the water is lower. However, an 80 km winter ski trip through wilderness terrain combined with a wet cave makes this a serious undertaking. Plan accordingly.

Description The entrance consists of a 5 m shaft leading into a stream passage taking a strong inward draught. Penguin-shaped ice stalagmites about a metre high sit on narrow bases on either side of the streamway, and the roof is decorated with needle and platelet hoarfrost. The initial section of passage requires bridging above the stream, awkward because of sections of verglas. Further in, the passage widens and becomes a rift some 20 m high, the water meanwhile having disappeared down a flat-out passage. The rift eventually becomes a low crawl and leads to a pool which is probably a sump. Another side lead also ends in a choke. The strong draught indicates the possibility of more cave, but because this point is close to the resurgence, it is probably not worth pushing.

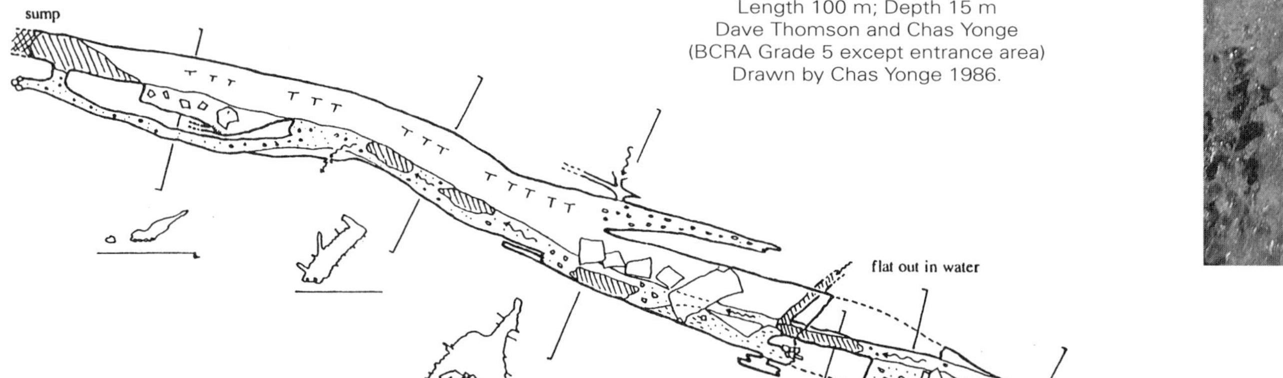

Chas Yonge in Alnus Creek Cave.
Photo Dave Thomson.

Mad Dog Cave

Map 83D/8 344120
Jurisdiction Hamber Provincial Park
Entrance elevation 2300 m
Number of entrances 1
Length, Depth 272 m, 87 m
Dscovered 1985, by DaveThomson

Location One of the sinks under Lick Peak. A small stream disappearing under the snow.

Entrance

MAD DOG CAVE

Alnus Creek, British Columbia

ICCC Survey 1986 to BCRA gd5c

Length 272m Depth 87m

true N N mag

dry inlet

duck

Wass Rift

-87m

Description Number six of nine sinks, Mad Dog Cave did not initially look promising, but once the stream was diverted around the cave entrance, a crawl under the snow accessed a small descending rift passage. This led to a low squeeze in one metre of icy water. Beyond this was 30 m of steeply sloping, very tight passage, the 'Wass Rift', which in turn led to another passage that just got too small as the water cut back along the bedding plane strike. The resurgence for these sinks was found several-hundred metres lower, in the same band of Cathedral limestone. The sinks appear to be juvenile karst features, not large enough to be entered for much of their length.

Warning The optimum period to visit coincides with the movement of grizzlies up into the alpine meadows to graze on roots and bulbs, and to dig up marmots. In 1986 the Imperial College Caving Club encountered five different grizzlies in one week.

Top: Sink #1, Alnus Creek. Photo Dave Thomson.

Bottom: Tim Flack in Mad Dog Cave. Photo Dave Thomson.

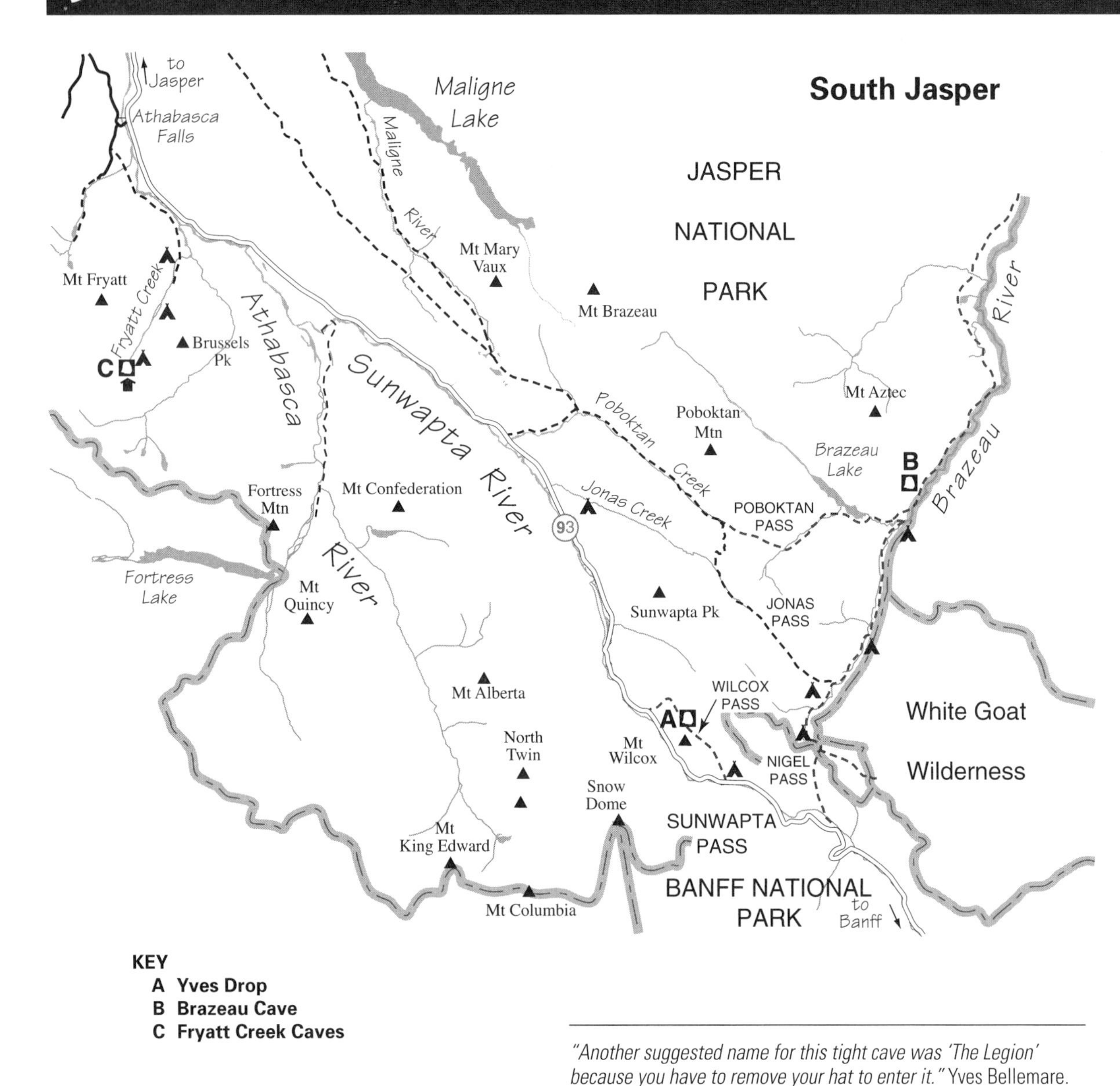

"Another suggested name for this tight cave was 'The Legion' because you have to remove your hat to enter it." Yves Bellemare.

Yves' Drop

Map 83 C/3
Jurisdiction Jasper National Park
Entrance elevation Approximately 2425 m
Number of entrances 1
Depth 45 m shaft
Discovered August 1995, by Yves Bellemare & Jasper ASS

Access Hwy. 93 (Icefields Parkway) just south of Sunwapta Pass. From the Wilcox Pass Campground follow the Wilcox Pass trail 4 km to the pass (360 m elevation gain). Yves Drop is located about 500 m down the other side, to the right (east) of the trail.

Location Yves Drop is a fissure sink that takes water during spring runoff. The wider part of the fissure entrance is too tight at 10 m depth. However, at its other extremity where a good natural belay is located, the constriction is negotiable. The best time to visit is late summer, early fall when the area is free of snow and water is at a minimum.

Area Exploration Cavers first visited the Wilcox Pass in 1967, and although it was noted as a good karst area, the cave entrances checked were too small to enter. The lake at the pass, for instance, although fed by two different creeks, drains underground through a fissure that so far has proved too wet and tight. In 1995, Jasper cavers made a more determined effort to push some of the more obvious fissure sinks.

Description Yves' Drop is a tight active sink that descends to a ledge at 25 m before ending at the bottom of a shaft where a pool drains into a fissure too tight to enter.

Warning This cave is for small cavers who can perform SRT in confined spaces.

Brazeau Cave

Map 83 C/7
Jurisdiction Jasper National Park
Entrance elevation 1760 m
Number of entrances 1
Length, Depth 176 m, +53 m
Discovered Unknown

Access Hwy. 93 (Icefields Parkway). Start 8 km south of Sunwapta Pass at the Nigel Pass trailhead. The trail (also the South Boundary Trail) is followed for 33 km: over Nigel Pass to the Brazeau River, past the Jonas Pass trail at Four Point Junction, and past the trail to Brazeau Lake (campground). Continue past the warden cabin and in about 2 km come to springs that consist of two fist-size holes out of which water jets as if from a fire hose.

Location The impresssive entrance (not visible from the trail) lies directly above the springs.

Exploration & Description The cave entrance is very obvious from across the Brazeau River and must have been noticed by outfitters operating in the area. In the summer of 1974 outfitter Tom Vinson Sr. reported the cave to the warden service and in October warden Ab Lowen and park naturalist John Steele surveyed the cave for 50 m to a blockage of packrat debris.

Thirteen years later, in 1986, ICCC expedition members refound the cave entrance, and after digging through pack-rat material explored a small passage for 120 m to a rising 20 m climb where the passage ended. A small low sump was found in a side passage about 100 m into the cave.

In the mid 1990s the entrance area was investigated by a Parks Canada archeologist, but nothing of interest was found.

The mountain immediately above Brazeau Cave has tiers of cliffs separated by terraces and talus slopes. Attempts to traverse high on the flanks of the mountain in search of further cave entrances was stopped by steep cliffs.

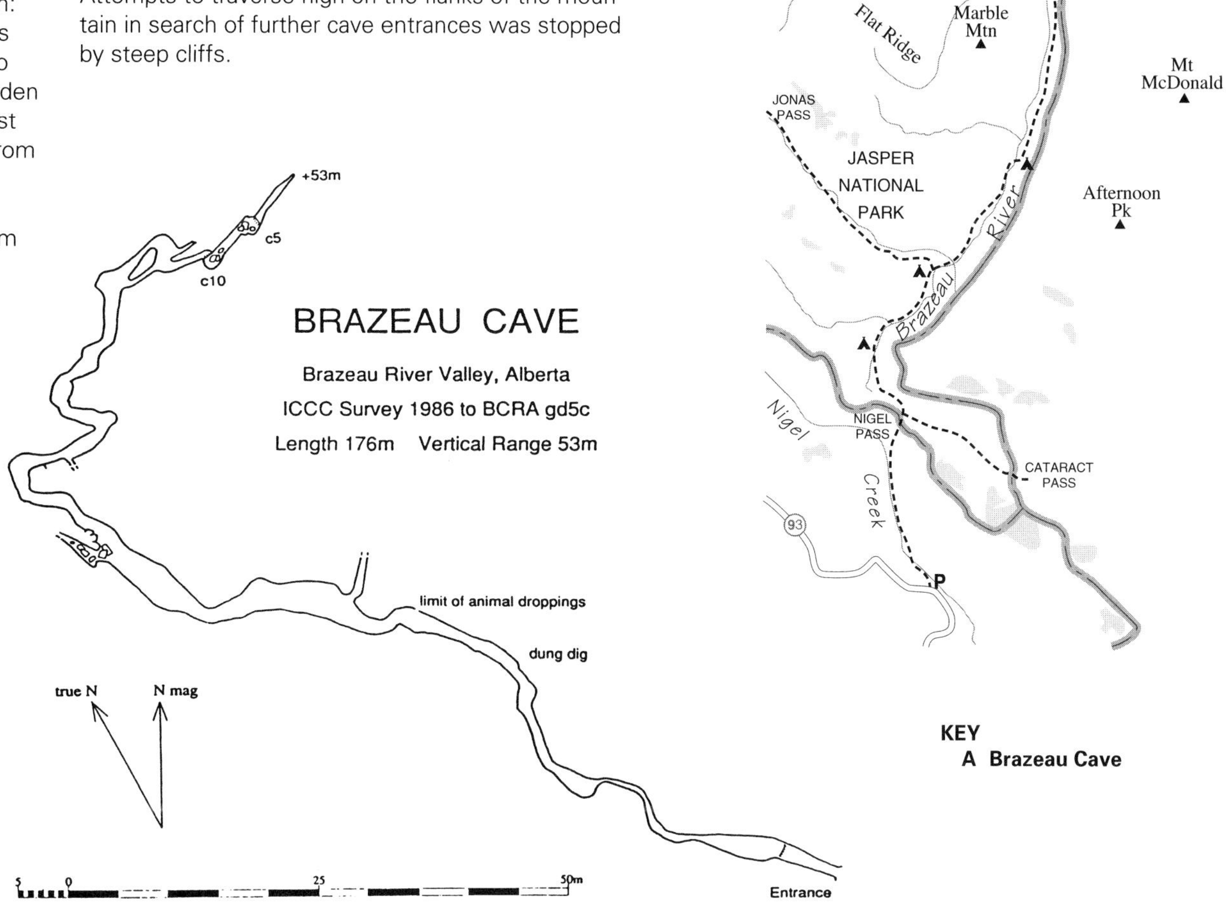

Fryatt Creek

This narrow hanging valley perched above the Athabasca River provides access to some splendid mountains, notably Mt. Brussels (3161 m), a high impregnable-looking peak not climbed until 1948. Just above the headwall at the mouth of Fryatt Valley the Alpine Club of Canada has a cabin (not the cabin marked on the topographical map). Two caves are located near the cabin and are wet in summer, taking water from Fryatt Creek.

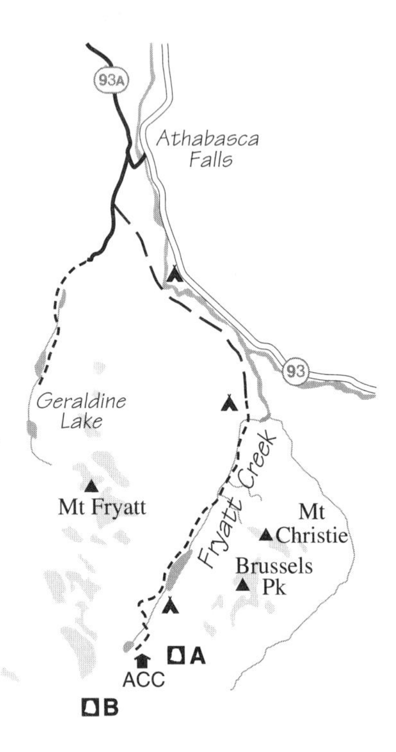

KEY
A **Headwall Cave**
B **Fryatt Creek Cave**

Access to all Caves Hwy. 93 (Icefields Parkway). From Athabasca Falls head north on Highway 93A for 1 km to the Geraldine Lakes Road. Turn south and follow it for 2 km to where a fire road turns off to the left (east) and park at the Fryatt Creek trailhead.

Hike the fire road for approximately 7 km along the Athabasca Valley. Near white clay cliffs, the trail swings away from the Athabasca River and enters the Fryatt Creek valley. After passing Fryatt Lake the trail ascends a steep headwall, keeping right of the falls. The ACC hut and the caves lie just above the headwall. A full day is needed for the 22 km hike (760 m elevation gain).

In winter it may be possible to cross the Athabasca River on ice opposite the Fryatt Creek valley (use caution!), thus reducing the approach by 7 km. The trail to the cabin is well-travelled and usually negotiable in winter.

Headwall Cave

Map 83 C/12
Jurisdiction Jasper National Park
Entrance elevation 2000 m
Number of entrances 2
Length, Depth 151.6 m, 10.0 m
Discovered October 1985, by Rick Blak

Location Just across Fryatt Creek from the cabin. The entrance drifts over with snow in the winter and is probably best vistited in fall when water levels are low.

Description This short cave provides a fine example of a lens-shaped, phreatic bedding-plane passage with well-formed scallops on the floor and ceiling. The sound of a roaring stream in the distance, alas, turns out to be a small underground tributary to Fryatt Creek which sumps downstream. In winter ice-curtains form along the ceiling and are banded in alternating clear and opaque layers.

Two other shorter caves similar to Headwall have been found higher in the valley below the upper lake.

HEADWALL CAVE
FRYATT VALLEY, AB

SURVEYED TO BCRA GRADE 5
FEBRUARY 1, 1986
SURVEYED BY: DON RUMPEL
BRIAN REUM
SURVEYED LENGTH: 151.6 m
SURVEYED DEPTH: 10.0 m

ALBERTA SPELEOLOGICAL
SOCIETY

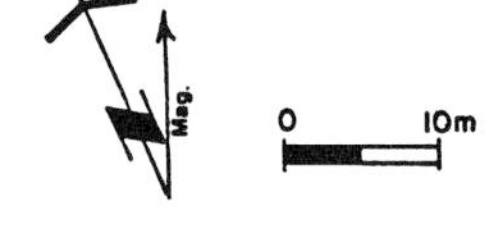

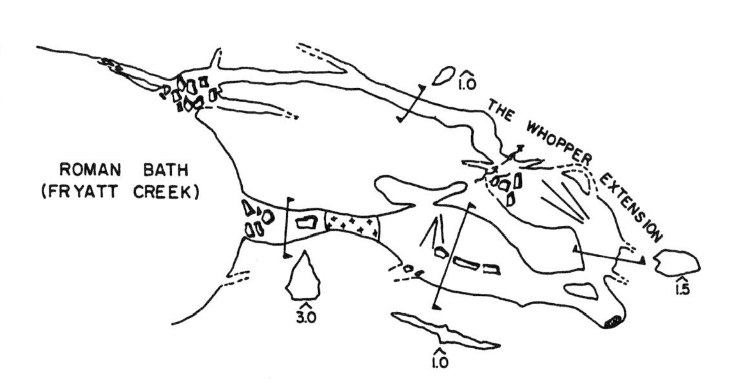

Fryatt Creek sink.
Photo Dave Thomson.

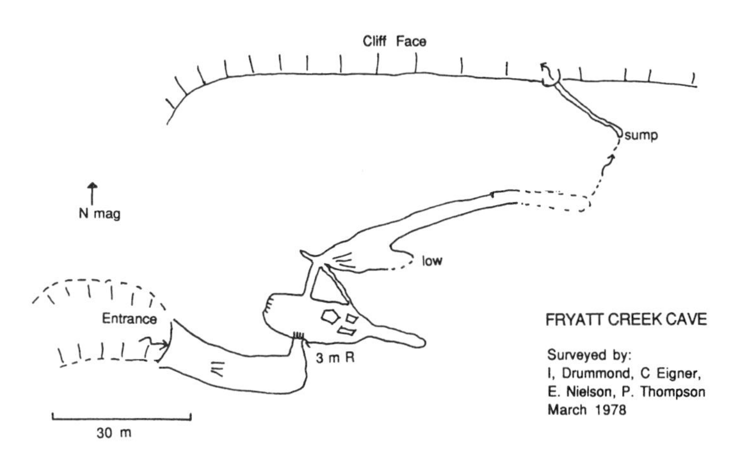

Fryatt Creek Cave

Map 83 C/12
Jurisdiction Jasper National Park
Entrance elevation 2000 m
Number of entrances 2
Length c. 120 m
Discovered Unknown

Location From the cabin follow Fryatt Creek upstream for about 0.5 km. You first come to the resurgence and then the massive sink entrance to Fryatt Creek Cave located 100 m distant in a direct line.

Although well worth visiting in summer, this cave is only enterable following a long cold spell in the winter.

Description The upper sink is one of the larger cave entrances in the Rockies, and in winter contains some spectacular ice formations. The 3 m entrance pitch forms a frozen waterfall leading down into a large chamber with a smooth ice floor. Water dripping from fissures in the ceiling creates massive hanging ice columns. From this chamber a passage leads towards the resurgence. Even in the depths of winter water still flows beneath the ice as unwary explorers have discovered when they found themselves swimming in bottomless slush. The cave usually sumps, however, following a long cold winter spell and in these conditions a through trip has reportedly been made.

The resurgence cave, a pleasant bedding plane crawl with some short side-passages, usually ends quickly when ice (in winter) meets the ceiling. Just above, a pleasant, dry, phreatic tube soon becomes too tight.

Warning Beware of falling through the ice.

Fryatt Creek resurgence in full flow.
Photo Dave Thomson.

Central Jasper

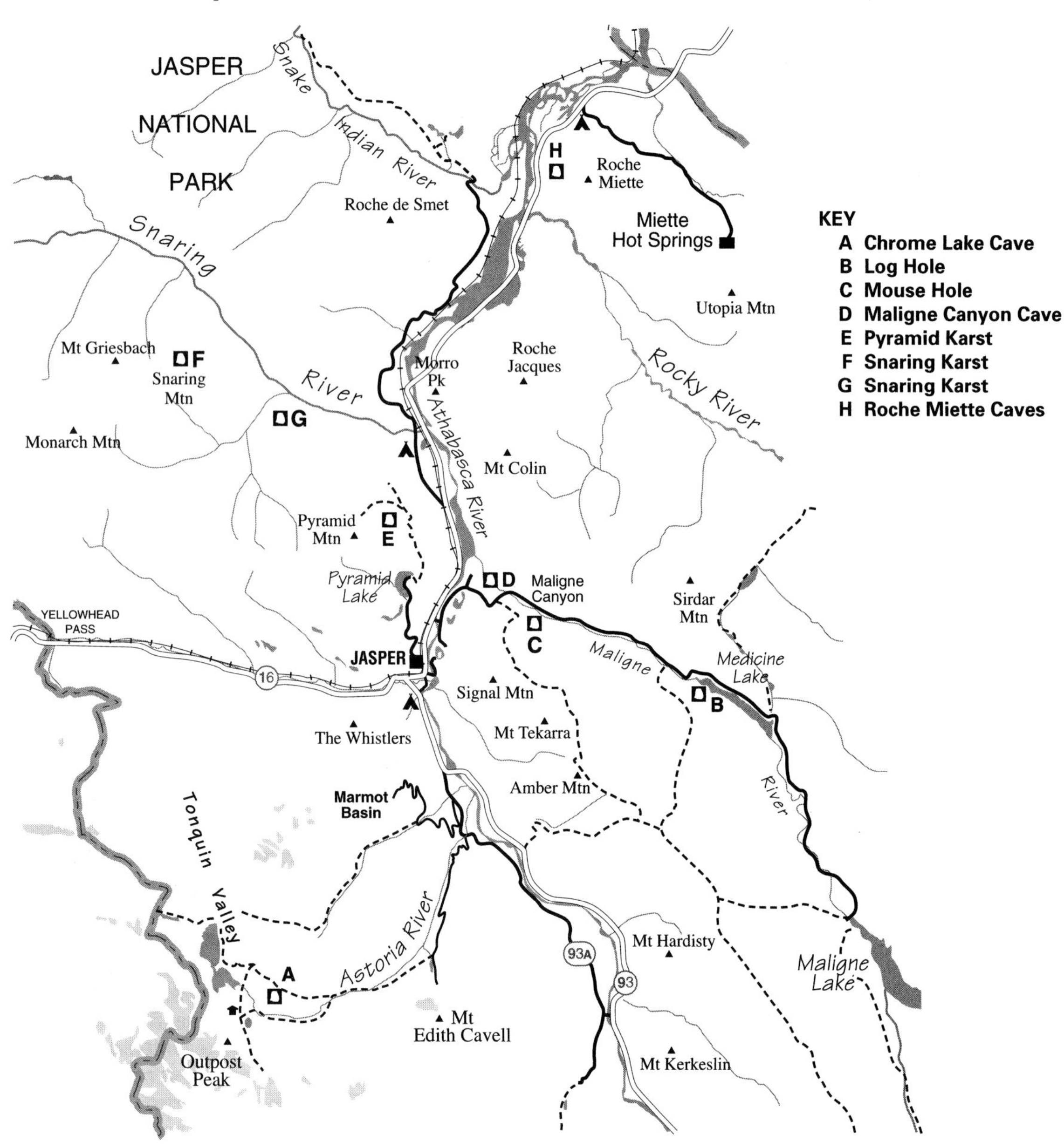

Tonquin Valley

Chrome Lake Cave

Map 83 D/9
Jurisdiction Jasper National Park
Entrance elevation 1800 m
Number of entrances 1
Length, Depth 15 m, 10 m
Discovered 1985, by Rick Blak

Access The Edith Cavell Road off Hwy. 93A. Start at the Astoria River to Tonquin Valley trailhead just before the bridge at the outlet of Cavell Lake and opposite the hostel. The trail descends and follows the Astoria River.

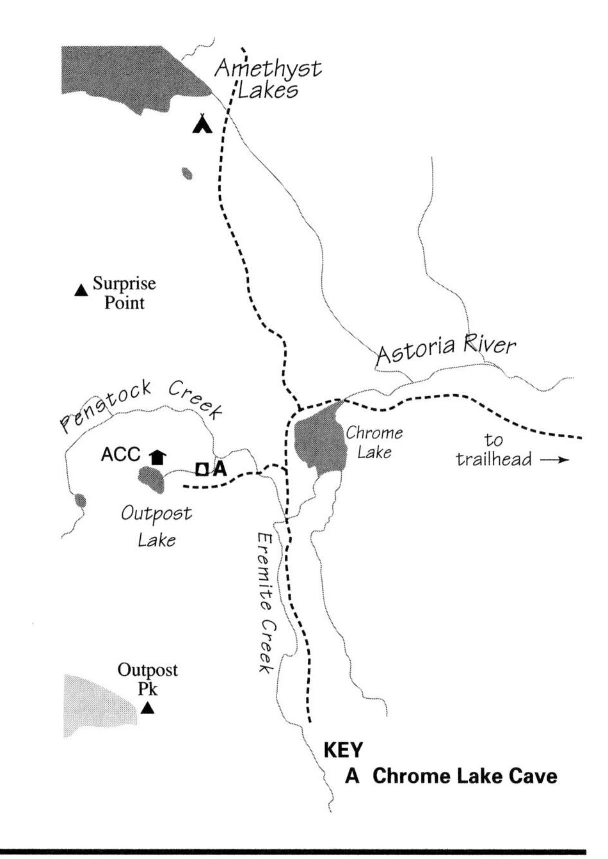

After about 8 km, at a fence just before the remains of the Oldhorn warden cabin, cross the bridge to the south side of the Astoria River and follow the trail to Chrome Lake — another 6.5 km.

Follow the lake inlet stream up towards Outpost Lake. The small entrance to this cave is located next to the top of a modest waterfall between Chrome Lake and Outpost Lake.

From Chrome Lake a trail also leads to Outpost Lake and the ACC hut — the Waites Gibson. The nearest campground is at Surprise Point near Amethyst Lakes.

Description The cave descends steeply to a sump at base level, essentially paralleling the adjacent falls.

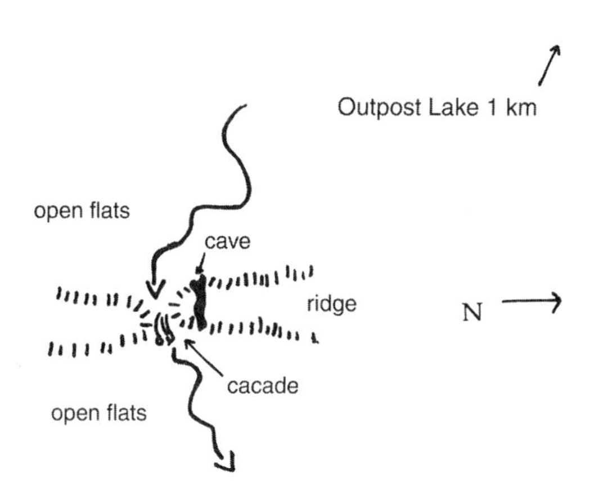

CHROME LAKE CAVE
From a sketch by Ian McKenzie 1992.

The Maligne System

"The Maligne System remained inviolate despite several thousand dollars worth of geophysics and drilling, and much watching of water." Chris Smart.

Area Exploration Since 1965 many attempts have been made to find the cave system thought to link Medicine Lake with a series of springs at the foot of Maligne Canyon. Running through Palliser Limestone, the straight-line distance between these two points is 15 km. By analyzing data from a series of dye-traces it has been estimated that a single large passage approximately 7 m in diameter and largely filled with water must exist (Chris Smart).

In 1967 a team of 13 cavers directed by Derek Ford spent five weeks systematically searching the area for caves without success. In 1970 a resistivity survey of the valley was commissioned by Parks Canada, and a large cavity was apparently detected 45 m down. The following year seven holes were drilled, all exceeding 45 m, but no cave was found.

The meager consolation prize for all this exploration was the discovery of three caves: Mouse Hole and Log Hole (two diminutive caves near Medecine Lake), and the somewhat longer but very tight and miserable Maligne Canyon Cave.

Recent attempts to drill a well for Maligne Canyon Teahouse encountered a void and resulted in a mixing of sulphur-laden waters with the Maligne Spring waters!

"The surveyors forgot to bring a notebook so we recorded the data on several flat rocks which we carried through the cave." Ian McKenzie, referring to Log Hole Cave.

Log Hole

Map 83 C/13
Jurisdiction Jasper National Park
Entrance elevation 1500 m
Number of entrances 1
Length, Depth 48.8 m, + 9.1 m
Discovered 1967, by Mike Goodchild

Access Hwy. 16 (Yellowhead Hwy.) 1.7 km east of Jasper townsite east junction, turn right onto the Maligne Lake Road and follow it to the northwest end of Medicine Lake (21 km). Park in a gravel parking area with washrooms.

Cross the dry river bed and hike along the southwest side of the lake for about 1 km to a canyon that contains the cave.

Location The large, south-facing entrance is reached by a 20 m rappel.

Exploration & Description Log Hole consists of a crawling passage leading off from a large chamber with some interesting fossils at the end. The cave was dug and extended in 1984 by a group of Jasper ASS members.

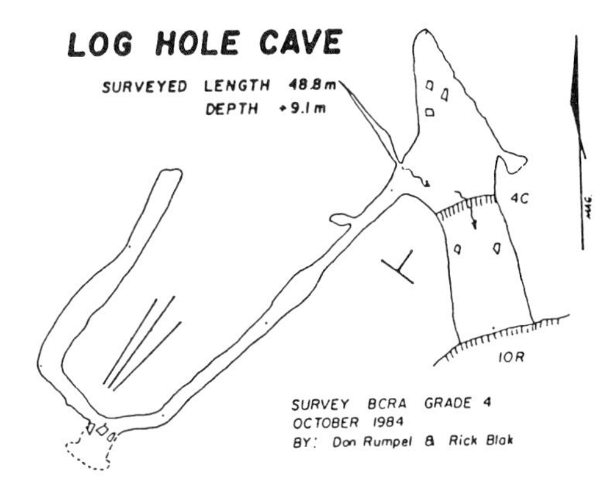

Mouse Hole (Spider Cave)

Map 83 C/13
Jurisdiction Jasper National Park
Entrance elevation 1220 m
Number of entrances 1
Length c. 30 m
Discovered 1967, by Mike Goodchild

Access A Hwy. 16 (Yellowhead Hwy.) 1.7 km east of Jasper townsite east junction, turn right onto the Maligne Lake Road and follow it for 8 km to where it crosses the Maligne River at the teahouse. Park here.

Hike back down the road across the bridge. Then turn left and head upstream along the right (south) bank of the Maligne River for about 4 km. There is no trail and several diversions have to made to avoid mud slides. Once the canyon proper is encountered, look for the large cave entrance at the base of the right canyon wall. Continue above the canyon edge to a point above the cave where two skylights are located.

Entrance to Mouse Hole at right of photo.
Photo Jon Rollins.

Access B Continue driving beyond the teahouse up the Maligne Lake Road for about another 4 km to a gravel pull-off on the south side shortly before Two Valley Creek bridge. Park here.

Hike steeply down hill, crossing an old road bed shortly before encountering the Maligne River. Head downstream to an impressive canyon. Just upstream of the cave (located on the south side of the canyon) is a bridge of collapsed limestone blocks. For those with climbing skills the bridge can be accessed by a narrow ledge. A rope is recommended for the short climb up the other side of the canyon. The cave is now below you and can be viewed through two narrow skylights.

During high water the cave can only be entered using a 20 m rope (bring SRT gear) via the skylights.

Warning Access B involves extreme hazard from loose rock and the possibility of drowning. Do not fall into the canyon!

Description The cave contains skylights and in winter serves as a harvestman hibernaculum, which is its only feature of interest.

Cave Softly Winter visits are not recommended. Don't light fires in the cave as some thoughtless people have done in the past.

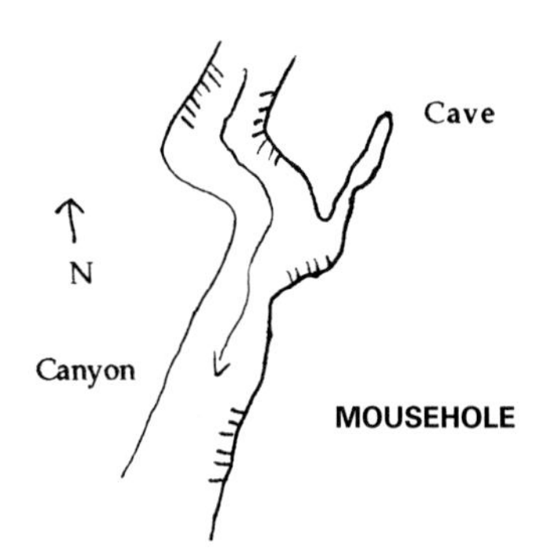

Maligne Canyon Cave

Map 83 D/16
Jurisdiction Jasper National Park
Entrance elevation 1060 m
Number of entrances 1
Length 373 m
Discovered 1974, by Chris Smart

Access Hwy. 16 (Yellowhead Hwy.) 1.7 km east of Jasper townsite east junction, turn right onto the Maligne Lake Road and follow it for about 8 km to the Maligne Canyon parking lot at the teahouse. Walk down the Maligne Canyon Trail.

Location The cave entrance is visible just below 4th Bridge from the railed walkway on the right (east) side of the canyon. In summer the entrance is a sizeable spring and not enterable. In winter the entrance is often covered by a thick layer of icicles.

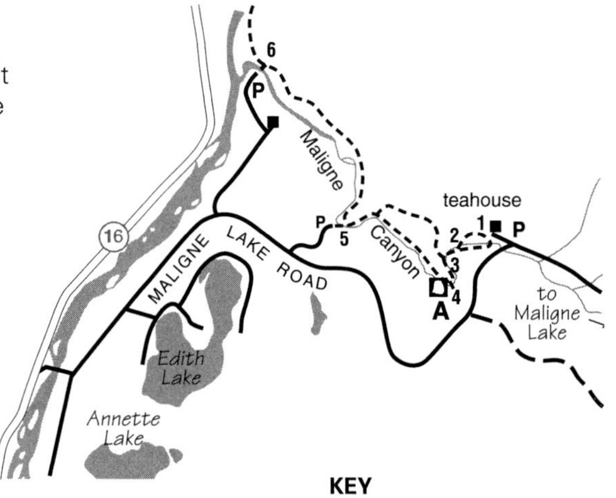

Maligne Canyon Cave entrance in full flow.
Photo Jon Rollins.

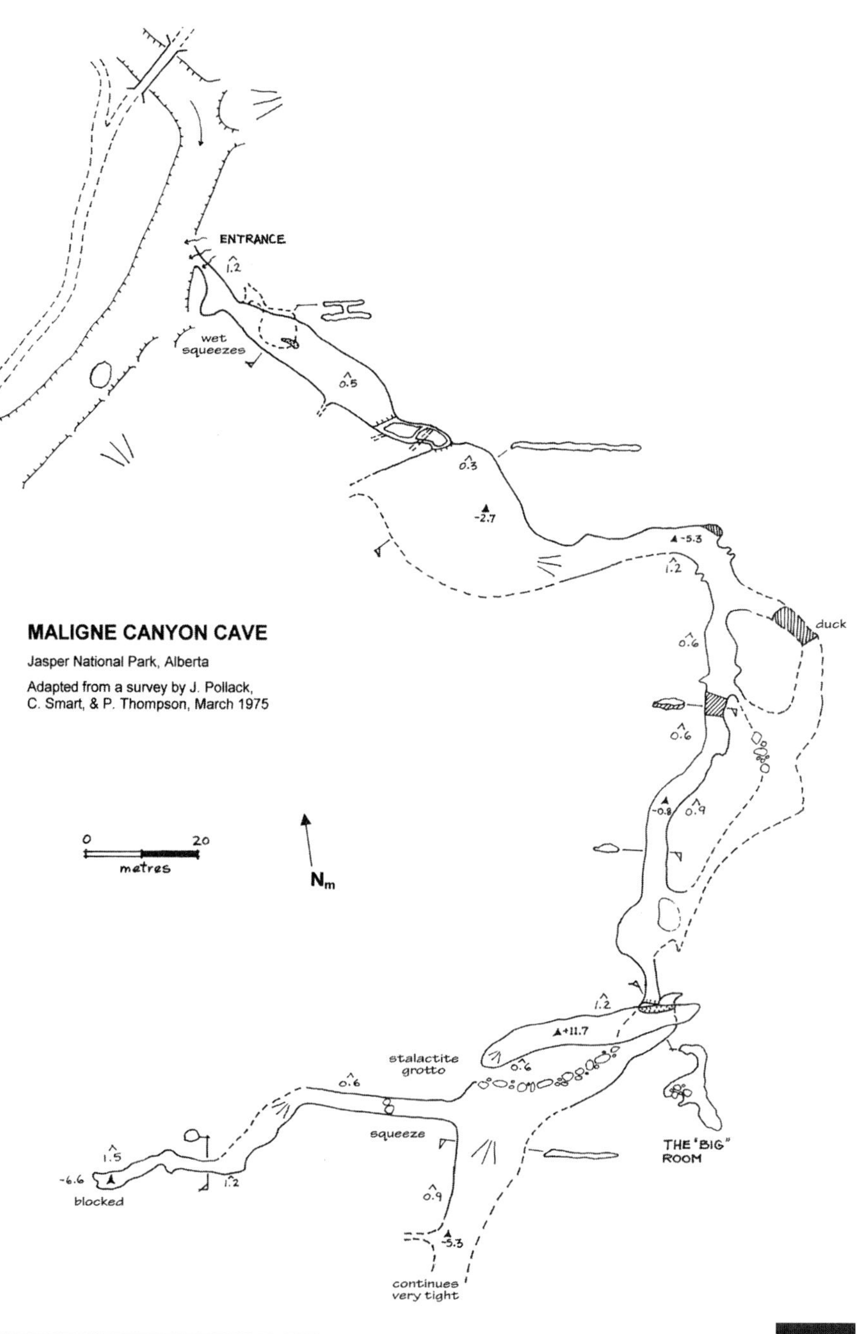

Description Only dreams of finding the "Maligne Master Cave" could have inspired the discovery and exploration of such a tight and unpleasant cave. For much of its length it's a bedding plane crawl with frequent squeezes between breakdown blocks. Occasional pools are impossible to avoid, one leaving only 15 cm of air space. The "Big Room," an enlarged joint, is one of the few points in the cave where you can stand up. The cave eventually ends in breakdown out of which the water presumably rises.

Maligne Canyon Cave is currently the second longest cave in Jasper National Park.

Warning The cave should only be visited by experienced and particularly determined cavers.

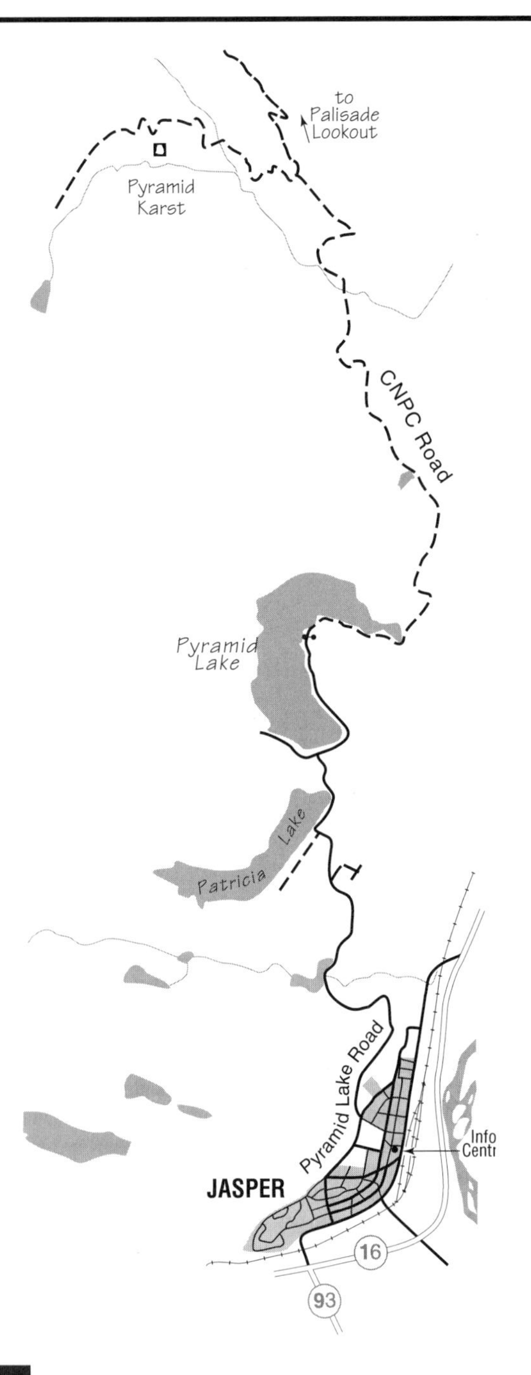

Pyramid Karst

Just below the CNCP Telecommunications Road, where it winds across the benches on Pyramid Mountain, is a small area of forested karst. A creek bed (usually dry) runs across the bench, becoming a canyon shortly before plunging down a line of cliffs 30 m to 70 m high. Some small sinks and shafts are located on the top of the bench. The attractive canyon contains some sections of oxbow passage and a delightful keyhole cave through which the creek passes. Several cave entrances — Ice Hole (partially blocked by ice), Pink Hole, Goober Cave and Lost Light Cave — can be seen lower down in the cliff face.

Access to all Caves From Jasper drive 8 km along Pyramid Drive to Pyramid Lake. From the end of the road hike the CNCP Telecommunications Road (gated). The road forks after 7.5 km, the right-hand option going to Palisade Lookout. Keep left. The Pyramid karst is located 500 m farther along below the road. Hike down from the road to a cliff with several obvious entrances.

Cave Softly All caves contain fragile mineral formations.

Pink Hole

Map 83 D/16
Jurisdiction Jasper National Park
Entrance elevation 1600 m
Number of entrances 1
Length, Depth 109.9 m, 7.8 m
Discovered 1988, by Ron Lacelle, Gille Roy, Craig Wagnell and Goober the dog

Location A two-hole cliff-side entrance in pink limestone.

Exploration & Description A dig accessed a 10 m aven and a small keyhole passage. This passage then divides into some small decorated crawls.

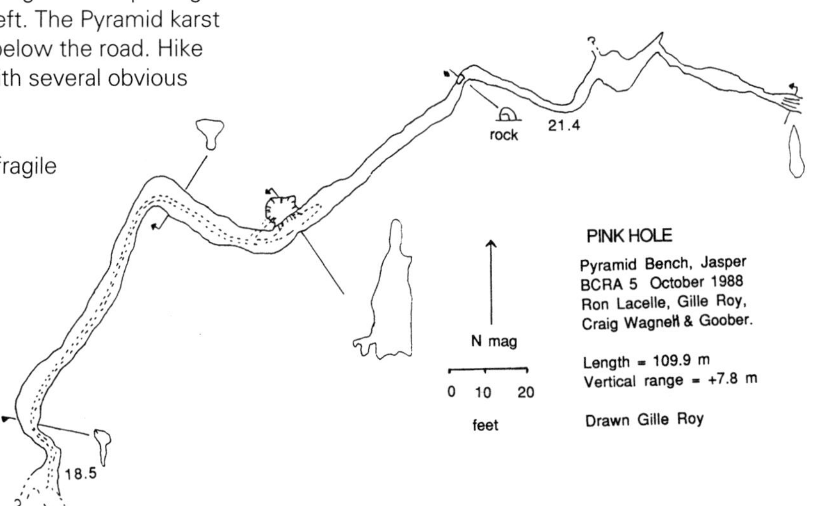

Lost Light

Map 83 D/16
Jurisdiction Jasper National Park
Entrance elevation 1600 m
Number of entrances 1
Length, Depth 785 m, 77m
Discovered 1999, by Ron Stanko

Location In the cliff face just above Goober Cave. The cave is normally frozen shut from December until mid-summer.

Exploration & Description The 10 m entrance excavation — a five-year project for Ron — revealed the longest cave in Jasper National Park. The ongoing passage is initially small, forcing you down into the water, but following a squeeze and down-climb, it becomes a high rift passage with a wet 10 m pitch. This section of the cave is well decorated with straws, carrots and helictites. Following a stream inlet from the left, the passage becomes a low tube before opening up again to 200 m of walking passage at the end of which the water runs down into a large slot. An 8 m pitch can be circumvented to farther dry passage and a deep pool that can be bridged. Beyond is the largest room in the cave, The Thread Room. From here, a down-climb (hand-line) accesses 30 m of passage leading to a sump.

Warning This is an active cave with a small stream and parts of this cave flood, as evidenced by pine needles in the ceiling.

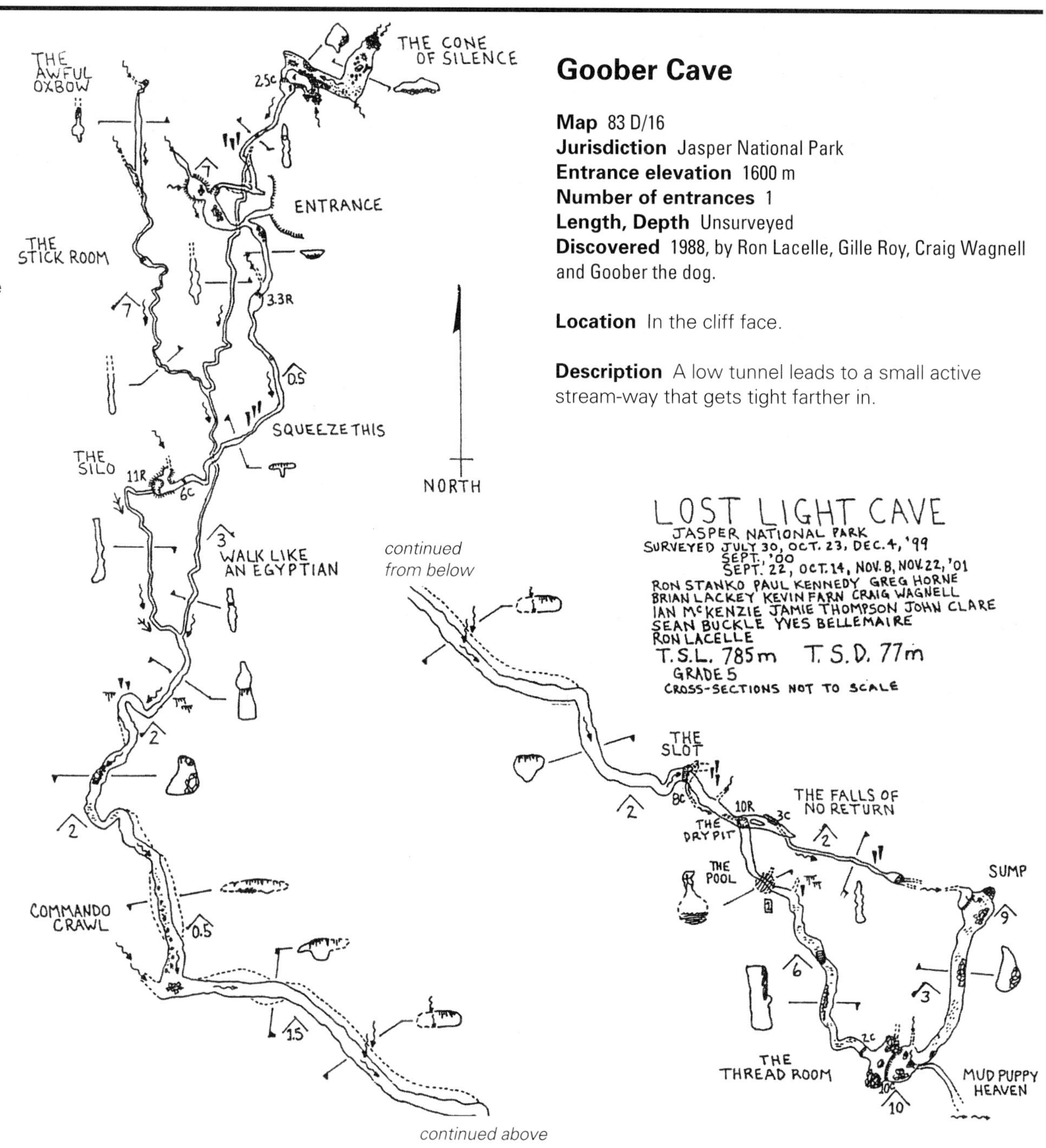

Goober Cave

Map 83 D/16
Jurisdiction Jasper National Park
Entrance elevation 1600 m
Number of entrances 1
Length, Depth Unsurveyed
Discovered 1988, by Ron Lacelle, Gille Roy, Craig Wagnell and Goober the dog.

Location In the cliff face.

Description A low tunnel leads to a small active stream-way that gets tight farther in.

The Snaring Karst

Area Exploration Long-time ASS caver John Donovan first searched for caves in this area when he walked around Mt. Robson in 1973. He was unable, however, to find a way through the cliff bands up onto the karst plateau above the Snaring River. Some years later, while on a ski trip up the Snaring River, Ben Gadd noted that the geology of the area looked promising, and a subsequent examination of aerial photographs confirmed this impression. This led to a three-day reconnaissance in August, 1982, when Ben, Will Gadd and Ian McKenzie flew in courtesy of a Parks Canada helicopter. During this trip one cave was discovered: Double Rubble Pit. Crooked Eye, Goat's Lair and Trundle Cave were noted, but not entered.

The locality was encouraging enough that a one-week expedition involving eight ASS members was organized for August, 1983, funding for a helicopter once again provided by Parks Canada — a unique co-operative exercise in Canadian Rockies caving not seen since the McMaster explorations headed by Derek Ford in the 1960s. A subsequent report was compiled by Ben Gadd from which the majority of this information was obtained.

In 1983 two main karst areas were explored totaling 19 km^2. These were the plateau to the east of Snaring Mountain and the northwest face of Buttress Mountain. All the explored caves were found here, contained in Palliser limestone. The Oliver cirque north of Mt. Oliver and east of Mt. Grisbach was also explored. But while this area of Lynx Group limestone had good surface karst features, no caves were found.

Ben Gadd identifies the Snaring surface karst as being amongst the most extensive in the Canadian Rockies, comparing it favorably with the Dezaiko Range to the north and the Castleguard karst to the south. Although no major cave systems have yet been found in the Snaring Karst, it is obvious that such features must underlie the area, and perhaps in the future more passage will be accessed by digging through the glacial debris and breakdown that currently blocks all the relatively short explored caves.

Area Geology Regarding the Oliver cirque area, Ben Gadd makes the following remarks: "Perhaps the most interesting thing about the cirque — and about the entire Oliver area — is that the karst is developed in interbedded limestones, siltstones, and dolostones of the Lynx Group, an upper Cambrian unit not previously known as karstic. (In the Snaring massif next door, we found karst only in the Palliser Formation, a massive Devonian limestone famous for karst development in other parts of the Rockies). Bedding in the Lynx is fairly thin. Most layers are 10 cm to 1 m thick, and the silty and dolomite layers are not as easily dissolved by groundwater as the limestone ones. Thus, the limestone beds show well-developed karst, while the silty and dolomitic beds show little or none. The rock colour, texture, and karst development vary from bed to bed in any exposed sequence, giving it a striking, banded look. The geology of the karst at Oliver is quite different from the rest of the Snaring karst and perhaps unique in the Rockies."

Hydrology Three major sinks in the Oliver cirque seem to reappear only a kilometre to the north where the creek draining the cirque rises. A dye test will be needed to confirm this. Two springs at 1980 m and 2135 m, just below the northeast edge of the Snaring Karst, appear to drain the area. Goat's Lair at an elevation of 1830 m is obviously a resurgence point of some significance.

Cave Softly The Snaring Karst is a pristine wilderness area — leave no sign of your passing. Ben Gadd believes that "Cavers, especially in organized groups, are extremely preservation-minded and safety-conscious. Beyond the activities of organized groups such as the Alberta Speleological Society, little serious caving is done in the Rockies and I doubt that any but competent cavers would take the trouble to reach the isolated Snaring area. Therefore, I don't think it's necessary to issue some sort of order placing the caves off-limits to park visitors."

Access for all Caves Without the use of helicopters (park-sponsored trips only), the Snaring Karst is hard to reach. Ben Gadd notes that "the ground approach to the area remains untried. Perhaps it will be easier than anticipated." A suggested route through the seemingly impenetrable headwall above the Snaring River is via a small cirque at GR 154764 on map 83 E/1. During exploration of Goat's Lair Cave this proved to be the least precipitous route up into the karst (short of entering from the Oliver end). This would require 1200m in elevation gain from the Snaring River to the ridge crest at 2255 m, then a drop of 150 m to reach the Sinking Ponds campsite. In 1983, Kim Smallwood and Will Gadd accessed the area from a south fork of the Snaring River via a col at GR 142751. The hike/climb took 12 hours and required about 50 m of easy 5.5 technical climbing on the final headwall below the col.

The Swing and Slim Pickins can be accessed on foot from the Snaring River via a hike up the northwest ridge of Buttress Mountain.

"James Hector, leader of the 1858 Palliser Expedition, named Snaring Mountain after the Snaring Indians who obtained their food by snaring rabbits etc. in pits with loops." from *Place Names of the Canadian Alps.*

Ice Trap

Map 83 E/1
Jurisdiction Jasper National Park
Entrance elevation 2165 m
Number of entrances 1
Length, Depth Unknown
Discovered August 1983, by Will Gadd

Location On the eastern edge of the Snaring Karst is a small raised plateau named The Deck. Ice Trap is located at the base of the cliff marking the south wall of the cirque immediately northwest of The Deck. From The Deck a steep goat trail descends the east side of a creek into the cirque.

Description Ice Trap has a small entrance leading to an ice-covered shaft with an aven. The cave has not been fully explored or surveyed and is a good prospect for a future visit.

Sand Crawls

Map 83 E/1
Jurisdiction Jasper National Park
Entrance elevation 2135 m
Number of entrances 2
Length A few metres
Discovered August 1983, by Eric von Vorkampff & Ben Gadd

Location The Sand Crawls are contained in a patch of tightly folded, unstable rock below the northwest side of The Deck. Follow a steep goat track that leads from The Deck to a depression on the north side. The Sand Crawls are just east of the trail.

Description & Geology Although these caves are short, Ben Gadd included these tubes in his survey of Snaring caves owing to the pale orange Gog Quartzite sediments found in them (also found in Cliffside Cave). "Because the nearest Gog Group rock is on the west side of a thrust fault, ten kilometres to the west, these sediments had either been carried into the area or were deposited before being stripped from the area by erosion. Either way, they provide important clues as to the geomorphological history of the area, and the speleogenesis of the caves."

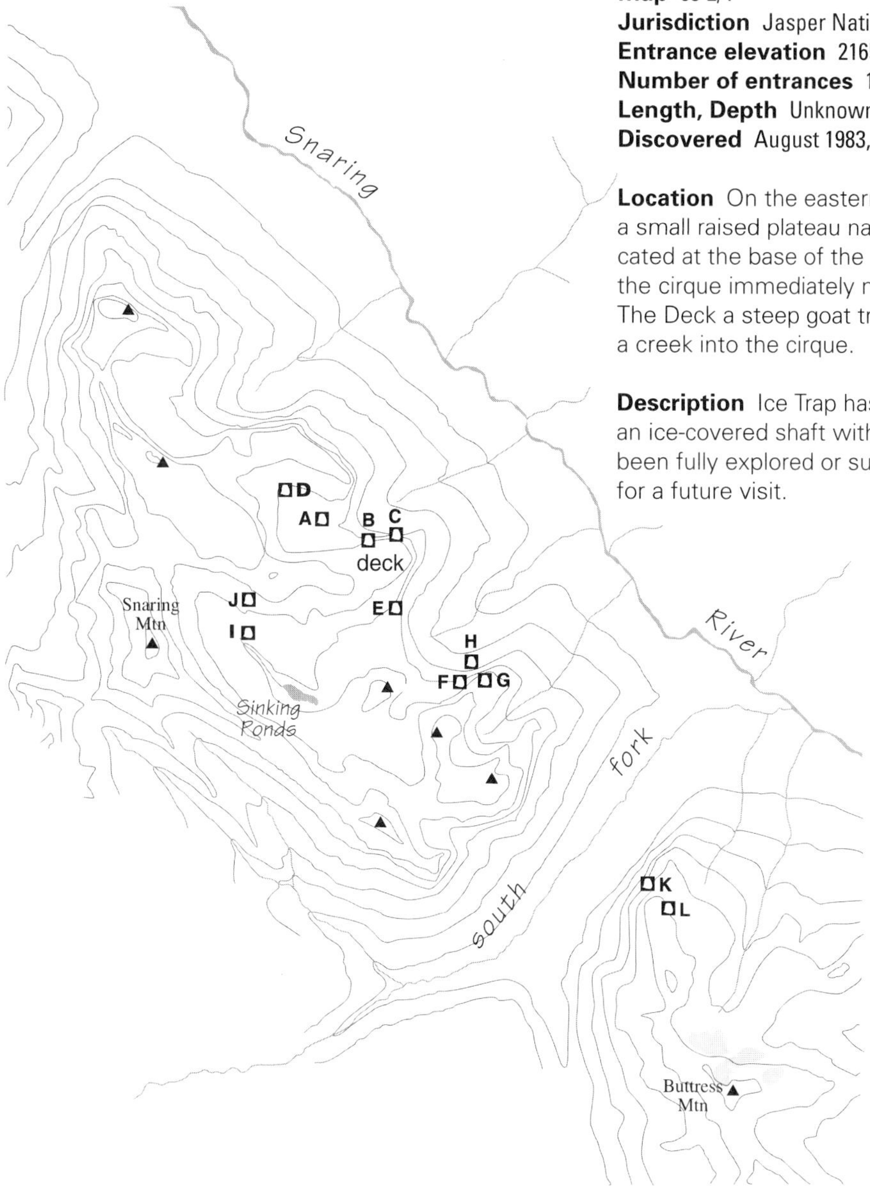

KEY
- A Ice Trap
- B Sand Crawls
- C Cliffside Cave
- D Ice Flower Cave
- E Afterthought Cluster
- F Crooked Eye Cave
- G Trundle cave
- H Goat's Lair
- I Double Rubble Pot
- J Ca-Ca Cave
- K The Swing
- L Slim Pickins

Cliffside Cave

Map 83 E/1
Jurisdiction Jasper National Park
Entrance elevation 2350 m
Number of entrances 1
Length 75 m
Discovered Spotted and photographed in 1982.

Location The dramatic entrance to Cliffside Cave, 20 m x 20 m, is located in the 500 m-high cliff face along the north side of The Deck.

It is most easily reached from the northeast via a traverse along a ledge followed by an 85 m rappel.

Exploration & Description Cliffside was first explored in 1983 by Eric Von Vorkampff and Ben Gadd. After noting the entrance contained an ice-floored lake, they climbed over the blocks of breakdown inside and "found that the passage remains wide but becomes lower. It ends at a blockage of rubble and ice about 75 m back. That this is a major cave passage is shown by its size and by the presence of numerous shafts in the ceiling. Eric climbed into a couple of the more accessible ones; they narrowed upward. Near the entrance, the smaller shafts were plugged with Gog-derived sand. There are a couple of cliff-swallow nests there, at the high elevation of 2350 m."

Cliffside Cave. Photo Ben Gadd.

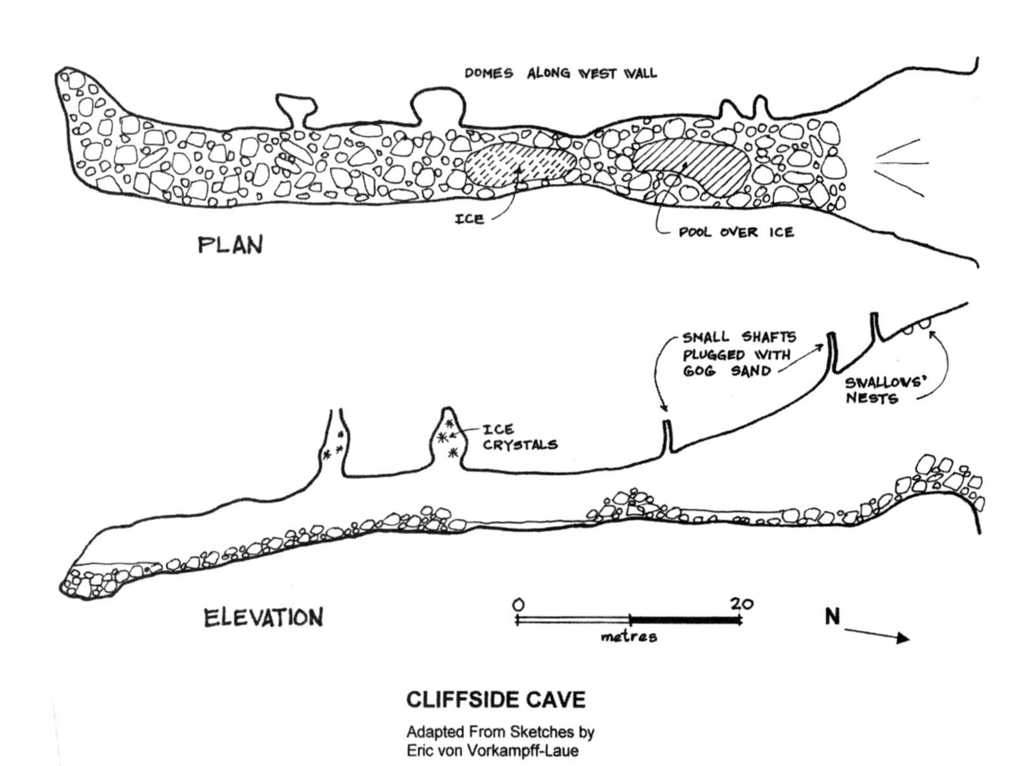

CLIFFSIDE CAVE

Adapted From Sketches by
Eric von Vorkampff-Laue

Ice Flower Cave

Map 83 E/1
Jurisdiction Jasper National Park
Entrance elevation 2257 m
Number of entrances 4 or 5
Length 200 m
Discovered August 1983, by Eric von Vorkampff

Location Ice Flower Cave is located in the top of a bench above the cirque northwest of The Deck. From The Deck a steep goat trail descends the east side of a creek into the cirque. Cross the cirque on karst pavement to just west of a large rock slide.

Description The longest cave yet discovered in the Snaring Karst, Ice Flower has four or possibly five entrances depending on snow conditions. It consists of four joining phreatic tubes, passage size varying from stooping to walking. Large (5 cm across) ice crystals were observed on the ceiling.

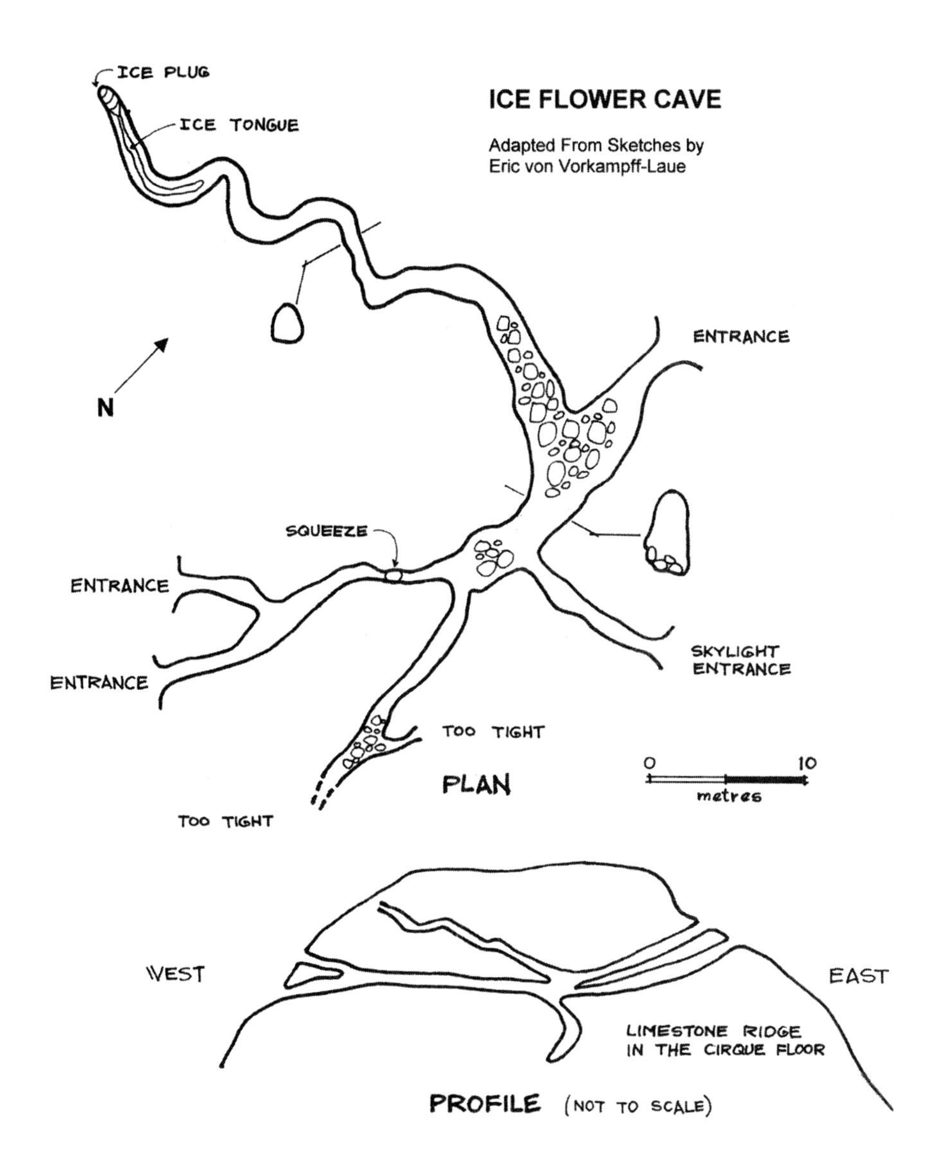

Afterthought Cluster

Map 83 E/1
Jurisdiction Jasper National Park
Entrance elevation 2287 m
Number of entrances Upper cave 1, lower cave 1
Length 33 m & 20 m
Discovered August 1983, by Marg Saul, Eric von Vorkampff, Gordon Fraser & Ben Gadd

Location Of several entrances found in the cliff just below the southern edge of The Deck, these are the two largest. These caves can be seen while approaching in a helicopter. They cannot be seen from above. The upper cave is approached by following a sloping ledge from the east. Accessing the lower cave requires an easy 20-m down-climb.

Description Both the upper and lower caves are formed in the same fault, the upper cave having good-sized passage, but ending after 33 m in an ice plug.

Three other entrances 20 m to the east were not checked, but one looks like a good prospect.

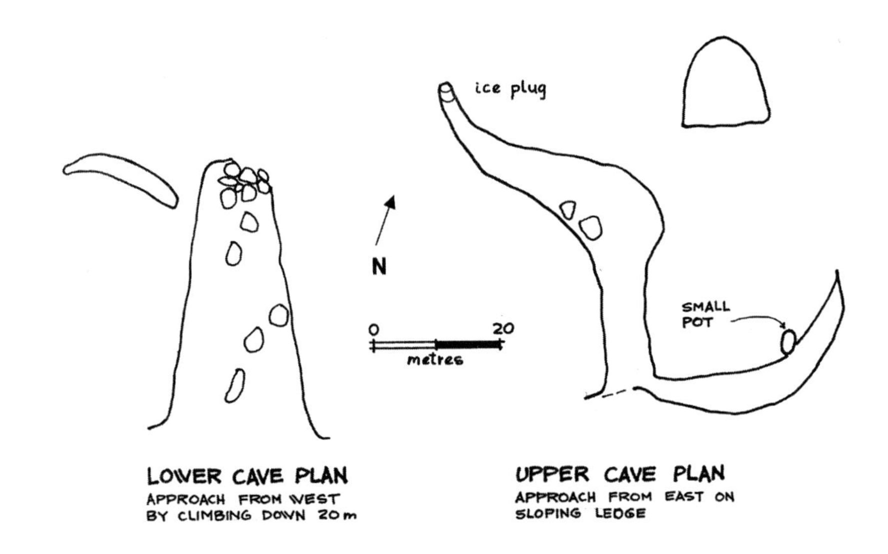

AFTERTHOUGHT CLUSTER

Adapted from sketches by Ben Gadd

Crooked Eye Cave

Map 83 E/1
Jurisdiction Jasper National Park
Entrance elevation 2013 m
Number of entrances 5
Length Approximately 130 m
Discovered August 1983, by Eric von Vorkampff & Will Gadd

Location This cave is located 1 km southeast across the cirque from Afterthought Cluster and faces north. It can be reached by a short rappel from the cliff top.

Description Ben Gadd writes: "The surveyed length was 129 m, all in dry phreatic tubes filled to within one or two metres of the roof with silt and sand (including more Gog material, according to Eric). The deepest passage ends at a silt blockage, which Eric felt could be removed with only a little effort and would likely open to what he thought might be a very large cave system beyond."

CROOKED EYE CAVE

Adapted from a survey by Eric von Vorkampff-Laue and Will Gadd, August 19, 1983

Trundle Cave

Map 83 E/1
Jurisdiction Jasper National Park
Entrance elevation 2013 m
Number of entrances 1
Length 38 m
Discovered August 1983, by Eric von Vorkampff & Will Gadd

Location A little to the east, and in the same rock layer as Crooked Eye Cave. Approach as for Crooked Eye and make a short rappel from the cliff top.

Description Trundle Cave consists of 38 m of gently descending phreatic passage ending in a mud and ice plug.

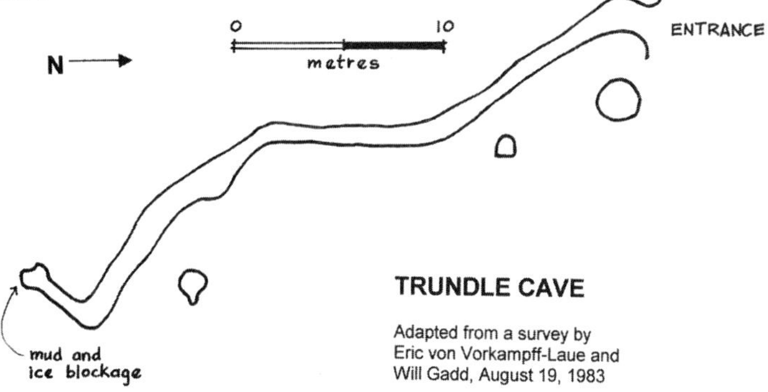

TRUNDLE CAVE

Adapted from a survey by Eric von Vorkampff-Laue and Will Gadd, August 19, 1983

Goat's Lair

Map 83 E/1
Jurisdiction Jasper National Park
Entrance elevation 1830 m
Number of entrances 1
Length c. 30 m
Discovered 1983, by Eric von Vorkampff

Location As for Crooked Eye Cave. Goat's Lair is situated at the base of the headwall beneath Crooked Eye and Trundle caves. The cave is reached from above by way of a very steep goat path (a 300 m descent not for the faint-hearted). It can also be accessed from the Snaring River via the small cirque at GR 154764.

Description Goat's Lair is interesting because it occurs nearly at the base of the Palliser Formation and about 300 m lower topographically than any other cave found in the Snaring karst. Furthermore, it carries a small stream out of the cliff.This was the headwall that had stopped John Donovan on his 1973 reconnaissance.

Eric von Vorkampff describes the cave thus: "The passage is of the rift type, angling gently upward at about 15° in shaley limestone. There is a climbable three metre step in the cave about 20 m in, and the passage contains a small stream. The cave becomes too tight at the end, but can be seen to continue."

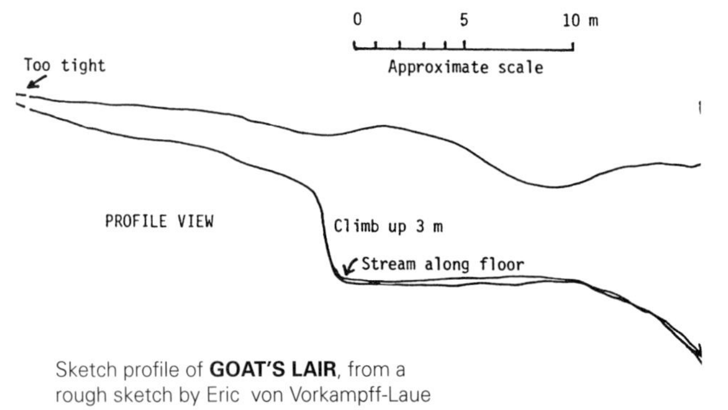

Sketch profile of **GOAT'S LAIR**, from a rough sketch by Eric von Vorkampff-Laue

Double Rubble Pit

Map 83 E/1
Jurisdiction Jasper National Park
Entrance elevation 2196 m
Number of entrances 1
Depth 60 m shaft
Discovered August, 1982, by Will Gadd

Location Double Rubble Pit is located towards the western edge of the plateau, across from and below the summit of Snaring Mountain. From the Sinking Ponds (site of 1983 camp), follow an ephemeral drainage channel northwest for about 1 km, looking for a narrow rectangular pit.

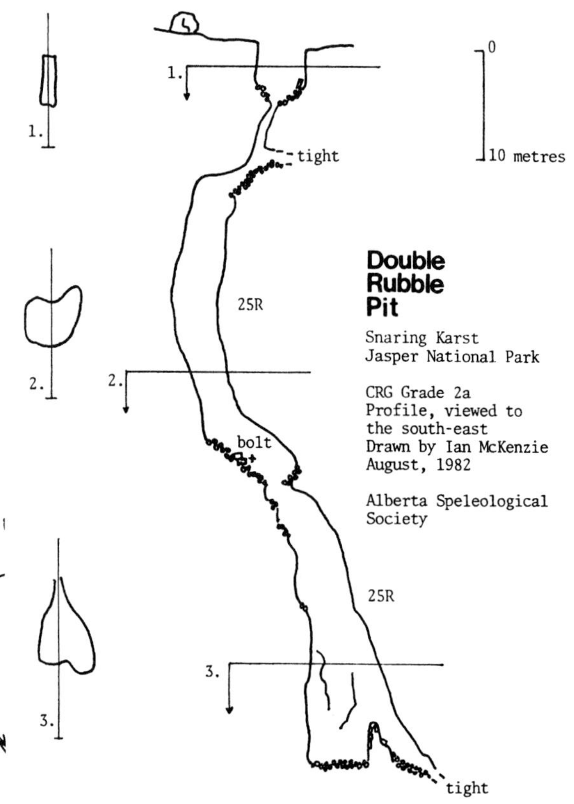

Description Double Rubble Pit was the only cave mapped during the three-day reconnaissance trip to the Snaring karst in 1982. (Four other pits were descended but all choked at about 15 m to 25 m depth.)

Ian McKenzie wrote in 1982: "The narrow, rectangular pit didn't look too promising, located in thin, alternating beds of limestone and siltstone dipping at about 60° and severely chipped and shattered at the surface. But when the very tight squeeze at the bottom produced 16 seconds of ricocheting rock fall, we began to wonder..." Eventually, after two 25 m pitches and a break for lunch, the pit ended at a disappointing flat choke.

Double Rubble, with some squeezing and removal of debris achieved the respectable depth of 60 m.

Ca-Ca Cave

Map 83 E/1
Jurisdiction Jasper National Park
Entrance elevation 2257 m
Number of entrances 1
Length 10 m
Discovered August 1983, by Marg Saul & Pam Burns

Location A prominent rift-like entrance located 500 m north of and 60 m higher than Double Rubble Pot. Approach as for Double Rubble Pot from Sinking Ponds, then continue north. Reportedly, the entrance was full of snow during the 1992 visit.

Description Named for the packrat scat found on the floor, Ca-Ca Cave consists of a short length of rift passage disappearing into a clear ice plug. Not surveyed.

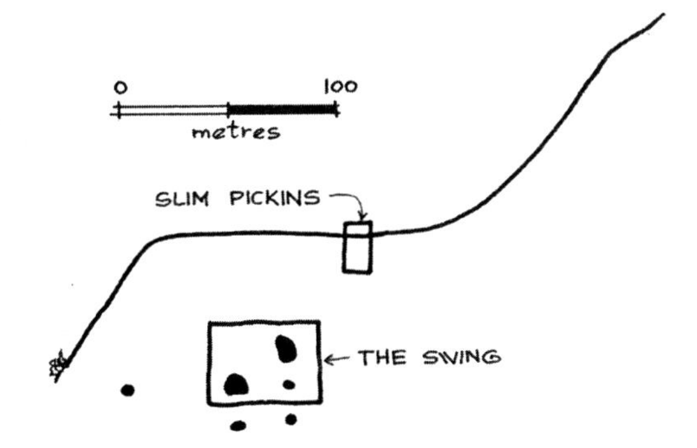

CLUSTER ARRANGEMENT
VIEW FROM NORTH

CAVES IN BUTTRESS MOUNTAIN

Adapted from sketches by Kim Smallwood, Pam Burns & Marg Saul

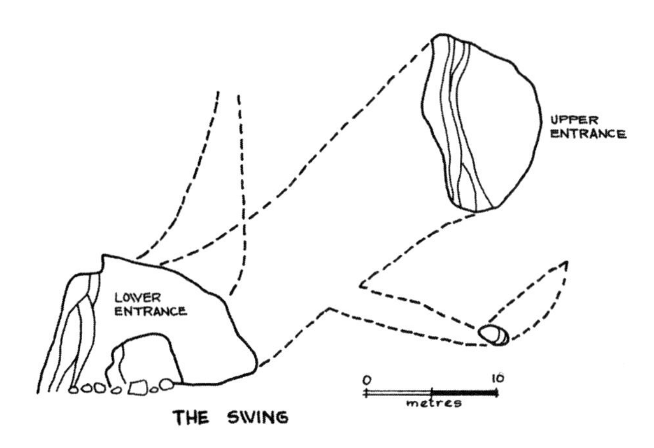

The Swing

Map 83 E/1
Jurisdiction Jasper National Park
Entrance elevation 1982 m
Number of entrances 2 (+1 skylight)
Length, Depth 30 m, 20 m
Discovered Located and photographed in 1982.

Location The northwest ridge of Buttress Mountain, which is separated from the Snaring karst by a south fork of the Snaring River. Either helicopter in or hike up the ridge from the Snaring River. The cave is located on west side overhanging cliffs and requires an 85 m free rappel from the ridge above. Finding the entrance from above is difficult. During the first exploration in 1983, the use of radios and a spotting scope enabled Pam Burns, located across the valley, to direct Kim Smallwood and Will Gadd to the largest of the holes. Even so, on the first attempt Kim found himself dangling 6 m to the west and 8 m out from the entrance with 600 m of space beneath him! A second attempt by Will Gadd with a rock tied to the end of a rope enabled them to pull into the entrance.

Exploration & Description From inside the cave a lower entrance and window was located. The only lead, a 20 m+ aven, heads up from just inside the lower entrance and looks climbable.

Slim Pickins

Map 83 E/1
Jurisdiction Jasper National Park
Entrance elevation 2135 m
Number of entrances 1
Depth 30 m shaft
Discovered Located and photographed in 1982.

Location High on the northwest ridge of Buttress Mountain, west of the rappel point for The Swing. Either helicopter in or hike up the ridge from the Snaring River to a flat area close to tree line.

Exploration & Description A few metres from where Will Gadd and Kim Smallwood camped in 1983 for their exploration of the entrances in Buttress Cliff, they found a tight-branching shaft that they explored to a depth of about 30 m.

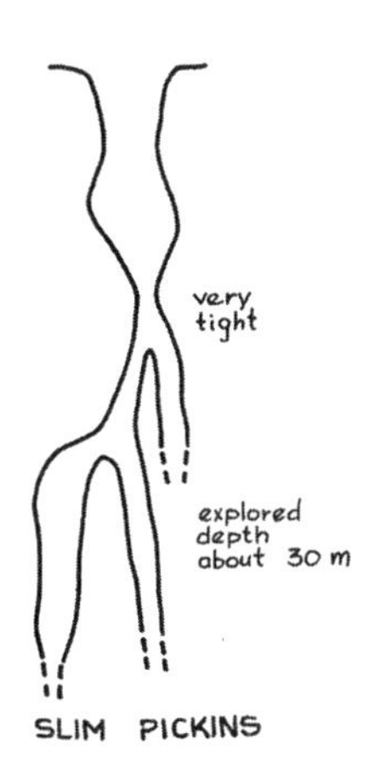

Roche Miette Caves

Named after the French corruption of "myatuck," Cree for Bighorn Sheep, this prominent mountain (2316 m) is formed of Palliser limestone that contains four caves. The two shafts on the benches below Roche Miette (Disaster Point Caves and Procrastination Pot) are fossil sink points. Caveat and Twin Holes are passage remnants exposed by glacial activity along the Athabasca Valley.

These caves are best visited when the area is snow free, as the approaches for all but the Disaster Point Caves require steep scrambling.

Access for all Caves Hwy. 16 (Yellowhead Hwy.), approximately 27 km north of Jasper townsite and 4 km north of Talbot Lake. Shortly after crossing the Rocky River you cross a second bridge just before a road-cut bluff on the right side. Park at a pull-off on the right just after the bluff, site of an old quarry and former Alpine Club of Canada shelter.

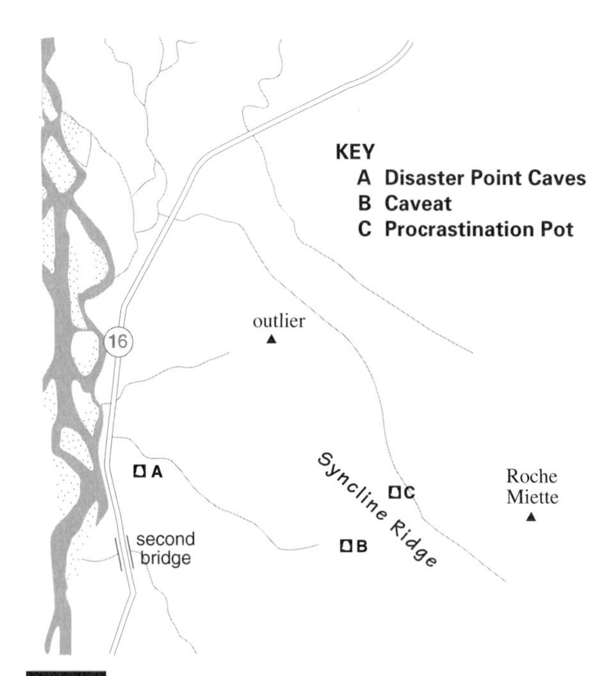

Disaster Point Caves

Map 83 F/14
Jurisdiction Jasper National Park
Entrance elevation 1082 m
Number of entrances 2
Length, Depth 673 m, 50 m
Discovered Unknown

Location The caves are located 90 m above the highway and about 200 m east on a dry rocky slope surrounded by scraggly White and Englemann spruce, Douglas fir and juniper — a favorite location for bighorn sheep and mountain goats.

Walk back along the highway ditch for a short way, then climb steeply to the top of the bluff. Follow the open ridge southeast and uphill for about 15 minutes, keeping the highway visible below you on the right. Look for a faint trail leading into trees on your left. The closer of the two entrances is located in a dense clump of trees about 10 m to the left of the open ridge.

The Climb out of Disaster Point Caves.
Photo Jon Rollins.

Exploration & Description Disaster Point Caves are fossil sink points, consisting of two steeply descending ice-lined tubes that join and end in wet sediment. The McMaster Karst Research Group was made aware of it by "a young Jasper resident" in 1968. In the same year they explored and surveyed the cave after having removed a large hanging icicle from the entrance by the use of explosives.

Just off the main shaft at about 30 m depth is a small chamber containing some nice ice crystals. The ice in the cave was sampled by Bill MacDonald in 1991 and found to be derived from snow that has drifted in and become compacted into ice. As with other ice caves in the Rockies, the ice in Disaster Point Caves appears to be going through a cycle of melting and reforming; whether it is due to human interference or to natural causes is unclear. Ice has currently (1998) reformed in the main entrance shaft, rendering it impassable.

Fauna The large mammal bones found in the sediments at the bottom of the shaft indicate the cave traps animals. A tiger salamander was seen on a visit to the cave in 1991.

Warning Although short, the steep inclination of the passage and the presence of ice make bottoming the cave a less than straightforward exercise. Be familiar with SRT before attempting this cave.

Cave Softly Avoid disturbing bones at the bottom of the cave.

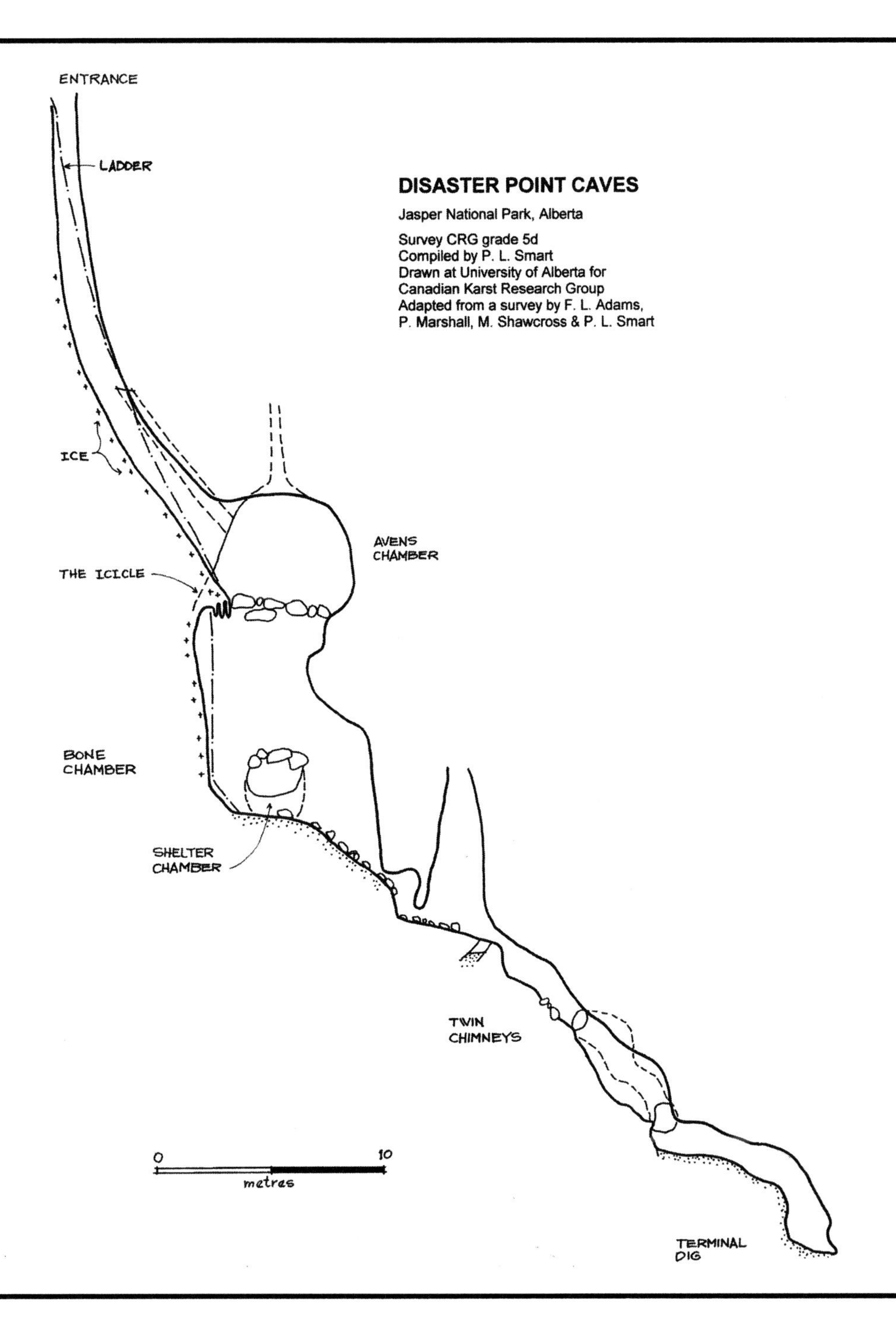

Twin Holes

Map 83 F/4
Jurisdiction Jasper National Park
Entrance elevation 1433 m
Number of entrances 2 caves, each with 1 entrance
Length Cave #1 15.5 m, Cave #2 35.1 m
Discovered 1982, by Jamie Thompson & Ron Lacelle

Location Park as for Disaster Point Caves. From the pull-out hike northeast? across an intermediate drainage , then through a gap between Syncline Ridge and an outlying hill. Angle east? below the north-facing slope of Syncline Ridge. The entrances are located just above treeline on the northwest face of Roche Miette and can be seen from the highway.

Description These two caves are large (2 m in diameter) phreatic tubes that provide pleasant walking passages to rubble and mud chokes.

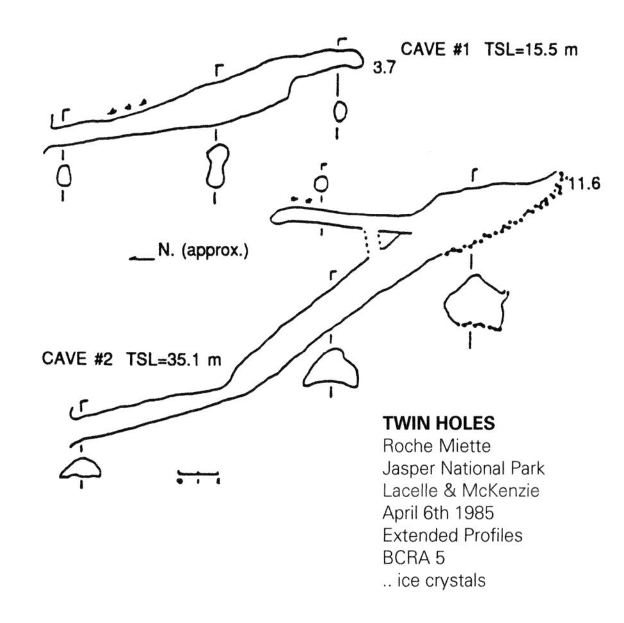

Caveat

Map 83 F/4
Jurisdiction Jasper National Park
Entrance elevation 1281 m
Number of entrances 1
Length c. 20 m
Discovered 1988, by Ian McKenzie

Location This tiny cave is located in the west face of Syncline Ridge, about 1.5 km northeast ??? of Disaster Point Caves. It is visible from the highway just north of the Rocky River crossing as a black dot near the bottom of a diagonal fault.

Access as for Disaster Point Caves, but park at a pull-out on the left (west) side of the highway opposite the bluff (plaque commemorating Jasper House).

Cross the highway and heading right, follow the old road-bed between the cliffs and a pond until it heads back towards the highway. Climb the steep slope on your left and follow a treed ridge above for about 2 km, keeping open views of the Rocky River and the highway on your right. The ridge climbs steadily with patchy burn areas in the trees. Upon reaching a larger burn area, climb steeply up to your left, looking for a distinctive orange-coloured rake on the cliff with a white patch towards the top. Continue right along the base of the cliff until the small round cave entrance comes into view. Allow about 1 hour.

Description A short section of promising phreatic passage intersects a passage that quickly becomes too tight up to the left, and that has been excavated a short distance to the right. 10-20 m higher along the diagonal fault are tiny water-worn fissures and scallops, and two other holes. The uppermost is enterable but immediately closes down. A piton is located in a bulge between the two lowermost entrances.

The entrance to Caveat Cave at left centre.
Photo Jon Rollins.

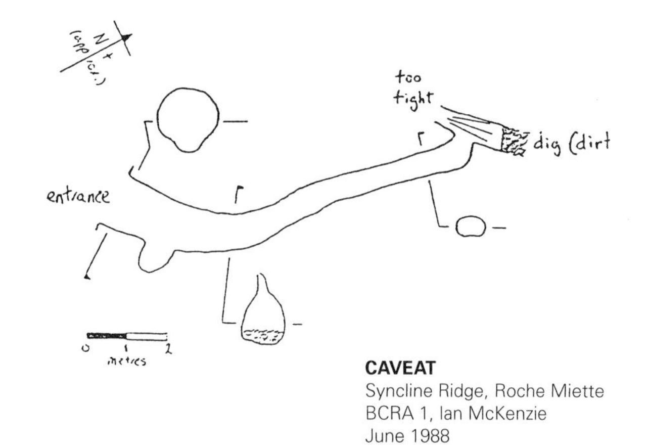

CAVEAT
Syncline Ridge, Roche Miette
BCRA 1, Ian McKenzie
June 1988

Procrastination Pot (NDP)

Map 83 F/4
Jurisdiction Jasper National Park
Entrance elevation 1700 m
Number of entrances 3
Length, Depth 159 m, 56 m
Discovered 1975, by Myron Oleskiw

Location Park as for Disaster Point Caves. Then climb up Syncline Ridge, looking for a depression on a bench. The two impressive shaft entrances (the third is small and unsurveyed) are hidden by trees.

Warning The approach requires a short section of steep scrambling. Be conversant with SRT before attempting this cave.

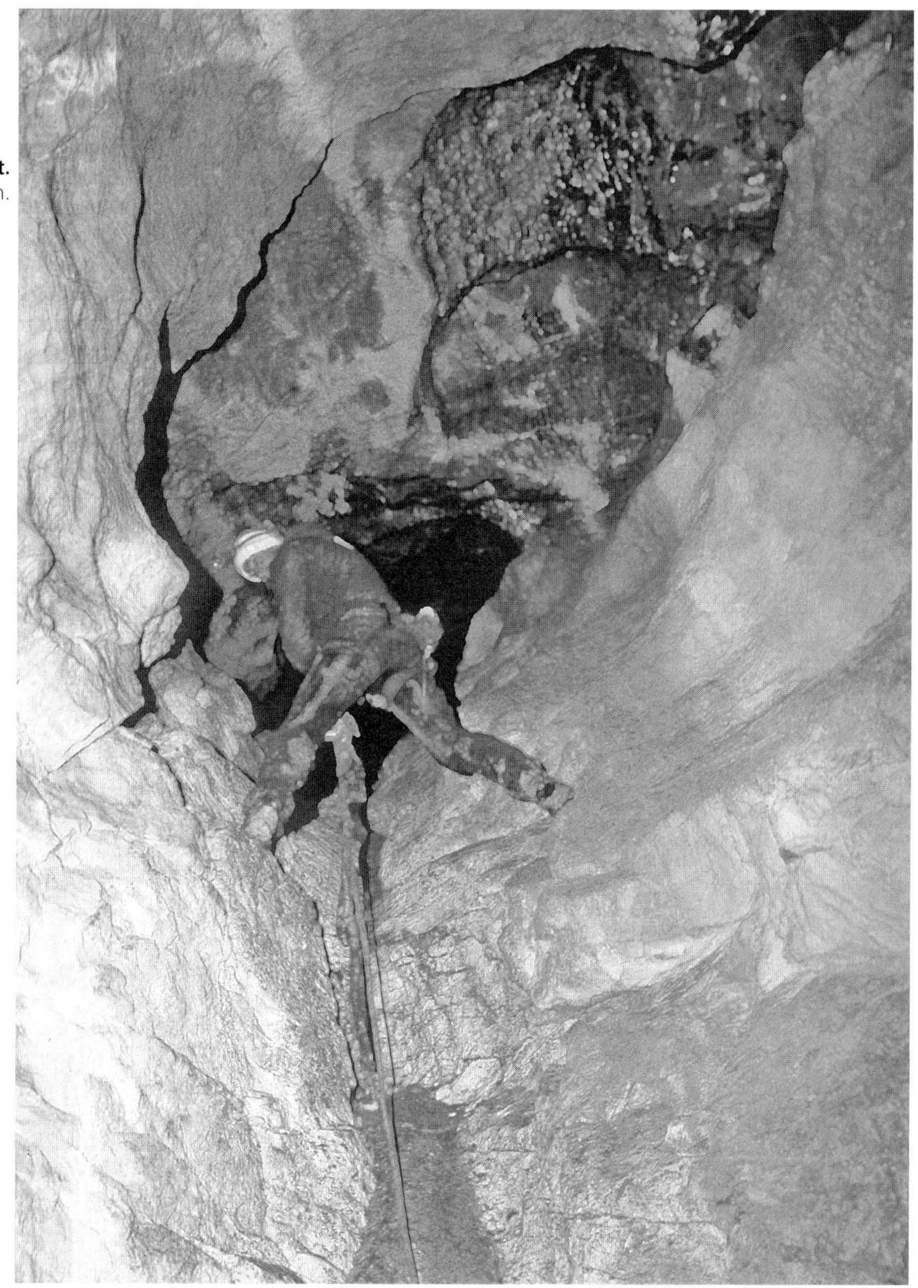

Procrastination Pot. Photo Dave Thomson.

Looking along Syncline Ridge at the approximate location of Procrastination Pot. Photo Jon Rollins.

Exploration & Description Procrastination Pot was at first thought by its discoverer to be the Disaster Point Caves, hence the interim name NDP (Not Disaster Point). Although similar in structure, it is in fact grander than Disaster Point Caves, being much deeper and having larger passages. The two large phreatic entrance tubes are particularly impressive with well-developed scallops.

The cave consists of four pitches leading to a drafting boulder choke that was dug on a later visit and lengthened the cave by a few more metres. There is a good possibility of reaching more cave, but the digging in breakdown is awkward.

Fauna During exploration large numbers of Little brown bats were seen roosting in the bottom of the cave. Large deposits of bat bones throughout the cave (some calcified) suggest that it has been used as a roost for a very long time. A bear skull was also found just before the head of the last pitch.

Cave Softly Do not visit this cave during winter months when Little brown bats are hibernating.

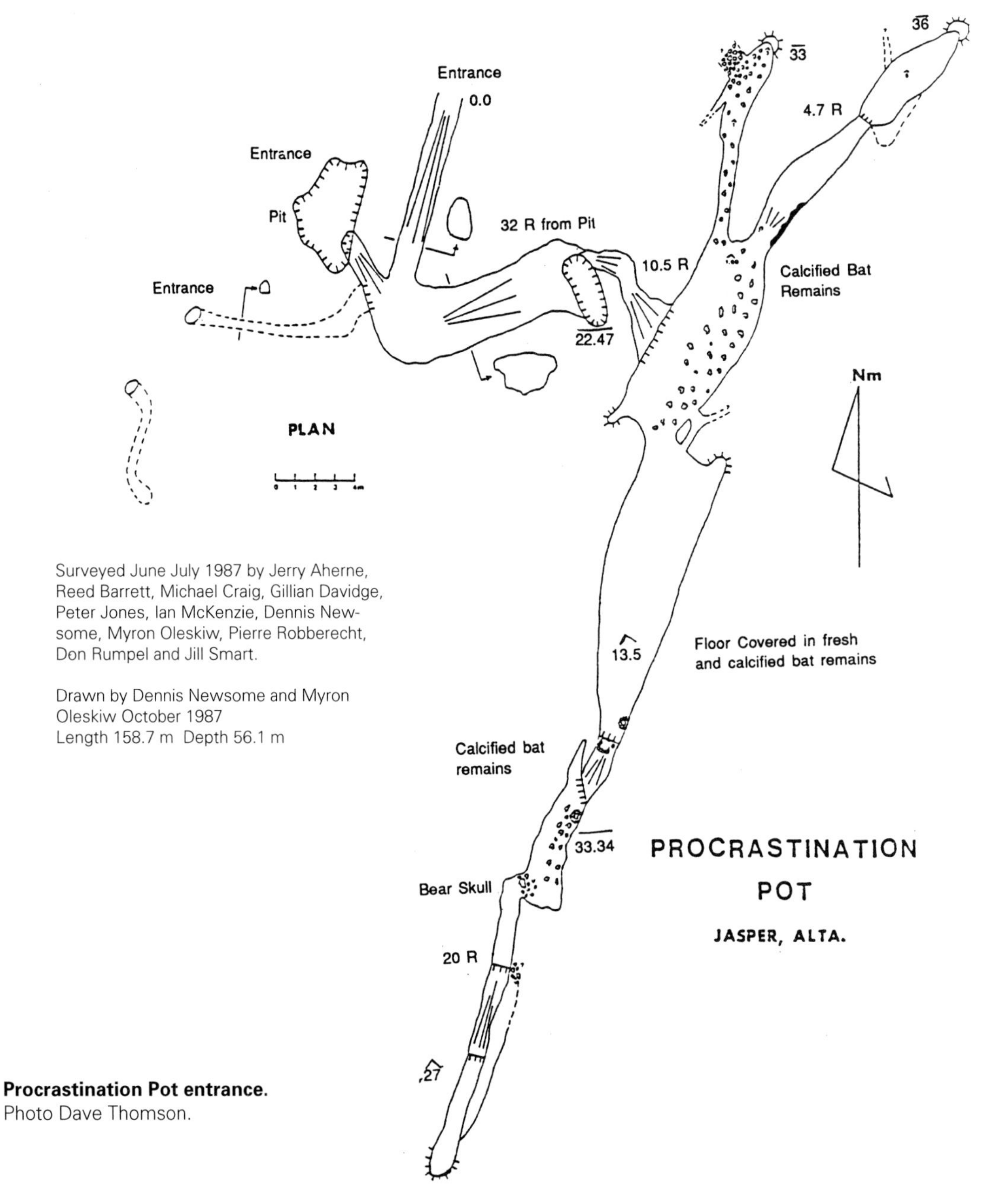

Surveyed June July 1987 by Jerry Aherne, Reed Barrett, Michael Craig, Gillian Davidge, Peter Jones, Ian McKenzie, Dennis Newsome, Myron Oleskiw, Pierre Robberecht, Don Rumpel and Jill Smart.

Drawn by Dennis Newsome and Myron Oleskiw October 1987
Length 158.7 m Depth 56.1 m

Procrastination Pot entrance.
Photo Dave Thomson.

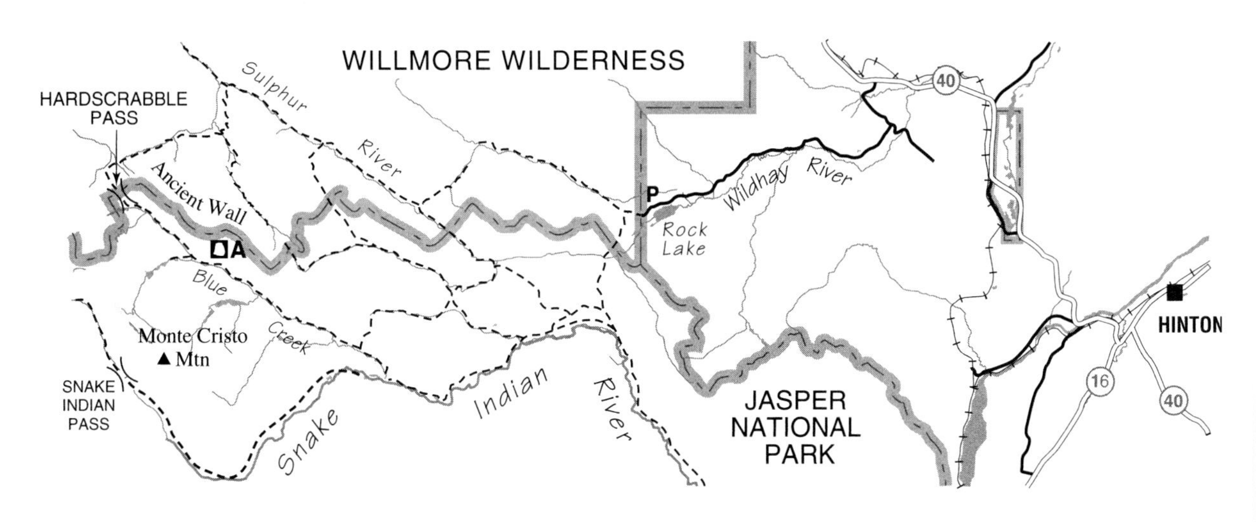

Key
A Ancient Wall Caves

Ancient Wall Caves

Map 83 E/7
Jurisdiction Jasper National Park
Entrance elevation 2135 m
Number of entrances Several small tubes at the base of the Natural Arch
Length, Depth 10 m
Discovered Unknown, the Natural Arch has been known about for over a century.

Access 4 km west of Hinton on Hwy. 16, turn north along Hwy. 40. Follow Hwy. 40 for 40 km then turn left (west) onto the Rock Lake Road. In 2 km turn left and follow the main road for 27 km to Rock Lake Provincial Recreation Area with campground. The road continues along the north shore of Rock Lake to a parking lot and trailhead.

From Rock Lake it is about 65 km to Natural Arch backcountry campground via the Willow Creek, North Boundary and Blue Creek trails (2 1/2 days hiking on trails one way).This probably rates as the longest hike to the shortest caves in the Canadian Rockies.

Description The Ancient Wall is a 30 m-long ridge along the northeast side of Blue Creek. Marked on the topo map is the Natural Arch and a number of closed depressions. The arch is impressive, being about 30 m across and 20 m high. At its base are a cluster of phreatic tube remnants explored by ASS members Rick Blak and Craig Wagnell in 1984. Two were pushed for about 10 m.

Natural Arch. Photo Gillean Daffern.

MOUNT ROBSON PROVINCIAL PARK

Mount Robson Provincial Park contains Arctomys Cave, which at just over half a kilometre deep is not only Canada's deepest cave, but also the deepest cave north of Mexico. It is fitting that the deepest cave should be adjacent to Mt. Robson which at 3954 m is the highest mountain in the Canadian Rockies.

The same Mural limestone that contains Arctomys also outcrops a few kilometres to the south where the Goat Valley caves are located, and to the west of the Robson syncline where the Small River karst contains many fine caves. Two small caves have also been found on the north side of Mt. Robson.

Area Exploration In August of 1982 and 1983, a small team of English and Canadian cavers made a lightweight expedition around Mt. Robson following the Mural limestone. Starting from the Valley of a Thousand Falls the team crossed Robson Pass and went up Calumet Creek, over Moose Pass and then back down the Moose River. Apart from the Goat Valley area just south of Arctomys Valley, no new karst areas were found. However, bad weather and a lack of time limited exploration and the Robson area still holds good prospects for those willing to hike long distances.

Area Geology The geology map reveals the Mt. Robson area to be a major syncline with its axis trending approximately northwest-southeast. The effects of faulting and thrusting have modified the simple synclinal structure causing the repetition of the Mural and associated beds in some areas.

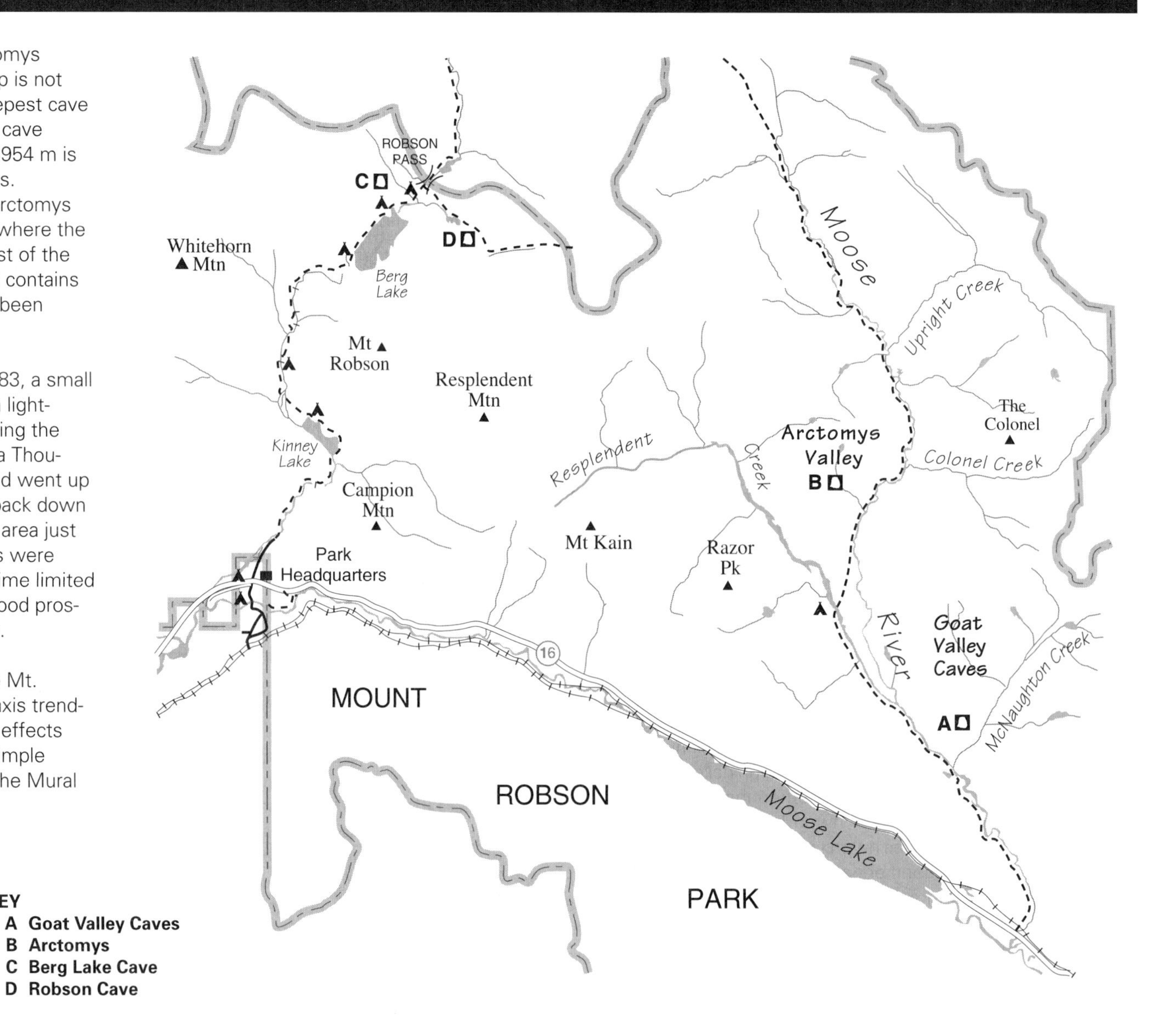

KEY
- A Goat Valley Caves
- B Arctomys
- C Berg Lake Cave
- D Robson Cave

Goat Valley

Located 1000 m above the Moose River is a valley named by cavers after its hoofed inhabitants. It is a 2 km long, and dry with a talus-filled floor. On the west side are inaccessible ramparts of shale and limestone rising to 2300 m, and on the east side is an extensive area of steeply dipping karst pavement composed of the pinkish, speleogenic Mural limestone. Deserted except for hoary marmots and mountain goats, it is an imposing location, offering some fine views of Mt. Robson to the west.

Area Exploration Goat Valley presently contains four caves. Calvados was explored by the 1983 ACRMSE expedition and Chartreuse Pot, Robson View Cave and Meteor Rift by the 1986 ICCC expedition.

The potential for cave development seems good, but it is unclear where the main resurgence for the valley is. A small lake at the head of the valley at 2348 m has an outlet stream that sinks, but is not enterable. Several other sinks were located but were also found to be blocked with debris.

It was while hiking down McNaughton Creek that the large sink, Thor's Mouth, was discovered, with a smaller resurgence 200 m below. The sink (estimated flow 2 to 3 m^3/sec) is in boulders and does not appear enterable. A massive cave entrance in a cliff-face high on the north side of the valley (777735) was checked and found to be the "world's largest frost pocket."

Other valleys in the vicinity remain to be checked for Mural limestone outcrops, and for those willing to climb through steep bush, the likelihood of finding more caves is good.

Area Geology The major sinks in Goat Valley occur in the Lower Carbonate Member of the Mural limestone, which forms a karst dip slope on the east flank of the valley. Along much of the valley length, the Upper Carbonate Member is present in the cliffs forming the steep western valley side. Where this outcrop steadily descends to cross the valley, close to its hanging culmination in the southeast are a number of minor streams sinking into choked holes. Analogous with the Arctomys Valley area, it seems most likely that the Goat Valley water follows the dip of the Mural Limestone to resurge by the major stream (a tributary of the Moose River) in the valley below the end of Goat Valley. (Adapted from a report by Deej Lowe, ACRMSE 1983/84.)

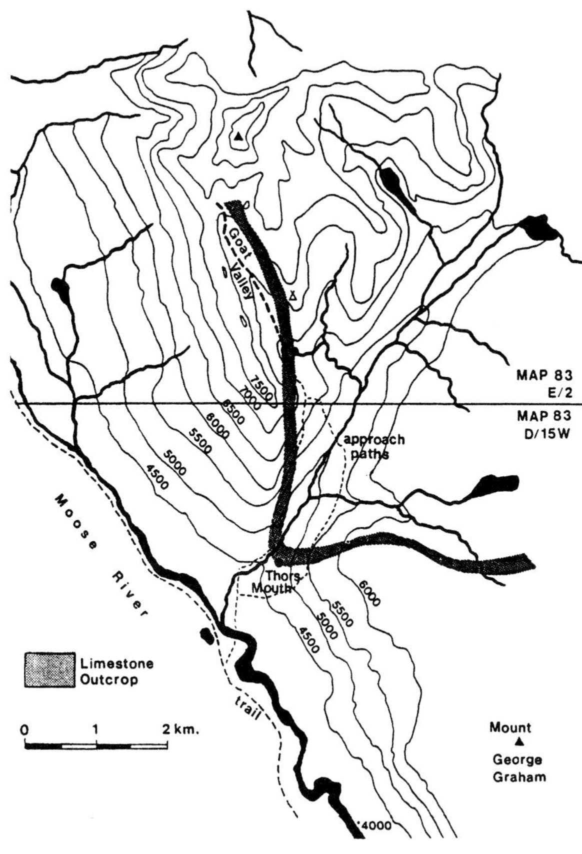

Goat Valley Area (Harry Lock et al 1986).

Access to all Caves Hwy. 16 (Yellowhead Hwy.) 29 km west of the Alberta/B.C. boundary and 2 km before Moose Lake, turn right (north) on a gravel turnoff that runs a short distance to a parking area just below the railbed.

Cross the tracks and follow Moose River trail for 6 km, first over a small shoulder and then along the west bank of Moose River. When level with an obvious forested valley (McNaughton Creek) on the opposite side of the river, cross the Moose River, here conveniently braided into two channels. In summer the water is waist deep (care needed).

A relatively easy hike through forest cut by many channels where McNaughton Creek meets the Moose River leads to Thor's Mouth Sink and resurgence separated by a short, dry valley. The remainder of the route to reach the shoulder at the southern end of Goat Valley (781736) is steeply uphill through thick deadfall bush with occasional game trails and across steep, vegetated talus slopes (total elevation gain 2200 m from the trailhead). Allow 6-8 hours.

Checking for pits in Goat Valley.
Photo Dave Thomson.

Calvados

Map 83 E/2 776755
Jurisdiction Mount Robson Provincial Park
Entrance elevation 2272 m
Number of entrances 1
Depth 43 m shaft
Discovered 1983, by the ACRMSE

Location In the Mural benches on the northeast side of Goat Valley is a small hole in a strike depression.

Description The small hole leads into a 43 m icy shaft beautifully decorated with ice flows and ice flowers. In 1983, the shaft was plugged with ice, but ice chippings could be heard falling at least another 10 m below. The cave entrance was also visited in 1896 during the ICCC expedition, but entry was not possible owing to ice.

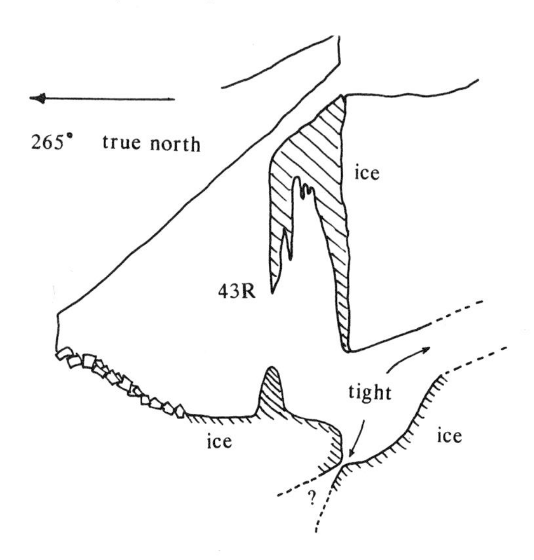

Chartreuse Pot

Map 83 E/2 777750
Jurisdiction Mount Robson Provincial Park
Entrance elevation 2133 m
Number of entrances 1
Length, Depth 88 m, 47 m
Discovered 1986, by the ICCC

Location The lowest elevation cave discovered so far in Goat Valley, this icy shaft is located 500 to the southwest of Calvados Cave and 130 m lower. In fact, Chartreuse Pot was first entered in the mistaken belief it was Calvados.

Exploration & Description An icy 25 m shaft followed a 10 m down-climb on ice was descended using a combination of ropes, descenders and ice axes. It bottomed at –40 m, hence the confusion with Calvados.

From a gravel-floored icy chamber at the foot of the pitch, various sloping bedding plane passages led off, all ending in constrictions. Two hours of digging removed one constriction, and another 15 m of passage went to a definite end just past a beautiful ice column.

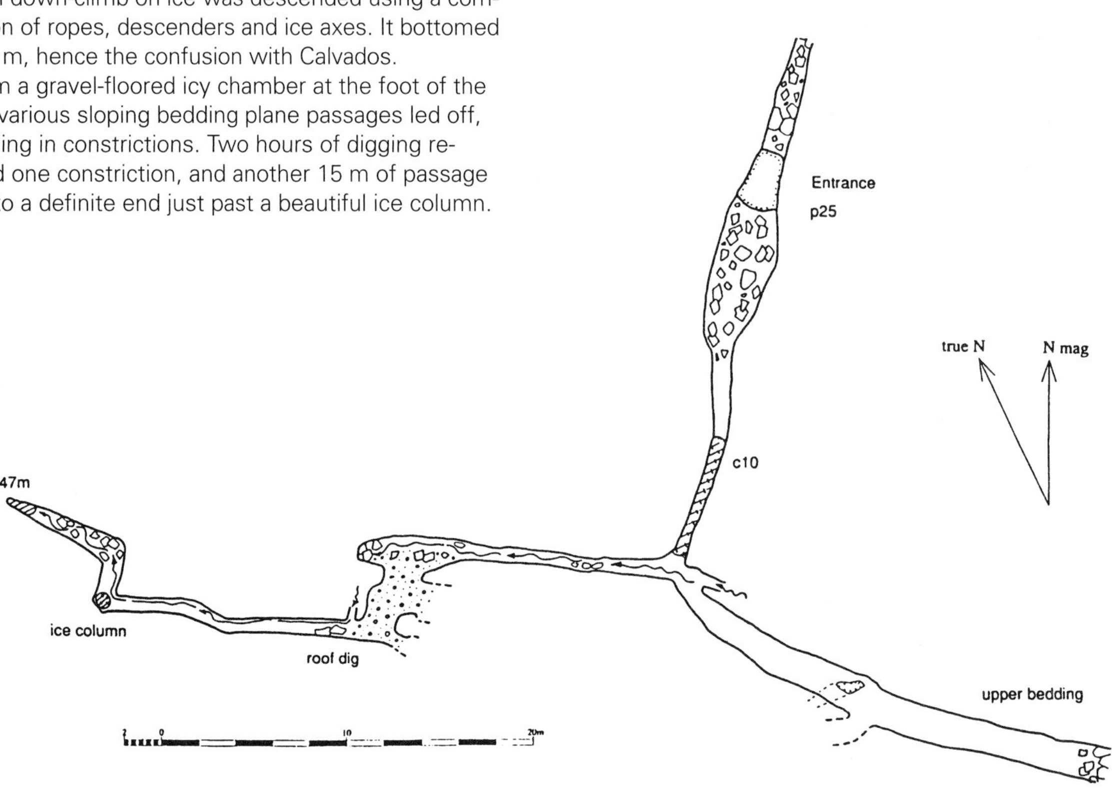

Meteor Rift

Map 83 E/2 769763
Jurisdiction Mount Robson Provincial Park
Entrance elevation 2280 m
Number of entrances 1
Length 60 m
Discovered August 1986, by the ICCC

Location High in Goat Valley, a short distance below the lake sink and on an obvious joint.

Description Meteor Rift takes a small stream that flows to a forbidding green sump pool. Upstream was dug and pushed a short distance to another blockage. The cave has not been surveyed.

Robson View (The Pit)

Map 83 E/2 775760
Jurisdiction Mount Robson Provincial Park
Entrance elevation 2318 m
Number of entrances 2
Length, Depth 168 m, 77 m
Discovered August 1986, by Jon Rollins

Location Situated just below the lake sink is a round pit choked with large boulders.

Description A squeeze beneath boulders leads to a pleasant rift passage taking a stream. Where the stream starts to descend steeply, use a bypass that rejoins the stream lower down in a chamber. Below this, the passage continues descending, becoming tight and icy before emerging in a surface rift 77 m below the entrance.

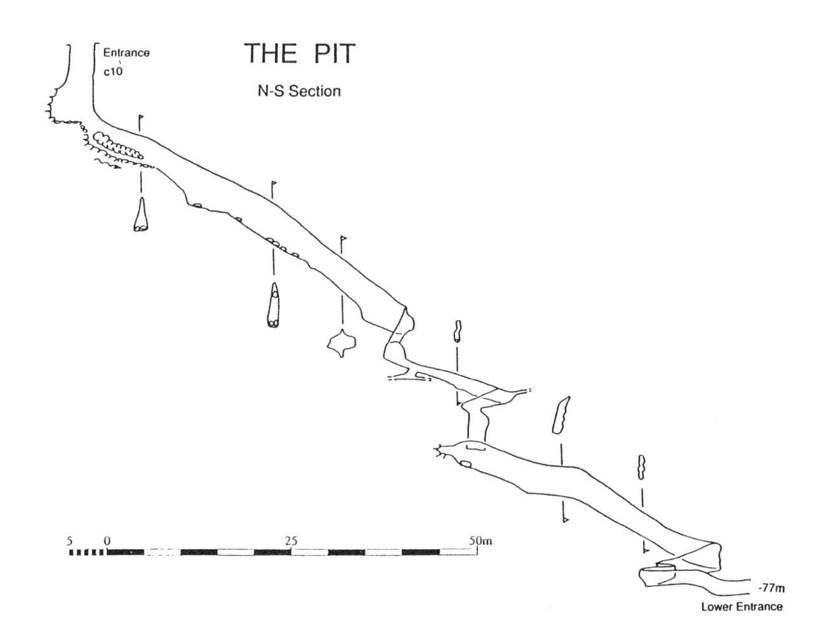

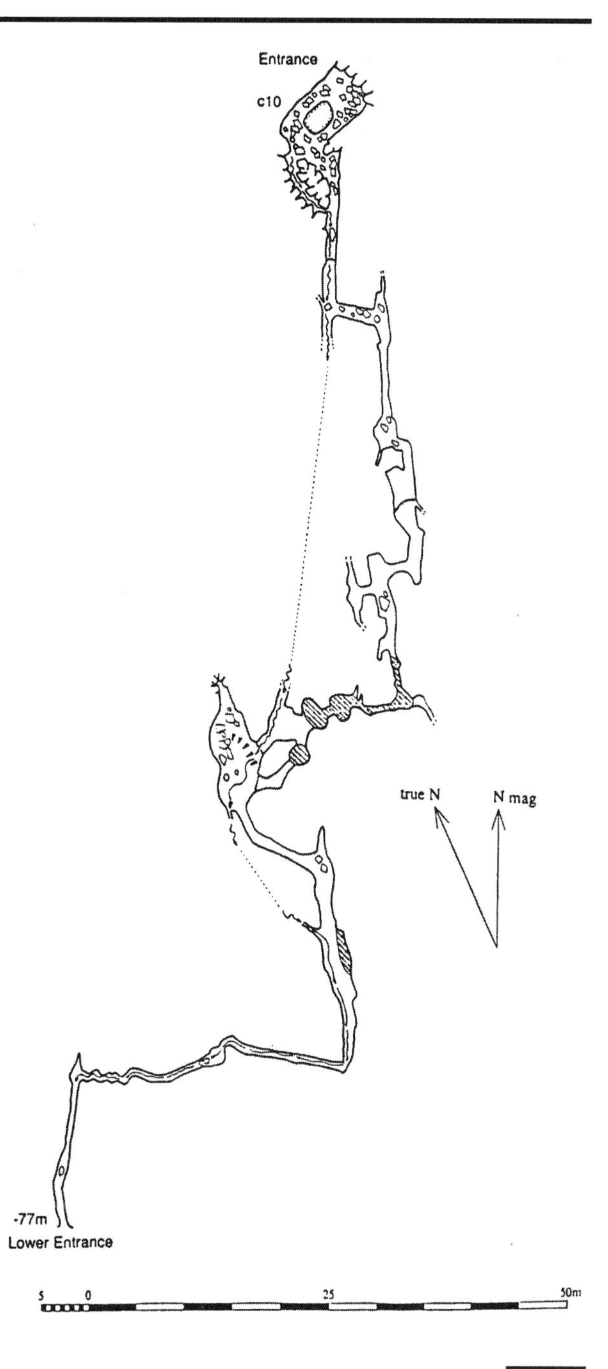

Arctomys Valley

A description of the valley is best left to the discovers — George Kinney and Conrad Kain in 1911. "The valley is a delightful one of the 'hanging' type, with charming alp-land surroundings; scattered wide-spreading spruce trees, open slopes carpeted with pink heath and white heather and brilliant with many species of alpine flowers. Rushing crystal streams join and flow through a small rock canyon, characteristic of the valleys here, and patches of snow create a truly alpine effect. Immediately on our appearance, the shrill resounding whistle of numbers of hoary marmots (*Marmota sibila*) greeted us from all sides as they sent forth their notes of surprise, indignation and warning as we topped the crest; and there is no sound that gives a more eerie feeling than this same long dawn whistle heard unexpectedly in the solitudes of the high mountain valleys. It was named Arctomys (Greek for hoary marmot) Valley in their honor. (*Canadian Alpine Journal*, 1912 Vol IV.)

Access Hwy. 16 (Yellowhead Hwy.) 29 km west of the Alberta/B.C. boundary and 2 km before Moose Lake, turn right (north) on a gravel turnoff that runs a short distance to a parking area just below the railbed.

Cross the tracks and follow Moose River trail for 24 km, first over a small shoulder, and then along the west bank of Moose River/Resplendent Creek. Ford Resplendent Creek at the site of an old horse camp (km 17), where there are several channels to negotiate. Continue east to where the trail rejoins the Moose River. In about 2 km the trail crosses Arctomys Creek.

Follow Arctomys Creek up the steep mountainside (600 m elevation gain) to where the gradient eases.

Although it can be done in one long day, most parties take two days, camping at Resplendent Creek crossing en route.

While the valley is located in a wilderness conservation zone, the Resplendent and Moose valleys are designated "wilderness recreation." This means helicopter access may be permitted to designated sites. Contact the Mount Robson Office at (250) 566-4325.

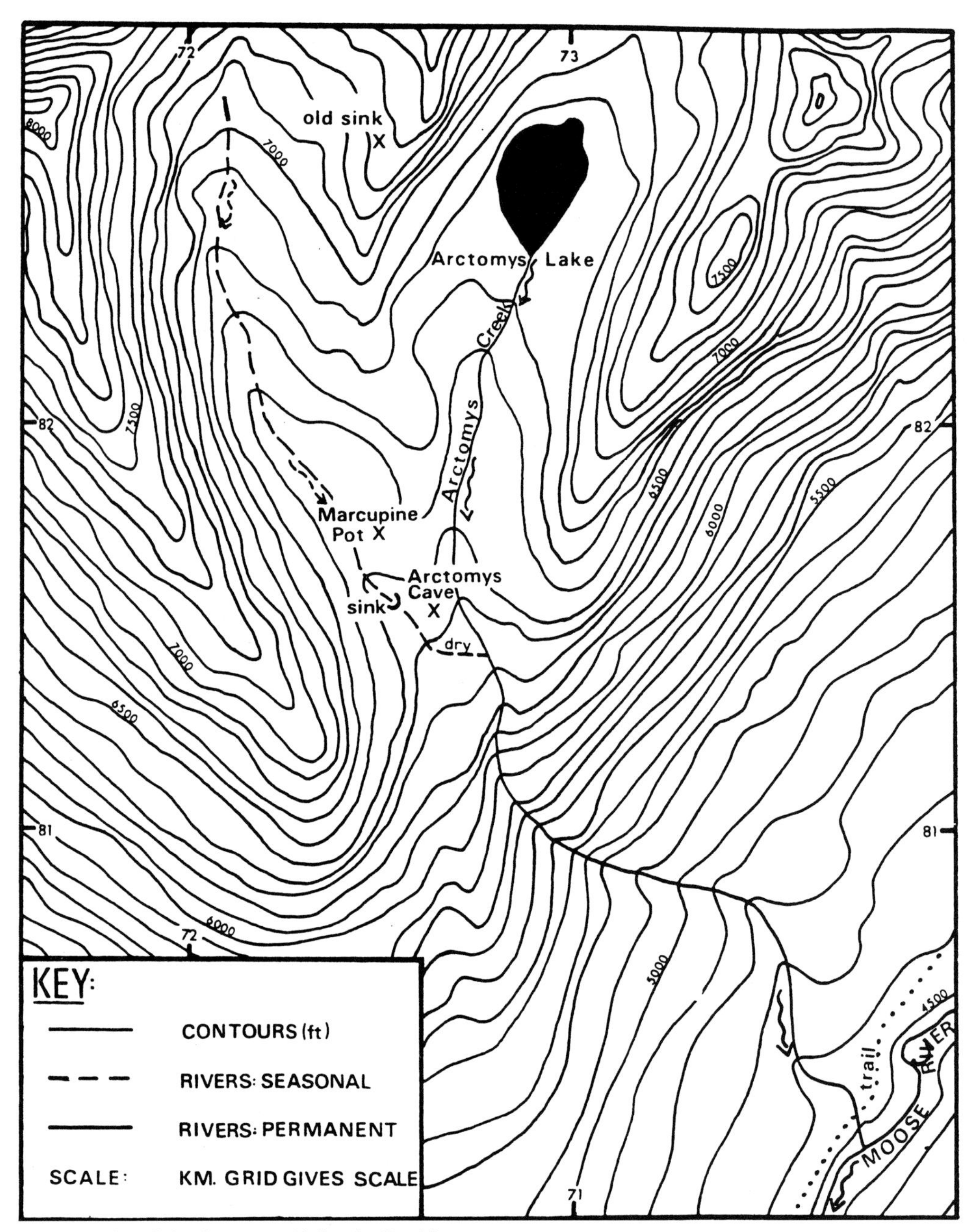

Topography of the Arctomys Valley by Deej Lowe.

Arctomys Valley. Photo Ian Drummond.

"Surely only a dedicated caver would walk 24 km carrying a fifty pound pack, swim/wade a 16-30 m wide, fast flowing river fed by glacial meltwater, climb 600 m up a mountain beating off flies and mosquitoes while stumbling through a maze of deadfall and knee-deep moss only to reach a 3 m by 1.5 m hole in the ground surrounded by stunted trees, limited firewood and snow. Only a die-hard caver would do this and love every minute of it! Such a person would probably have no problems with Arctomys or any other cave. The challenges and risks involved are, for the most, the essence of caving while the self-satisfaction which follows a successful trip is the reward. Any other attitudes or motives are dangerous."
Peter Thompson.

Geology Arctomys Valley is a classic example of a strike-controlled valley, running more or less north south, with the solid sequence dipping westward at about 35°. A great thickness of McNaughton formation quartzites form the eastern part of the area, though much of the outcrop is obscured by talus. Both the Lower and Upper Carbonates are present in the floor of the western arm of the valley and in the ridge between here and Arctomys Lake, the total thickness being somewhat less than that seen to the west of the syncline. Similarly, the Middle Shale Member is relatively thin, more of a parting than a significant bed, and even this appears to thin southward, though detailed exposure is lost in the steep forested slopes above the Moose River. Siltstone of the overlying Mahto Formation forms an impressive cliff to the west of Arctomys, with a 10 m exposure of a carbonate band about 100 m above the top of the Mural Formation.

Over the ridge to the west, the steep slope down towards Resplendent Creek consists of overlapping steeply-dipping slabs of Mahto siltstone and quartzite. As in the Hanging Valley area of Small River, and the Goat Valley area to the south, the Mural dip slopes are strongly karstified and surface karren formations are in evidence. Arctomys Cave is contained in the Upper Mural limestone running alternately by dip and strike passages, until the stream sumps and resurges at a similar altitude by the Moose River (adapted from a report by Deej Lowe, ACRMSE report 1983/1984.)

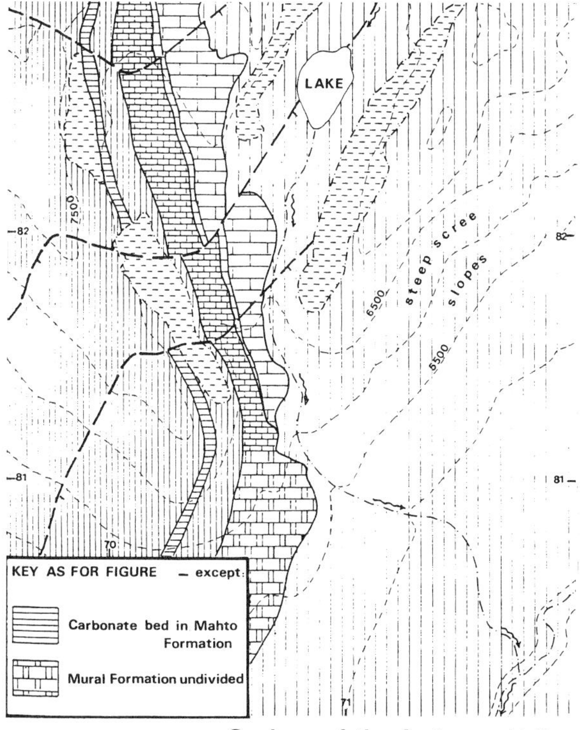

Geology of the Arctomys Valley. Drawn by Deej Lowe 1984.

Arctomys Cave

Map 83 E/2 726816
Jurisdiction Mount Robson Provincial Park
Entrance elevation 1967 m
Number of entrances 1
Length, Depth 3496 m, 536 m (-523, +13)
Discovered 1911, by George Kinney & Conrad Kain

Location The cave is located in a small hanging valley just above treeline, about 1 km south of Arctomys Lake and about 90 m north of Arctomys Creek. The uninspiring slot entrance, 3 m x 1.5 m, lies in a small depression and is marked on the most recent topographical map.

Exploration The following is an account of the discovery by the Rev. George Kinney and Conrad Kain on July 19, 1911: "...Immediately above the timbered ridge, high on the end of the great spur separating the west and main Moose valleys could be seen a series of very picturesque waterfalls. There are seven distinct leaps. In the valley where the stream has its origin, Kinney discovered a rock opening which seemed to lead into the bowels of the earth. It is evidently a rift in the strata, which here dips steeply. This through the action of water, has been widened into a deep shaft leading down, goodness knows how deep.

"The shaft, it can hardly be called a cave, was examined with candle, rope and barometer. The opening, a slit in a shallow depression, at one time undoubtedly furnished a water exit to lower levels. It is only large enough to admit one at a time. We descended 250 feet by barometer measurements to a point where a small stream of water tumbles through a tributary crack [they were only 20 feet out in this measurement]. Beyond that the going is wet and the exploration was not carried farther, as there was no change in the character of the subterranean shaft. Kinney claimed that, at the time of his discovery, he had gone some distance beyond the fall. The crack descends at an angle of about 65° or 70° from the horizontal. The rock is hard and rough, and affords good hand and foot holds. In places the width is ten to twenty feet and minor cracks lead off here and there. The walls are a dark limestone, dun coloured on the outside surface from seepage of the lime. There were no stalactites of more than two or three inches in length, and, generally speaking, it was unattractive. It appeared to be one of these subterranean waterways that are frequently encountered in mountains of a limestone formation" (*Canadian Alpine Journal 1912, vol IV*).

In 1972 a boot print bearing tricouni nail marks was found at the extent of George Kinney's exploration, mute testament to the sheltered solitude of the cave environment.

Arctomys Cave was not visited again until 1971 when Mike Goodchild rediscovered the cave. He descended a short distance with a flashlight and confirmed that it was, indeed, a good prospect.

Arctomys Cave entrance.
Photo Ian Drummond.

Formations in Arctsomys Cave.
Photo Ian Drummond.

The cave was first bottomed in 1973. Since then, trips to Arctomys have averaged one a year, with written records of 23 people having reached the final sump. The true number is probably close to twice this, although many turn back before reaching the bottom.

Description Arctomys Cave is best attempted in the fall before the first snowfall, the majority of trips having taken place in September. Participants in a couple of November trips have remarked on how much lower the water is. But even at low water, expect to get soaked.

Bottoming trips by experienced cavers typically take between 11 and 14 hours. The inexperienced should first cut their teeth on shorter caves before attempting to add Arctomys to his or her list. Despite, or probably because of the demanding nature of this cave, a trip to the final sump is one of the most enjoyable caving trips in the Canadian Rockies.

One of the most remarkable characteristics of Arctomys from a caver's perspective is the necessity of having to rig less than 50 m of pitches in order to bottom a 536 m-deep system. (Substantially less tackle than this is needed if all bypasses are utilized.) The whole cave is a series of down-climbs, straightforward when you are fresh and dry and on your way down, but a lot more serious on the way out when you are cold, tired and wet.

Initially the cave is dry to about -70m, where a junction is reached with a stream entering from the roof. Downstream is a wet 4 m drop, followed by walking passage down an inclined rift and farther down-climbs. Following a 6 m drop, the cave takes a sharp turn to the right and heads down dip in a narrow fissure where a traverse with hand-line leads back down to the stream.

At a point 450 m into the cave, the main way on apparently ends. However, a low sandy roof tube leads on to the top of Webster's Pit, a 15 m rappel. Below the pit the cave continues for 280 m as a high narrow fissure with breakdown blockages that require more short climbs. Here, a large waterfall (the 1000 Waterfall) ended exploration in 1972. Peter Thompson described it as, "a thunderous roar, we could only approach to within 15 feet of the waterfall before the carbide lamps were blown out." This explains why later trips were attempted in the fall when the water was lower.

The high, narrow rift passage continues down a series of wet and dry climbs to the Refresher — a notorious 9 m pitch for which it is hard to arrange a dry rig. Just below the Refresher is the Elbow, below which the cave changes character.

Above the Elbow the cave consists of almost continual climbing in a narrow, high rift passage — referred to as The Endless Climb. Below the Elbow the passage flattens out, and although still rift-shaped, has a lower ceiling. The passage remains fairly horizontal until the Straw Gallery, after which three pitches of 8, 5 and 5 m bring you to the 1700 Foot Room and the end of the cave.

"The tourist trip was extremely enjoyable even though The Endless Climb virtually finished off some of the team. Variety was the key to the quality of the trip, with some fine formations, extensive roomy walking stream passages, tricky climbs, wet and dry passages and even some archaeology [deer bones] to look at. Upstream was quite a contrast as the passages were almost dry."
Deej Lowe, 1983.

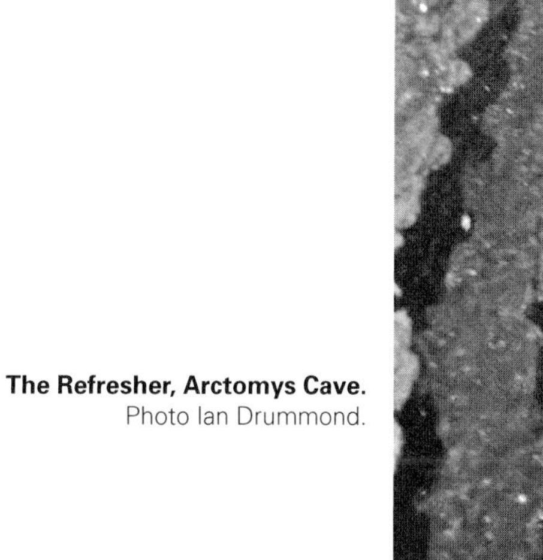

The Refresher, Arctomys Cave.
Photo Ian Drummond.

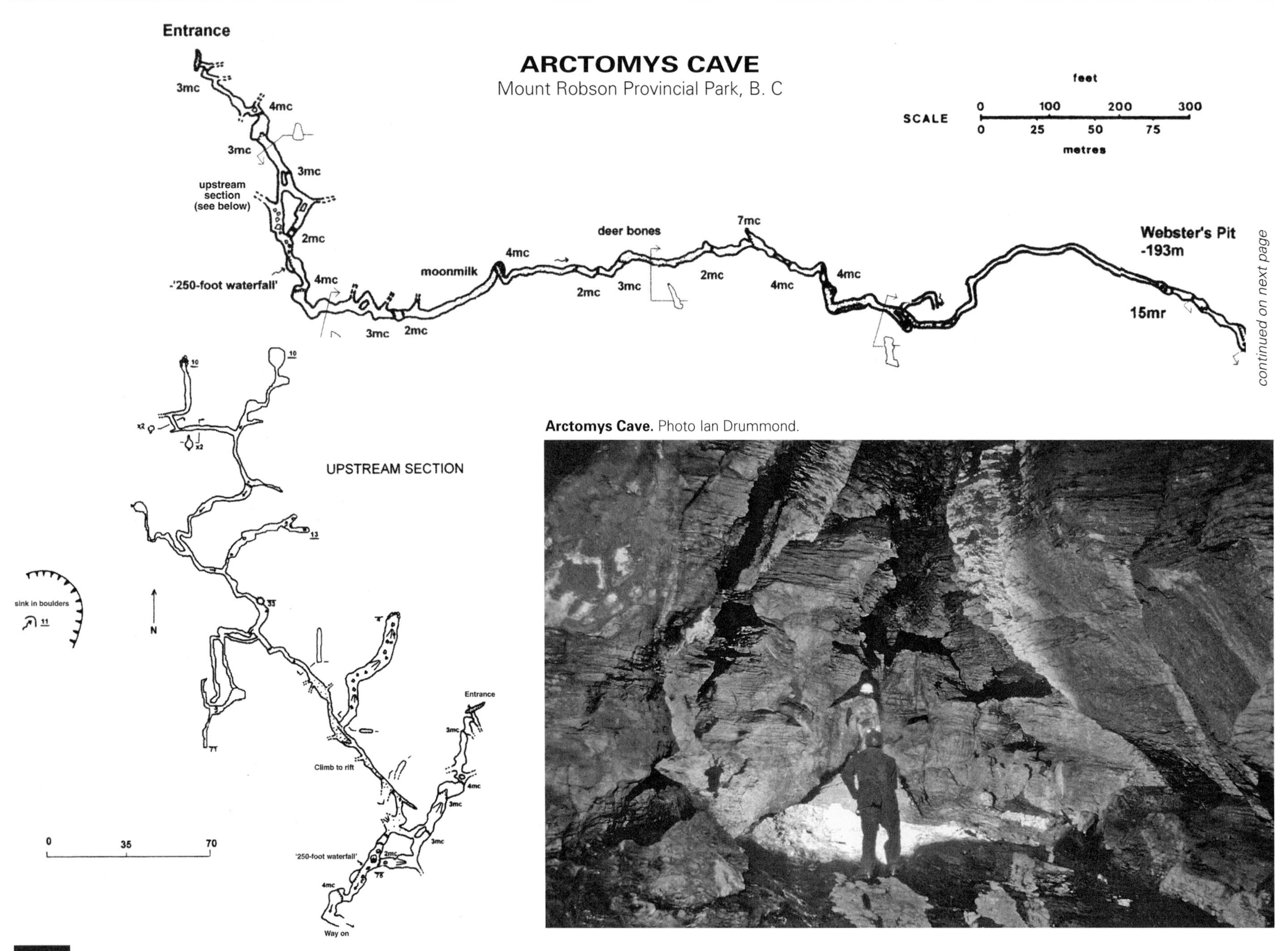

continued on next page

Arctomys Cave. Photo Ian Drummond.

continued from previous page

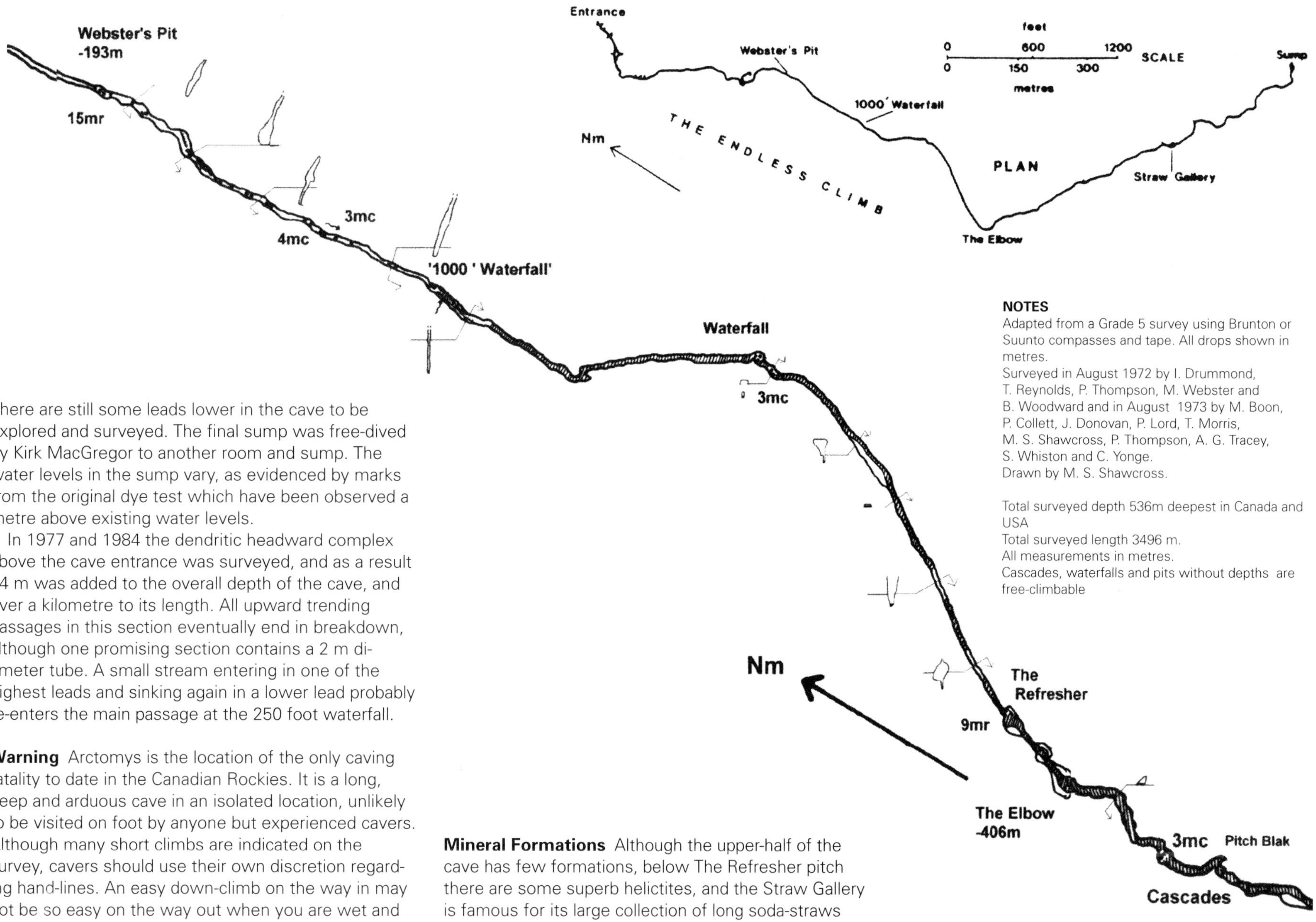

NOTES
Adapted from a Grade 5 survey using Brunton or Suunto compasses and tape. All drops shown in metres.
Surveyed in August 1972 by I. Drummond, T. Reynolds, P. Thompson, M. Webster and B. Woodward and in August 1973 by M. Boon, P. Collett, J. Donovan, P. Lord, T. Morris, M. S. Shawcross, P. Thompson, A. G. Tracey, S. Whiston and C. Yonge.
Drawn by M. S. Shawcross.

Total surveyed depth 536m deepest in Canada and USA
Total surveyed length 3496 m.
All measurements in metres.
Cascades, waterfalls and pits without depths are free-climbable

There are still some leads lower in the cave to be explored and surveyed. The final sump was free-dived by Kirk MacGregor to another room and sump. The water levels in the sump vary, as evidenced by marks from the original dye test which have been observed a metre above existing water levels.

In 1977 and 1984 the dendritic headward complex above the cave entrance was surveyed, and as a result 14 m was added to the overall depth of the cave, and over a kilometre to its length. All upward trending passages in this section eventually end in breakdown, although one promising section contains a 2 m diameter tube. A small stream entering in one of the highest leads and sinking again in a lower lead probably re-enters the main passage at the 250 foot waterfall.

Warning Arctomys is the location of the only caving fatality to date in the Canadian Rockies. It is a long, deep and arduous cave in an isolated location, unlikely to be visited on foot by anyone but experienced cavers. Although many short climbs are indicated on the survey, cavers should use their own discretion regarding hand-lines. An easy down-climb on the way in may not be so easy on the way out when you are wet and exhausted after 14 hours of caving.

Mineral Formations Although the upper-half of the cave has few formations, below The Refresher pitch there are some superb helictites, and the Straw Gallery is famous for its large collection of long soda-straws and flowstone formations.

continued at top of next page

continued from previous page

continued from far right

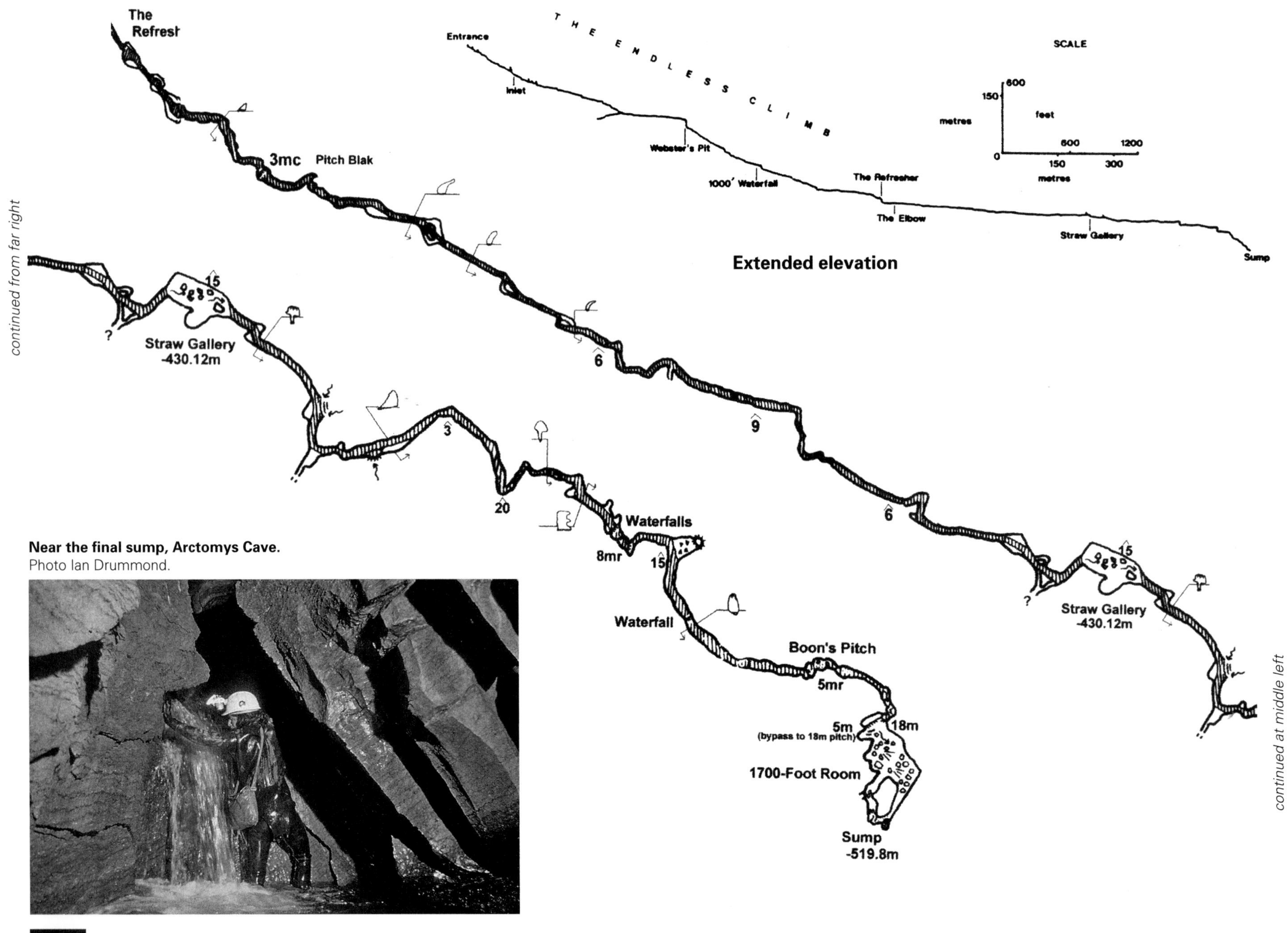

Near the final sump, Arctomys Cave.
Photo Ian Drummond.

continued at middle left

Marcupine Pot

Map 83 E/2 724818
Jurisdiction Mount Robson Provincial Park
Entrance elevation 2013 m
Number of entrances 1
Length, Depth 305 m, 76 m
Discovered 1973, by Pete Lord, Chas Yonge & Odile Renault

Location A choked rift about 100 m above Arctomys Cave and just over 250 m distant.

Description During a trip to Arctomys Cave to complete the survey, several short caves in the valley were checked out. It was hoped that Marcupine would provide a higher entrance to Arctomys and thus add to its depth.

A choked rift was opened, and led to a 4 m climb down into a large passage. Following a farther 15 m climb down through boulders, a squeeze accessed a large strike passage. However, all leads beyond this were either heading back to the surface or were too tight. Attempts to extend Marcupine by digging proved unsuccessful.

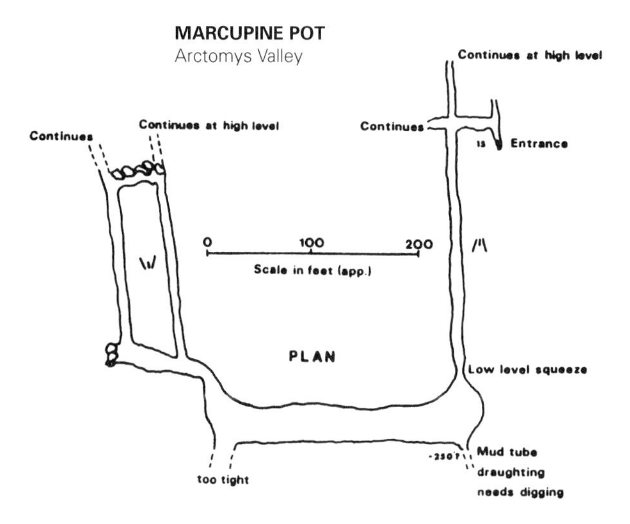

Berg Lake Area

Access to all Caves Hwy. 16. (Yellowhead Hwy.) at the Mount Robson service area, 58 km west of the Alberta/B.C. border. The Berg Lake trailhead is 2.5 km along a gravel road. Hike 19.6 km to Berg Lake campground. Contact the Mount Robson Area Office at (250) 566-4325 for reservations.

Berg Lake Cave

Map 83 E/3 555930
Jurisdiction Mount Robson Provincial Park
Entrance elevation 2100 m
Number of entrances 1
Length 111.6 m, 15.2 m (-2.4 m, +12.8 m)
Discovered Unknown

Location The cave is located 500 m above Berg Lake in a small limestone cliff. It can be reached from Berg Lake campground in a one hour walk along a well-signposted trail leading past Toboggan Falls.

Description Developed on prominent joints, the cave consists of walking or squeezing passages in a rectangular pattern. The cave appears to have been initiated by an opening of the joints with some subsequent solutional activity.

Another small cave about 15 m long is located just to the south.

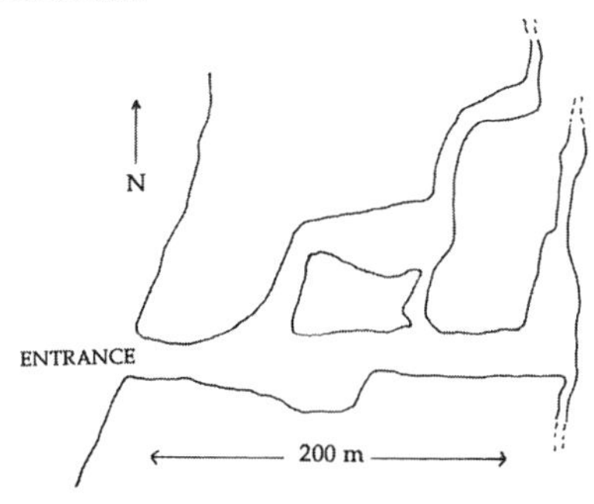

Grade 1 Elevation , Julian Coward

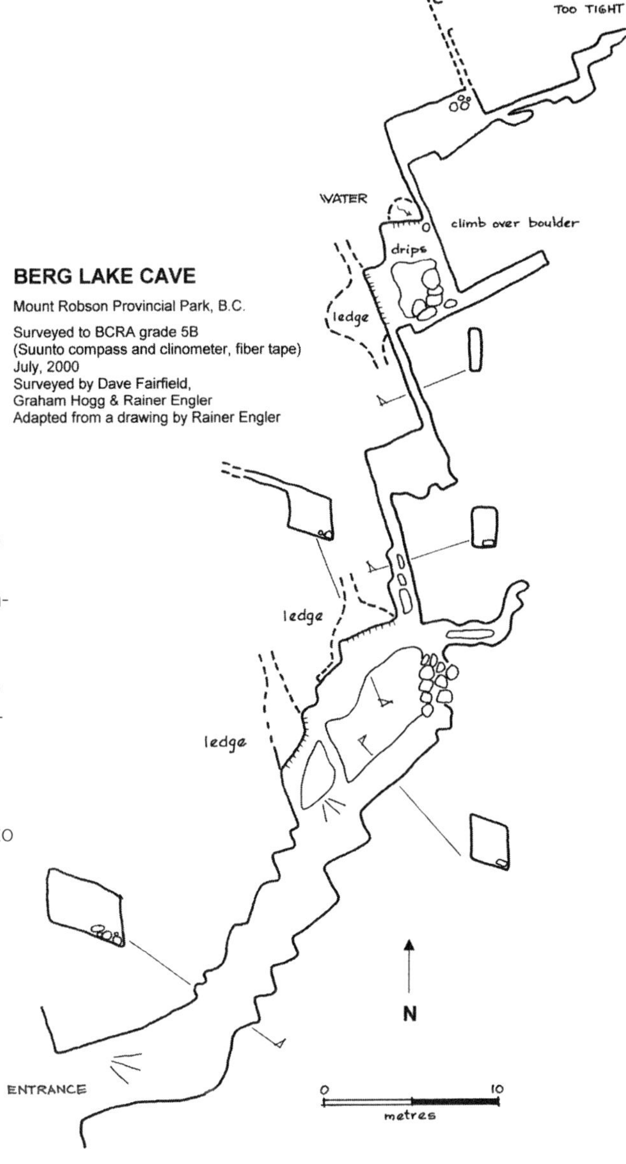

Robson Cave

Map 83 E/3 597906
Jurisdiction Mount Robson Provincial Park
Entrance elevation 1921 m
Number of entrances 1
Length c. 100 m
Discovered 1986, by Julian Coward

Location From Berg Lake campground take the Robson Glacier Trail that branches southeast from the main trail 1 km west of Robson Pass summit. Follow the trail for about 3 km up the lateral moraine on the east side of the glacier (cairns). The entrance is located a few hundred metres from the dry valley floor.

Description Robson Cave dips fairly steeply. Initially it is roofed by boulders, but farther in the cave is contained by bedrock.

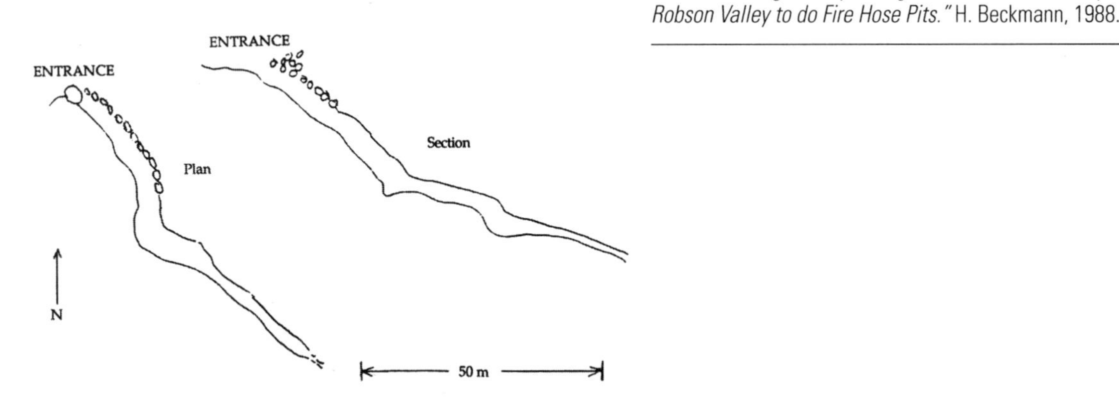

ROBSON CAVE
Grade 1 sketch, Julian Coward

Robson Valley

Fire Hose Pits

Map 83 D/14, location uncertain
Jurisdiction (Headwaters)
Entrance elevation Uncertain
Number of entrances 2
Length, Depth 52.4 m, 17.0 m
Discovered 1970, Unknown

Access Start from Hwy. 16, about 1.4 km west of the junction of Hwy. 16 and Hwy. 5. Follow the creekbed leading to Little Lost Lake. Located about 400 m in from the highway, the caves are fissures in insoluble rock in the creekbed.

"We all had a good day caving and I recommend a day in beautiful Robson Valley to do Fire Hose Pits." H. Beckmann, 1988.

Discovery & Description Many rumours circulate about caves that upon investigation prove impossible to substantiate. The discovery of Fire Hose Pits was one of the more unusual stories which proved true. The discovery was made when a forest fire was burning on the side of Mt. Chamberlin, and a firefighter, following the creek that led to Little Lost Lake, almost fell into the upper entrance. It was allegedly descended for 'a hundred feet' on fire hoses before the intrepid explorer finally decided he had had enough.

The ASS first heard about the cave in 1970, and four ASS cavers surveyed it in April of 1988.

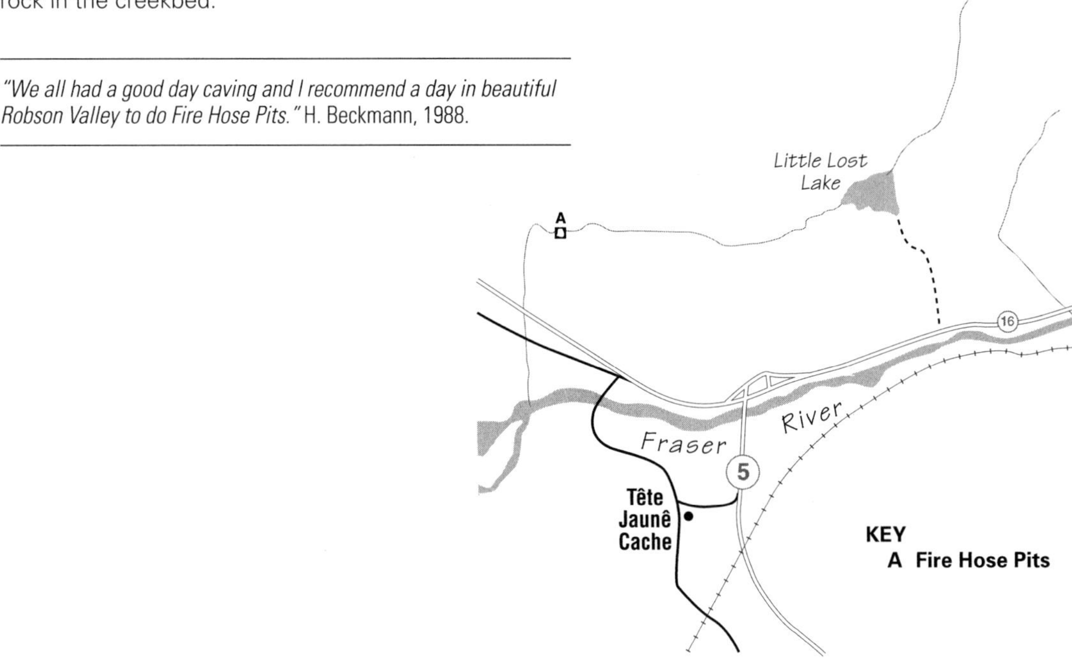

THE SMALL RIVER KARST

The Lower Cambrian Mural Limestone outcrop that contains Arctomys Cave is part of a syncline that runs under Mt. Robson and appears again in the Small River area. With numerous streams running off high altitude impermeable rocks there is good potential for cave development between the alpine karst and the valley bottom resurgences.

The caves were discovered in two main areas just north of Small River, extending up to a small glacier (Small River Glacier) not shown on topographical maps. The westernmost area, known as the Top Area (the average elevation of the caves is 1980 m), is divided into three locations arranged from south to north: the Far Karst, Middle Karst and North Karst. Hanging Valley is located across a valley to the east. In an adjacent valley just to the west of Hanging Valley is a large closed despression (Manic Depression) investigated by the 1984 ACRMSE expedition. No significant caves were found.

In 2002 the area became the 1848 ha Small River Caves Provincial Park. For up-to-date information on access and cave visitation contact the Prince George office at (250) 565-7086.

Access to Top Area (Middle, Far, North Karst) From Hwy. 16 (Yellowhead Hwy.) 14 km northwest of Tete Jaune Cache turn right onto a logging road that heads north along the Small River. Drive to the 18 km marker just south of the creek that drains the Small River Glacier. The access trail starts here.

Initially crossing overgrown logging roads, the rough trail climbs 800 m up the west side of the Small River valley to eventually emerge into the Top Karst area. Allow 3 hours from the logging road to Porcupine Cave.

Although caving has taken place in the winter when the road is occasionally ploughed, the snow can be deep in the bush, making the approach up the snow-covered creek bed a miserable experience.

KEY
A **Top Area**
B **Hanging Valley**

Mt Longstaff
River
Small
To McBride
16
Mt Goslin
to Jasper

Access to Hanging Valley Area As for Top Area to the 18 km marker on the logging road.

Then drive a further 2 km along the road to where an intermittent creek that drains from the Hanging Valley intersects Small River at GR 378937.

Follow the ridge to the west of the intermittent creek (no trail) to Hanging Valley. It is a steep hike (elevation gain is 800 m over 4 km) and takes a long day.

Warning Maintenance on the logging road is spasmodic, usually occuring during logging activity when the road is repaired and the bridges, which are subject to frequent washouts, are replaced. The 1983 expedition had great problems with porcupines chewing through the rubber hoses on a van parked at the trailhead. Chicken wire weighted with rocks was placed around the base of the vehicle to prevent a repetition of the incident. However, cattle grazing in the area must have been working in league with the porcupines, for they removed the rocks and for the second time a brakeless trip into Valemount had to be made for repairs.

The Small River Glacier.
Photo Chas Yonge.

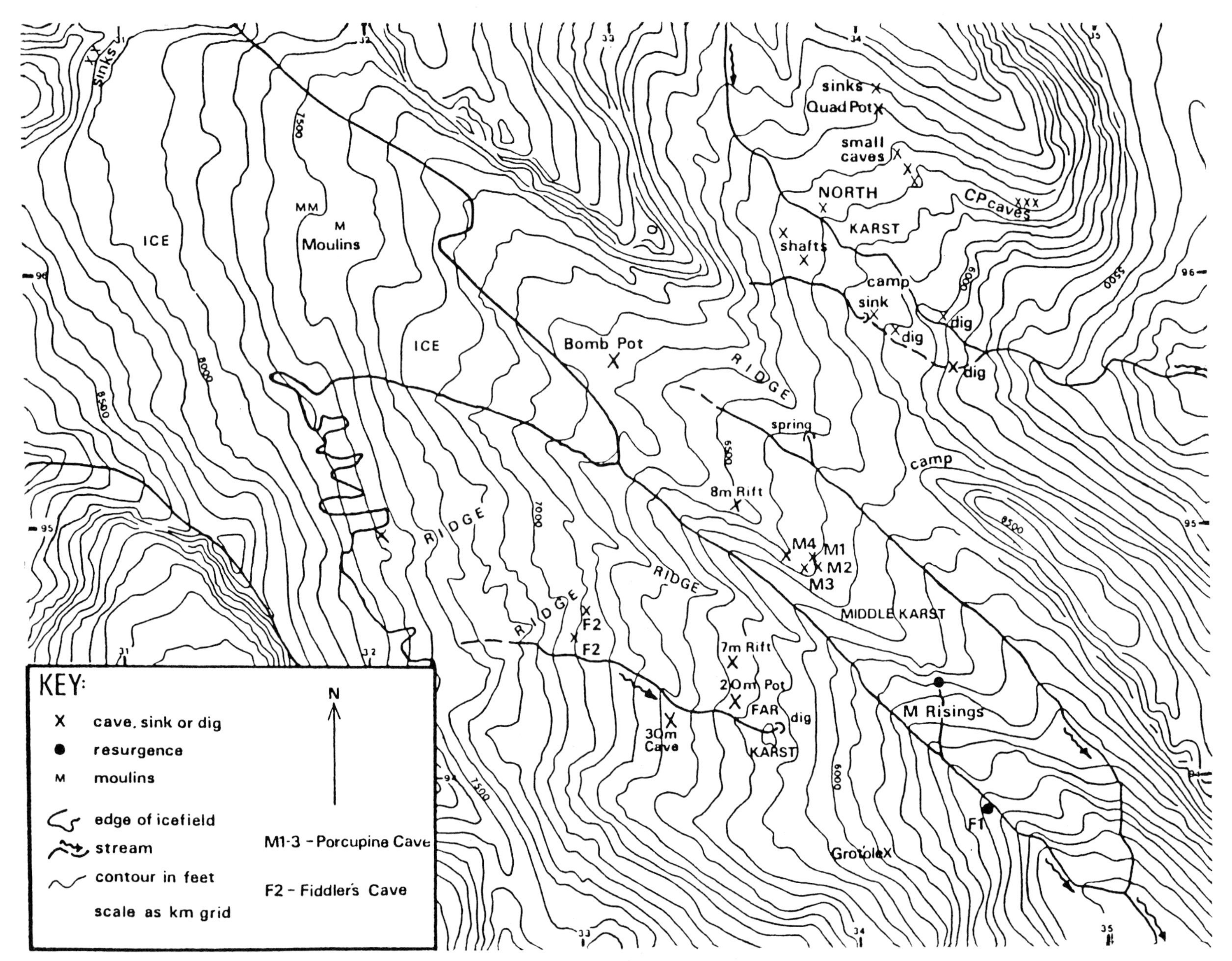

Topography of the Top Area — Far, Middle, North Karst (Deej Lowe, 1984).

Area Exploration In August of 1982 and 1983, a small team of English and Canadian cavers (ACRMSE) spent up to a month checking the Mural limestone, not only in the Goat Valley area of Mount Robson Provincial Park, but also in the Small River area. In 1983 the majority of the caves were discovered and explored. In 1984, the ACRMSE expedition returned to Small River, but owing to heavy snow conditions, not much cave exploration was possible, and the group soon moved on to the Bocock Peak area in northern B.C.

A few caves such as Stump Cave and Porcupine Cave have been extended substantially since the ACRMSE expeditions. In the case of Porcupine Cave, a glacier covers much of the catchment area, and so a moulin was descended during the ACRMSE explorations in the hope of finding a higher entrance. After descending for 30 m, the cavers found it too wet to continue. Subsequent winter trips in November, 1983, and November, 1984, were unable to locate the moulin which was under snow. However, Porcupine Cave was extended and Gorge Hole was discovered. In 1986, the Imperial College Caving Club (ICCC) visited the area briefly from London, England. They extended Stump Cave and pushed Grot 'Ole for a further 50 m of tight unpleasant passage. In 1990 more passage was discovered in the lower sections of Porcupine.

The North Karst has received less attention than the other areas, but has yielded a number of smaller caves that were not pushed.

Since 1987, Chris Smart of the University of Western Ontario has been conducting research in the Small River karst, directed at understanding the hydrology of alpine karst mantled by glaciers. Recent fieldwork (1999) indicates that new cave entrances are being exposed as the glacier retreats, including yet another entrance to Porcupine Cave.

Geology The geology map reveals the Small River (and Mt.Robson area) to be a major syncline with its axis trending approximately northwest-southeast. The effects of faulting and thrusting have modified the simple synclinal structure causing the repetition of the Mural and associated beds in some areas.

The General Geological Sequence in the Small River Area

Middle Cambrian and younger beds	(Hota Formation)		(84 m)
Lower Cambrian	(Mahto Formation) (Mural Formation) (McNaughton Formation)	Gog Group	(459 m) (245 m)
Pre-Cambrian	(Upper) (Middle) (Lower)	Miette Group	Very thick

The Mural Formation is comprised of three distinct divisions:
Upper Carbonate Member (limestone)
Middle Shale Member (shale)
Lower Carbonate Member (limestone/dolomite)

Small River, Hanging Valley area. Photo Chas Yonge.

The beds underlying the Mural Formation (the McNaughton Formation) are the distinctive massive blocky-weathering quartzites that typically form boulderfields with a flourishing lichen flora that is a distinctive pale green. Lakes impounded on the quartzite have an abundant algal growth while those on the carbonate beds are notably sterile. Above the Mural Formation is a mixed sequence of quartzite, siltstone and shale, with one significant band of limestone 10 m thick, comprising the Mahto Formation.

As indicated in the geology map on the next page, the Far Karst is formed in the Lower Carbonate Member whereas the Middle and North Karst are in the Upper Carbonate Member. The Middle Shale Member is exposed in Porcupine Cave where it forms the slippery floor of the stream passage. The quartzite of the McNaughton Formation forms a wide outcrop above the Hanging Valley area where there are extensive lichen-covered block-fields. The upper part of the Hanging Valley is composed of massive pale-coloured carbonates with well-developed karren on the dip slopes, similar to that of the Goat Valley area near Arctomys Cave. Rillenkarren, rinnenkarren, kluftkarren, meanderkarren and trittkarren are all represented here. Fossils found in the locality include fragmental Olenellid trilobites (rare), the trace fossils *Skolithus* and *Cruziana semplicata*, Archaeocyathids (very common) and primitive Echinodermata (common). (Adapted from Deej Lowe, 1984.)

Speleogenesis Most of the Small River karst, particularly the Top Area, has only relatively recently been exposed from beneath the ice. Evidence for this is quite strong; the freshness of the morainic landforms, neoglacial pavements with sharp striae still intact and unweathered, and the presence of sub-glacial calcite still preserved on exposed rock surfaces (Chris Smart, 1989).

The end of the Late Wisconsin glacial advance was only about 10,000 years ago, and it would appear impossible for such large cave systems to have formed from scratch in such a short time. It can, therefore, be assumed that most of the Small River caves predate this event. Further evidence lies in the large amounts of old cave that have been removed by glacial action (i.e. Shattered Illusions) and in the blockage of the abandoned dry sections of caves such as Stump and Porcupine, by glacial debris. The dating of flowstone in Shattered Illusions at > 350,000 years and up to 1.2 million years ago also supports this premise (Chris Smart, 1989).

Deej Lowe, a member of the ACRMSE expeditions, postulated the following geological history for the Small River caves:

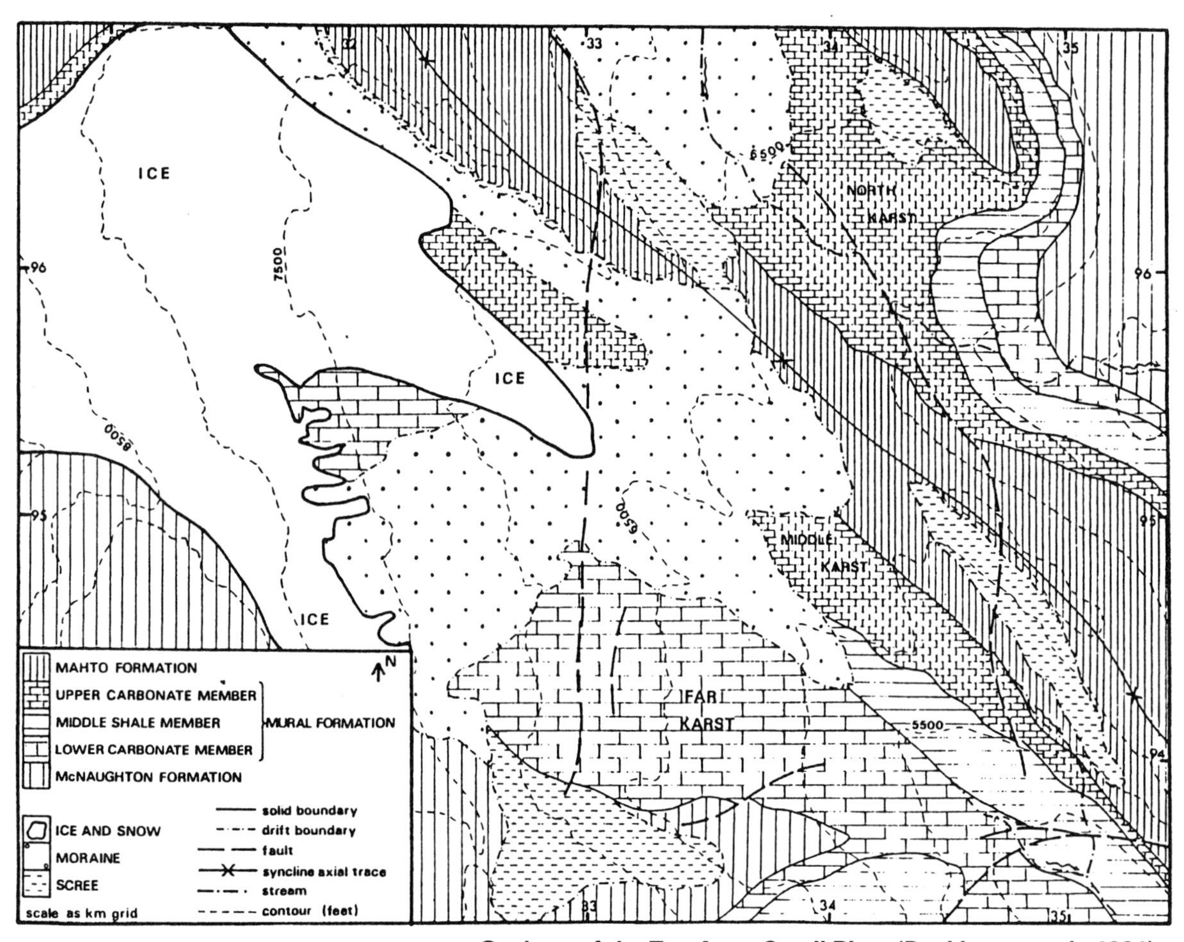

Geology of the Top Area, Small River (Deej Lowe et al., 1984).

1. Pre-Late Wisconsin (this very loose term could include any time between pre-Pleistocene glaciation and the commencement of the Late Wisconsin Advance): deep phreatic drainage routes formed. In order that a hydraulic gradient and viable sink-rising systems could be set up it seems likely that a proto Small River valley must have existed, and, therefore, an interglacial date is preferred for the commencement of this phase.

2. Late Wisconsin: Advancing ice removed much bedrock, de-roofing many phreatic passages and possibly scraping away entire drainage systems. Concurrently, some of the holes exposed must have been plugged by ground moraine, though not necessarily to any great depth.

3. Post-Late Wisconsin: As retreat began, surface drainage must have increased dramatically with debris-laden streams flowing down surface valleys. Any open holes would be rapidly invaded and rapid vadose downcutting would ensue if drainage were impeded. Where old underground conduits were blocked by moraine or where outlets were moraine- or ice-dammed, ponding would lead to further blockage by silt deposits. If an alternative drainage route was initiated the silt-filled tubes would be left abandoned at high level. At the same time, minor phreatic enlargement of surface fissures probably began across the entire area of glacially-scoured pavement revealed by the retreat. Eventually, a second underground drainage system would develop and in some cases take over surface drainage to form immature vadose cave passages. In turn some of these probably intersected fragments of the older system, as in Porcupine Cave, and in some cases might have commenced removal of clastic fill.

4. Present Situation: Fragments of the old system are still extant, and in some cases partly or completely active. The passages of the new system are still generally immature with perched sumps, nick-points and unmodified phreatic tubes. The large vadose canyons cut below the old phreatic tubes (as in Fossil Cave) by debris-rich streams are now generally abandoned, or fed by under fit streams, except at the time of the annual spring thaw.

Hydrology Meltwater from the Small River Glacier flows over the moraine before sinking at various points in the karst. Dye tracing in a moulin on the glacier indicates most of the water does not enter bedrock sinks, although a possible link with F1 resurgence (Stump Cave) was made. A lower sink on the glacier was, however, connected to the Porcupine 'M' resurgence. Of all the streams flowing onto the karst, only the largest cross the full width of the Mural Limestone, although the situation is different during spring melt. No local resurgence is recognized, and water sinking into the two limestone bands remains independent (Chris Smart, 1988).

Caves in the Far Karst drain to F1, those in the Middle Karst to the M rising through Porcupine Cave. In the North Karst a major surface stream crosses the limestone outcrop but does not sink. Minor streams in the area do sink, but because no resurgences have been located, it is assumed the water flows to a resurgence by the logging road in the valley floor.

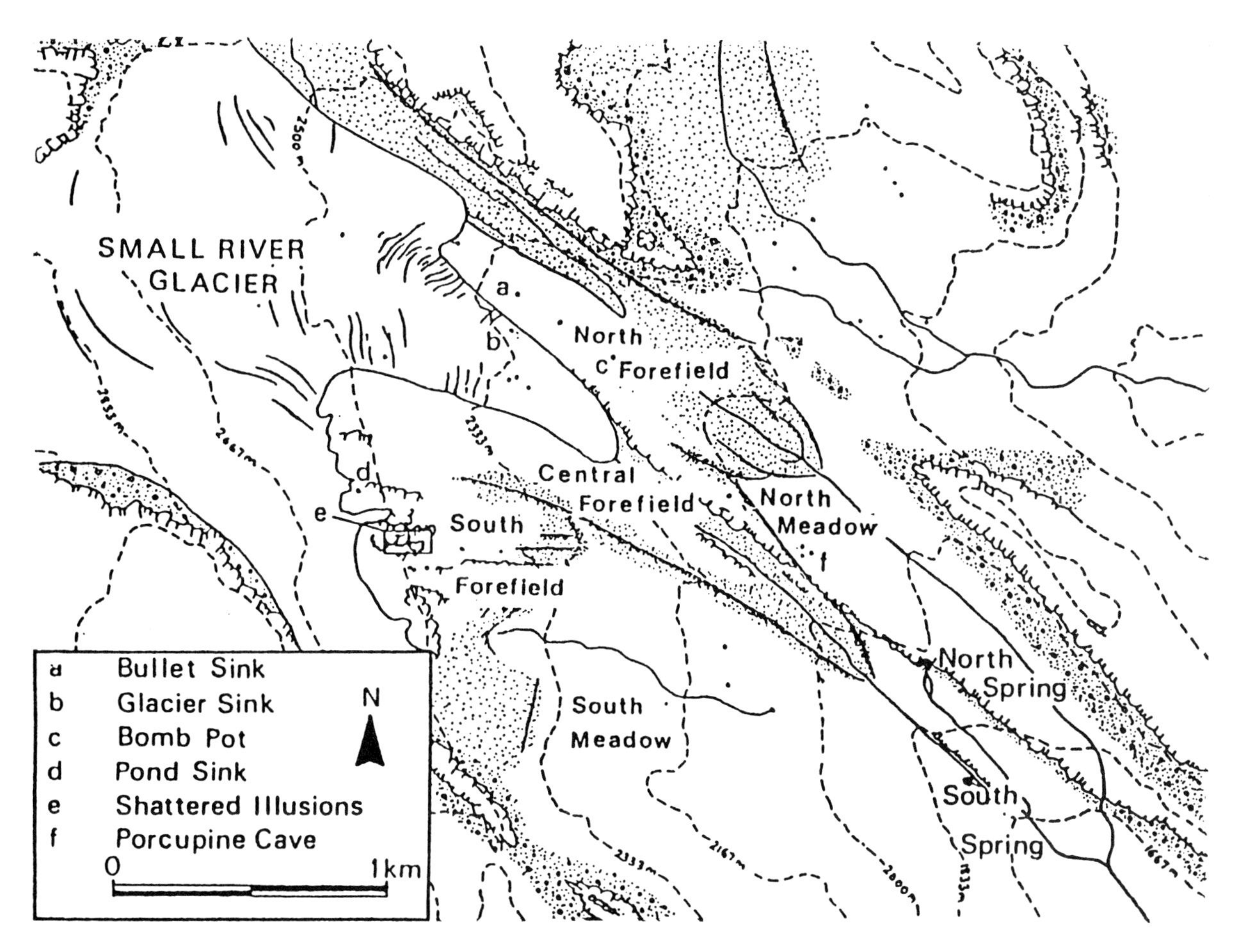

Hydrology of the Small River Glacier Karst adapted from drawing by Chris Smart, 1988).

A similar situation exists in the Hanging Valley Area: no impermeable rock-bands out-crop on the slopes below, and so it is assumed the water flows down-dip to the logging road resurgence, or to a smaller resurgence nearby.

Dye tracing has been carried out by Chris Smart on a number of sinks at the margins of Small River Glacier. From Pond Sink (d), a very fast flow-through time to F1/South Spring (trace peaked 2 hr. 30 minutes after injection) of 1000 m/h indicates a well-developed drainage connection. As expected, a trace from Bomb Pot (c) went to the M rising/North Spring, presumably via Porcupine Cave, a slow flow-through time indicating drainage constrictions above the cave. Of interest was a trace made from Glacier Sink (b) to the North Spring. This suggests a situation analogous to the M3 entrance to Porcupine exists under the lobe of the glacier.

The Middle Karst

Gorge Hole

Map 83 E/3 353930
Jurisdiction Small River Caves Provincial Park
Entrance elevation 1464 m
Number of entrances 2
Length 160 m
Discovered November 1983, by Jon Rollins & Chas Yonge

Location Gorge Hole is the lowest elevation cave discovered at Small River and the first cave you come to when hiking up the creek that drains the Small River Glacier. Look for three small apertures in the west wall of the same canyon into which Stump Cave resurges 230 m higher up.

Description The upper opening is a small resurgence; the lower two lead to some 150 m of walking passage up to 5 m high in places.

Stump Cave (F1 Resurgence)

Map 83 E/4 344941
Jurisdiction Small River Caves Provincial Park
Entrance elevation 1690 m
Number of entrances 1
Length, Depth 570 m +, 30 m
Discovered 1983, by the ACRMSE

Location The prominent entrance is low on the south side of the main access valley to the Top Area, and often has a large stream issuing from it (1 m³/s in August) — probably the entire drainage from Far Karst.

Description A large passage (10 m x 8 m) leads past a partial boulder choke to an upstream sump under the right wall. Dry passage continues up-dip, becoming low and narrow before a cross-passage is reached. To the right, a passage increases in size and leads to a breakdown chamber with a stream below, which can be followed in either direction along a canal to sumps. To the left of the chamber a rift leads up to a loose boulder slope. To the right, a squeeze through boulders accesses a large (15 m high) breakdown chamber.

From here three different leads have been explored. 1. A drafting boulder choke leading to a small chamber with a muddy crawl on the left ending at a high rift that becomes tight. 2. A hole in the floor that leads to a series of small chambers and a possible continuation through a "flaggy" hole. 3. A further sump that the ICCC cavers pushed through to find a further 30 m of stream passage (the Boulder Streamway). A waterfall issuing from a hole in the wall stopped progress. Attempts to locate this point in the cave in winter, when the flow of water would be reduced, have so far been unsuccessful. With 800 m vertical distance to the sinks above, there is potential for finding more passage in this resurgence, but it is not surrendering its secrets willingly.

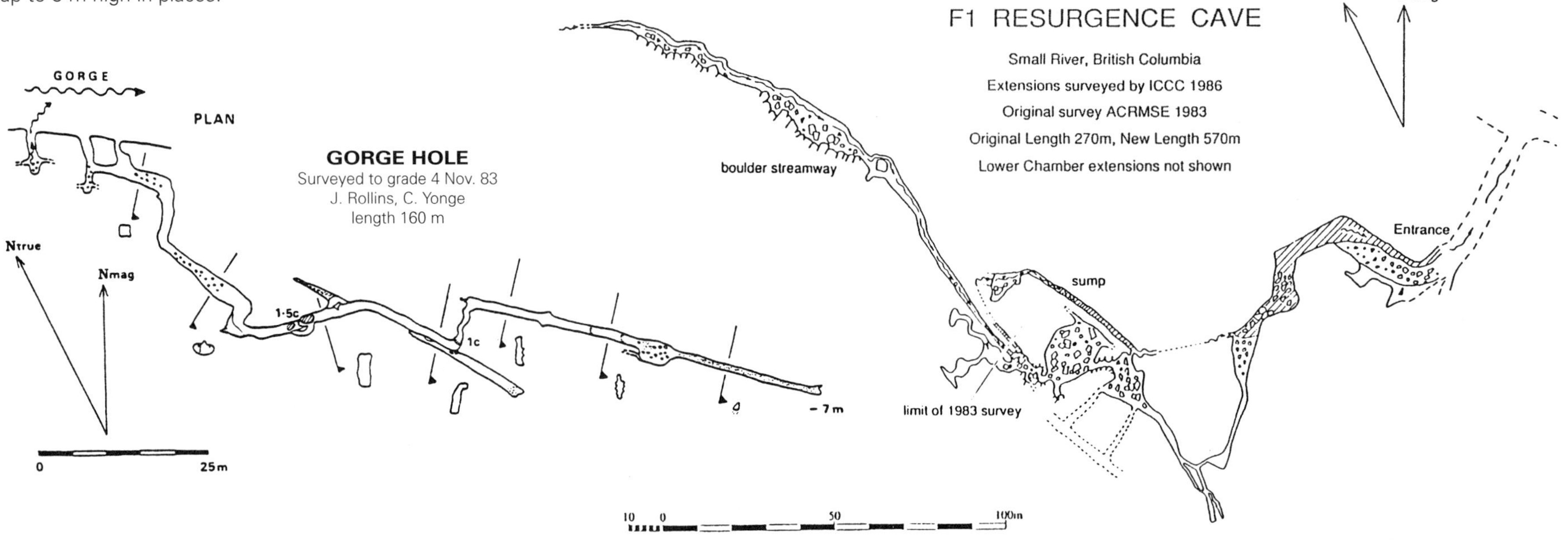

Porcupine (M1, M2, M3 & Latecomers Entrances)

Map 83 E/3 336941
Jurisdiction Small River Caves ProvincialPark
Entrance elevation 1920 m
Number of entrances 4
Length, Depth 1660+ m, 100 m
Discovered 1983, by ACRMSE

Location See M1, M2 and M3 on the Top Area location map. M3, a small hole in the cliff face on the limestone-shale boundary, provides the easiest way into Porcupine Cave.

Exploration & Description Porcupine has mostly pleasant walking passages along winding rifts. There are occasional formations and some interesting sediment deposits. The rushing stream-way provides for exciting caving, the shale floor being very smooth and slick in places.

Exploration of Porcupine Cave commenced from the M2 (Rat Hole) Entrance, from which the easier M3 Entrance was located. Subsequently, M1 was also found to join the system. Excitement ran high when the sound of a roaring stream was heard from a high tube in the main cave. A climb gave access to a sub-parallel active system, but sadly, the stream-way was found to sump upstream, while down-stream a short sump was bypassed, only to lead to another, apparently terminal sump. Porcupine Cave proved to be the longest and deepest system explored during the ACRMSE expedition of 1983.

The first of many return trips by ASS members was made on the November, 1983, long weekend. Latecomers Entrance was discovered by climbing 30 m of pitches from the inside after entering the cave from the M2 Entrance.

Thirteen ASS members attempted to visit the cave on the long weekend in November of 1984. After eight hours of difficult ascent in heavy snow conditions, only four persons made it underground. Finding a way round the terminal downstream sump made the trip

Staying out of the water near the end of Porcupine Cave. Photo Dave Thomson.

M3 Entrance to Porcupine Cave. Photo Jon Rollins.

worthwhile, another 300 m of passage being found. Exploration stopped at the point where the way on involved negotiating a canyon above deep water (Remembrance Passage).

Early in 1985 four ASS members returned on skis, and managed to locate and dig their way into the M3 entrance. After surveying the 1984 discoveries, they traversed 30 m along the canyon, but were stopped by a swimming section. A return trip was made using snowmobiles in March of 1985. Donned in wetsuits, the cavers swam down the canyon which quickly became too tight. Tantalizingly, the canyon could be seen widening only 5 m farther on. A strong draft indicated more cave ahead, especially as the resurgence (the M rising) was still 0.5 km away and 50 m lower down.

In 1990, members of the ASS from Jasper found a side passage that has not yet been surveyed, and during field work in 1999 students from the University of Western Ontario found a fourth and higher entrance (Latecomers Entrance) believed to be recently exposed by the retreating glacier.

PORCUPINE CAVE
Middle Karst — Top Area
Small River, British Columbia
Grid ref. LJ83819486 MI Alt. 1920 m
Length 1660 m Depth 100 m
ACAMSE 1983 & 1984 to BCRA gd 5c
By S. Worthington, J. Miller, C. Pugsley, C. Yonge, O. Whitwell, I. McKenzie, P. Langdon, E. Von Vorkampff, A. Bennett, C. Roberts.
Latecomer's Entrance by J. Rollins, C. Yonge.
Remembrance Passage by C. Yonge.

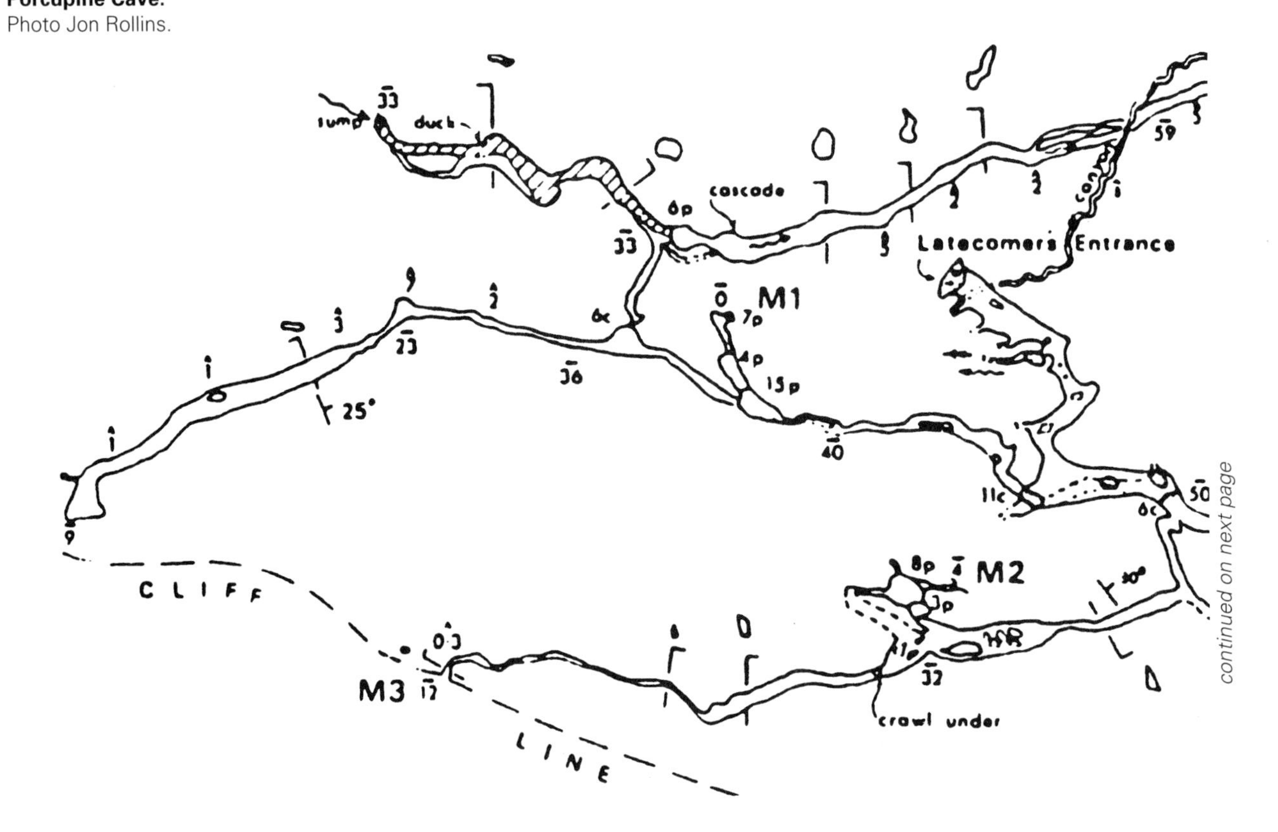

continued on next page

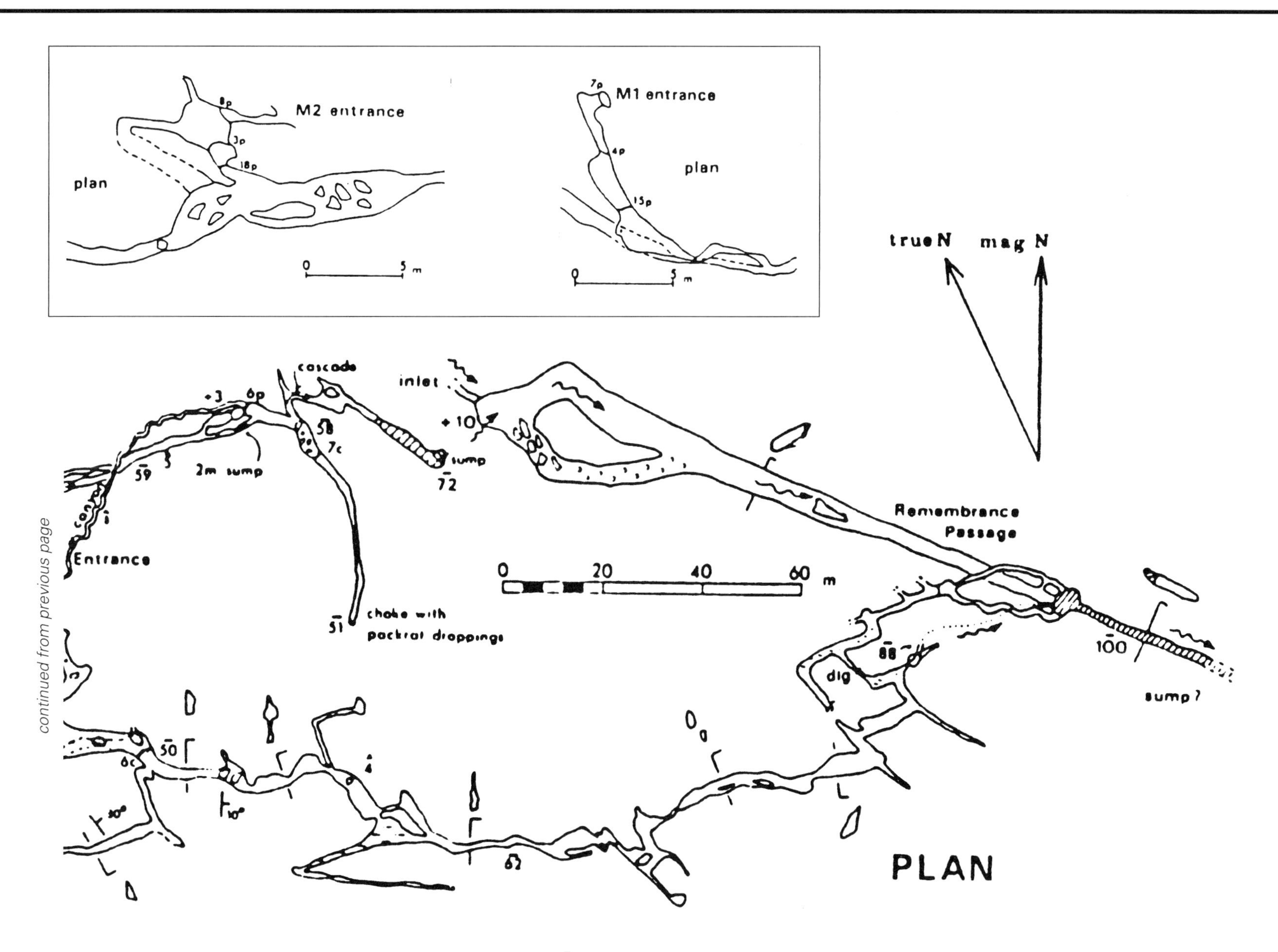

continued from previous page

M4

Map 83 E/3 336953
Jurisdiction Small River Caves Provincial Park
Entrance elevation 1946 m
Number of entrances 1
Length 60 m
Discovered 1983, by ACRMSE

Location Just above and slightly west of the M3 Entrance to Porcupine Cave.

Description A short drafting cave with a choked shaft. M4 is a good prospect for digging.

M4
Gd 1
Middle Karst

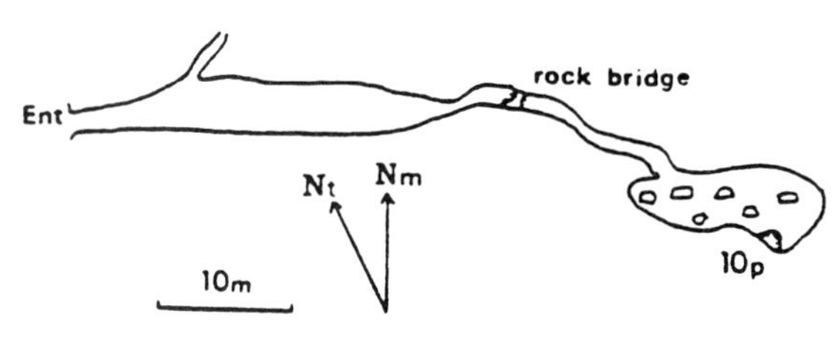

Far Karst

Grot'ole

Map 83 E/3 338939
Jurisdiction Small River Caves Provincial Park
Entrance elevation 1859 m
Number of entrances 1
Length 350 m
Discovered 1983, by the ACRMSE

Location 500 m due west and 170 m higher than F1 (Stump Cave) is an excavated sink in a dry valley.

Description Grot'ole consists of 350 m of generally unpleasant passage, progress being dependent on water levels and the use of wet suits. It was extended to its present length in 1986 by the ICCC.

GROT'OLE
Far Karst – Top Area
Small River, British Columbia
Grid ref. LJ 3380 9390 Alt. c.1859m
ACRMSE-83 survey to BCRA 2 by C Yonge

flat-out in water
stream appears
3c
stream appears & disappears
boulder ramp
2c
dry valley
sinks
mag N
true N
PLAN
0 30 60 90 m

Bomb Pot

Map 83 E/3 330957
Jurisdiction Small River Caves Provincial Park
Entrance elevation 2058 m
Number of entrances 1
Length 40 m
Discovered 1983, by the ACRMSE

Location 1 km northwest and about 100 m higher than Porcupine Cave, between the glacier and the prominent ridge. A precariously perched boulder in the entrance gives Bomb Pot its name.

Exploration & Description The cave ended after only 40 m, but is a promising dig.

BOMB POT
Gd 1
Far Karst

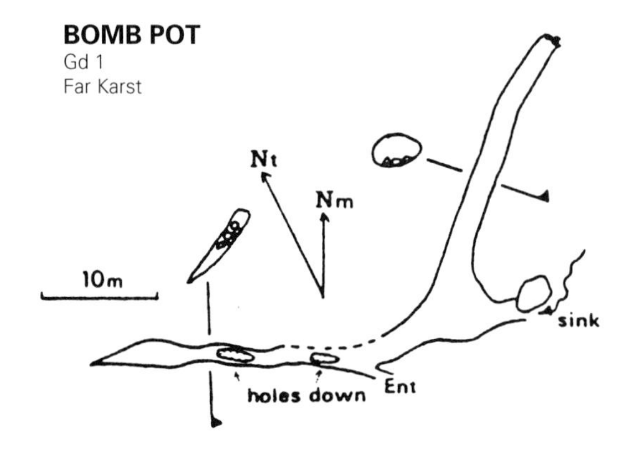

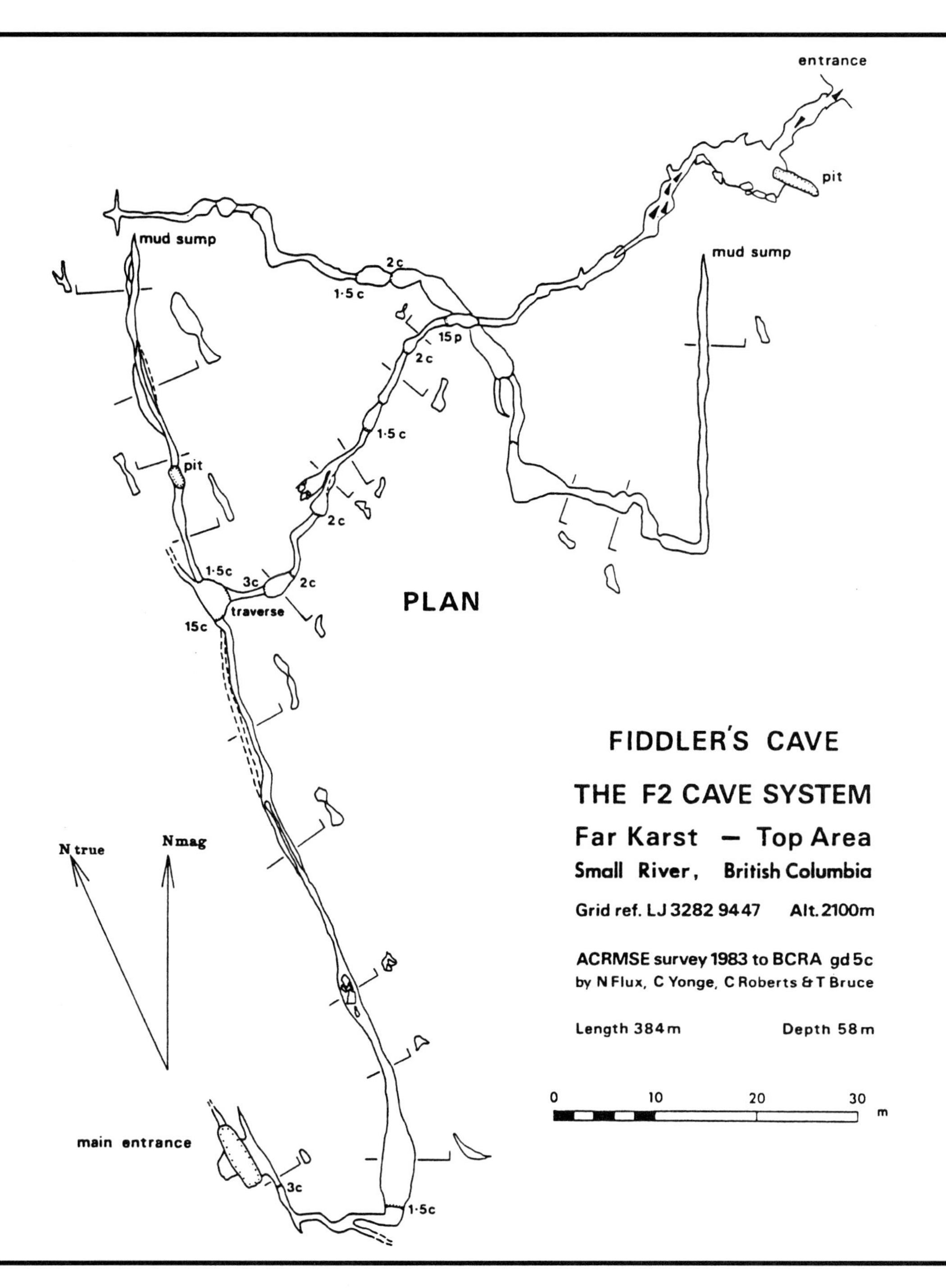

Fiddler's Cave

Map 83 E/4 328945
Jurisdiction Small River Caves Provincial Park
Entrance elevation 2100 m
Number of entrances 2
Length, Depth 384 m, 58 m
Discovered 1983, by the ACRMSE

Location The entrances are located just to the north of a stream that runs from the glacier above to a prominent sink below in the Far Karst.

Description A climb down the main entrance shaft accesses a small passage and a narrow descending rift that quickly chokes. Above this, another larger passage climbs steadily to reach a second higher entrance in a small depression. Upstream, small climbs and pools lead to a cross-rift choked by silt.

Shattered Illusions

Map 83 E/3 321949
Jurisdiction Small River Caves Provincial Park
Entrance elevation 2500 m
Number of entrances 1
Length 48 m
Discovered 1988, by Dave Huntley & Chris Smart

Location Close to the projecting ice lobe of the retreating glacier at the head of the Far Karst.

Exploration & Description Discovered during field work directed towards understanding the hydrogeology of the Top Area, Shattered Illusions consists of 48 m of cave passage and 50 m of unroofed canyon. It is a series of remnants of a much larger system which has been heavily dissected by glaciation. The only passages left occupy a few isolated bedrock knobs deeply carved by glacial ice.

A frost-shattered flowstone sample, recovered from rubble in the cave passage, has been dated at greater than 350,000 years, and up to 1.2 Ma (Chris Smart).

"The whole system is a marvelous pocket example of what ice has done to the former karst glory of the Canadian Rockies."
Chris Smart.

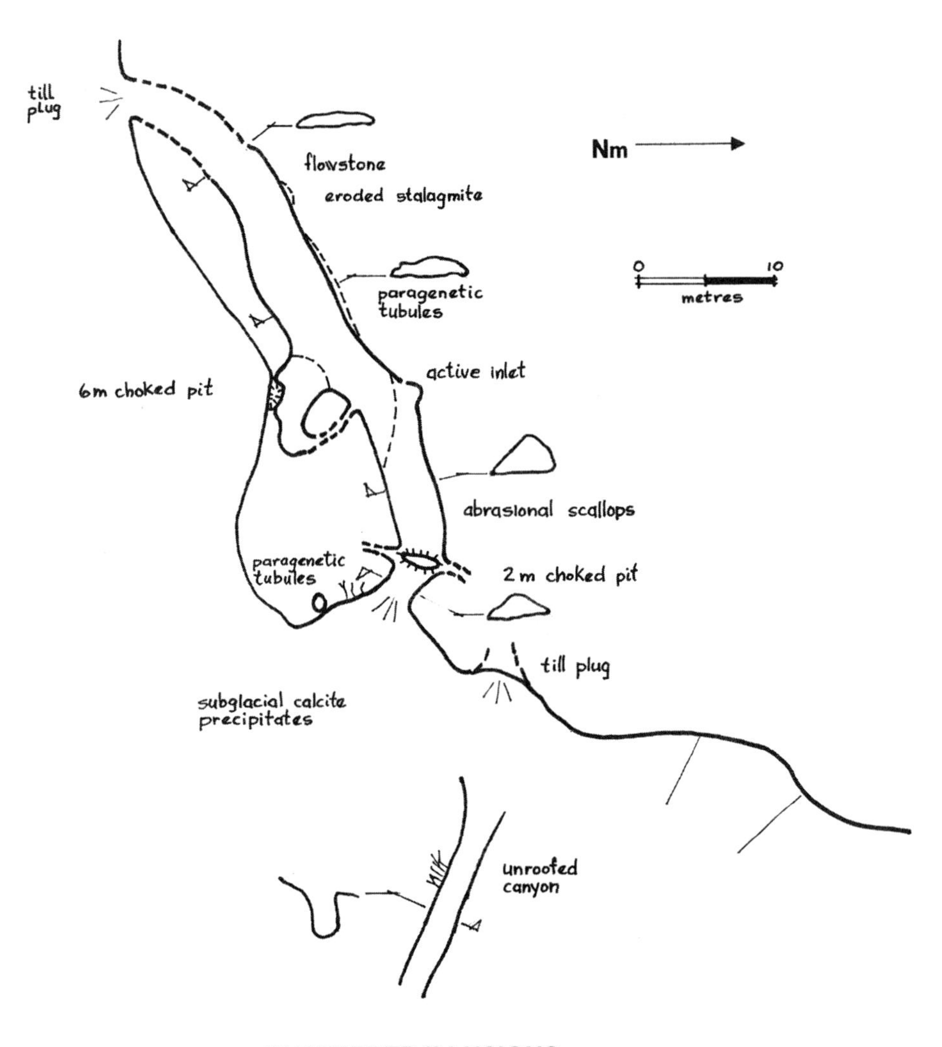

SHATTERED ILLUSIONS

Small River, British Columbia

Adapted from a Survey by: C. Smart & D. Huntley

The Moulin

Map 83 E/4 319962
Jurisdiction Small River Caves Provincial Park
Entrance elevation 2287 m
Number of entrances 3
Length, Depth c. 75 m, 50 m
Discovered 1983. by the ACRMSE

Location A dry moulin on the surface of the Small River Glacier.

Exploration & Description Descending early in the morning to avoid snowmelt, the explorers came to a stop where water from an active moulin close by was intersected at the top of an estimated 14 m deep pit.

Moulins, also known as millwells, are glacial sinkholes where meltwater flows from the glacier surface to beneath the ice. Because the catchment area for Porcupine Cave lies beneath the glacier above the Top Area, it was hoped that the sub-glacial bedrock sink-point might be found by descending one of these moulins. As these glacial karstic features usually take meltwater, descending them can be a very unpleasant task. Attempts to enter them in the winter when they could be dry have so far failed as they are buried under deep snow with an absence of landmarks. It is also probable they are temporary features, and that they move with the glacier. As little is known about subglacial hydrology and cave formation, the successful exploration of a moulin would offer a unique opportunity for direct observation. Cavers from Quebec have descended moulins in the Svalbard Glacier in Norway to a depth of 150 m. (Jacques Schroeder, 1990.)

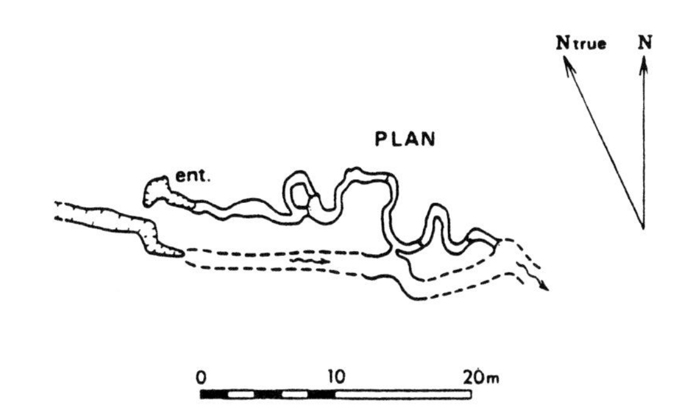

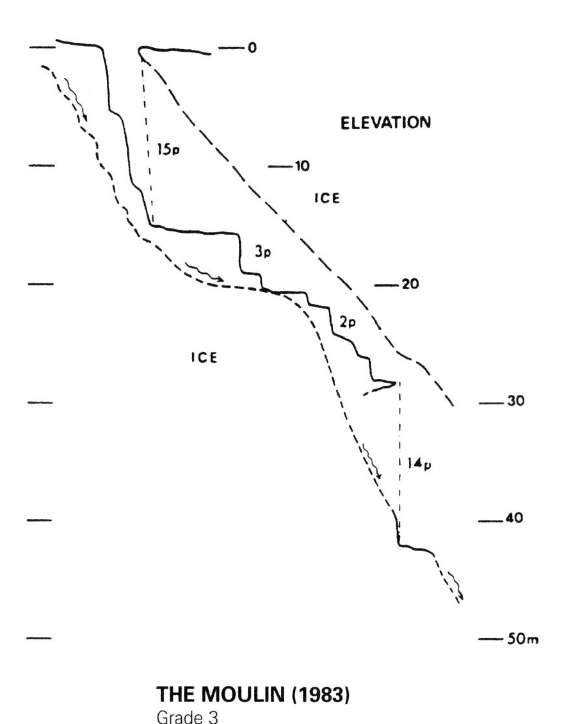

THE MOULIN (1983)
Grade 3
By S. Worthington

The North Karst

Quadruple Pot

Map 83 E/3 342966
Jurisdiction Small River Caves Provincial Park
Entrance elevation 1980 m
Number of entrances 4
Length 60 m
Discovered 1983, by ACRMSE

Location High in the North Karst.

Description Quadruple Pot is the longest and only surveyed cave explored to date in the North Karst. It has four shaft entrances, all climbable.

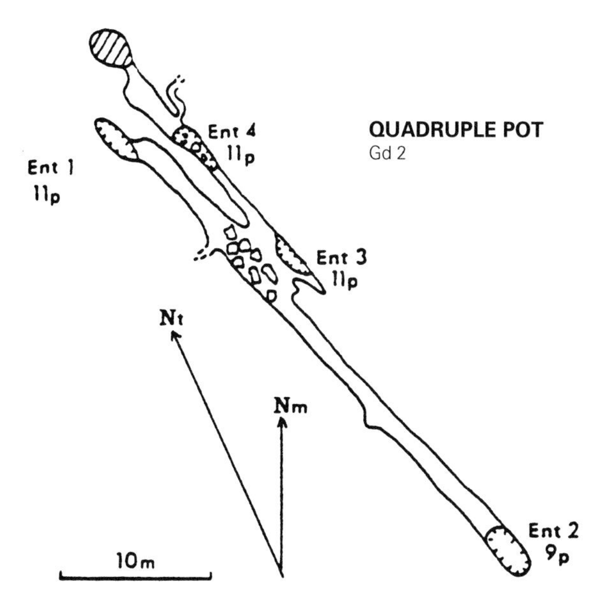

QUADRUPLE POT
Gd 2

The Hanging Valley

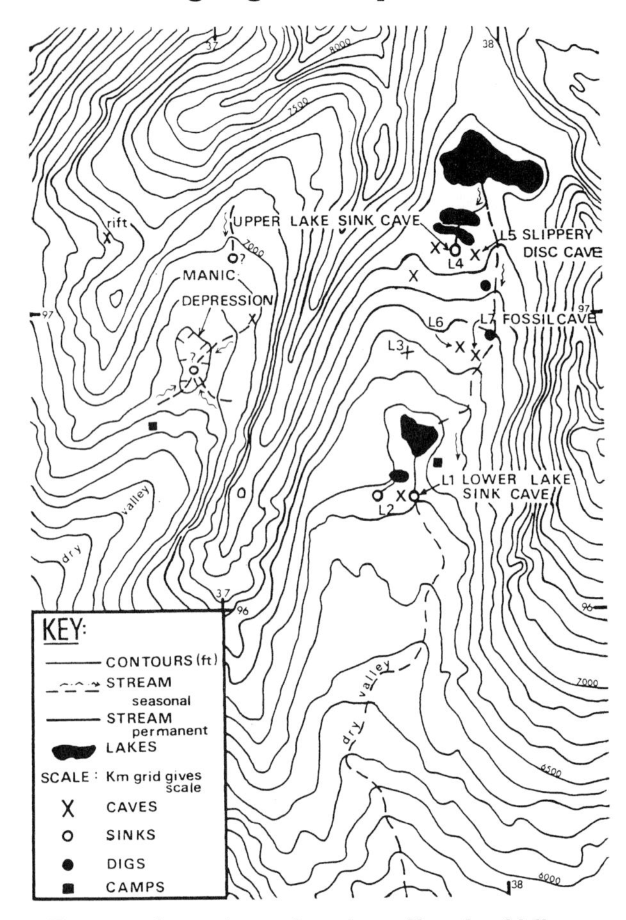

Topography and cave locations, Hanging Valley area (Deej Lowe et al., 1984).

Fossil Cave

Map 83 E/3 379969
Jurisdiction Small River Caves Provincial Park
Entrance elevation 2090 m
Number of entrances 1
Length, Depth 245 m, 62 m
Discovered 1983, by ACRMSE

Lcation Fossil Cave lies just to the west of the intermittent stream that flows from the upper to the lower lake. The small but well-defined entrance is located in highly fossiliferous, pale Mural limestone.

Exploration & Description An ice and snow slope descends at 30° into a chamber with a choked floor. A climb up to the left over boulders leads to a short descending traverse and a climb into a chamber (a bedding plane crawl bypasses the climb). The main passage continues, large and dry, with blocked tubes at various points. Several holes in the floor blocked by boulders are passed before a traverse over and down a rift leads to a mud-floored chamber and a small pool. Below the descent in the rift, an intricate route leads back to an unstable boulder area below the main passage. From here an impressive phreatic tube heads down-dip until filled to the roof with material. A tributary tube heading upwards also ends in a choke.

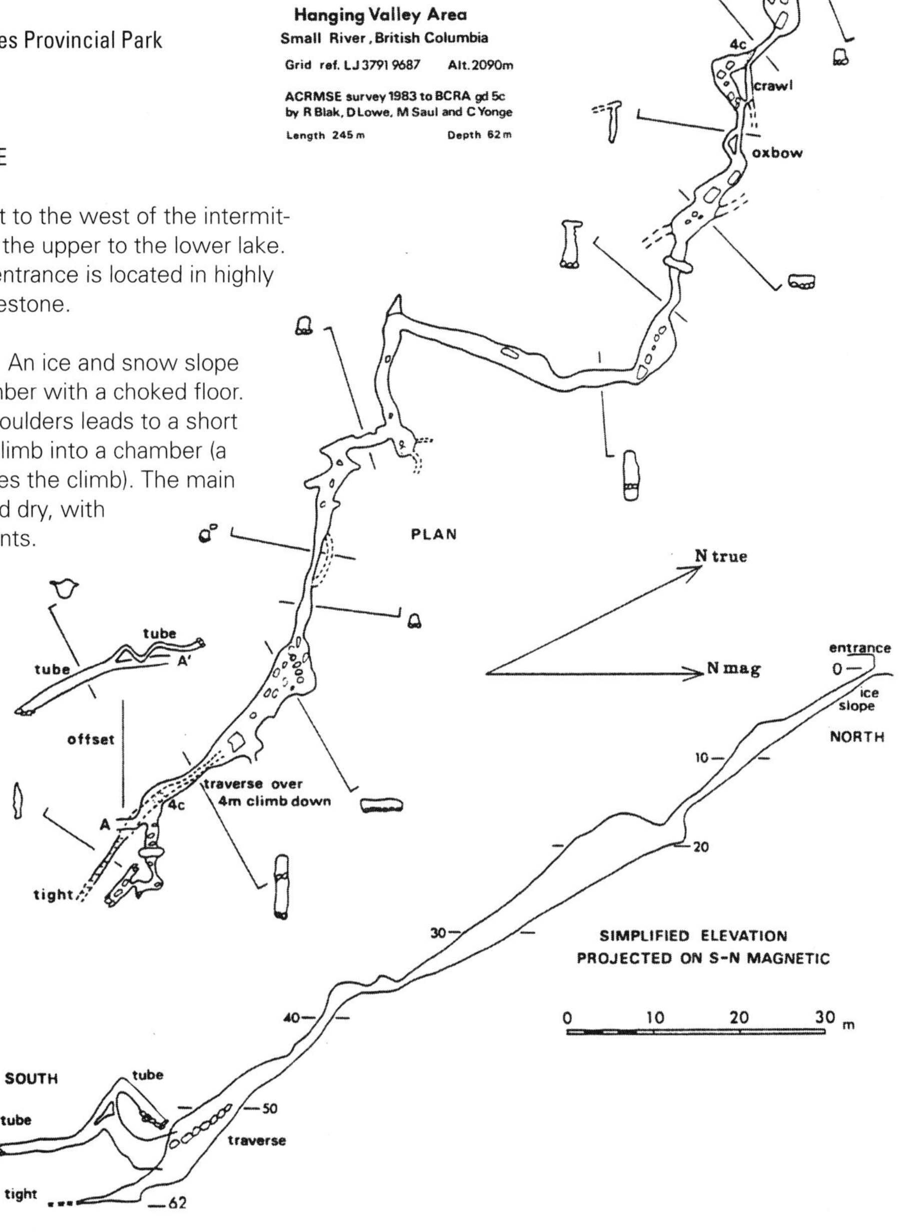

Lower Lake Sink Cave

Map 83 E/3 377963
Jurisdiction Small River Caves Provincial Park.
Entrance elevation 1982 m
Number of entrances 1
Length, Depth 190 m, 47 m
Discovered 1983, by ACRMSE

Location To the south of the lower lake, a stream passes through a small pond into a gully in an outcrop of Mural limestone, and sinks after a few metres. The entrance has been enlarged by hammering.

Description The cave entrance is tight and awkward and drops into a constricted stream-way, which is followed for about 5 m to a sharp bend leading to larger passage. The stream-way flows down dip, along a steep rock ramp to a 2 m climb to a chamber, where another stream joins. Beyond, the passage continues and descends steeply, becoming rifty and tight. The stream then sumps, probably on a fault. In wet conditions the lower section of this cave quickly sumps.

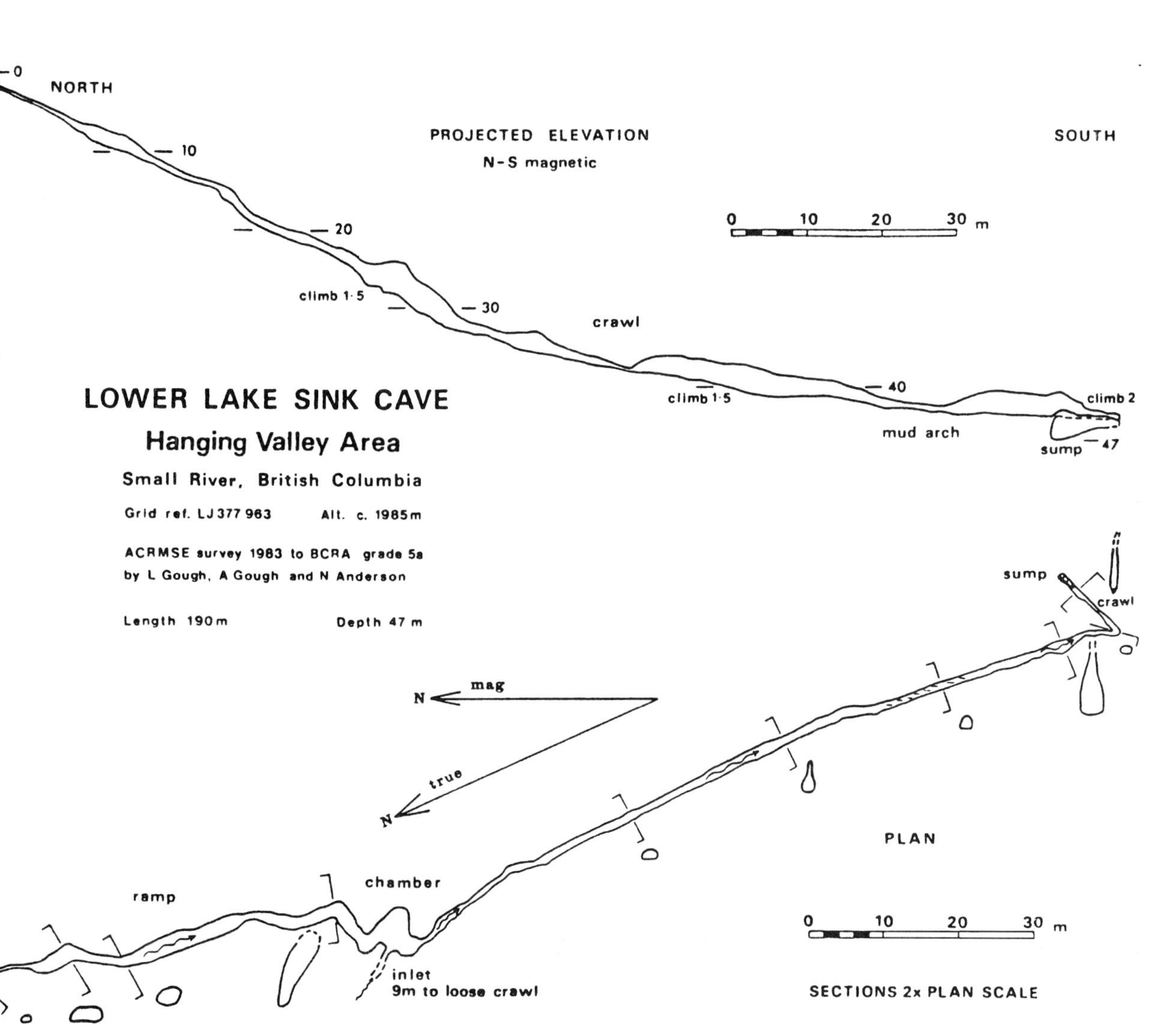

Slippery Disc Cave

Map 83 E/3 379972
Jurisdiction Small River Caves Provincial Park
Entrance elevation 2170 m
Number of entrances 1
Length, Depth 150 m, 50 m
Discovered 1983, by the ACRMSE

Location The entrance is an obvious open fissure in the Lower Mural limestone buttress to the east of Upper Lake Sink Cave.

Description Scramble down an open hole. A rock arch leads to a straight passage and a series of squeezes that were excavated during the original exploration. The passage eventually widens to a small chamber where it ends in a choke close to the end of Upper Lake Sink Cave.

Upper Lake Sink Cave

Map 83 E/3 378972
Jurisdiction Small River Caves Provincial Park
Entrance elevation 2186 m
Number of entrances 1
Length, Depth 330 m, 71 m
Discovered 1983, by the ACRMSE

Location As the name implies, this cave is the sink-point for a stream from the upper lake. The massive entrance gully has been formed by the breakdown of thinly bedded soft limestone where the descending stream encountered the basal part of the Lower Mural carbonates.

Description Several routes through the huge boulders unite in a large passage with a steep boulder slope following the dip downward. Shortly after the stream is met — minor inlets join at several points — the stream passage becomes too tight. (The passage just inside the entrance on the left is a rifty oxbow that rejoins the stream-way just before the tight section.) To the left of the constriction, a scramble over boulders followed by a crawl over a huge slab (The Grand Piano) gives access to a squeeze leading back to the stream-way.

For a short distance, the passage passes through orange-coloured limestone before the canyon narrows. Traversing along ledges makes progress easier. The rift then widens to form a chamber with dry abandoned routes leading off left. All of these are choked, but a stream is audible in one. The main stream-way continues to a 7 m pitch with an unstable boulder slope at the top that drops into a circular chamber. Beyond this, the stream-way continues in a high rift before appearing to sump after about 30 m.

This cave has a potential depth of 968 m to the valley springs.

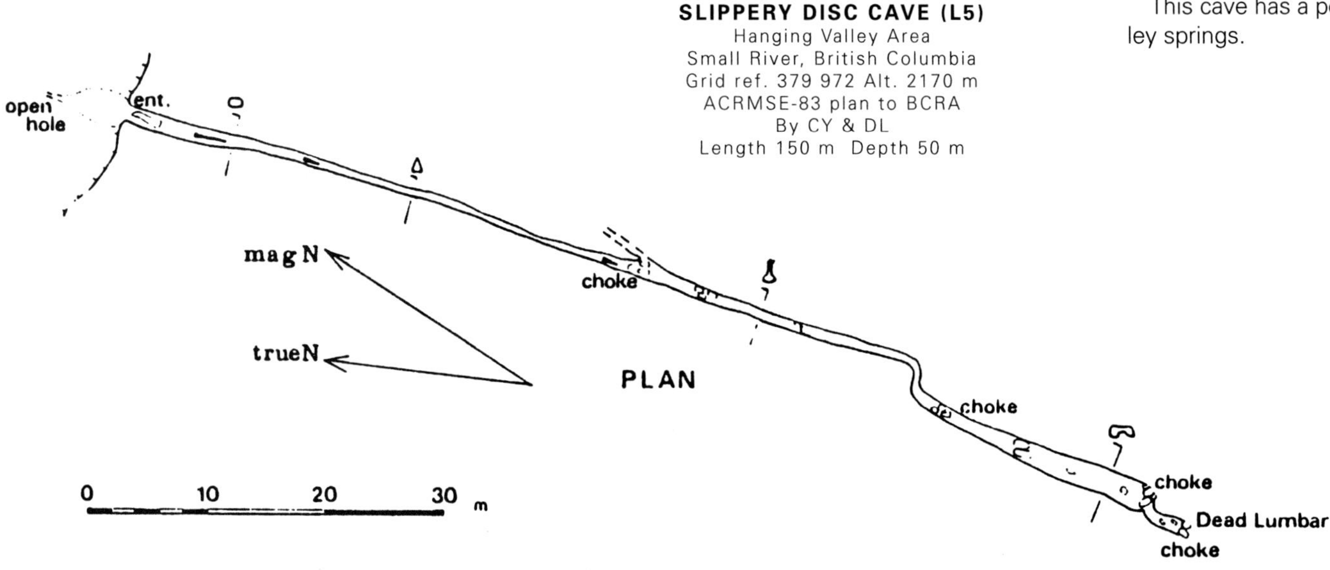

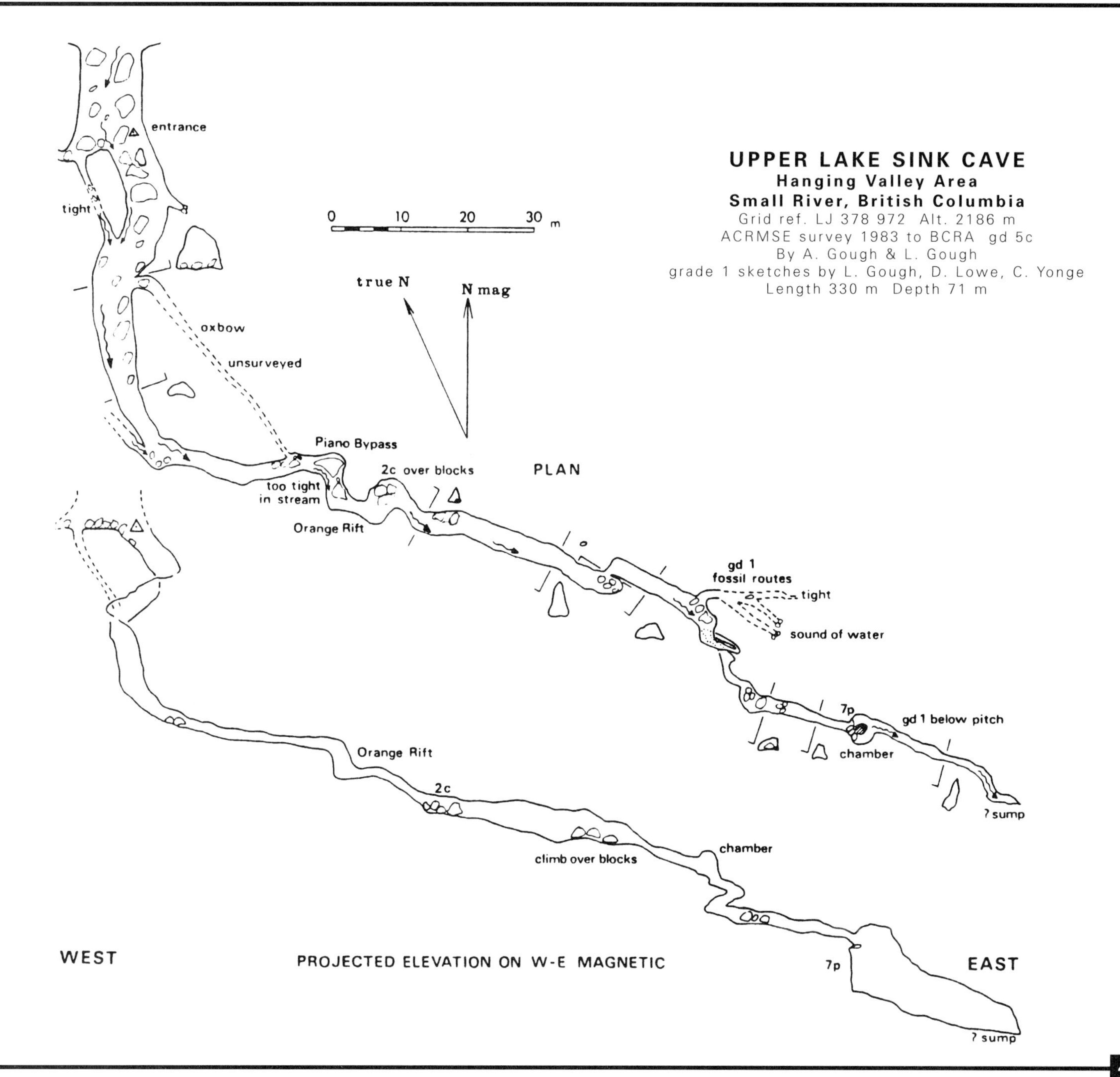
UPPER LAKE SINK CAVE
Hanging Valley Area
Small River, British Columbia
Grid ref. LJ 378 972 Alt. 2186 m
ACRMSE survey 1983 to BCRA gd 5c
By A. Gough & L. Gough
grade 1 sketches by L. Gough, D. Lowe, C. Yonge
Length 330 m Depth 71 m
0 10 20 30 m
true N
N mag
entrance
tight
oxbow
unsurveyed
Piano Bypass
too tight in stream
2c over blocks
Orange Rift
PLAN
gd 1 fossil routes
tight
sound of water
7p
gd 1 below pitch
chamber
? sump
Orange Rift
2c
climb over blocks
chamber
WEST
PROJECTED ELEVATION ON W-E MAGNETIC
7p
EAST
? sump

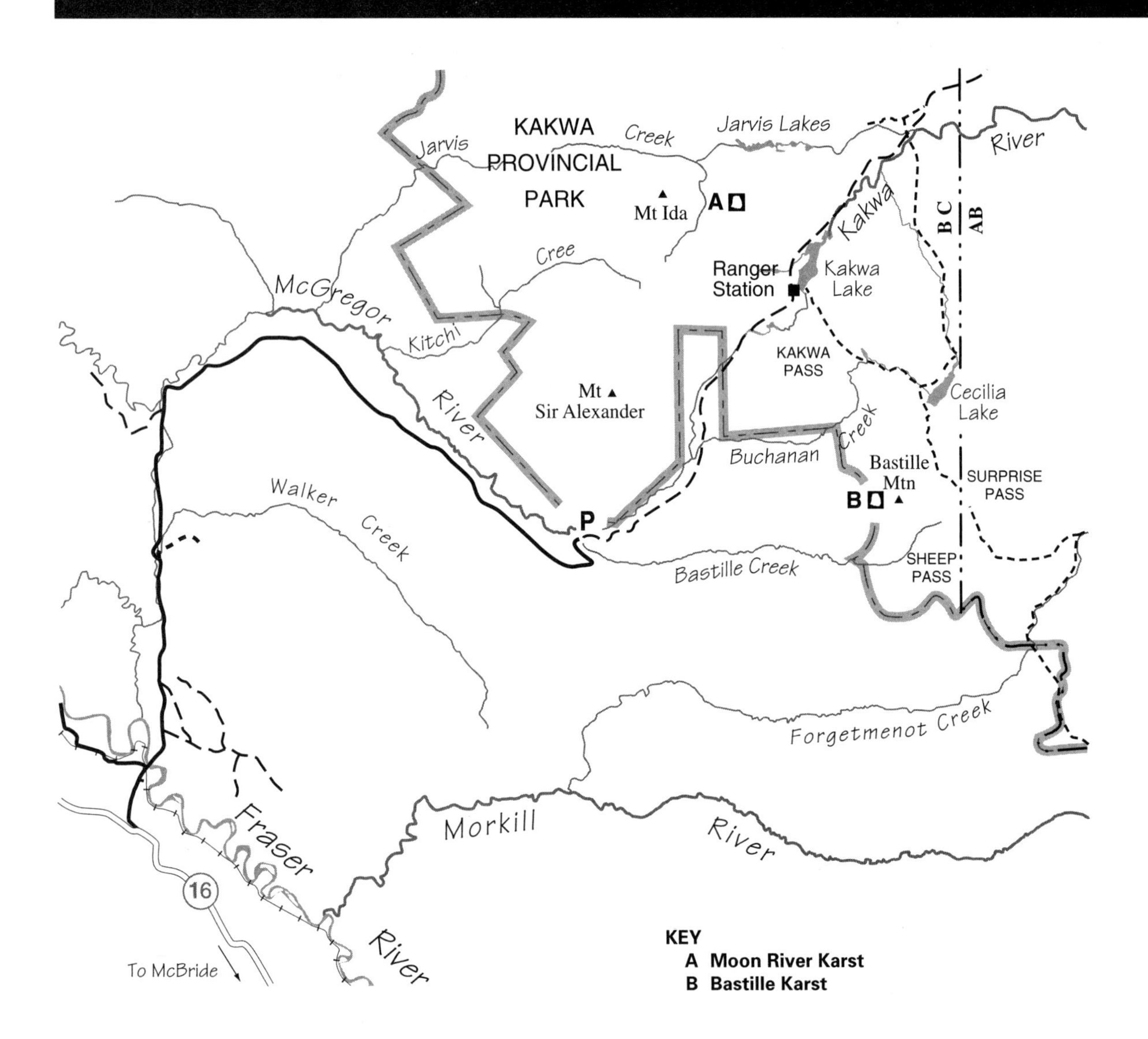

The recently created Kakwa Provincial Park, encompassing 170,890 ha, is located 70 km north of the town of McBride on Hwy. 16. It contains both the Moon River karst and the Bastille karst.

Moon River Karst

Moon River is a karst plateau with a limestone/impervious contact running through it approximately north -south. Besides Moon River Cave and Moon Valley Cave that are located on the contact, the plateau contains thousands of blocked sinks. Blocked resurgences have been found near Jarvis Lakes.

As with many Rockies caves, finding the window of opportunity with unfrozen lakes, minimal snow and low water seems to occur in the early fall. Finding the window of opportunity is especially critical with Moon River Cave. Attempts in August have been turned back by high water — despite the use of wet suits. October appears to be the best month, although weather forecasts should be checked. Winter attempts seem futile, with bitterly cold temperatures and deep hard-packed snow.

Access Different modes and combinations of transport can be used — helicopter, floatplane, 4x4 vehicles, packtrain.

A helicopter can take you right to the plateau (Yellowhead Helicopters based in Valemount (250) 566-4401). Taking a small floatplane to one of the lower lakes is cheaper, bearing in mind that floatplane access requires ice-free lakes. However, the floatplane charter that used to operate out of McBride is not available at the time of publication. Check with BC Parks, as fly-in access might be restricted.

Access by 4x4 vehicle to the trailhead, then on foot is possible, but carrying heavy loads of caving gear is a struggle. Consider using packhorses to carry the gear.

From Hwy. 16, 74.5 km north of McBride, turn right onto the resource road (Walker Creek Forest Road) leading to Kakwa Provincial Park. En route cross the Fraser River and at the junction in the McGregor River valley keep right. Park at Bastille Creek (the road beyond has been washed out in places by the heavy rains of July 2001).

From here walk along the "road" to Buchanan Creek (12 km, camping area), then over McGregor Pass towards the ranger cabin on the southern tip of Kakwa Lake (another 17 km, camping area at Mariel Lake en route). Before the ranger station head left on the continuing road, more of a track, that follows the west side of the lake (keep straight at Babette Lake junction). Leave the track as it heads down the Kakwa River and contour left (northwest) to Jarvis Pass. From Jarvis Lakes beyond the pass walk up to the plateau. The total hiking distance of 40 km takes at minimum three days with a 400 m elevation gain at the end up from Jarvis Lakes to the Moon River Plateau.

For the latest information on the state of the access road and cave visitation, call the Prince George Forestry Office at (250) 565-7086.

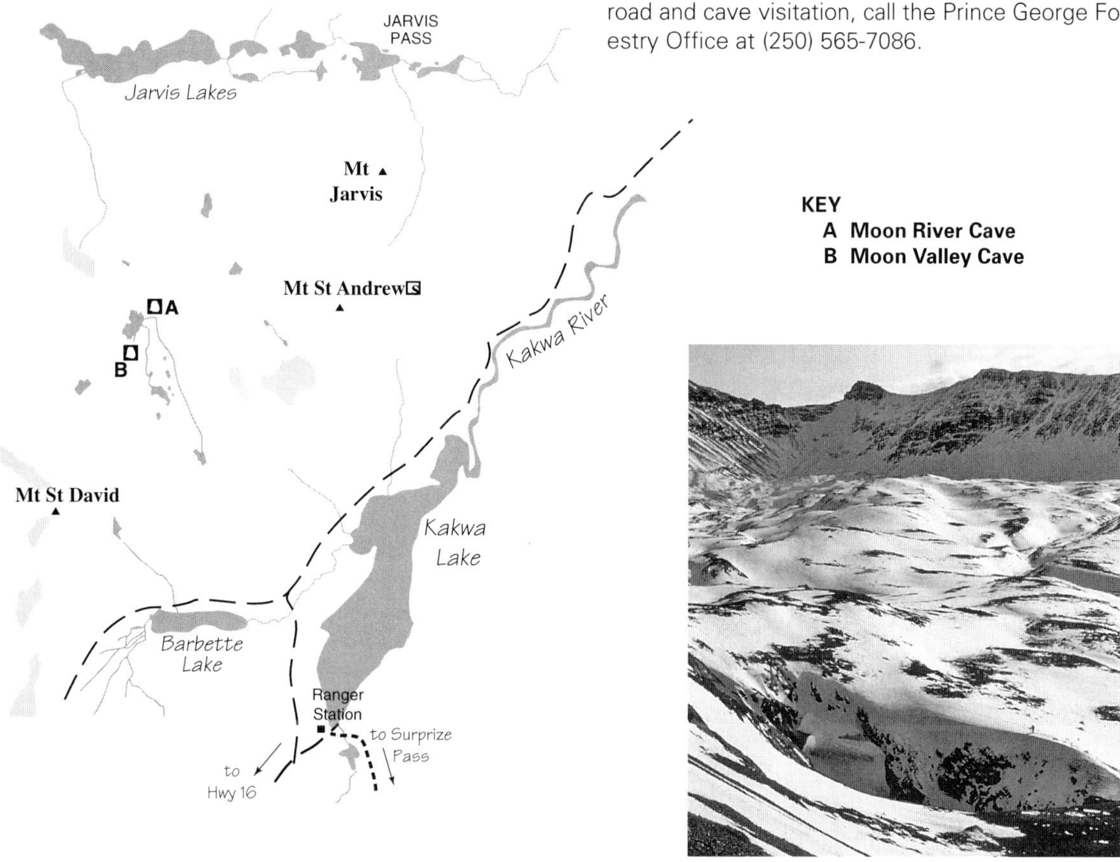

The Moon River plateau, showing the huge entrance sink of Moon River Cave. Photo Ian McKenzie.

Area Exploration Cavers first visited the Moon River area in August of 1982 before the area became a provincial park and there were access restrictions. Even then, the approach took three days and involved the use of a 4 wheel-drive pickup and dirt bikes, followed by 30 km of hiking. A group of outdoor enthusiasts from Prince George had helicoptered into the area several days earlier and poked around in a number of sinks, but lacking caving gear, were not involved in any of the underground exploration. During this first visit ASS cavers explored the obvious, huge snow-plugged sink, Moon River Cave, for 45 m to the third wet pitch. A smaller sink higher up the valley, Moon Valley Cave, was explored to a definite conclusion after 353 m.

A return helicopter visit in January 1983 suffered from a week of temperatures as low as -40°C, strong winds and lots of snow. Much digging in snow and a little skiing was accomplished, but most of the time the beleaguered party just tried to keep warm. No new passage was explored.

In October, 1983, a Beaver floatplane was used to fly to the lower lakes. The party of ASS cavers established a base camp at Pegasus Lake before moving over the col to the high karstic plateau to continue battle with Moon River Cave. This time, the cave being considerably dryer than in August of 1982, a low point of –90 m in the Big Shaft was achieved.

The last trip to date was made in October, 1987, when an ASS group arrived in a Cesna 185 float plane from McBride. During this trip Moon River Cave was explored to a surveyed length of 774 m, a second entrance to Moon Valley Cave was found, and the diminutive Pika Cave was explored.

Moon River Cave

Map 93 I/1 818933
Jurisdiction Kakwa Provincial Park
Entrance elevation 1860 m
Number of entrances 1
Length, Depth 774 m, 142 m
Discovered 1982, by a Prince George outdoors group

Location The outlet of a nearby ice-filled lake plunges in a 15 m waterfall into the 30 m-wide triangular doline formed on the limestone/quartzite contact. The doline is almost filled by a small glacier.

Exploration & Description It was aerial photographs of the large doline that marks the entrance of Moon River Cave that first attracted attention to the Moon River karst.

The small glacier almost filling the doline forces explorers into close contact with the sinking stream. Below the ice (at one point an estimated 25 m above the floor) a series of short pitches lead to the Main Shaft, an intimidating water-washed drop that terminated exploration during the October, 1983, trip. In October, 1987, lower water levels permitted the rigging of a devious traverse across the Main Shaft (the Moondance) and down a shorter, parallel drop, allowing the bottom of the Main Shaft to be finally reached through a small hole. The Moondance traverse was immediately rendered obsolete when a dry bypass was found (Every Day Way).

"I looked back at the ghostly white cascades simultaneously falling down and standing in place like a drunken bride, before crashing off into the black maelstrom below... then felt a shock of recognition at the black buttons of the bolt traverse we had placed in 1983 at the head of the main shaft. It was like being dead and reviewing your past life from above — man, we would have been nuts to try dropping the pitch from there." The view from the Moondance, Ian McKenzie.

From the passage adjacent to the base of the Main Shaft were two obvious leads: a large phreatic tube leading to a sump (the Full Moon Cascades) and a cobbled stoop leading to a large fossil passage (The Dark Side). At first the Dark Side passage was thought to end just past a 24 m pitch, but at the end of the expedition a crawl was found to access a hundred metres of phreatic passage partially filled with sand and cobbles (Partial Eclipse Passage). This was explored to a 7 m drop and a sump. A good lead around the corner at the top of the drop was not pushed owing to lack of time.

Warning Moon River Cave is probably one of the more serious caves in the Canadian Rockies, with complex route-finding, difficult ropework and danger from fast-flowing water and collapsing ice.

The entrance to Moon River Cave. Photo Dave Thomson.

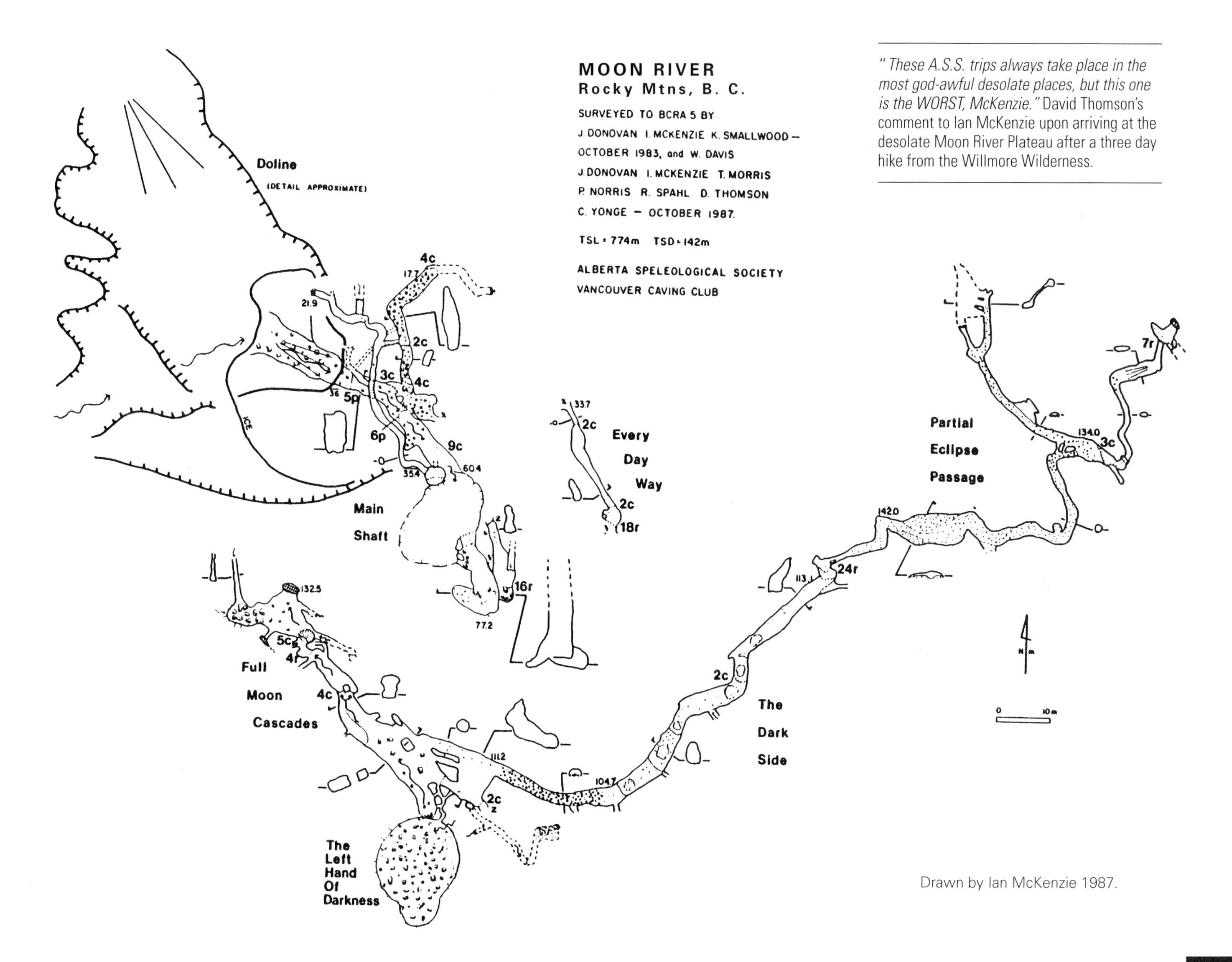

" These A.S.S. trips always take place in the most god-awful desolate places, but this one is the WORST, McKenzie." David Thomson's comment to Ian McKenzie upon arriving at the desolate Moon River Plateau after a three day hike from the Willmore Wilderness.

Drawn by Ian McKenzie 1987.

continued from bottom left

Moon Valley Cave

Map 93 I/1 815925
Jurisdiction Kakwa Provincial Park
Entrance elevation 2789 m
Number of entrances 2
Length, Depth 352 m, 134 m
Discovered 1982, by a Prince George outdoors group

Location A series of shakeholes trend upwards (south) from the main depression of Moon River Cave. One of these is the rather loose Moon Valley Cave.

Exploration & Description In August, 1982, the entrance was probed by a group of outdoor enthusiasts from Prince George who were unable to descend the first 8 m drop as they had no vertical gear. Two days later ASS cavers explored and surveyed the cave during two trips.

Descent of a snow plug leads to a boulder choke over the first 8 m pitch. This is followed by a series of damp down-climbs and pitches. Squeezes through a couple of boulder chokes lead to the Terminal Room with no apparent way on. Following a lower stream route down through a small hole in the floor accesses "a low roofed sandy little grotte with zero potential."

The second entrance to Moon Valley Cave (not on the survey) was located and explored during the 1987 ASS Moon River trip (no details available).

MOON VALLEY CAVE
British Columbia
CRC Grade 4
Surveyed by J. Donovan, D. Gilliatt, I. McKenzie August 1982.
Surveyed length: 1157 feet
Surveyed depth: 439.2 feet
All measured in feet
Sections not to scale
Alberta Speleologicsl Society

Drawn by Ian McKenzie 1982

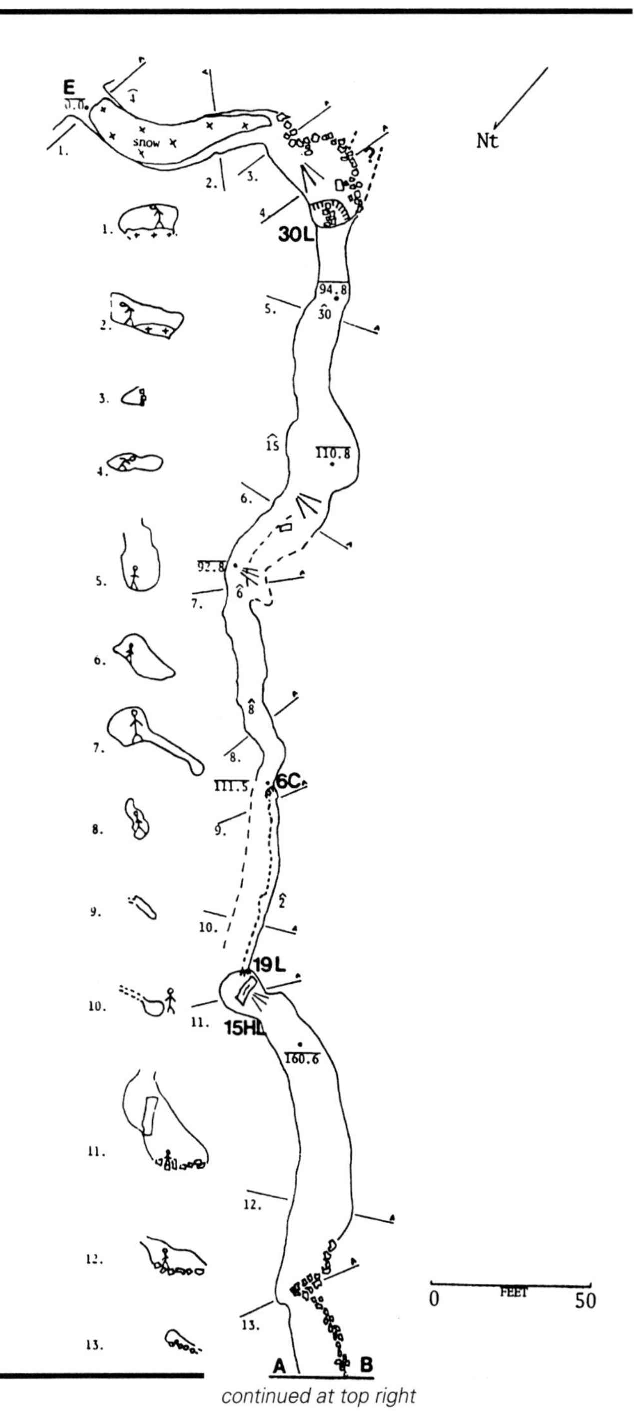

continued at top right

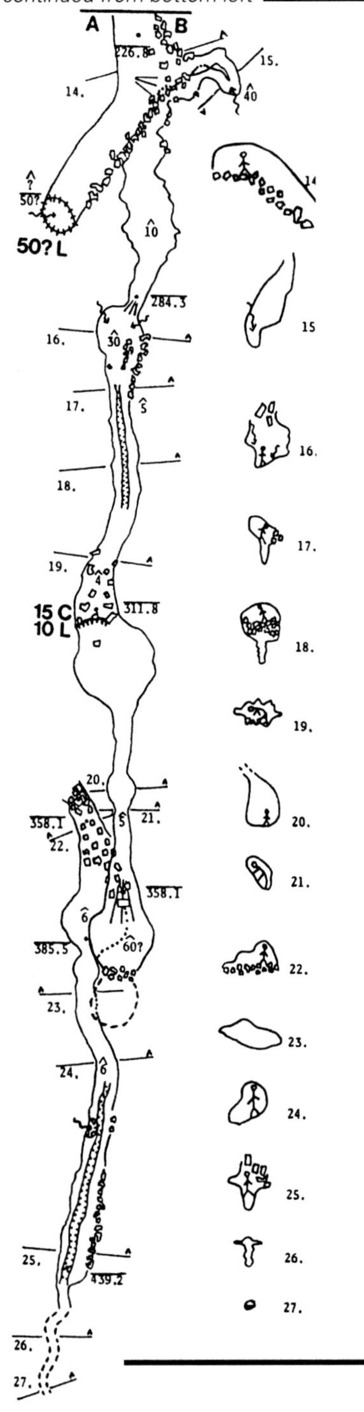

Pika Cave

Map 93 I/1 GR location uncertain
Jurisdiction Kakwa Provincial Park
Entrance elevation Unknown
Number of entrances 1
Length, Depth 20 m, 6 m
Discovered 1987, by Chas Yonge

Location A cave situated where the lake in the lower hanging valley drains into the ground — the water source for the 1987 expedition.

Description A diminutive cave.

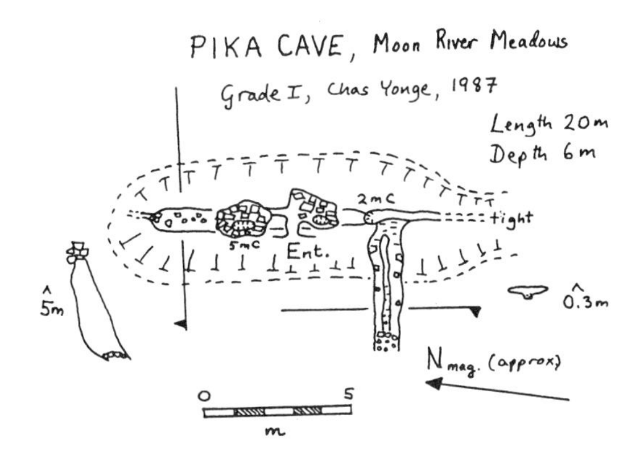

The Bastille Karst

The Bastille karst is located in the southeast extent of Kakwa Provincial Park to the northeast of Sheep Pass between Sheep and Bastille creeks.

Access to all Caves The only caving expedition to date flew in by Yellowhead Helicopters. Staging was along logging roads to the north.

Access by 4x4 vehicle and on foot is possible from near McBride. As for the Moon River karst, an approach on foot carrying heavy caving gear would be a hard undertaking.

From Hwy. 16, 74.5 km north of McBride, turn right onto the resource road (Walker Creek Forest Road) leading to Kakwa Provincial Park. En route cross the Fraser River and at the junction in the McGregor River valley keep right. Park at Bastille Creek (the road beyond has been washed out in places by the heavy rains of July 2001).

From here hike up Bastille Creek (no maintained trail) to its headwaters at Sheep Pass (about 25 km).

For information on the access road and cave visitation contact the Prince George Forestry Office at (250) 565-7086.

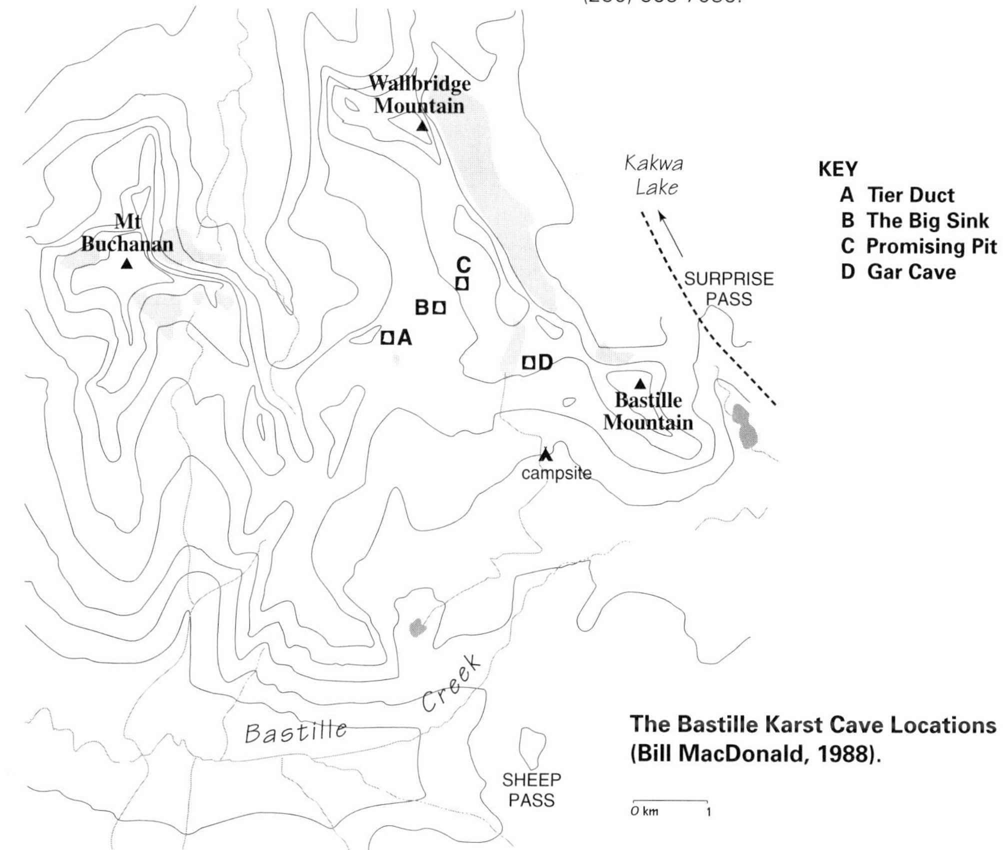

The Bastille Karst Cave Locations (Bill MacDonald, 1988).

Area Exploration Long-time ASS caver Tom Barton first noticed the promising looking plateau while studying topographical maps in the University of Calgary library. Dave Thomson walked the area on his epic hike out from the Moon River at the end of the 1987 expedition, and described it as "A vast apron of a plateau billowing up to 1970 m in the middle and tied down at the corners by Wallbridge, Buchanan and Bastille mountains." He made note of the many surface karst features, including large areas of bare limestone pavements, cave entrances, sinks and springs. Confirming the large extent of the karst using aerial photographs, ASS member Bill MacDonald organized an expedition, ultimately consisting of 13 cavers, that helicoptered in during August of 1988. Over five days of exploration (during which several attempts were made to climb Mt. Buchanan on days off from caving), four caves were explored and surveyed, the most promising being the very loose Tier Duct, in which exploration was cut short due to lack of time.

The Bastille karst area (named after Bastille Mountain 3390 m, first climbed on July 14th, Bastille Day), probably more than any other karst area in the Canadian Rockies is deserving of another caving expedition.

"Large, sub-cubical guillotines and grand pianos were slung raggedly on incredibly tiny rock points, held more by whatever faith rocks possessed in their own destinies than by any known laws of physics." Tom Miller.

"Out of hundreds of shafts, only two to three score, at most, had been thoroughly checked, and the cave was consuming nearly all available rope." Tom Miller, 1989, referring to Tier Duct.

Tier Duct

Map 93 H/16 920721
Jurisdiction Kakwa Provincial Park
Entrance elevation c. 2090 m
Number of entrances 1
Length, Depth 431 m, 92 m
Discovered 1988, by the Bastille Karst Expedition

Location The furthest west (towards Mt. Buchanan) of the caves. Look for a cairn. Originally named Jaws due to its propensity for eating equipment and cavers, this cave was later renamed Tier Duct owing to the distinctive tiers of limestone that surround the cave entrance.

Exploration & Description The cold, blowing cave entrance contains a large canyon heading down-dip to the first 8 m drop. Shortly afterwards, a step across a 30 m shaft leads to the top of a high rift, the roof of which is soon out of sight.

A large pile of jammed boulders sets the nature for the rest of the cave: loose, complex and scary. Squeezes over, under and through boulder heaps lead steadily downwards to 60 m depth. This is followed by a climb up a steep debris ramp to a vertical climb (10 m prussic) below the "Guillotine"—a series of large balanced boulders. A short, tight squeeze leads to a 21 m drop, and an 8 m climb back to the roof. The roller coaster nature of route finding continues with yet another climb to the ceiling in a narrow crevice, then another 22 m drop, and a free-hanging cleft named Bold Step Pit. Following yet another climb up collapse material and a squeeze through boulders, the passage abruptly widens to where piles of gypsum were noted (Gypsum Junction).

From here, a 12 m drop into a wide room accesses a narrow passage and a high terminating aven. The way on, the left-hand passage from Gypsum Junction, consists of a sinuous canyon (stream audible beneath), that is too wide to chimney. Out of time and tackle, the exploration team turned back at this point.

Warning Beware of the extremely loose nature of this cave.

The entrance to Tier Duct. Photo Dave Thomson.

Opposite near right: Tom Miller in Tier Duct. Photo Dave Thomson.

Opposite far right; Olivia Whitwell performing SRT in Tier Duct. Photo Dave Thomson.

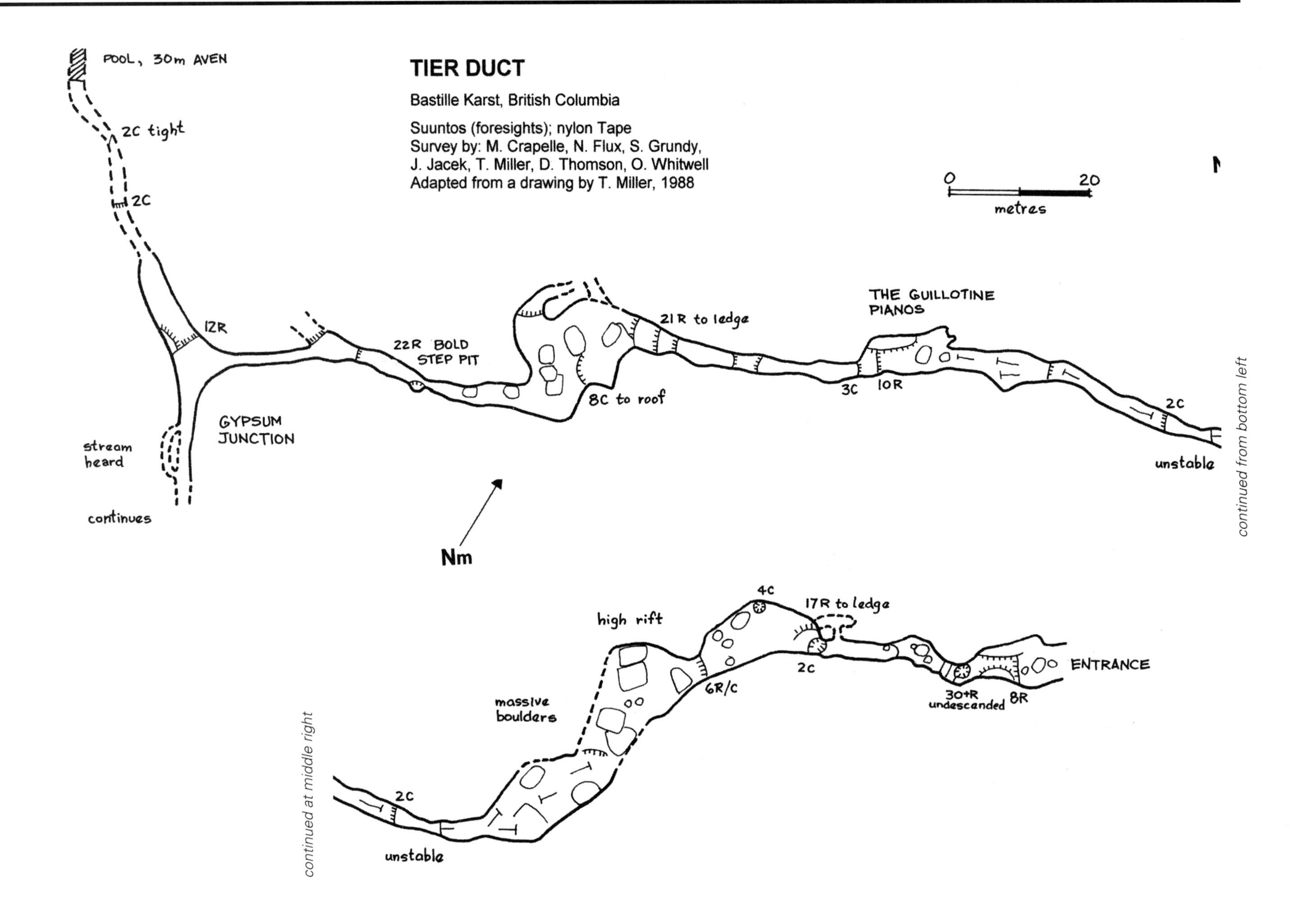

continued from bottom left
continued at middle right

The Big Sink

Map 93 H/16 928742
Jurisdiction Kakwa Provincial Park
Entrance elevation c. 2100 m
Number of entrances 1
Length, Depth c. 70 m, 27 m
Discovered 1988, Bastille Karst Expedition

Location A small lake drains into a large sink (9 m entrance pit).

Exploration & Description From the base of the pit, a stream was followed along a horizontal passage to where it disappeared in the floor of a small room. A squeeze past a flake of rock accessed a draughting, wet passage that became progressively tighter, with the sound of the stream thundering in the distance.

Surveying in The Big Sink.
Photo Dave Thomson.

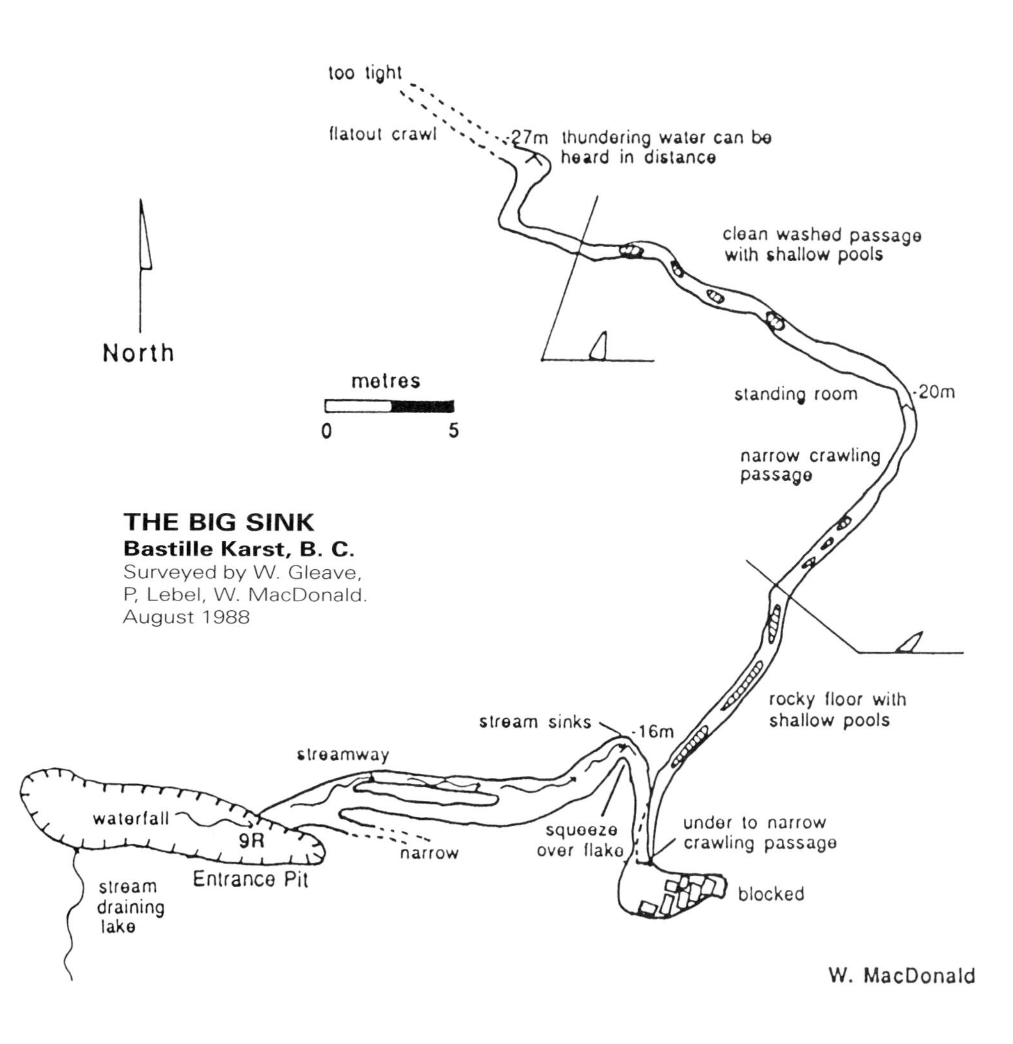

Gar Cave

Map 93 H/16 939733
Jurisdiction Kakwa Provincial Park
Entrance elevation c. 2150 m
Number of entrances 1
Length, Depth 214 m, 103 m
Discovered 1988, Bastille Karst Expedition

Location Look for a 6 m pit adjacent to a small ice-field just over 1 km west of Bastill Mountain. At the time of exploration the pit had a large mound of snow in its centre.

Description From the base of the pit a small drafting, vadose passage leads to the head of a series of wet interconnecting pitches, both ending in boulder chokes.

The name originates from the large number of cigars smoked by two degenerate British Columbia cavers while waiting for more rope to arrive at Close to the Ledge.

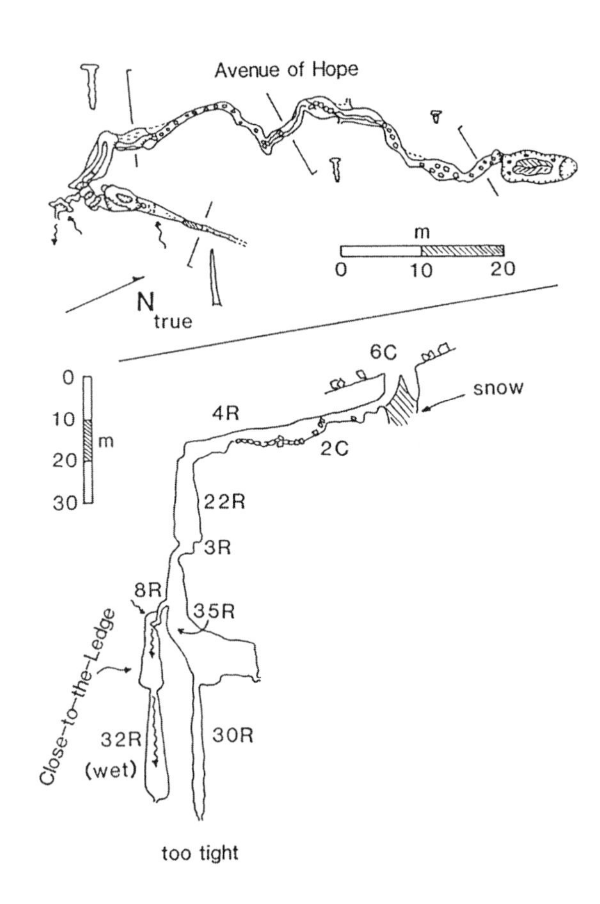

GAR CAVE
Bastille
Elevation: 2200 m asl
Depth: 103 m length: 214 m
Survey: Jim Jasek, Art Peters, Chas Yonge.

Promising Pit

Map 93 H/16 929744
Jurisdiction Kakwa Provincial Park
Entrance elevation c. 2200 m
Number of entrances 1
Length, Depth c. 100 m, 25 m
Discovered 1988, Bastille Karst Expedition

Location A few hundred metres northwest of The Big Sink. A 25 m-deep shaft contains a streambed that is blocked at both ends.

Description A round hole 7 m down the shaft leads to 75 m of upward trending phreatic passage ending in two narrow rifts, one heading up and the other down, both too tight to continue along.

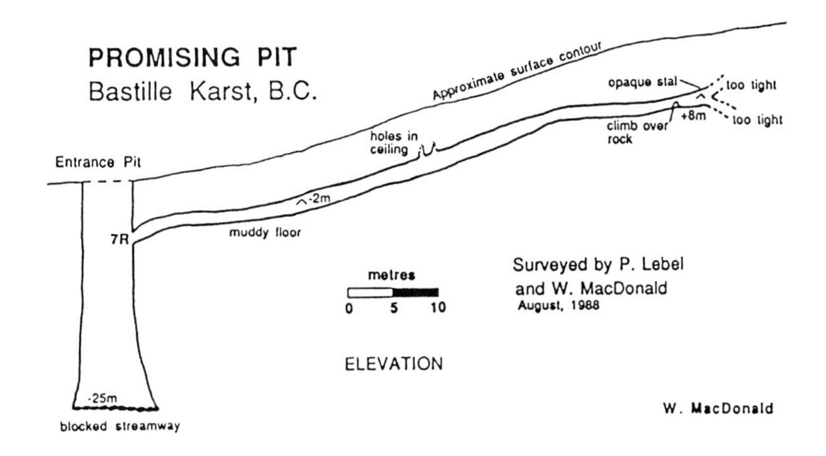

THE MCGREGOR RANGE

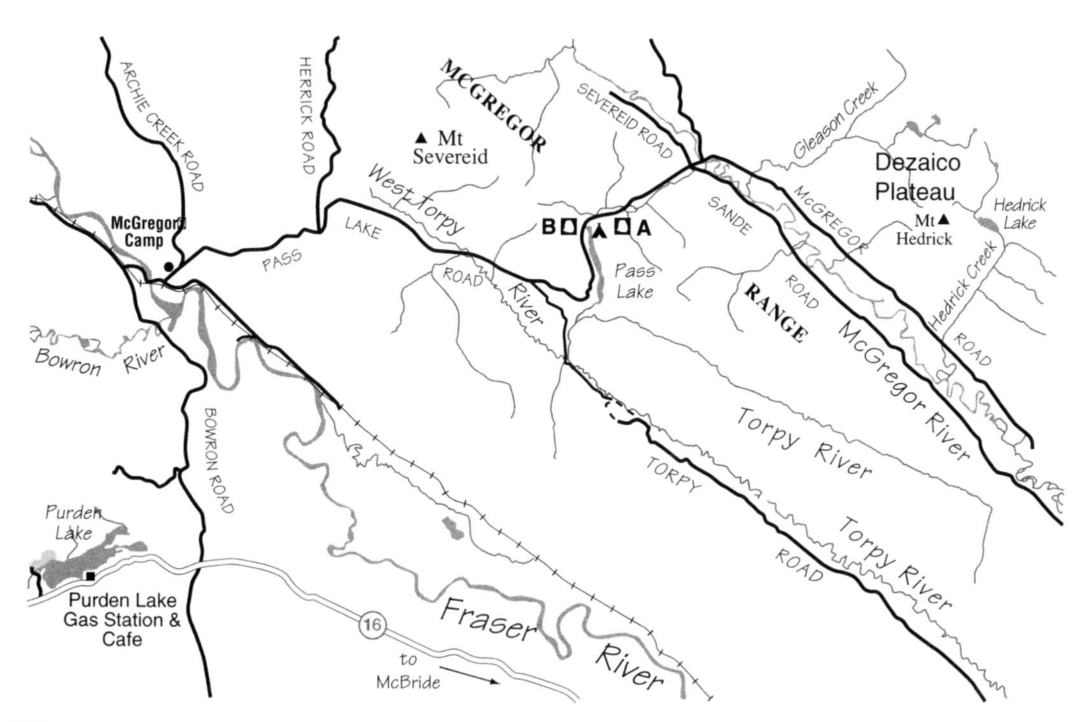

KEY

- **A Fang Cave, Meadow Cave, Tooth Decave, Window on the West, Useless Cave**
- **B Pass Lake Cave**

Southeast of Prince George, bordered by the Fraser River to the east and the Dezaiko Range to the west, are the caves of the McGregor Range. They are easily accessed via gravel logging roads from Hwy. 16 near Purden Lake.

In 2000, several new B.C. Provincial Parks were created in order to protect the caves from logging, mining and other commercial activities. In this area, the 1473 ha Evanoff Provincial Park protects Fang Cave.

Access to all Caves Make sure you have a full tank of gas. From Hwy.16 travelling north towards Prince George, turn right (east) just before Purden Lake Resort onto Bowron Road. Follow it over a bridge to the junction with a surfaced highway. Turn right and cross a shared railway bridge over the Fraser River. Then fork left through McGregor Camp onto Pass Lake Road. At the junction with Herrick Road bear right and continue past the junction with Torpy Road to Pass Lake (campground at far end).

For Fang Cave, Tooth Decave, Window on the West, Useless Cave and Meadow Cave turn right shortly before the first bridge crossing after Pass Lake. Park a few hundred metres along the overgrown road.

Fang Cave

Map 93 I/3 075950 (Upper Entrance)
Jurisdiction Evanoff Provincial Park
Entrance elevation Upper Entrance 1525 m, Resurgence Hall 1400 m
Number of entrances 4, Upper Entrance (sink), Middle Entrance (skylight), Trademan's Entrance, Resurgence Hall (via duck)
Length, Depth 3342 m, 247 m (from datum at Upper Entrance -143 m +104 m)
Discovered Known since the early 1970s, Fang Cave was first entered for any distance in 1981 by George Evanoff, Michael Nash et al.

Location The distinctive limestone pinnacle located above Fang Cave (and that gave the cave its name) can be seen on the drive in from southeast of Pass Lake.

From the parking area follow a well-used trail south up the drainage that emanates from Fang Cave (700 m elevation gain over 3 km).

Warning If approaching in winter, care should be taken on the final steep approach that may be avalanche prone.

Description Fang Cave is one of the finest caves in the Rockies. The cave system is active, water entering through three distinct sets of passages: the Upper Entrance sink, the Root Canal, and the Missing Lynx, before meeting at the head of the Queen's Gallery and resurging at the impressive Resurgence Hall. Water levels vary greatly; the resurgence entrance is often dry in the winter. At this time, Coliseum, Corkscrew and other passages in the cave contain impressive ice formations.

Currently, Fang Cave is the seventh longest and tenth deepest cave in Canada.

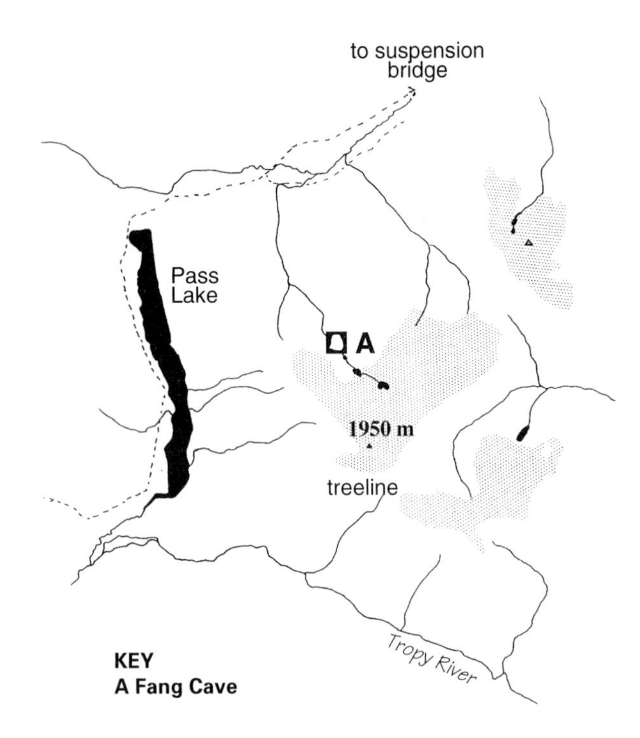

Exploration In September, 1981, George Evanoff, Michael Nash and three ASS members entered Fang Cave via the Resurgence Hall, carried on through The Plunge, eventually coming to a stop at the 5 m waterfall at the end of the Queen's Gallery. The next day they entered the impressive Upper Entrance sink and explored through the low 1 m-high Crab Crawl to a 17 m pitch, and on down into a cavernous chamber partially lit by a high skylight (Middle Entrance). With impressive dimensions — 65 m long, 30 m wide and 25 m high — the Coliseum is believed to be second in size only to Boggle Alley in Gargantua. The explorers were finally stopped by a lack of tackle at the descending Corkscrew passage. Next, they backtracked to the Entrance Hall, entered the Fumarole and went down the Tiger Tunnel to the top of a waterfall (Disconnection Falls), which they wrongly surmised was the same waterfall that had stopped exploration at the head of the Queen's Gallery. The trip back to the Upper Entrance was complicated by them getting lost, running out of lights and being fumigated by dense smoke from a fire at the entrance — hence the name Fumarole.

During later visits in 1981 they entered the cave from the skylight into the Coliseum, and using a handline, descended the Corkscrew, thus making the connection with the Ante Room and Queen's Gallery. A rough survey of the cave (Grade 5 with sections of Grade 1) made in 1982 incorrectly shows the Fumarole and Tiger Tunnel connecting to the Queen's Gallery. (This connection was not made until 1987.)

In 1983 Disconnection Falls was descended to a sump, verifying that a connection with the Queen's Gallery had not been made. The inlet beyond the Queen's Gallery was pushed via a tight squeeze, and the large (up to 10 m wide, 30 m high) Missing Lynx passage was explored to its conclusion at a waterfall issuing from a hole 10 m above the floor. A side lead off the Missing Lynx passage was explored through a complex of dirt crawls and short pits (upper passages) to a boulder choke. A tight squeeze revealed the Big Grin, a dry canyon 150 m long and 20 m high that ended in a boulder choke. A later visit found a way through the choke to more ascending passage that ended in another choke. A 40 m-deep pit in the upper passages, Deep Throat, was descended back to the Missing Lynx stream way, thus bypassing the first waterfall. Another 25 m+ pit was found that provided an easier connection. The Missing Lynx stream was then pushed for another 200 m past two more waterfalls to a small room beyond with a huge boulder choke (surveyed in 1984). The Root Canal passages, explored in 1982, were surveyed for 250 m to the impressive 30 m-high Twin Avens. By the end of 1983 Fang Cave was 2632 m long and 238 m deep, making it the ninth longest and ninth deepest cave in Canada.

In February, 1987, the cave was entered through the Plunge which was dry (the inlet to the right was still flowing) and a "dry pool" near the Ante Room was connected with Disconnection Falls. A side lead from this connection, a small tube, was explored to a choke beneath the Coliseum through which walking passage could be seen. In July, the Big Grin was pushed past the 1983 far point choke for another 74 m to a more definite end thought to be close to the surface.

In 1993, the UNBC Caving Club discovered a fourth, free-climbable entrance (Tradesman's Entrance) and in 1995 explored some small side passages just inside the Resurgence Hall. During the removal of a rock and gravel plug, a surge of water was released, washing two cavers a considerable distance down some tubes!

Upper entrance to Fang Cave. Photo Ian McKenzie.

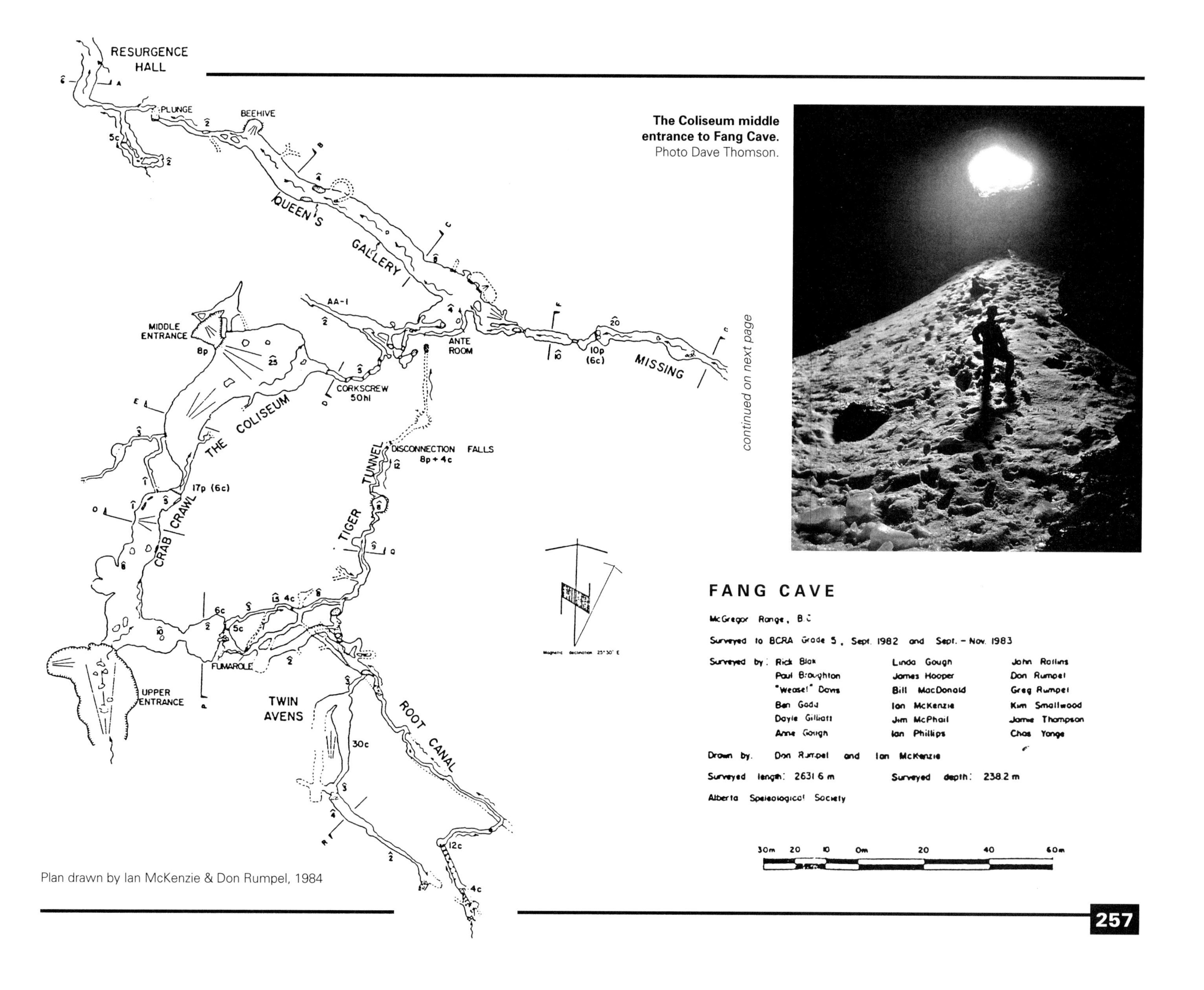

The Coliseum middle entrance to Fang Cave. Photo Dave Thomson.

Plan drawn by Ian McKenzie & Don Rumpel, 1984

continued on next page

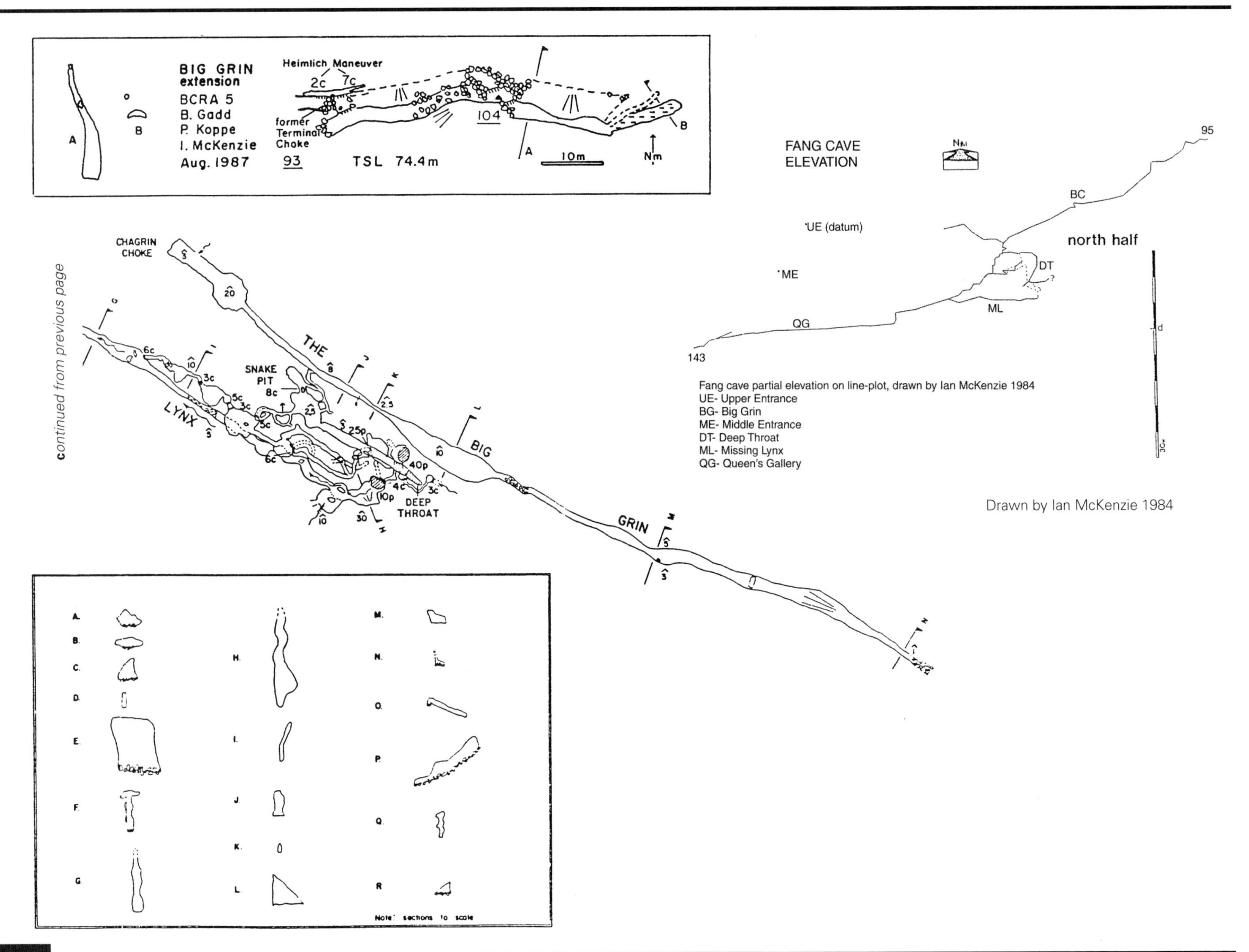
BIG GRIN extension
BCRA 5
B. Gadd
P. Koppe
I. McKenzie
Aug. 1987
Heimlich Maneuver
former Terminal Choke
93
TSL 74.4m
104
10m
FANG CAVE ELEVATION
95
BC
UE (datum)
north half
ME
DT
ML
QG
143
Fang cave partial elevation on line-plot, drawn by Ian McKenzie 1984
UE- Upper Entrance
BG- Big Grin
ME- Middle Entrance
DT- Deep Throat
ML- Missing Lynx
QG- Queen's Gallery
Drawn by Ian McKenzie 1984
CHAGRIN CHOKE
continued from previous page
SNAKE PIT
THE BIG GRIN
LYNX
DEEP THROAT
Note: sections to scale

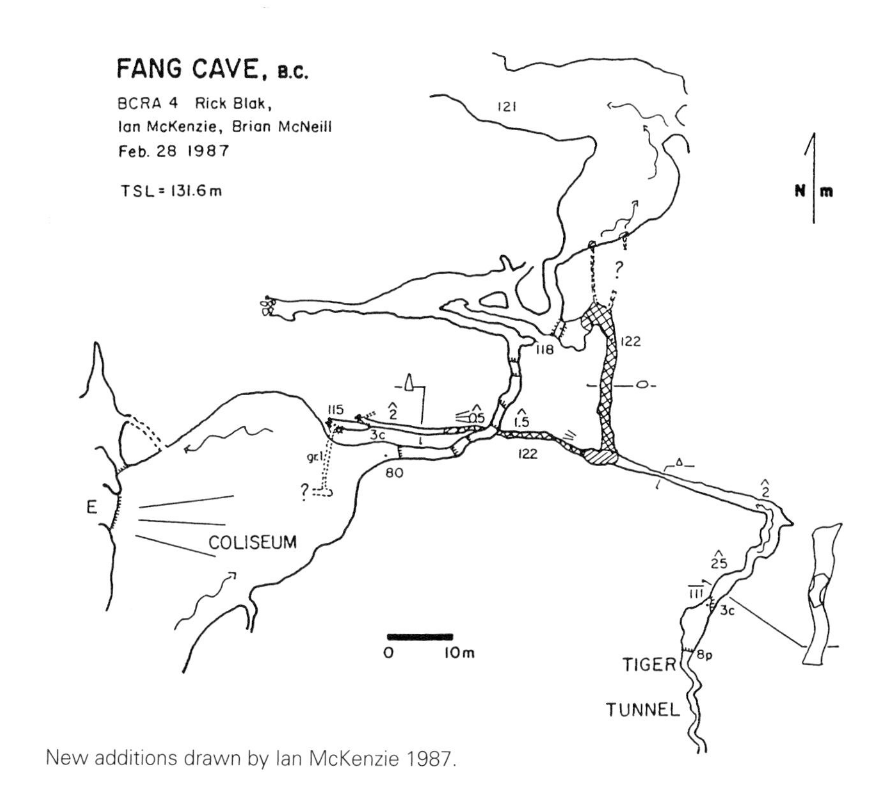

New additions drawn by Ian McKenzie 1987.

Corkscrew Passage in Fang Cave.
Photo Ian McKenzie.

Tooth Decave

Map 93 I/3 072952
Jurisdiction Evanoff Provincial Park
Entrance elevation 1490 m
Number of entrances 1
Length, Depth 215 m, 45 m
Discovered 1985, by Ian McKenzie, Greg & Don Rumpel

Location The small entrance is located at the base of the cliff west of Middle Entrance of Fang Cave.

Exploration & Description The cave was visited in 1994 by UNBC cavers who dug out crawl passages at the end of the cave. At one time Tooth Decave may have been part of the same system as Fang Cave. It is also possible that Window on the West is a blocked upper entrance to Tooth Decave.

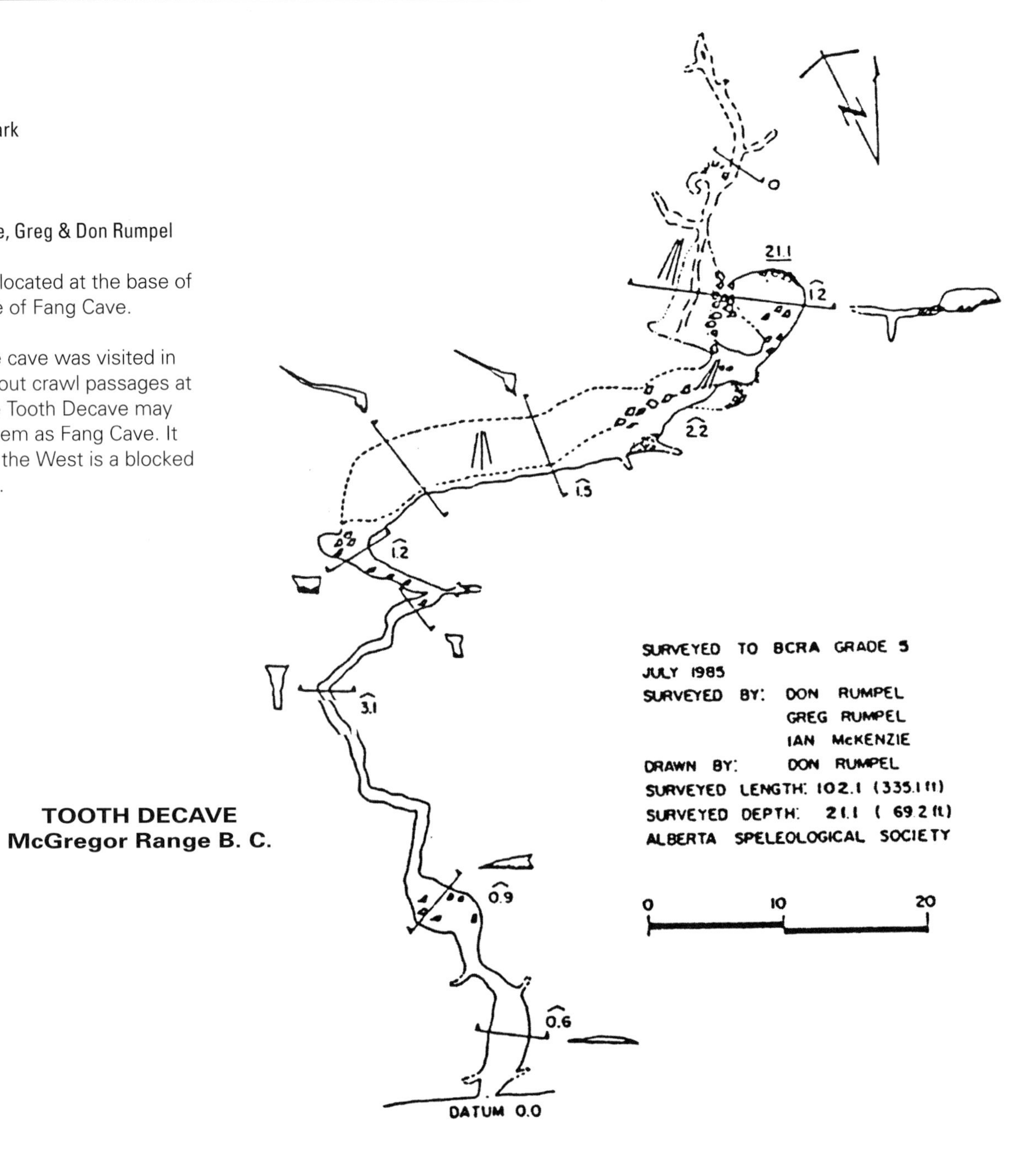

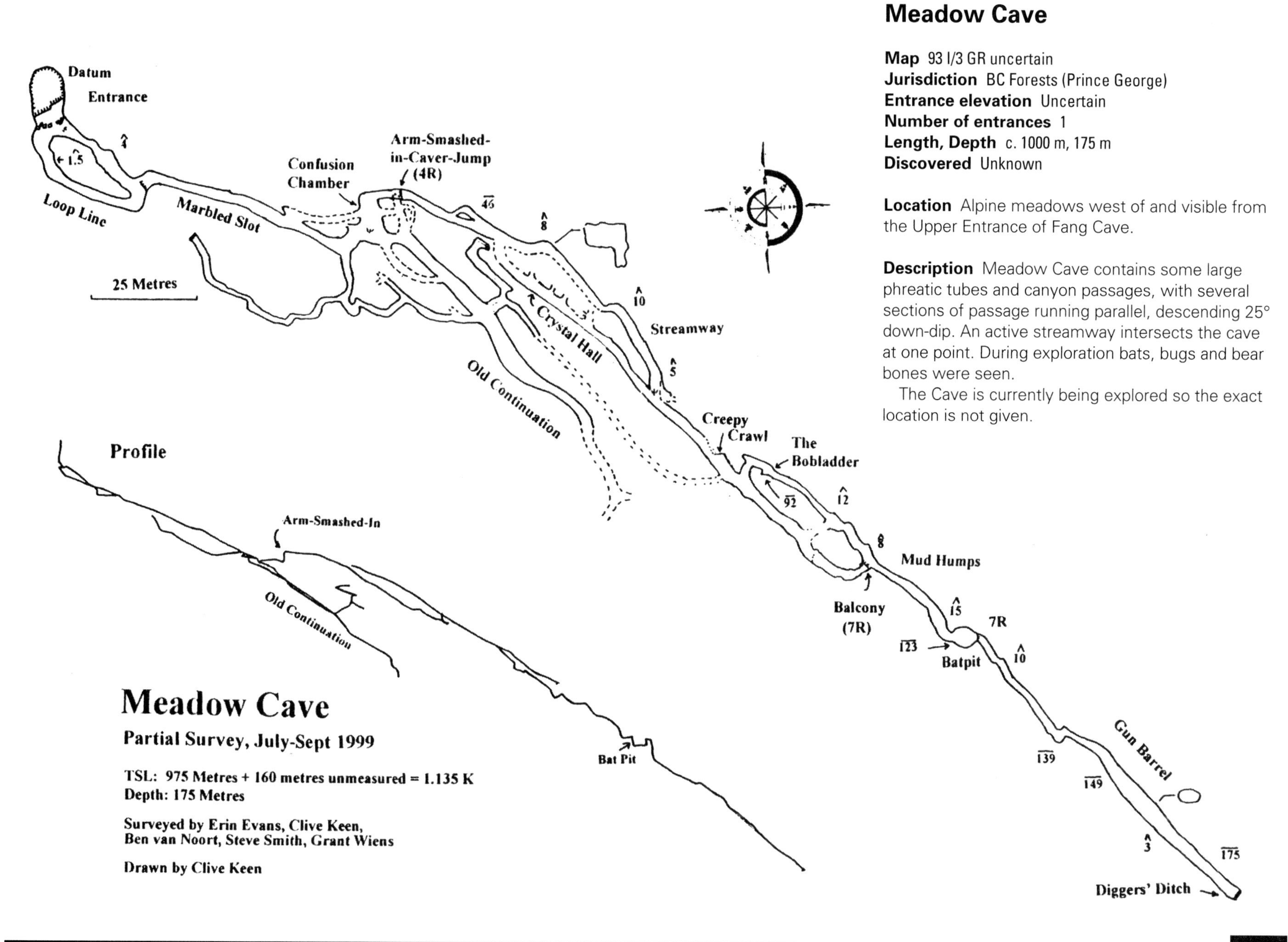

Meadow Cave

Map 93 I/3 GR uncertain
Jurisdiction BC Forests (Prince George)
Entrance elevation Uncertain
Number of entrances 1
Length, Depth c. 1000 m, 175 m
Discovered Unknown

Location Alpine meadows west of and visible from the Upper Entrance of Fang Cave.

Description Meadow Cave contains some large phreatic tubes and canyon passages, with several sections of passage running parallel, descending 25° down-dip. An active streamway intersects the cave at one point. During exploration bats, bugs and bear bones were seen.

The Cave is currently being explored so the exact location is not given.

Window on the West

Map 93 I/3 071954
Jurisdiction Evanoff Provincial Park
Entrance elevation 1525 m
Number of entrances 1
Length 12 m
Discovered 1985, by Ian McKenzie, Greg & Don Rumpel

Location A 3 m entrance above Tooth Decave

Description A short but intimidating climb up to the entrance is rewarded by a few metres of passage and a draughting boulder choke. It is possible this is an upper entrance to Tooth Decave. Half-hearted digging by various parties has been unsuccessful.

Useless Cave

Map 93 I/3 078950
Jurisdiction Evanoff Provincial Park
Entrance elevation 1585 m
Number of entrances 1
Length c. 35 m
Discovered 1987, by Ian McKenzie

Location Large keyhole entrance located in the meadows beyond Fang Cave's Upper Entrance.

Description The meadows beyond Fang Cave's Upper Entrance are an attractive hiking destination, so it is very likely the large entrance to this short cave has been investigated previously to 1987.

A phreatic roof-tube in the main room and the one small side lead indicate this is a small remnant of a larger system likely destroyed by glacial valley deepening.

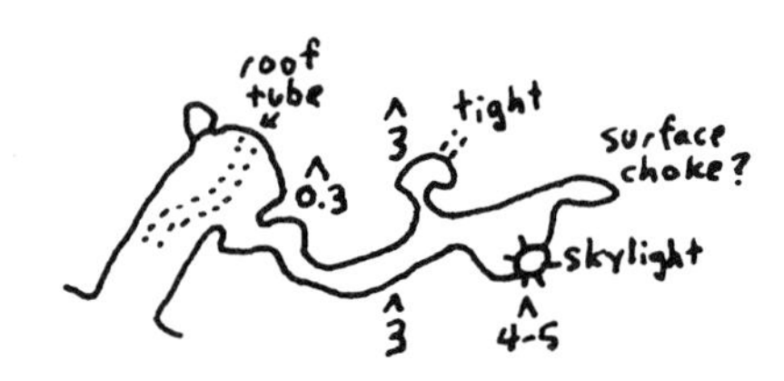

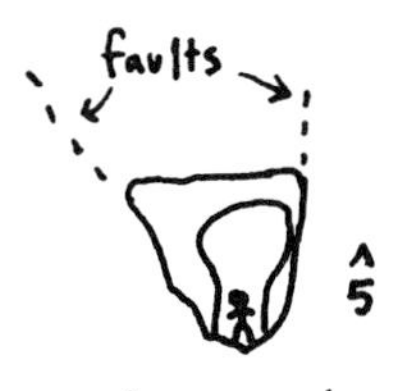

Pass Lake Cave

Map Gleason Creek 93 I/3 035968
Jurisdiction BC Forests (Prince George)
Entrance elevation 820 m
Number of entrances 1
Length c. 70 m
Discovered Labour day weekend, 1996, by Pat Shaw & Pete Norris

Location A small spring just across the road from Pass Lake.

Description A dive was made to some dry passage. No survey was made.

THE DEZAIKO RANGE

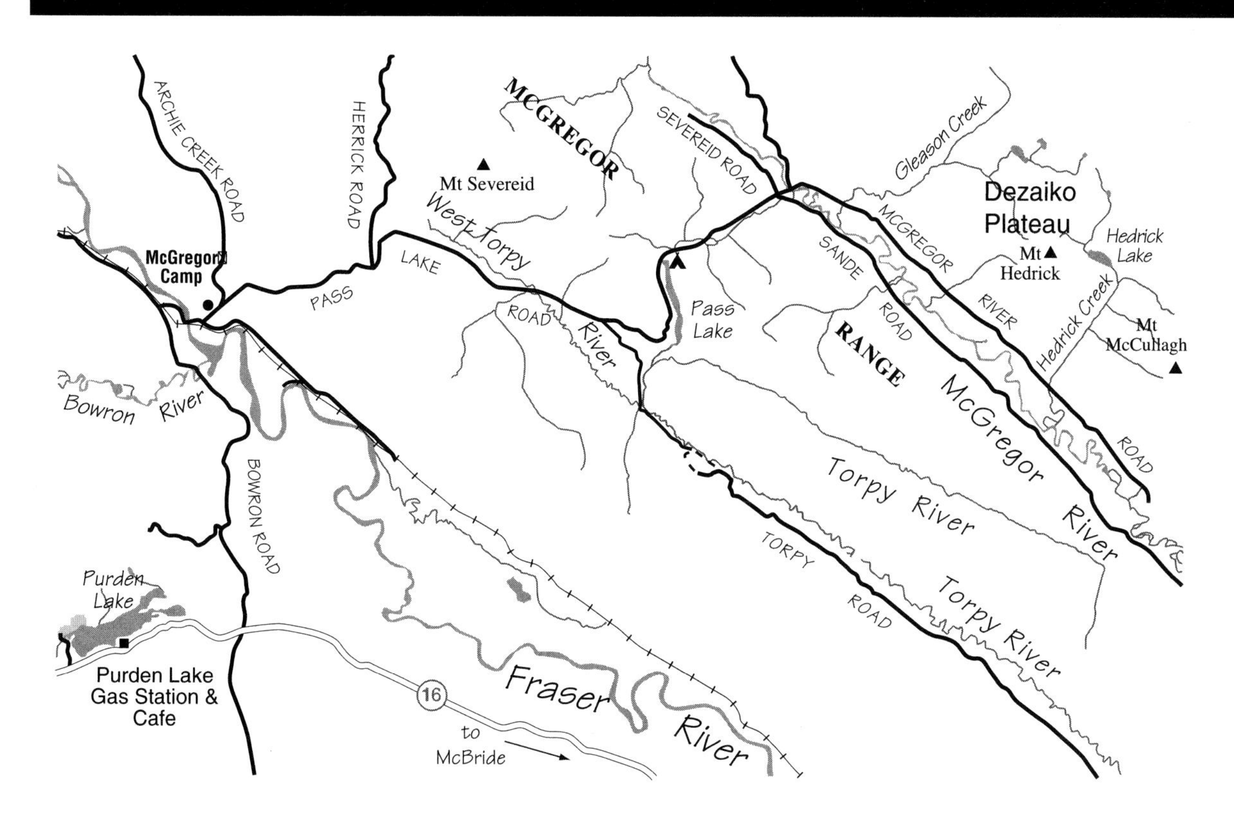

Area Exploration The Dezaiko Range, located northwest of Mt. Sir Alexander, contains several excellent-looking karst areas. During a reconnaissance flight in 1982, Paul Hadfield and John Pollack noticed one that looked exceptional. Located near Gleason Creek was a mountainous plateau occupying an area of 40 to 60 square kilometres. Visible were a number of sinks up to 1950 m elevation and two large resurgences. A reconnaissance up Gleason Creek indicated that ground access would be particularly difficult, and so in August, 1983, an expedition comprised of 19 assorted Alberta, B.C,, British and American cavers helicoptered in. Many sinks, pits and small caves were investigated, including Dooley's Drop, Chute of Many Crawlers, Gleason Creek Resurgence and Dezaiko Cave. A return helicopter trip in 1984 was cancelled at the last moment, and it was while walking the valleys below that a group of disconsolate cavers discovered Bluebell Cave.

With the aid of a grant from the Royal Canadian Geographical Society, 17 cavers made a return visit to the plateau in September of 1985. Despite awful weather, the exploration of Dezaiko Cave was completed. It was while establishing the 1985 Dezaiko base camp that the entrances of Close to the Edge and Twin Falls Resurgence were seen.

In 2000, several new B.C. Provincial Parks were created in order to protect the caves from logging, mining and other commercial activities. Close to the Edge Provincial Park (702 hectares) protects Close to the Edge, Twin Falls Resurgence and Bluebell Cave.

Much still remains to be explored on the Dezaiko Plateau; a return trip is long overdue.

Access to Valley Caves Make sure you have a full tank of gas. From Hwy.16 travelling north towards Prince George, turn right (east) just before Purden Lake Resort onto Bowron Road. Follow it over a bridge to the junction with a surfaced highway. Turn right and cross a shared railway bridge over the Fraser River. Then fork left through McGregor Camp onto Pass Lake Road. At the junction with Herrick Road bear right and continue past the junction with Torpy Road to Pass Lake (campground at far end).

Beyond Pass Lake, cross the suspension bridge over the McGregor River. Then turn right onto the McGregor River Road. For Gleason Creek Resurgence and Guttentite Cave turn left (north) after about 5 km and follow rough logging roads requiring a 4x4 up Gleason Creek. For Bluebell, Twin Falls and Close to the Edge continue along the McGregor River road. After about 21 km from the bridge park at the Hedrick Creek trailhead, identified by a corral on the left and a large lone cottonwood. Hike an established horse trail up the west side of Hedrick Creek.

Access to the Dezaiko Plateau Caves Access has been by helicopter in the past. However, as logging proceeds inexorably up each side valley, it could be that a route on foot up from Gleason Creek might eventually prove practical.

High water in Bluebell Cave.
Photo Ian McKenzie.

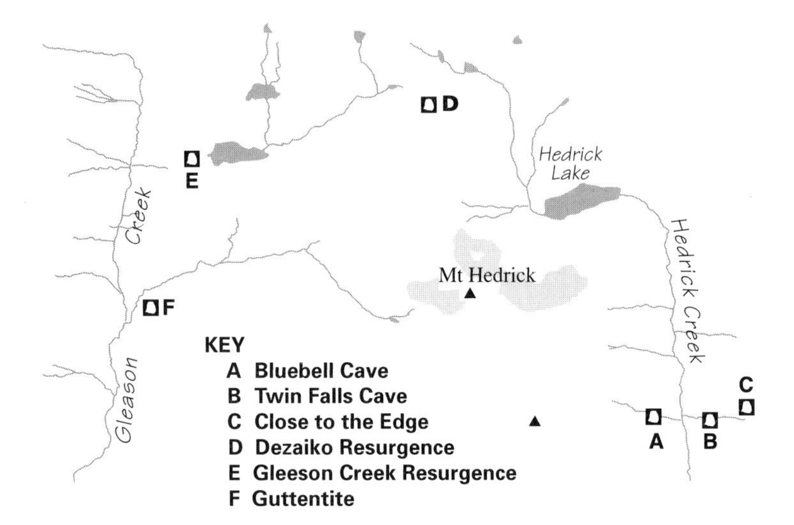

Valley Caves

Bluebell Cave

Map 93 I/3 281930
Jurisdiction Close to the Edge Provincial Park
Entrance elevation 1052 m
Number of entrances 1
Length, Depth 421.7 m, 8.6 m
Discovered 1984, by Paul Hadfield & John Pollack

Location Hike the horse trail up the west side of Hedrick Creek for approximately 3 km to where it crosses the resurgence stream. Follow the resurgence stream up steep blocky talus for about 200 m to the headwall. The entrance is mostly hidden behind a talus slope at the back of a huge alcove, and the large size of the triangular entrance only becomes apparent when you are right up to it. An excellent viewpoint for Bluebell is from Close To the Edge, situated 500 m higher up across the valley.

"Paul boasted for several years that he was one of the few cave divers who couldn't swim. This always puzzled us, and for a many years, we pondered what a hapless, non-swimming Brit would look like in a cold, out-of-depth alpine water cave. While kitting up at the entrance, Paul exhibited with great pride his secret weapon — a large inner tube — that he proceeded to inflate with a bicycle pump. Unfortunately the tube was holed, and we were treated to the sight of a barely floating cave diver who bobbed rapidly through the deep pools and bridged everything in sight." John Pollack.

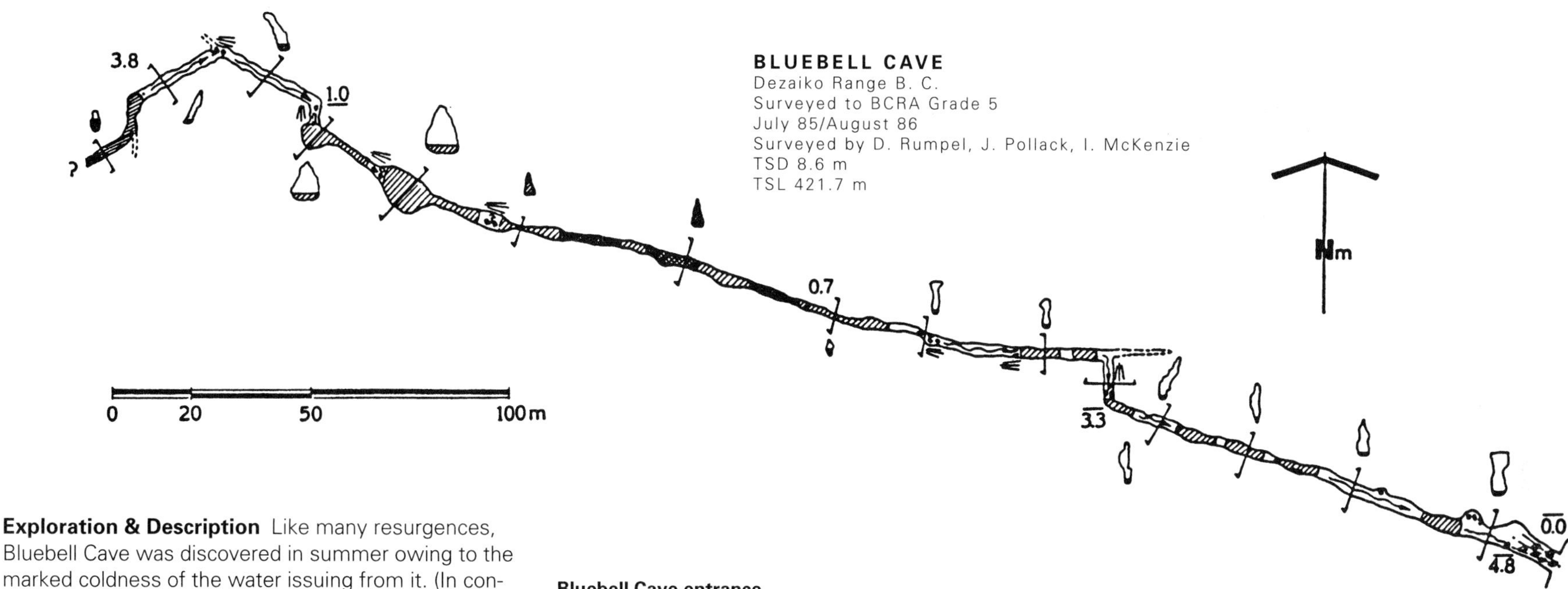

Exploration & Description Like many resurgences, Bluebell Cave was discovered in summer owing to the marked coldness of the water issuing from it. (In contrast, resurgence streams often feel relatively warm in winter.) The stream, with a varied flow between 0.5 and 2 m³/sec, was followed to a large cave entrance about 200 m above its confluence with Hedrick Creek. Inside, Bluebell consists of a large (up to 5 m wide by 15 m high) remarkably straight, gently ascending rift passage. Caver's lights can be seen 150 m from the entrance until the passage takes a small jog to the right (north). A large vibrant stream flows along the floor, making the cave very sporting in high water conditions when standing waves bridge the passage. After wading several deep pools, the caver reaches the first sump, approximately 200 m in from the entrance.

In 1985, the first three sumps were dived, and in 1986 the cave was surveyed to sump #4 at 410 m where the cave turned sharply to the left (south). Shortly after, sump #5 was reached. In 1997, sump #5 was dived and then sump #6, beyond which the passage divided, the small left-hand passage going 10 m to a blind aven, the right-hand passage quickly ending in a bedding plane squeeze.

Bluebell Cave entrance.
Photo Dave Thomson.

Notes for Diving
Sump 1 10-12m long <4m deep.
Sump 2 12-15m long <5m deep.
Sump 3 10m long <3m deep.
Visibility is reduced from >20m to >10m after diving, with minimal silting. By going deep and avoiding the current, the first three dives can be connected into one long dive of 60 m.
Sump 4 3 m long (free-dived).
Sump 5 Strong current in narrow 1m-high x 1.5 m-wide passage.

Warning Avoid entering this cave when the water is high and turbulent. Cave diving is extremely dangerous, and should only be attempted by experienced cave divers.

Geology Bluebell Cave is believed to be the major underground drainage for the southern part of the Dezaiko Plateau. Lying in the axis of a syncline, it is located in the Mural or Hota limestone with an associated underlying quartzite bed.

Twin Falls Resurgence

Map 93 I/3 290920
Jurisdiction Close to the Edge Provincial Park
Entrance elevation 1075 m
Number of entrances 1 (below the entrance is an impenetrable spring)
Length 212 m
Discovered 1985, by John Pollack & Ian Jepson

Location Follow the horse trail up the west side of Hedrick Creek for approximately 3 km to where an indistinct trail heads down to a cable crossing of the creek (use at your own risk). If the cliffs above Bluebell Resurgence come into view up to the left you have gone too far.

On the far side of Hedrick Creek, the steep, rough trail to Close to the Edge starts just left of the Twin Falls resurgence stream. Follow it as it parallels the creek until a rough flagged trail leads off to the right (southwest).

Exploration & Description Located 600 m below Close to the Edge, Twin Falls Resurgence was also discovered on the 1985 flight into Dezaiko base camp. Although established by dye trace to be the resurgence for Close to the Edge, a caving connection seems highly unlikely, and so this cave has been written up separately.

It consists of a 1.5m-high by 5 m-wide overflow passage with clean-washed cobbles that descends to a quiet sump pool after only 20 m. This was dived in September of 1987 through a low 20 m sump (a side branch off Jepsump remains unchecked) to an ascending room and cascades where the water emerges from a slot. A further air-filled rift passage was followed for 160 m to an 8 m x10 m room. Several leads from this room were unchecked due to lack of time.

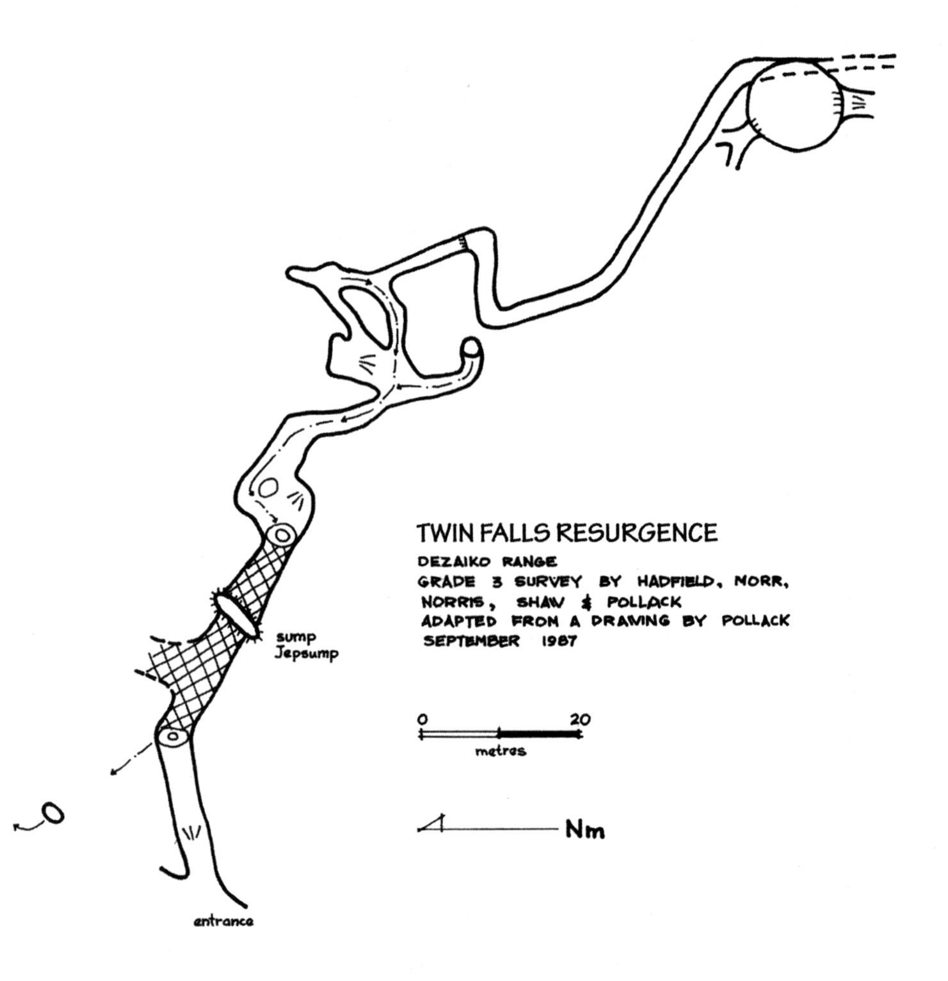

Entrance to Close to the Edge.
Photo Ian McKenzie.

"One steamboat, two steamboat, three steamboat, four steamboat, FIVE steamboat, and there it went again, the deep sonorous, rolling thunder. Suddenly, none of us wanted to stand close to the edge. They looked at me, envious yet relieved about not being first down this pit. How can I get out of this?"
Ian McKenzie.

Close to the Edge

Map 93 I/3 299918
Jurisdiction Close to the Edge Provincial Park
Entrance elevation 1554 m
Number of entrances 1
Length, Depth 967 m, 472 m
Discovered 1985, by Ian McKenzie

Location Follow the horse trail up the west side of Hedrick Creek for approximately 3 km to where an indistinct trail heads down to a cable crossing of the creek (use at your own risk). If the cliffs above Bluebell Resurgence come into view up to the left you have gone too far.

On the far side of Hedrick Creek, a steep, rough trail starts just left of Twin Falls resurgence stream. Sometimes flagged, it climbs directly up through a burn area, gaining 670 m (skirt cliffs to the left) to a flat camping area with no water. The hike with a large pack is hard and seems endless.

The cave is located slightly to the right of and slightly below the camping area. The cave entrance consists of a large cliff-face window opening into the top of a shaft, and thus is not visible from either above or below. From the camping area a trail leads down to the edge of the cliff where a fixed 20 m hand-line protects the descent to ledges that are easily traversed to the entrance window.

Exploration & Description The entrance to this huge shaft was first noticed from a helicopter when establishing the Dezaiko base camp during the 1985 expedition. Ian McKenzie first visited the cave on foot in October of 1985. He quickly established the 70 m of rope he had carried up was not sufficient by dropping stones down the shaft and estimating its depth as between 50 and 80 m.

On the next visit in 1986, the longest ASS rope of 100 m was carried up and the shaft descended to an estimated –83 m. On the third visit, using the 100 m with another 40 m rope tied on, a small ledge at –120 m was reached just above the large halfway ledge.

The Big Shaft, Close to the Edge. Photo Dave Thomson.

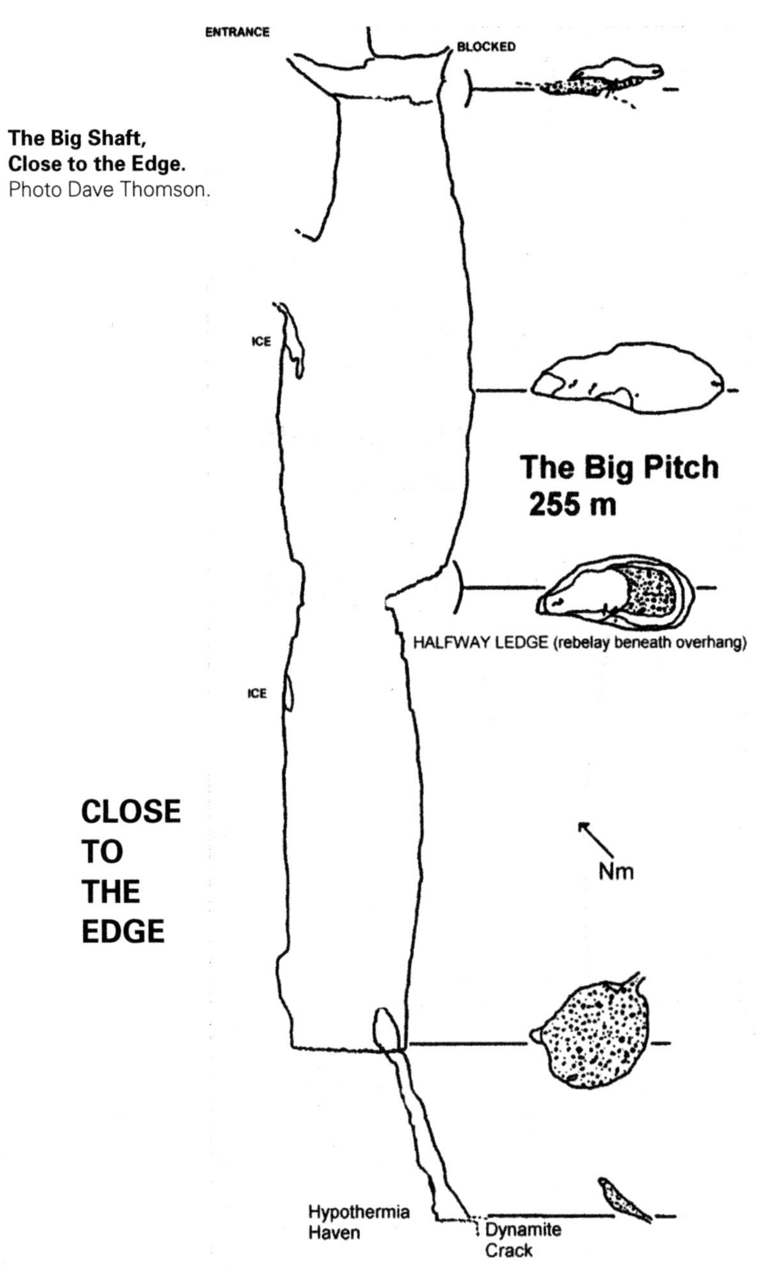

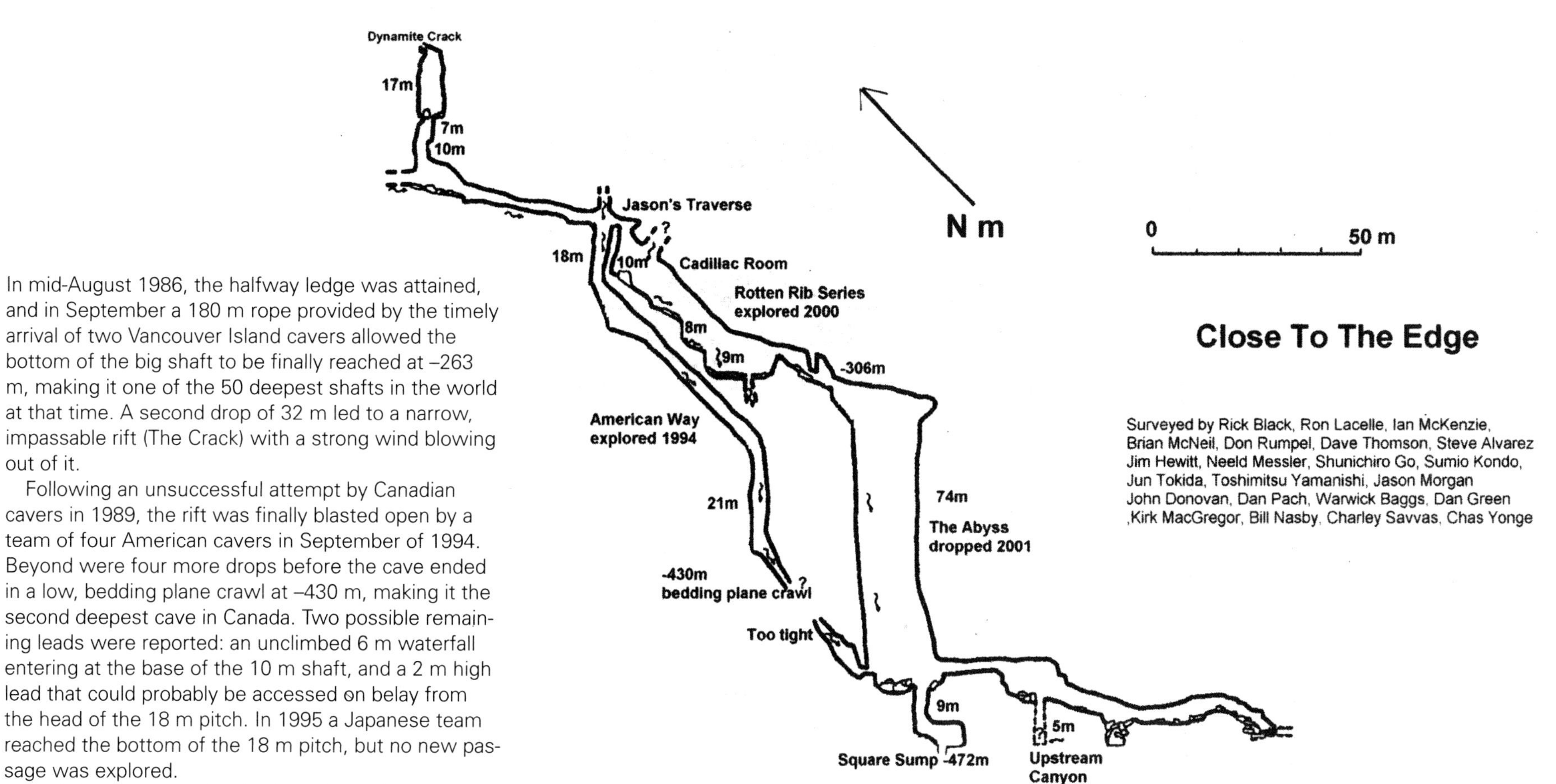

In mid-August 1986, the halfway ledge was attained, and in September a 180 m rope provided by the timely arrival of two Vancouver Island cavers allowed the bottom of the big shaft to be finally reached at –263 m, making it one of the 50 deepest shafts in the world at that time. A second drop of 32 m led to a narrow, impassable rift (The Crack) with a strong wind blowing out of it.

Following an unsuccessful attempt by Canadian cavers in 1989, the rift was finally blasted open by a team of four American cavers in September of 1994. Beyond were four more drops before the cave ended in a low, bedding plane crawl at –430 m, making it the second deepest cave in Canada. Two possible remaining leads were reported: an unclimbed 6 m waterfall entering at the base of the 10 m shaft, and a 2 m high lead that could probably be accessed on belay from the head of the 18 m pitch. In 1995 a Japanese team reached the bottom of the 18 m pitch, but no new passage was explored.

On Labour Day weekend, 1998, Dave Chase and Jason Morgan, part of a group of five ASS cavers, bottomed the cave. They found the terminal bedding plane crawl wet and unpromising. On the way back up, Jason Morgan traversed into the lead across the head of the 18 m pitch that proved to be the way on. Using the rope from the 18 m pitch Jason descended another 10 m pitch into a room containing a car-sized boulder (the Cadillac Room). He then negotiated a series of down-climbs before getting to the next pitch for which he had no rope.

A year later the ASS were again down the big shaft. Two more short pitches (8.2 m and 8.5 m) and some short down-climbs bought them to the head of a chute leading off into blackness (The Abyss). Jason got the honour of descending, only to reach the end of the remaining rope an estimated 35 m from the floor, or possibly, he thought, a ledge. The new lead was now at approximately –400 m and still going. In September, 2001, the Abyss (74 m) was descended and a further 9 m shaft that ended at a sump at –472 m — a frustrating 64 m short of breaking the Canadian depth record (Arctomys is –536 m).

Although there are several good leads, they all seem to sump or be heading upwards. The two most promising leads are a climb through a waterfall up from the Cadillac Room, and the Upstream Canyon at the bottom of the cave, reached using a 5 m hand-line down a mud bank.

The Halfway Ledge on the Big Pitch, Close to the Edge.
Photo Dave Thomson.

Warning & Considerations A number of hazards face parties attempting Close to the Edge. Loose rock and ice in the big shaft are a very real danger. Both seem to be at a minimum in September, probably the optimum time to attempt the cave.

Pack rats living at the top of the shaft have a voracious appetite for caving gear, and the rope should be lowered below the initial ledge (using some old rope or wire hawser) when the pitch-head is unattended. The original rigging created two rubbing points at the top of the first and second pitches in the big shaft. In 2000 the hangs were much improved with the addition of further bolts, eliminating the need for any rope protection.

The long ascent up the big shaft averages two hours, which can create a considerable wait for larger parties who may already be wet through from water deeper in the cave, and from water that sprays the pitch in the vicinity of the half-way ledge. If the cave is wet (and it usually is), carry gear to avoid hypothermia (stove, sleeping bag). A good rope-walking system that the user is thoroughly familiar with greatly speeds up the ascent.

Dynamite Squeeze at the top of the third pitch can be challenging to larger cavers. Packs should be shuttled through this awkward squeeze, as they jam if hauled on a leash.

The rock is particularly suspect lower in the cave and bolts are hard to place in the alternately hard and rotten limestone.

Due to the large distances involved, a whistle signalling system is useful.

Don's Baby

Map 93 I/3 298919
Jurisdiction Close to the Edge Provincial Park
Entrance elevation c. 1550 m
Number of entrances 2
Length 50 m
Discovered 1987, by Don Rumpel

Location Just above the entrance to Close to the Edge.

Description A short phreatic loop with two entrances and a draft.

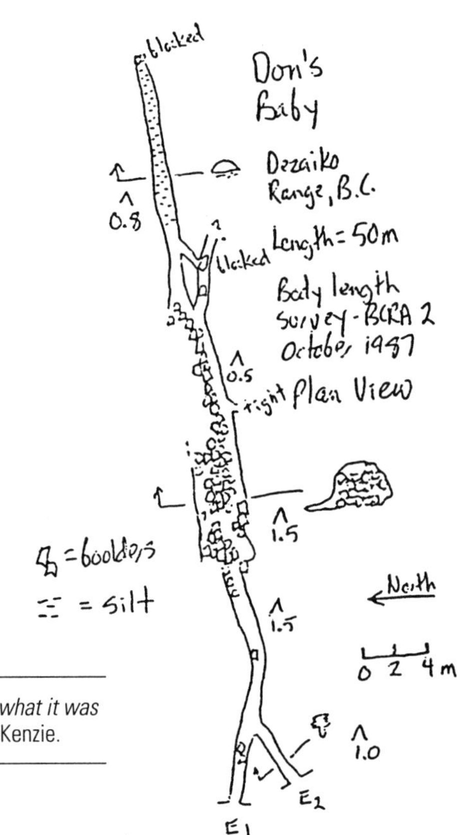

"Rick Blak inadvertently named this cave. When asked what it was called he replied 'I don't know; it's Don's baby.'" Ian McKenzie.

Guttentite Cave

Map 93 I/3 208029
Jurisdiction BC Forests (Prince George)
Entrance elevation c. 1000 m
Number of entrances 1
Length, Depth 105 m 11.6 m
Discovered 1993, by the UNBC Caving Club

Location In a small depression only 50 m from the end of the 4x4 logging road up Gleason Creek (in1993) and about 150 m above Gleason Creek Resurgence.

Exploration & Description The "good and tight" entrance was excavated to access a descending 35° chute. Contained in Mural limestone, this short cave was possibly once connected to Gleason Creek Resurgence.

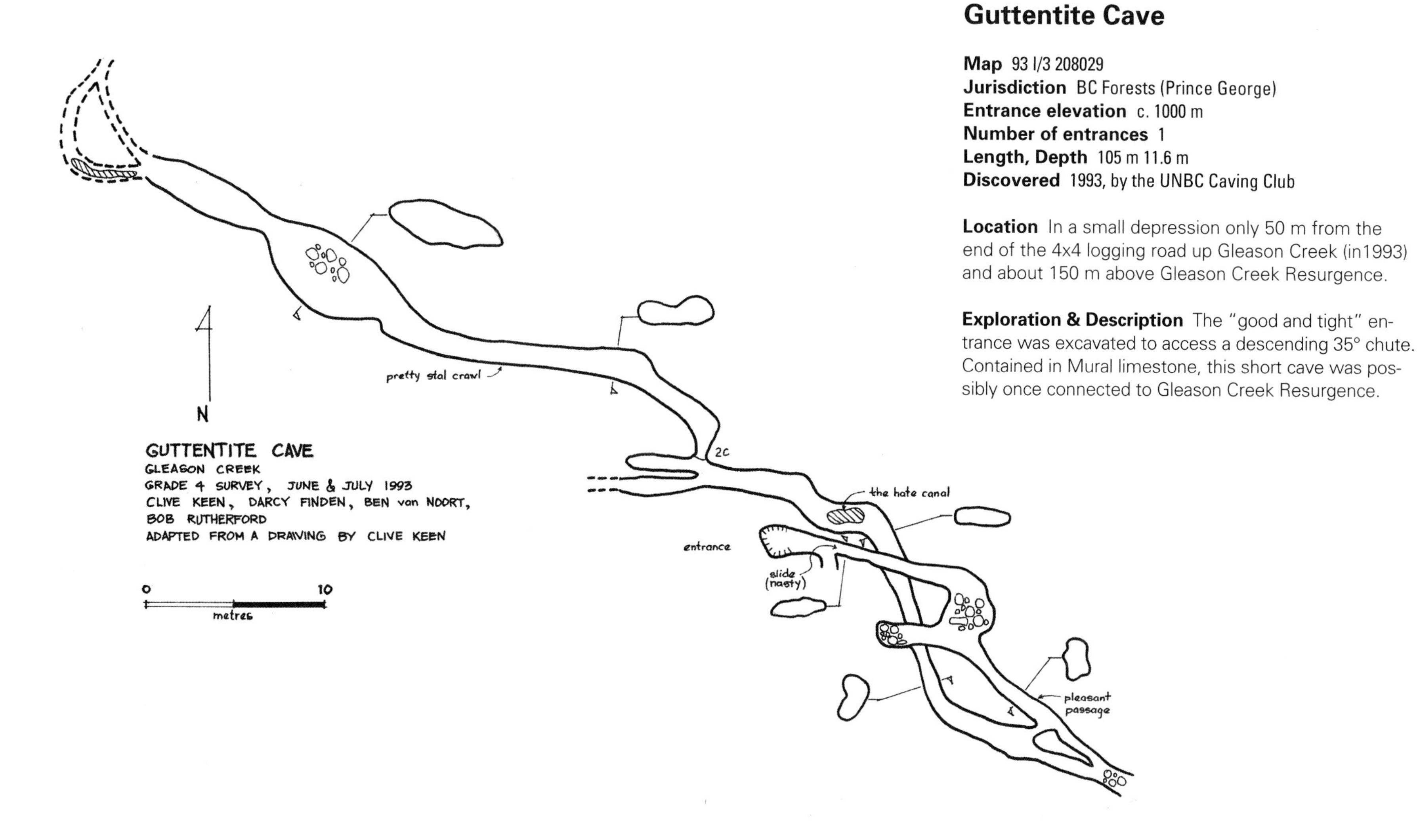

Dezaiko Resurgences (Upper Main Resurgence, Middle Resurgence & Hedrick Resurgences)

Map 93 I/3
Jurisdiction BC Forests (Prince George)
Entrance elevation Various
Number of entrances 1 each
Length, Depth N.A.
Discovered 1983 to September 1985

Six significant resurgences were observed or investigated during the 1983 and 1985 Dezaiko expeditions. The three with surveyed passage — Bluebell Cave, Twin Falls Resurgence and Gleason Creek Resurgence (formerly Lower Main Resurgence) — are covered separately. Although not enterable by cavers, the other three are major karst features that indicate the tremendous cave potential of the Dezaiko area. They are included here as a matter of record.

Upper Main Resurgence 290990 Members of the 1983 expedition investigated this cliffside resurgence above the north fork of Gleason Creek by jumping off the skids of the hovering helicopter onto the steep slopes below. The 7 x 7 m entrance led to 20 m of passage before it sumped.

Middle Resurgence 235046 Although not investigated at close range, this resurgence is likely water-filled and almost certainly drains the small lake on the other side of the ridge. This lake has no surface outlet and is known to sink through lake-bottom "blue holes" and a shoreline drain.

Hedrick Resurgences Location uncertain. These inaccessible resurgences pour out of the huge cliffs above Hedrick Lake, and were photographed from a helicopter in 1985.

Gleason Creek Resurgence (Lower Main Resurgence)

Map 93 I/3 location uncertain
Jurisdiction BC Forests (Prince George)
Entrance elevation c. 850 m
Number of entrances 1
Length, Depth 412.4 m, 23.1 m (-11.75 & +11.32)
Discovered 1983, by John Pollack & Paul Hadfield (the Dezaiko Expedition)

Location 150 m below Guttentite Cave, 20 m above the winter resurgence.

Exploration & Description When first visited in August, the cave consisted of a 1 to 2 m^3/sec resurgence that sumped a short way in. A later visit in February of 1988 revealed over 400 m of passage. The winter sump location is a good prospect for diving. However, skis or snowmobiles are needed for access at this time.

Hedrick Resurgence.
Photo Chas Yonge.

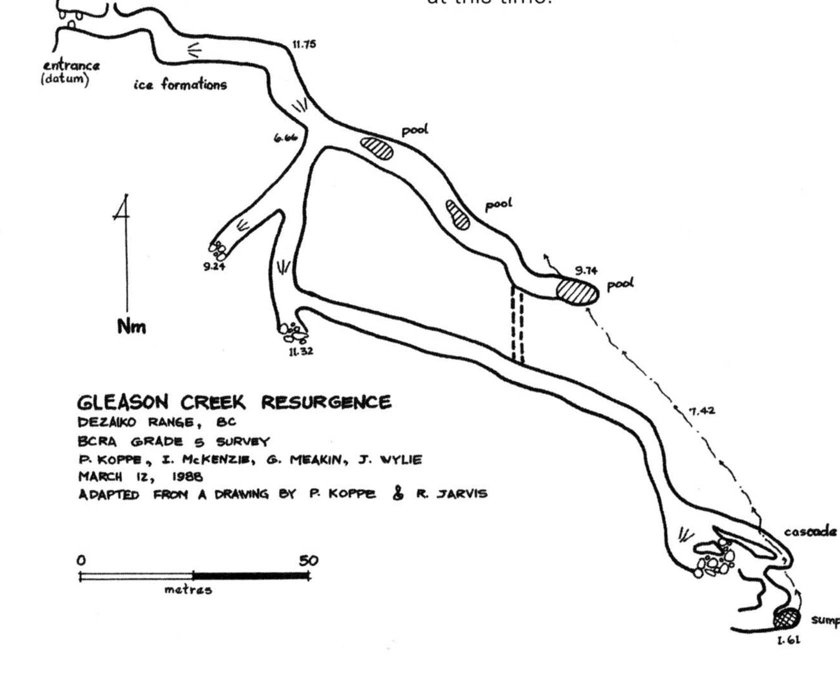

Dezaiko Plateau Caves

Dezaiko Cave

Map 93 I/3 293011
Jurisdiction BC Forests (Prince George)
Entrance elevation 1905 m
Number of entrances 1
Length, Depth 710 m, 253 m
Discovered 1983, by Erik von Vorkampff & Chas Yonge (the Dezaiko Expedition)

Location High on the side of a glaciated bench, the double-barrelled shaft is protected from filling by a large boulder erratic.

Exploration & Description The longest and deepest cave so far discovered on the Dezaiko plateau, Dezaiko Cave was explored to the bottom of the Spy pitch in August of 1983, and then to its conclusion in four very tight crawlways "only a marmot could attempt" during the awful weather of the September, 1985, expedition. Many pitches were rigged off natural features, or by using nuts and Friends which sped up exploration. The cave has not been entered since the bottoming expedition of 1985.

The entrance consists of a double-barrelled 9 m shaft (Tinker) protected from filling by the large boulder erratic. After a further 7 m down-climb the cave appeared to end for the first explorers, but digging revealed a dry fossil rift that led down-dip via another 3 m climb to a small cascading stream and to the head of the 5 m Tailor pitch.

From the base of this pitch the streamway was followed through boulders and past the Zimmer von Oozen Scroozen break-down room. The dip passage continued before almost shutting down at the Strike I Grovel, followed by the even more miserable low, wet, muddy passage of the Strike II Grovel.

The cave then turned abruptly counter-dip before swinging back into a nice rift passage decorated with long soda-straws. The 4 m Soldier Pitch was followed by the pleasant but slippery Shale Run down-dip to a beautiful elliptical shaft, the 18 m Spy pitch — the extent of exploration during the 1983 expedition.

Just beyond the Spy pitch is a further 13 m drop, The Bypass, then a series of small drops called The Cascades. These are followed by a large (estimated 40 m) pitch that in 1983 was rigged to a ledge at 11 m, from which a dry, high-level passage was followed to a 19 m pitch. The Meanderings consist of a narrow, winding passage which soon picks up the streamway again before a series of four wet pitches are encountered: the 10 m, 12 m, 13 m and 10 m. The end of the cave is reached at -253m. When it was bottomed, Dezaiko shared with White Hole the position of ninth deepest cave in Canada.

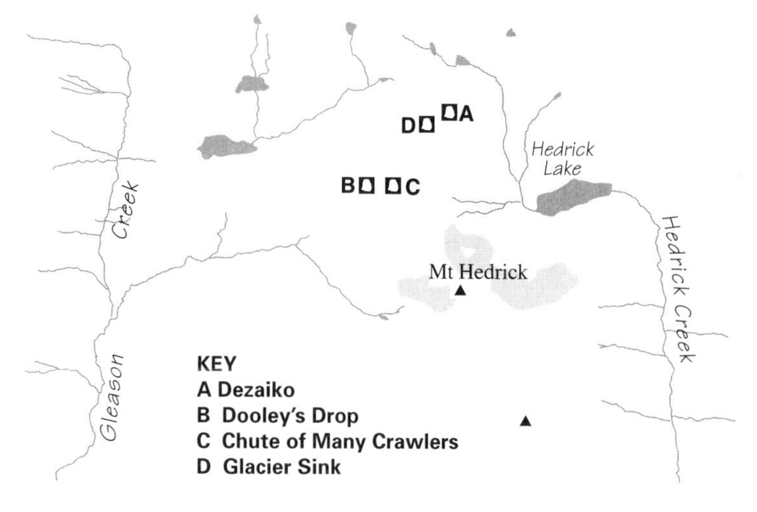

A short way in Dezaiko Cave.
Photo Dave Thomson.

"The cave itself was consistently demanding. A dozen vertical drops — some up to 20 m and some washed by the spray of waterfalls — were linked with steep canyon-like passages that required continual climbing. It was also miserably cold. The combination of wetness and temperatures barely above freezing drained away body heat whenever we stopped moving. We began referring to the cave, only half joking, as the worst in Canada." Dave Thomson.

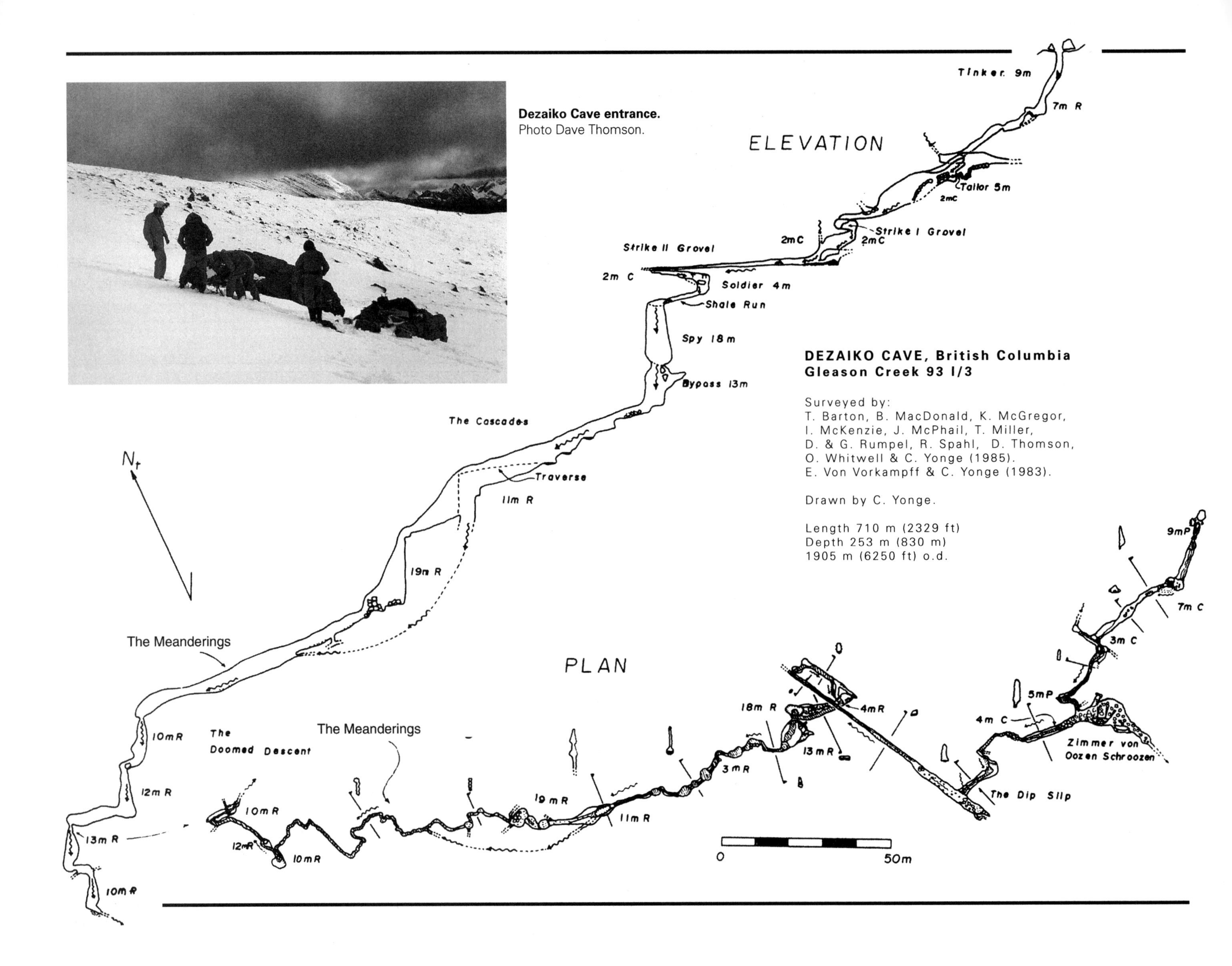

Dezaiko Cave entrance.
Photo Dave Thomson.

Glacier Sink

Map 93 I/3 289009
Jurisdiction BC Forests (Prince George)
Entrance elevation c. 1646 m
Number of entrances 1
Length, Depth Unknown
Discovered 1985, by four members of the Dezaiko Expedition

Location This small active sink is located at the edge of a small snowfield immediately below the col leading to Dezaiko Cave.

Exploration & Description Using chemical and mechanical persuasion (i.e. explosives and a hammer), a smallish slot was opened sufficiently to allow a 2 m drop to be negotiated. This accessed a small rift that was explored for a short distance in both directions before becoming constricted. This cave was not surveyed.

Chute of Many Crawlers

Map 93 I/3 271009
Jurisdiction BC Forests (Prince George)
Entrance elevation c. 1560 m
Number of entrances 1
Length, Depth 48.9m, 28.8m
Discovered 1983, by the Dezaiko Expedition (Ian McKenzie)

Location About 250 m southeast of Dooley's Drop. Metal-tagged B17.

Description This descending bedding plane cave contains several creative squeezes. Exploration was finally abandoned due to a dwindling carbide light.

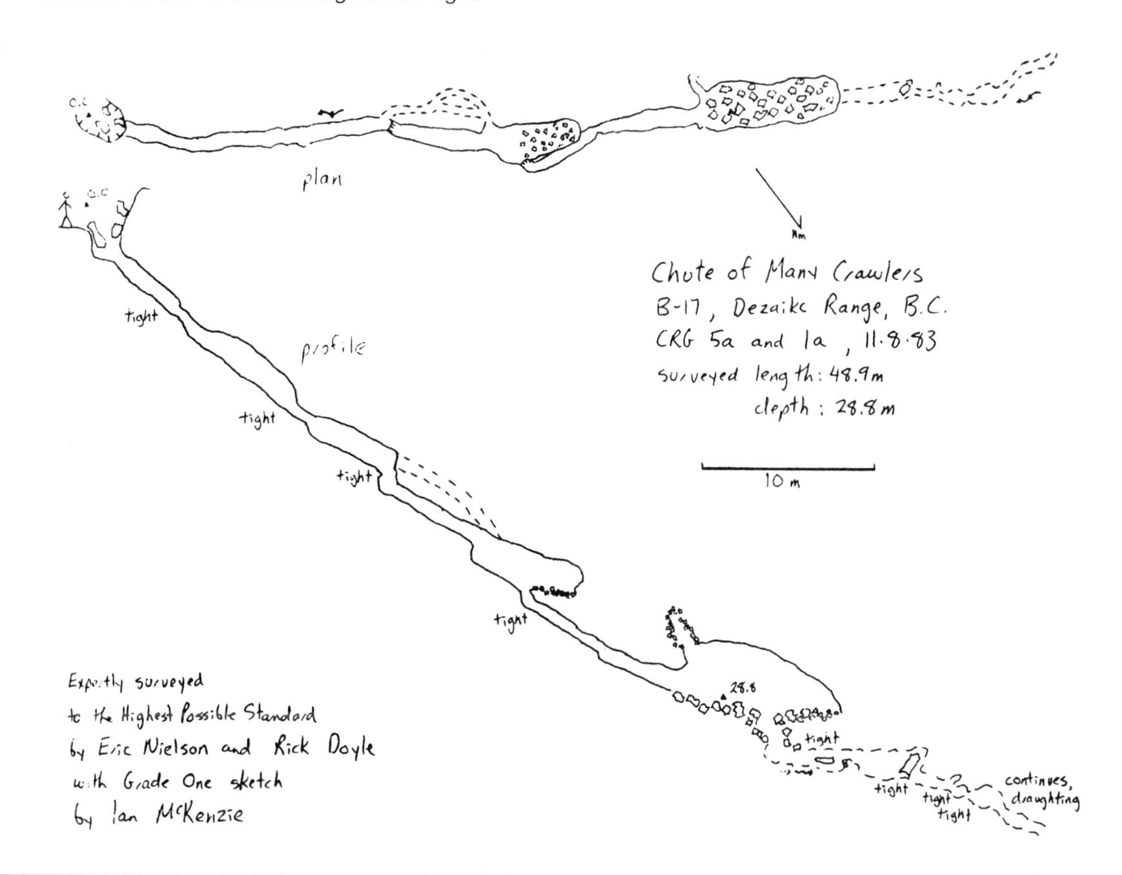

Unnamed Caves (B-13, B-15)

Map 93 I/3 250 to 295, 300 to 030 ??
Jurisdiction BC Forests (Prince George)
Entrance elevation 1850 - 1950 m
Number of entrances Typically 1
Length Less than 50 m
Discovered August 1985, by Ian McKenzie, John Pollack & Chas Yonge (the Dezaiko Expedition).

Location Believed to be located in the northern part of the plateau. Metal-tagged B13 and B15.

Exploration & Description Over a two week period, 52 enterable sinks and shafts on the Dezaiko Plateau were investigated and metal-tagged. The more significant ones have been named and documented, but the others are recorded only in the tattered field notes and fading memories of the participants. One caver's notes included these two sketches of his best finds of the day.

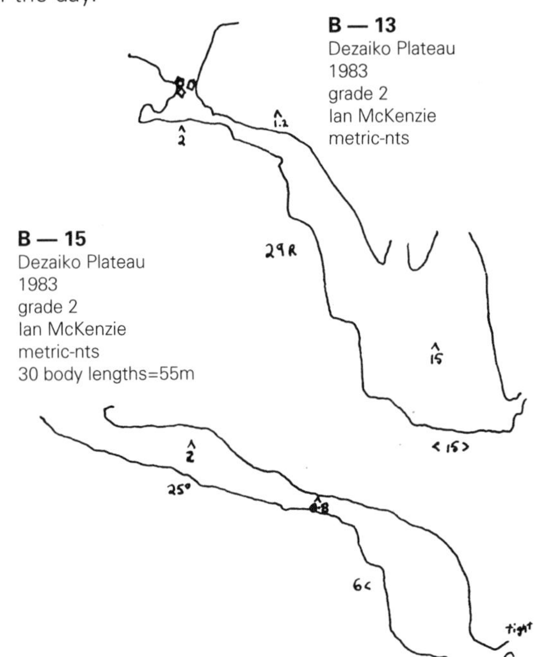

The Lake Cave

Map 93 I/3 Uncertain
Jurisdiction BC Forests (Prince George)
Entrance elevation c. 1585 m
Number of entrances 1
Length, Depth c. 20 m, 1 m
Discovered 1985, by Ian McKenzie, John Pollack & Chas Yonge (the Dezaiko Expedition).

Location South of the Dezaiko Plateau is Mt. Hedrick with flanking limestone benches that were checked during brief helicopter landings on the last day of the 1985 expedition. Several shafts were noted amongst well-developed surface karst on the south bench, but none were descended. A pond on a northerly bench drained directly into a small cave.

Exploration & Description Lake Cave was followed to the limit of daylight where the passage height dropped to about 1 m. A return with lights two years later resulted in an additional 2 m of exploration (!) to a sump. It is possible that The Lake Cave resurges at Gleason Creek Resurgence.

The Lake Cave entrance. Photo Dave Thomson.

Dooley's Drop

Map 93 I/3 269011
Jurisdiction BC Forests (Prince George)
Entrance elevation c. 1560 m
Number of entrances 1
Length, Depth 92 m, 72 m
Discovered 1983, by Ian McKenzie, John Pollack & Chas Yonge (the Dezaiko Expedition).

Location 2 km due west of Desaiko Cave, approximately 500 m west of a large closed depression.

Description Once tied with Raven Lake Pit (Vancouver Island) as the second deepest shaft in Canada, this fine 72 m vertical shaft starts as a rift and enlarges to 24 m x 15 m at the bottom. A steeply descending passage floored with rubble descends for an additional 20 m to a boulder-jammed rift that could not be passed.

THE MOUNT BOCOCK KARST

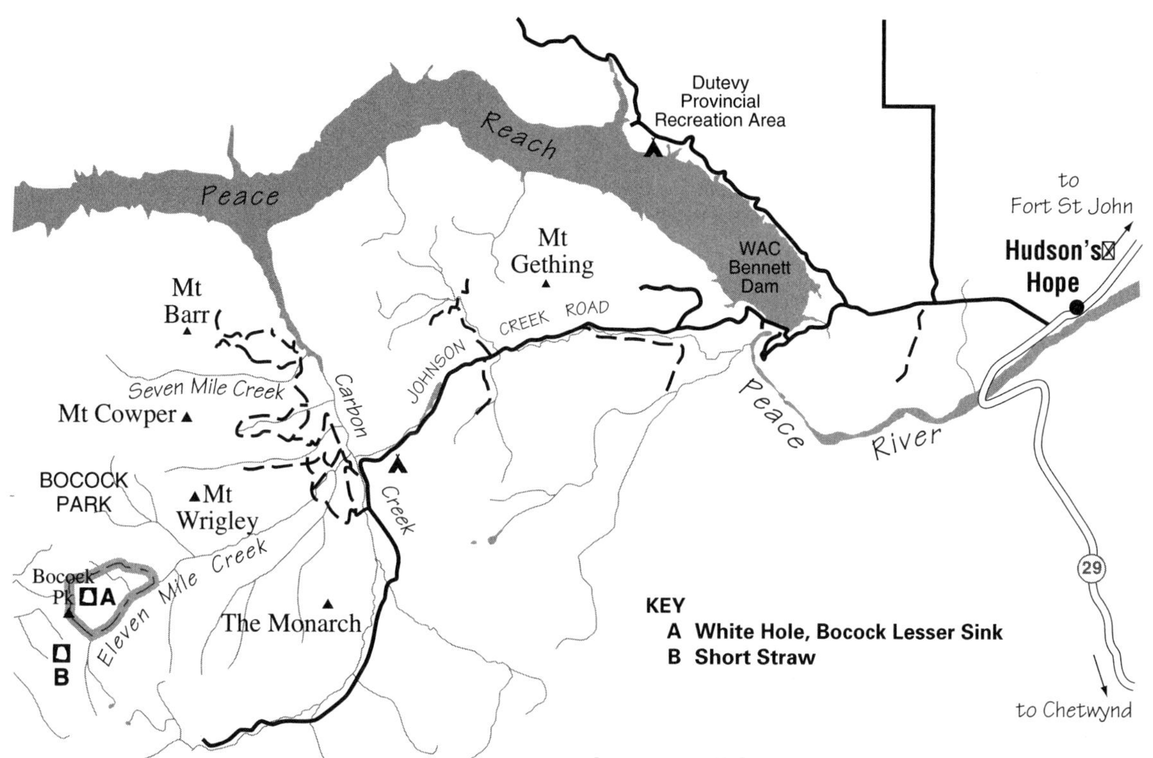

In contrast to the precipitous, rugged, bedrock terrain that typifies most karst areas in the Rockies, the Bocock area consists of rolling pastoral hills and broad tree-filled valleys. Two caves (White Hole, Bocock Lesser Sink) are located in the recently designated 1170 hectare Bocock Peak Provincial Park, which has no facilities or maintained trails. Just 25 km to the north is Williston Lake, which was created by the Bennett Dam built on the Peace River in 1967. Williston Lake became the largest reservoir in the world when it submerged a 360 km stretch of the northern Rocky Mountain Trench.

Access to all Caves In the past the caves have been accessed by helicopter flying out of Chetwynd and using the Carbon Creek trailhead as a staging area. Since the area has become a provincial park, helicopter use is being restricted to protect Bighorn sheep in the area. For current information contact Fort St. John Forestry Office at (250) 787-3562.

If you are on foot, drive from Hudson's Hope across the Bennett Dam. Follow the Johnson Creek forestry road to its end in a cutblock up Eleven Mile Creek — the current trailhead. From here, accessing the caves requires a 7 hour bushwhack with about 400 m of elevation gain. Above treeline movement is a lot easier across alpine meadows with occasional krummholz.

Area Exploration In 1984, Andy Legun, geologist for the Peace River District, saw a large sink adjacent to Bocock Peak. In early June he set up a helicopter reconnaissance that allowed four ASS members to investigate the caving possibilities of the area. The main sink (White Hole) was entered via two separate routes, but high melt-water prevented further exploration. Another sink (Bocock Lesser Sink) was likewise too wet for more than a cursory visit. In August, British and Canadian cavers from the 1984 ACRMSE expedition flew into the area for 10 days, making use of a Bell 212 helicopter (one flight for 11 people and gear) that just happened to be stationed at the head of the forestry road. Three significant caves — White Hole, Short Straw Cave and Bocock Lesser Sink — were explored as a result. Lacking the funds to fly out, the heavily-laden group walked out over a period of 13 hours.

The area was visited for a third time in August, 1986, by three ASS members and two members of the Imperial College Caving Club Expedition. Less snow than in August of 1984 enabled a number of minor shafts (between 5 m and 10 m deep) close to the main depression to be checked, as was the ridge containing Short Straw. Leads in the bottom of White Hole were visited, and Bocock Lesser Sink was pushed for a further very tight 45 m.

"We flew the 40 km up Eleven Mile Creek to the White Hole Depression in just five minutes (the same distance took the 1984 ACRMSE expedition 13 hours of desperate bush-whacking during their hike out), proving once more that helispelunking is the best method of cave exploration in the Canadian Rockies." Harry Lock et al.

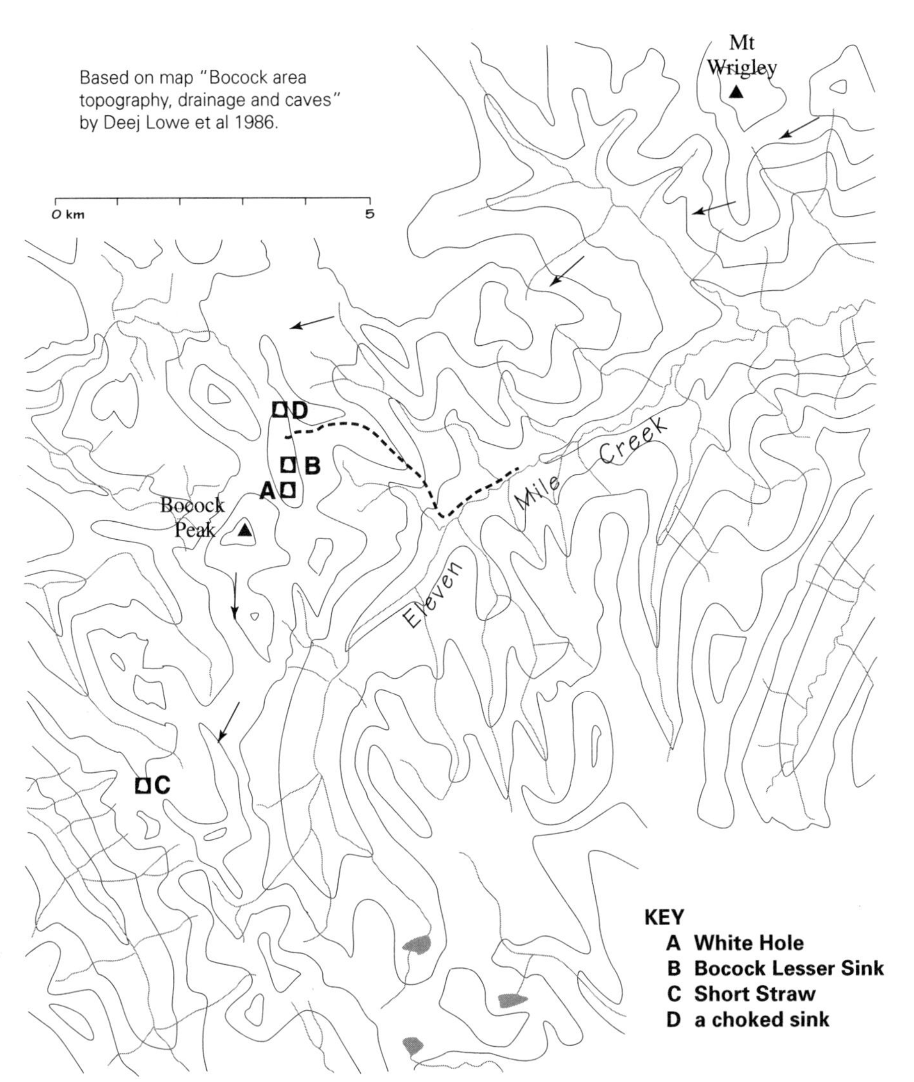

The Bocock karst. Photo Chas Yonge.

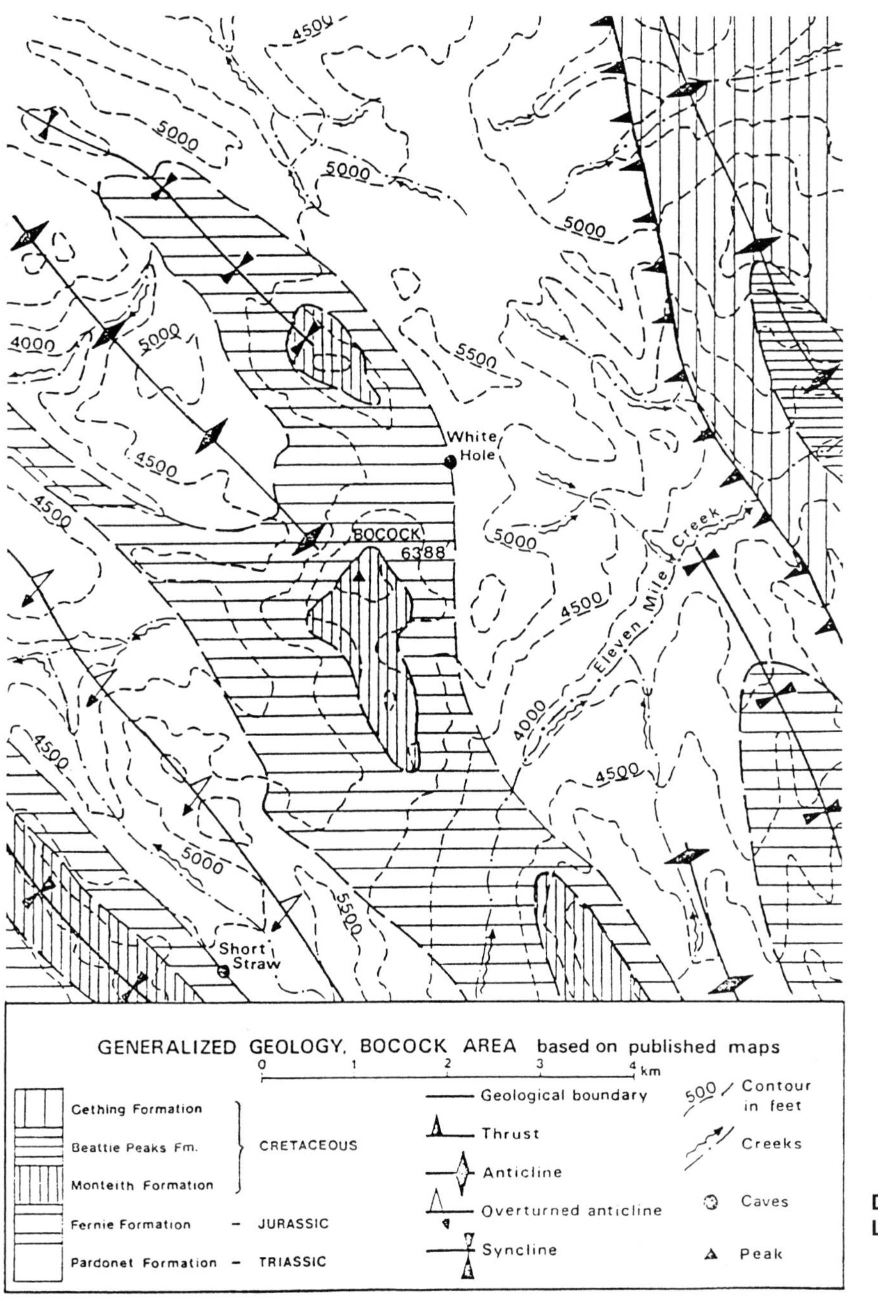

Drawn by Deej Lowe et al (1984).

Geology of the Area Bocock Peak to the west and the lesser summit to its north are synclinal remnants, presumed to be Jurassic sediments. The fold axes following the two ridges are not continuous. Bocock Formation limestone, in a bed about 45 m in thickness, dips westward beneath Jurassic rocks. The junction is assumed to be an unconformity and that slip-movement has taken place along the junction. Eastwards from the depression is a complex series of limestone dip slopes and scarps, some of which might be formed in inverted beds.

The general sequence in the Bocock region is:

Fernie Formation	Jurassic
Unconformity	
Bocock Formation	Triassic
Minor unconformity	
Pardonet Formation	Triassic
Baldonnel Formation	Triassic

(Adapted from the ACRMSE Expedition Report by Deej Lowe 1983/84.)

White Hole

Map 93 O/15 043908
Jurisdiction Bocock Peak Provincial Park
Entrance elevation 1600 m
Number of entrances 2 entrances to 2 caves: White Hole & White Dwarf Series (not connected).
Length, Depth 1320 m, 253 m
Discovered 1984, by Andrew Legun

Location White Hole is located in a spectacular depression above Eleven Mile Creek, dominated to the west by Bocock Peak (1947 m) and a smaller unnamed peak (1829 m). The sink is very prominent from the air, and is indicated on the Carbon Creek topographic map as a sinking stream in the middle of a closed depression. The first edition of the topographical sheet actually has the word "cave" written against the main sink in the depression.

Exploration & Description The original ASS reconnaissance of June 1984, left little doubt that a major cave must lie beneath this sink, but as is often the case in cave exploration, realizing this potential was more difficult than originally thought. Although snow did not completely block the entrances during the reconnaissance, melt water from the snowpack was a strong deterrent, and exploration in the Main Entrance of White Hole ended at a depth of 25 m.

The ACRMSE expedition in August found much of the snow melted and three icy shafts plunging into darkness. Attention initially focused on two shafts on the east side — the White Dwarf Series — that closed down at –83 m. The deepest shaft ended in a "dangerous but promising dig with a draft." Exploration in the Main Entrance was halted at the Neutron Collapse, a 3 m-wide breakdown-filled half-tube. No way on could be found, so somewhat despondent, most of the group headed off to check out the sink that later produced Short Straw. However, one last trip was made into White Hole.

Deej Lowe desribes what happened next. "Pete, Andrew and Chas returned to the Main Entrance, but with little hope of extending the cave via the Neutron Collapse dig; more so since there had been torrential thunderstorms all the previous night. The route to the half-tube was very wet and the dig area was really uninviting. Chas crawled up the tube to have a look around and on returning he noticed a strong draft emerging from a section of wall that was jig-sawed together. Pulling out a rock revealed a section of clean-washed wall behind. The three crow-barred the rest of the pieces out and were confronted by a solutional tube some 0.3 m in diameter (The Wormhole). The tube was tight but negotiable, and after about 10 m gradually increased in size to a small canyon very similar to those in Bocock Lesser Sink. Other small vadose trenches joined from the direction of the original dig, and the canyon, now 1 m high, continued down-dip to the northwest, carrying a small trickle of water.

"The distant boom of a waterfall was heard and after some 80 m of tortuous passage they descended two or three cascades to enter a canyon 4 m high and 1.5 m wide carrying a large stream. The water was warm and it was immediately assumed that its source was the main sink, with its pond of sun-heated water. After venturing only a short distance down stream the three turned back, not wanting the cave to "bomb-out" too soon. Back at camp the news was broken to the others over lunch, who naturally assumed it was a joke."

Following lunch, exploration continued. From The Nebula, a series of four interlinked shafts led on down: Stal Way, Dry Way, Wind Way and Wet Way. At the end of Wet Way was an impressive 60 m-deep stepped rift (the site of Olivia's near demise) that bottomed at –253 m, at the time making White Hole the ninth deepest cave in Canada. The Dry Way ended at –176 m (The Schwarzchild Limit), but was not pushed hard as the Wet Way was still going. The expedition report indicated that both the Dry and Wet Ways could yield more passage, exploration in both ending at drafting chokes.

The 1986 ICCC expedition visited the Schwarzchild Limit but found the way on too tight. They also attempted to push the White Dwarf Series which they found was "impossibly blocked with boulders."

Warning This cave contains all the usual hazards associated with large, partially ice-plugged active sinks in an isolated location.

Geology The White Hole commences in the uppermost Bocock Formation dip slope at a point where several faults intersect. The cave cuts down through the full thickness of the Bocock limestone, eventually running down-dip on top of the underlying argillaceous beds. Vertical drops in the main cave are guided by fracture zones, a number of which are high-angle reverse faults. Passages such as Quantum Jump and the Event Horizon are guided by these fractures and are locally floored by argillaceous limestone from the Pardonet Formation. In the Black Hole section of the cave, the shaft passes through marbled limestone below the argillaceous band, and it is possible that this junction represents a low-angle section of a thrust plane. This would imply that the cave re-enters the Bocock Formation at depth, a possibility that fits well with the surface exposure where no significant massive limestone is recorded in the upper part of the Pardonet Formation. (Adapted from the ACRMSE Expedition Report by Deej Lowe, 1983/84.)

"As Steve weighted the rope in preparation to descend, a rebelay broke and he fell a short distance before being brought to rest with a frightening jolt. The pair then traversed to an alcove away from the water in an attempt to get a dry hang. Suddenly, Olivia slipped on the treacherous surface and was over the edge of the pitch. Steve, who was fortunately tied in, instinctively dived forwards and got his hands to her quickly pulling her back onto the ledge."
Deej Lowe et al., 1984, describing an incident in White Hole.

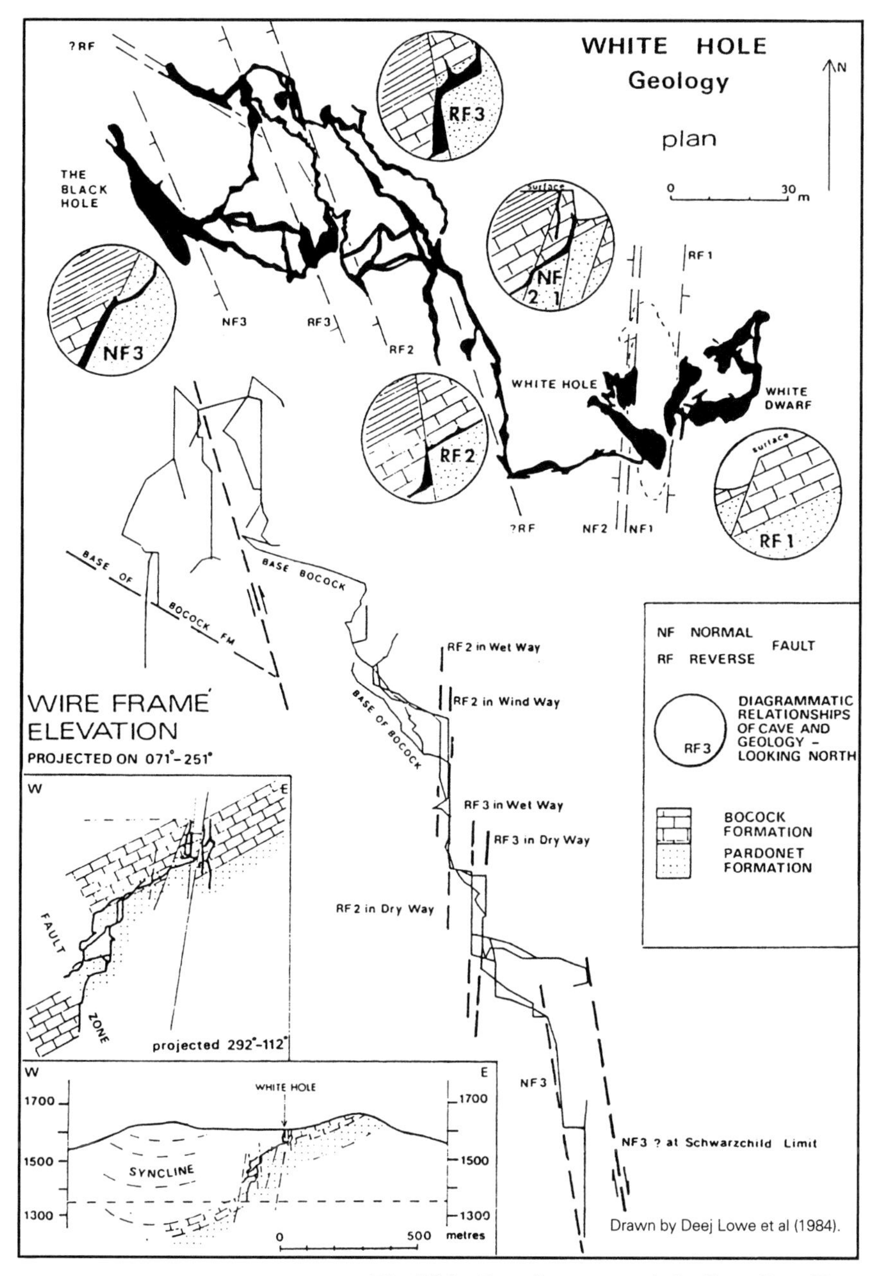

The White Dwarf entrance shaft. Photo Dave Thomson.

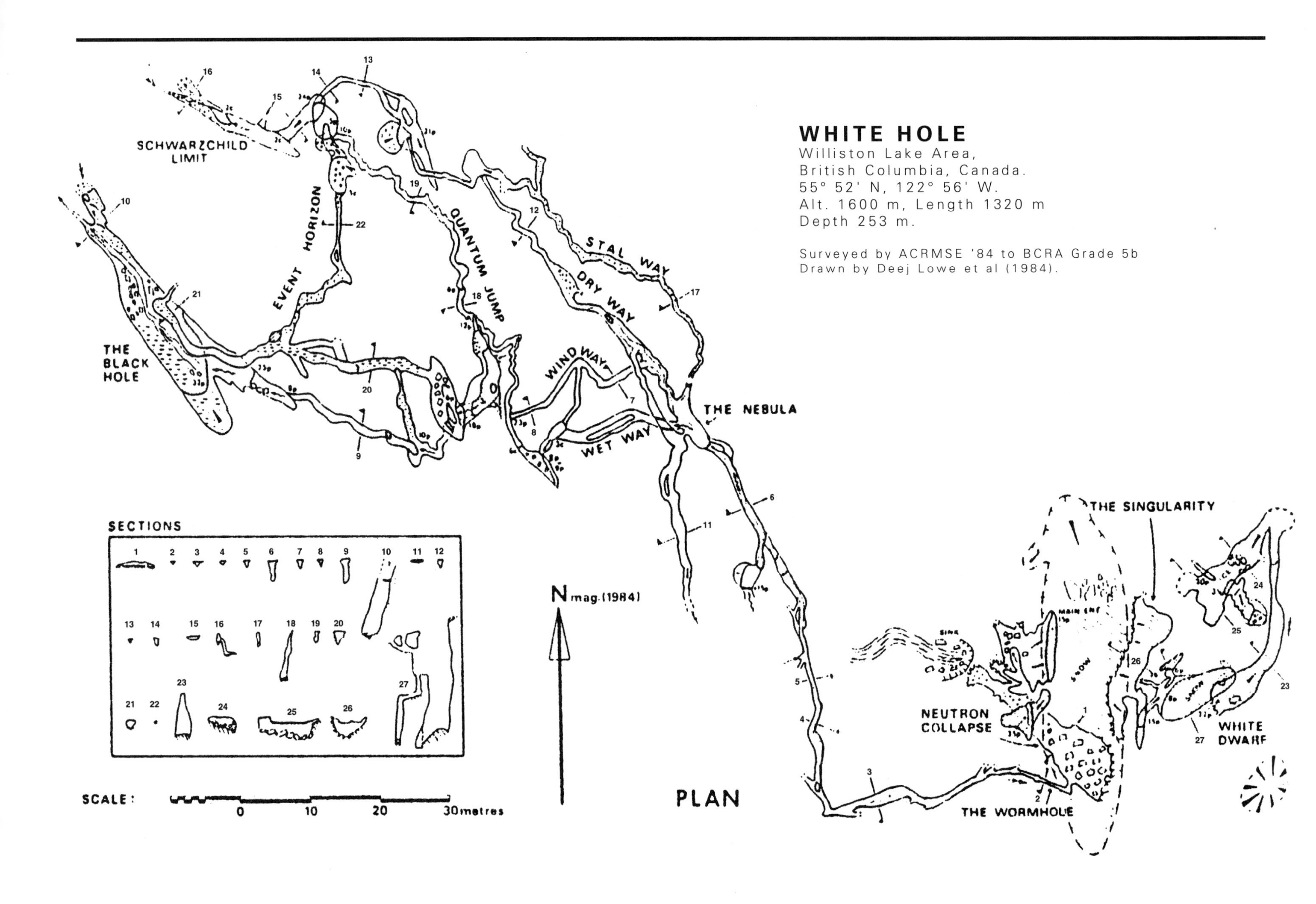
WHITE HOLE
Williston Lake Area,
British Columbia, Canada.
55° 52' N, 122° 56' W.
Alt. 1600 m, Length 1320 m
Depth 253 m.
Surveyed by ACRMSE '84 to BCRA Grade 5b
Drawn by Deej Lowe et al (1984).
SCHWARZCHILD LIMIT
THE BLACK HOLE
EVENT HORIZON
QUANTUM JUMP
STAL WAY
DRY WAY
WIND WAY
WET WAY
THE NEBULA
THE SINGULARITY
NEUTRON COLLAPSE
THE WORMHOLE
WHITE DWARF
SECTIONS
N mag.(1984)
SCALE: 0 10 20 30metres
PLAN

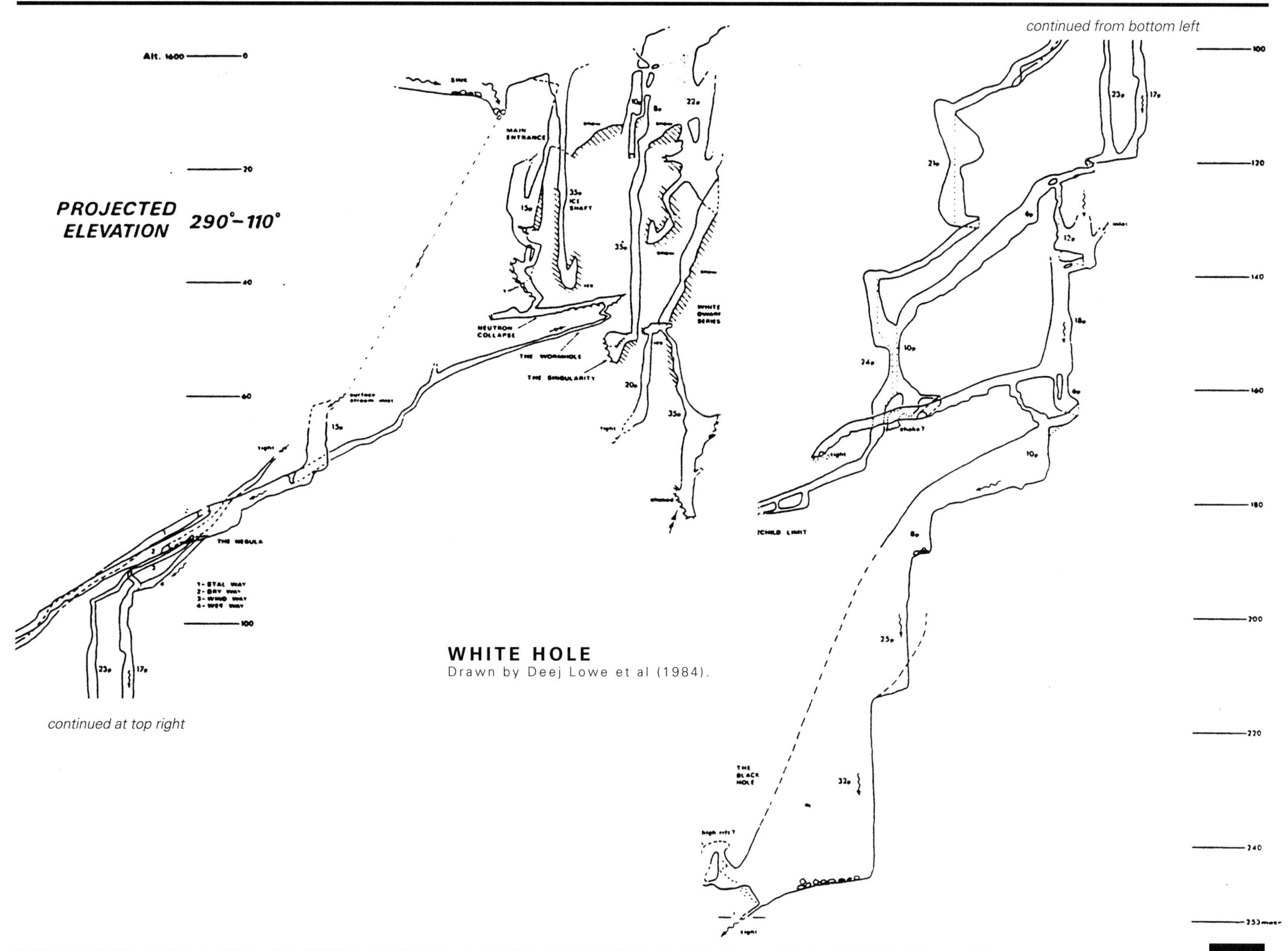

continued at top right

continued from bottom left

Bocock Lesser Sink

Map 93 O/15 043904
Jurisdiction Bocock Peak Provincial Park
Entrance elevation Approximately 1600 m
Number of entrances 1
Length, Depth 185 m, 10 m. Length includes a 45 m extension gained by the 1986 ICCC expedition. It is not known if this increased the depth.
Discovered 1984, by the ASS

Location Bocock Lesser Sink lies 500 m to the south of White Hole and is closest to the southern extent of the large depression below Bocock Peak.

Exploration & Description Pam Burns and Chas Yonge first entered this cave during the ASS reconnaissance in early June of 1984, but high water terminated exploration after only about 30 m.

Later in August of the same year, Chas and Linda Gough (ACRMSE) returned to push it for a very tight 140 m. Deej Lowe writes that, "Linda and Chas returned to the 'terrifying' crawl and where a violent stream had raged a month before, only a placid trickle remained. The rocking boulder blocking the wet continuation was attacked and removed with some difficulty. The passage dropped less than a metre and continued flat-out and tight. They were able to force a way through, extruding gravel to the sides, and round an awkward right-angle bend to gain access to a small vadose canyon cut through beautiful grey massive limestone. The canyon was followed to a junction with another canyon, the passage gaining about a metre in height at this point. The water dropped down a 3 m slot and turned through 180 degrees to follow a sinuous duct for 10 m, at which point the passage was too tight. Over the top of the slot a fossil, bedding plane-like passage continued to a number of leads, which all ended in gravel chokes. The other canyon from the vadose trench junction was dry and up to 1.5 m high, ending in a choke close to an excavated shattered cave discovered just north of the Lesser Sink. Although no vocal connection was realized, this other cave (a 10 m pit) is probably a tributary of the Lesser Sink."

A final visit was made to the cave at the end of the August Bocock expedition when the cave was surveyed and an attempt was made to dig the most prominent choke.

In 1986, the ICCC Expedition pushed Bocock Lesser Sink in a determined effort over three days, gaining a hard won 45 m of tight rift (not shown on the survey).

Warning This cave contains all the usual hazards associated with large, partially ice-plugged active sinks in an isolated location.

Geology Bocock Lesser Sink is also formed in the upper beds of the Bocock Formation. Water flowing northwards over impermeable or armoured ground, crosses a fault and sinks at once into the limestone. The whole of the cave is formed in massive Bocock beds and the underlying Pardonet Formation is not reached.

Abandoned sinks between here and the White Hole are also in Bocock limestone, which is locally well jointed adjacent to faults that trend east west across the depression. (Adapted from the ACRMSE Expedition Report by Deej Lowe 1983/84.)

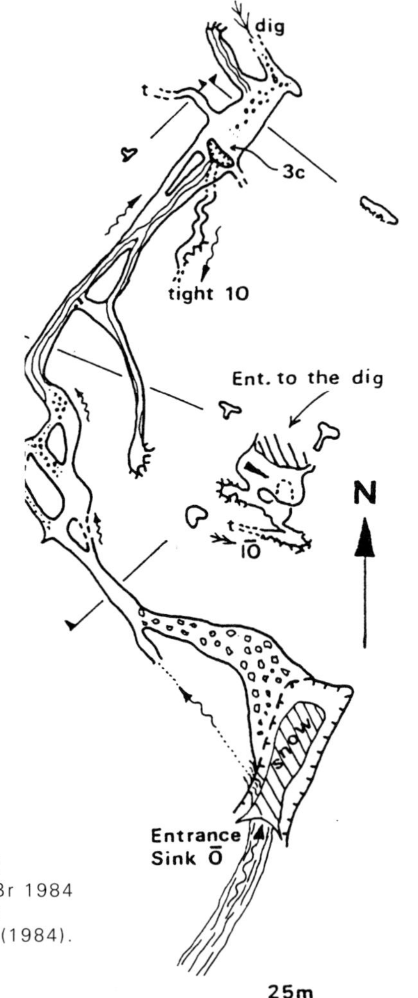

BOCOCK LESSER SINK
Grade 3 survey by CY & ABr 1984
Length 140 m Depth 10 m
Drawn by Deej Lowe et al (1984).

Short Straw

Map 93 0/15 019859
Jurisdiction BC Forests (Peace River)
Entrance elevation 1600 m
Number of entrances 1
Length, Depth 170 m, 107 m
Discovered 1984, by ACRMSE

Location Short Straw lies southwest of Bocock Peak. The sink, rather oddly situated atop a ridge, contains a stream that sinks through a boulder choke.

Exploration & Description After two frustrating days of unsuccessful exploration in the holes of the Bocock depression, forays were made further afield. Using binoculars, cavers discovered the sink containing Short Straw from an animal trail cutting through the bush on a ridge just east of Ducette Creek.

The stream sinks through a boulder choke to a steeply dipping wet passage. Fortunately, a parallel dry passage bypasses the first wet pitch. Three more drops are free-climbed to the head of the final 37 m drop. On the first exploration, straws were drawn to pick the two people who were to have the pleasure of dropping the final shaft, hence the cave's name.

The cave ends in a gravel-choked strike passage which holds some prospects for extension. The exploration team reported that the water, which makes the final pitch and gravel choke unpleasant, could easily be diverted down a narrow outlet passage at the head of the pitch by using some plastic sheeting.

Warning This cave contains all the usual hazards associated with large, partially ice-plugged active sinks in an isolated location.

Geology The ridge in which the cave is located is the upturned end of a steeply dipping, massive limestone band 30 to 40 m thick and assumed to be of the Bocock Formation. The dip is towards the southwest and the bed is probably not inverted. Here, as with the Bocock area, the regional strike and structural trend is NNW-SSE and it is assumed that the underground drainage will follow the most favourable hydraulic gradient towards Williston Lake to the north. (Adapted from the ACRMSE Expedition Report by Deej Lowe, 1983/84.)

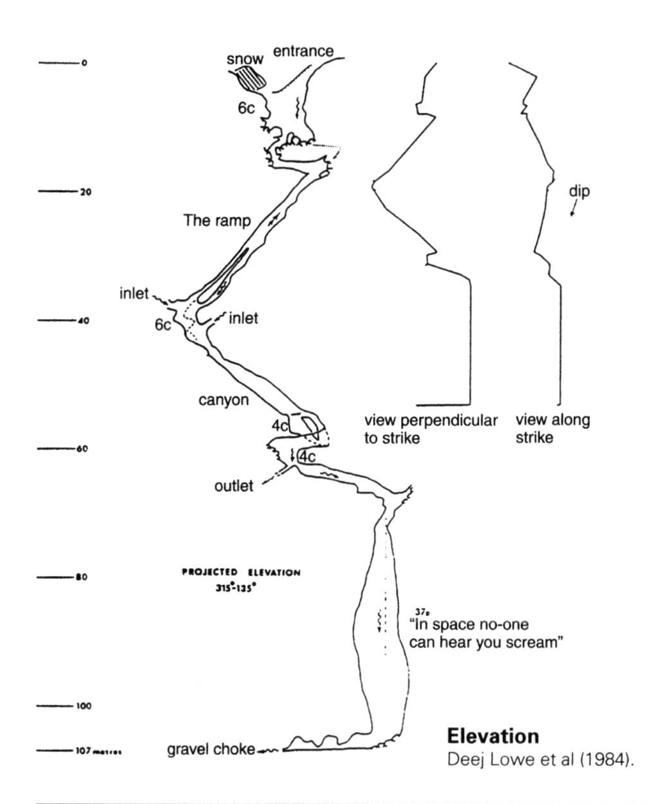

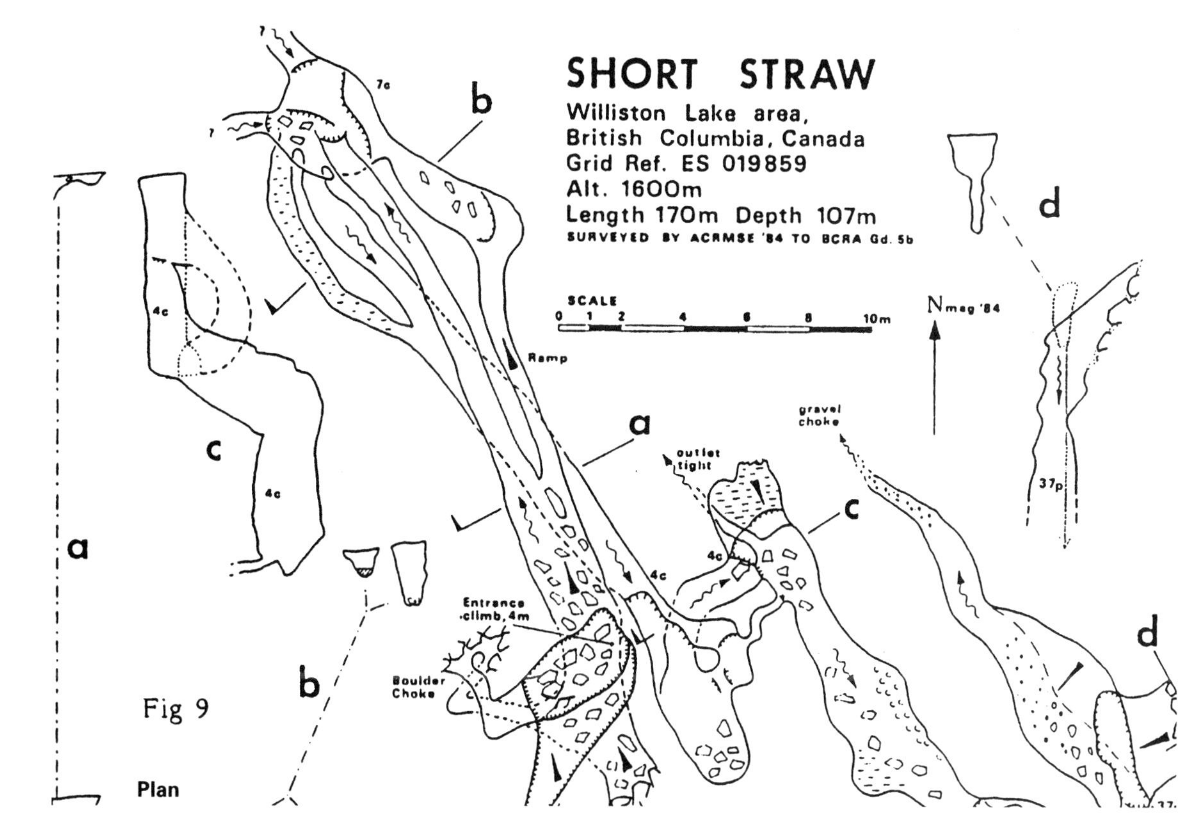

Elevation
Deej Lowe et al (1984).

CARIBOU RANGE

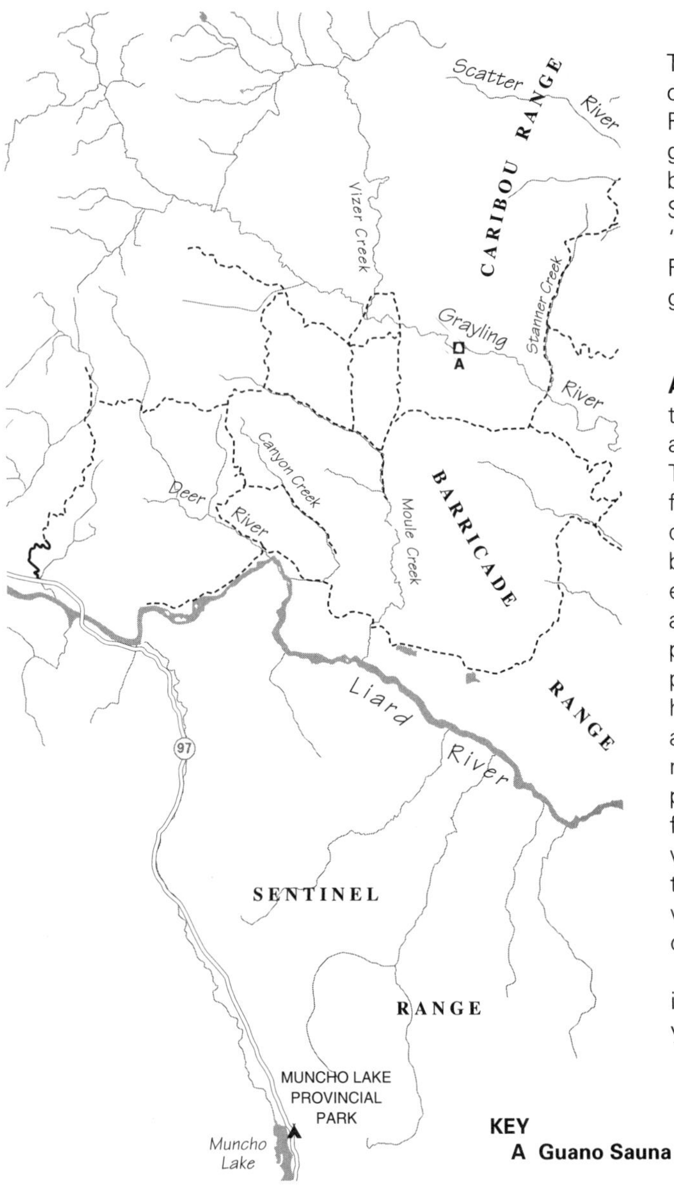

The Liard River at the northern extent of the Canadian Rockies in the Caribou Range has created the mother of all gaps, its confluence with Toad River being the lowest point in the Rockies. Steve Worthington in his PhD thesis, "Karst Hydrology of the Canadian Rocky Mountains," identifies the Liard gap as having great cave potential.

Area Exploration In 1980 Mike Boon travelled north to investigate a sinking and resurging stream adjacent to the Toad River, only to find it was a slide feature. In 1986 the Liard Gap was chosen as the destination for 13 members of a joint Anglo-Canadian caving expedition. Following a marathon approach involving 15 flights in a small plane (only capable of carrying one passenger at a time), three days on horseback and a day hike, the karst area was finally reached. Unfortunately, all sinks and cave entrances proved to be blocked with glacial fill, and despite much digging and a widening of the reconnoitered area up to 14 km from camp, the cavers left with only a small guano-filled gaseous chamber in their survey book.

The karst area to the north remains incompletely explored and may yet yield some caves.

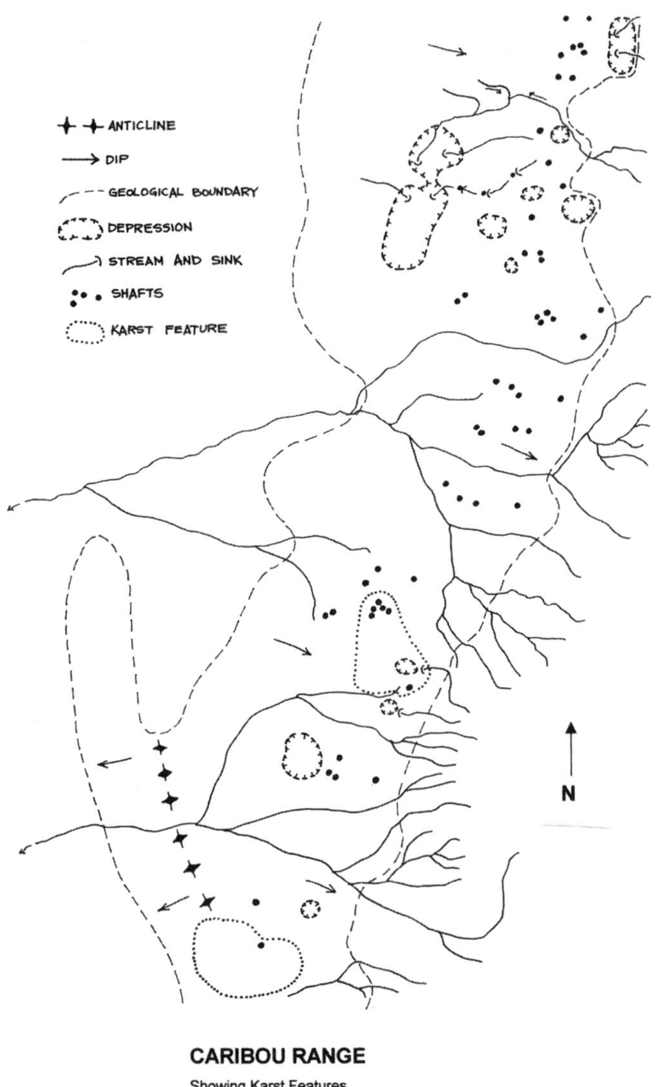

CARIBOU RANGE

Showing Karst Features

Adapted from a Drawing by
Steve Grundy & Paul Hatherley, 1986

Guano Sauna

Map 94 N/12 570110
Jurisdiction Grayling River Ecological Reserve
Entrance elevation 800 m
Number of entrances 1
Length, Depth 64 m, 35 m
Discovered Unknown

Access Guano Sauna is located on the Grayling River northeast of Liard Hot Spings Provincial Park on the Alaska Highway. Access is by helicopter, small plane or horseback. Contact outfitters in Fort Nelson.

Because the cave is located in an ecological reserve, permission to visit may be hard to obtain.

Location On the south bank of the Grayling River upstream of a series of hot springs. Above a 30 m-high tufa terrace.

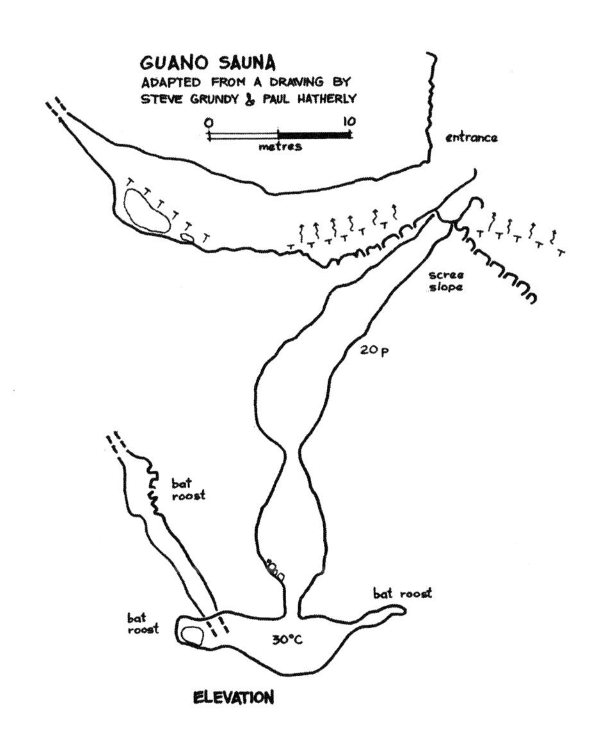

Exploration & Description According to Vancouver Island caver Steve Grundy, first impressions of the hot and steamy Guano Sauna "were of a descent into hell. Steam was venting around the entrance and a 30 degree slope led quickly to the head of a 30 m pit. As we descended the pit the temperature and smell increased dramatically, and with bats clinging to our oversuits, we landed in a heap of ammonia stenching guano in a small sweltering chamber." This is the total extent of the cave.

Warning Owing to the high temperature and extensive guano deposits, the air in Gauno Sauna is probably not conducive to good health.

Cave Softly Do not disturb the bats.

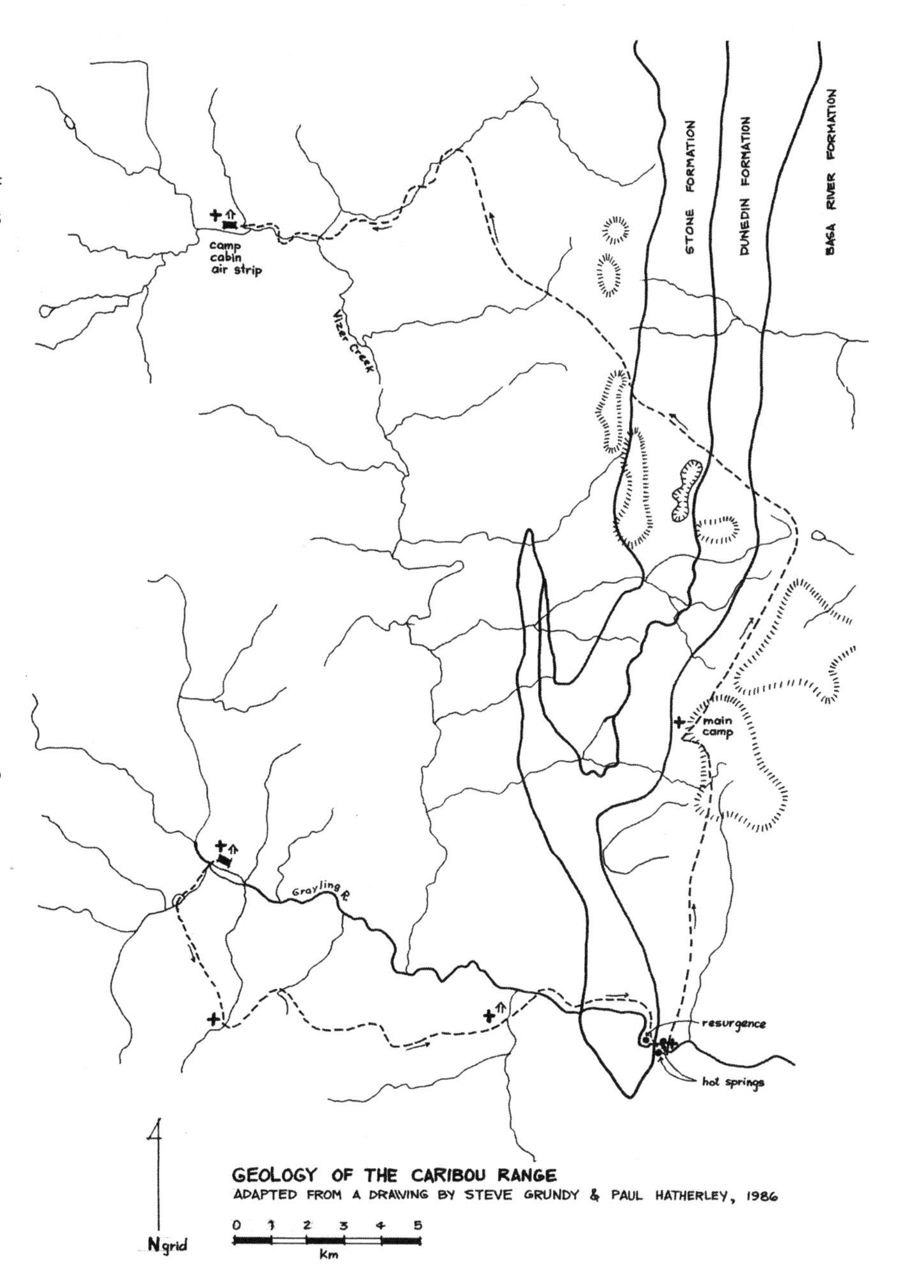

Nakimu Caves (Main Cave, Gopher Bridge & Mill Bridge caves)

Map 82 N/5
Jurisdiction Glacier National Park
Entrance elevation 1576 m (Main Entrance, others between 1580 m and 1708 m)
Number of entrances Numerous
Length, Depth 4500 m, 270 m
Discovered 1902, by D. Woolsey & W. Scott (first recorded visit)

Access In the past decade, management concerns have focused on grizzly activity in the Cougar Valley, and for this reason all access to the cave via the Cougar Valley trail and the old Stage Road has been denied. The only access — summer or winter — is now over Balu Pass, turning the former 3 km approach into an 18 km return hike or ski, with 1500 m elevation gain. Not surprisingly this has cut down on visitation, and currently only two or three groups a year visit the caves.

The Balu Pass trail starts behind Glacier Park Lodge, located just east of the summit of Rogers Pass. It is 5 km to Balu Pass at 2030 m (about two hours) where a campsite has been established for use of parties visiting the caves. From the pass the trail descends steeply, losing 450 m down to the gorge (approximately 4 km one way). From here, trails lead downstream to the Main Entrance, and upstream to Mill Bridge and Gopher Bridge caves.

The snow-free season is mid-June to mid-October, although snow usually lingers at Balu Pass until mid-July. The author has visited the caves in the winter. However, at this time the Main Entrance gate is usually frozen shut and buried, and the other entrances can be hard to locate.

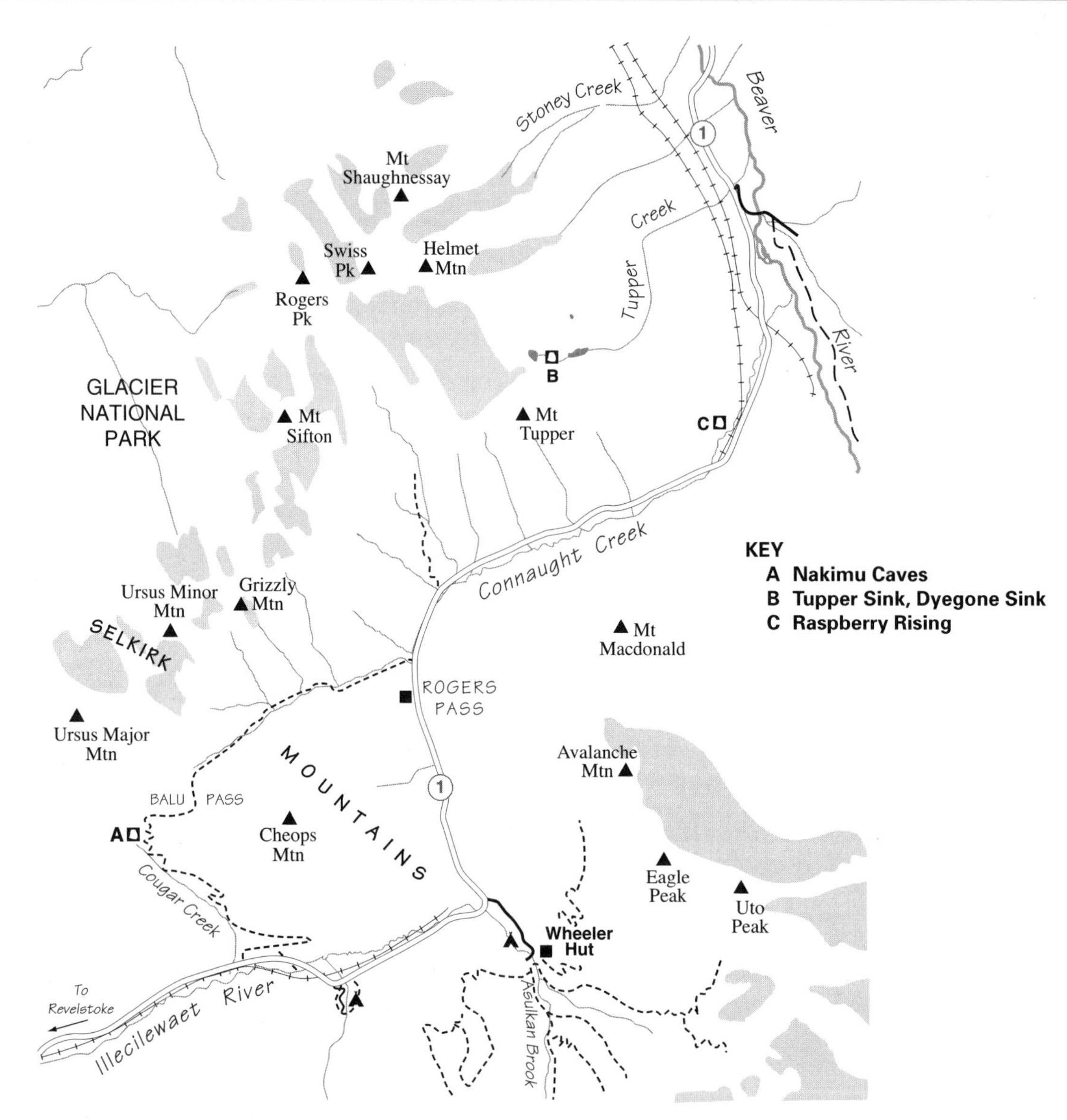

Warning It is recommended you travel in groups of five or more. The area around the caves is prime grizzly habitat (some of the best in the Rockies and Columbia Mountains), so keep a keen eye out for bears and bear signs at all times. In Cougar valley there were maulings in 1971 and a fatality in 1976. Parks often deny access to the caves at short notice.

Visiting the caves in winter while the bears are still hibernating could be a good idea. However, the avalanche hazard is often extreme, and closures are commonly in effect. The descent on skis from Balu Pass requires careful route finding to avoid avalanche terrain.

As with any active cave, water levels can be highly variable. Most of the lower parts of the caves are only accessible later in the summer or in the winter when water levels drop.

Exploration & Description The story begins with the railway, and the building of Glacier House, which opened in 1887 as a dining facility. Although the caves were located only 10 km from the hotel, it was not until 1902 that two prospectors from Nakusp, Walter Scott and D. Woolsey, descended into the gorge of Cougar Creek by climbing down a fallen tree. It is not recorded whether they entered the caves. In May of 1904, Charles Henry Deutchman was the first to enter the caves, and seeing their potential as a tourist attraction, filed a mineral claim in October of that year.

A. O. Wheeler was impressed with Deutchman's solo exploration of the Gorge series, which he descended into using a knotted rope. "The work of exploration he (Deutchman) has done without assistance shows a character utterly devoid of fear. The descent into the depths of blackest darkness, lighted only by the dim rays of a tallow dip, without a rope or other aid except in a case of direst necessity, requires more than courage; it requires strength of purpose and power of will far beyond the ordinary degree. For, added to the thick darkness, there was always the fierce vibrating roar of subterranean torrents, a sound most nerve-shaking in a position sufficiently uncanny and demoralizing without it."

This quote illustrates a notable attribute of Nakimu Caves, that of their active nature, carrying large volumes of water. This is in contrast to many Rockies caves which are relict and dry. It was the thundering roar of water that suggested to Deutchman the name of Nakimu, from the Cree for "spirit sounds," and prompted Derek Ford to note during the 1970's exploration of the caves that, "Nakimu is a fine example of a dynamic cave; in July when snowmelt enters the cave, the passages shake with the intensity of water flow."

In May 1905, a party of twelve visited the caves, including Howard Douglas from Banff (Superintendent of Rocky Mountains Park), W. S. Ayers, M.E. (an underground engineer) and R. B. Bennett (Associated Press correspondent). Because the caves were situated in the recently designated Glacier Park Reserve, the purpose of the visit was to report on their extent to the Dominion government.

Following this, a series of visits were made in order to survey the valley and caves (see 1905 survey by A.O. Wheeler), during which Deutchman, Ayers and Wheeler surveyed about 600 m of passage and explored a further 1200 m.

Using acetylene bicycle lamps, vastly superior to the previously mentioned tallow candles, and burning magnesium wire to get glimpses of the larger chambers, the trio explored the upstream Gopher Bridge Cave, which they named after the marmots seen in the vicinity. They first entered this section of cave by "wriggling and squeezing through narrow cracks over dirty rocks" (the Old Entrance). Later, Deutchman excavated a more user-friendly entrance (Gopher Bridge Entrance). This short section of cave and the waterfall it contained recalled for the party Dante's Inferno, and so they named a chamber Avernus.

The next section of cave downstream from Gopher Bridge Cave was named Mill Bridge, and the entering streamway, The Flume, after its resemblance to a millrace. Entering through a narrow cleft (Entrance No. 1), the party explored the Auditorium, Corkscrew and Pothole Passage where Deutchman and T. Kilpatrick (CPR Superintendent from Revelstoke) later placed ladders and a floating bridge to aid tourists.

Below Mill Bridge Cave lies an open section of Cougar Creek known as The Gorge, 30 m deep and spanned by two natural rock bridges. From here,

Deutchman's shack at Nakimu Caves. Photo courtesy Whyte Museum of the Canadian Rockies.

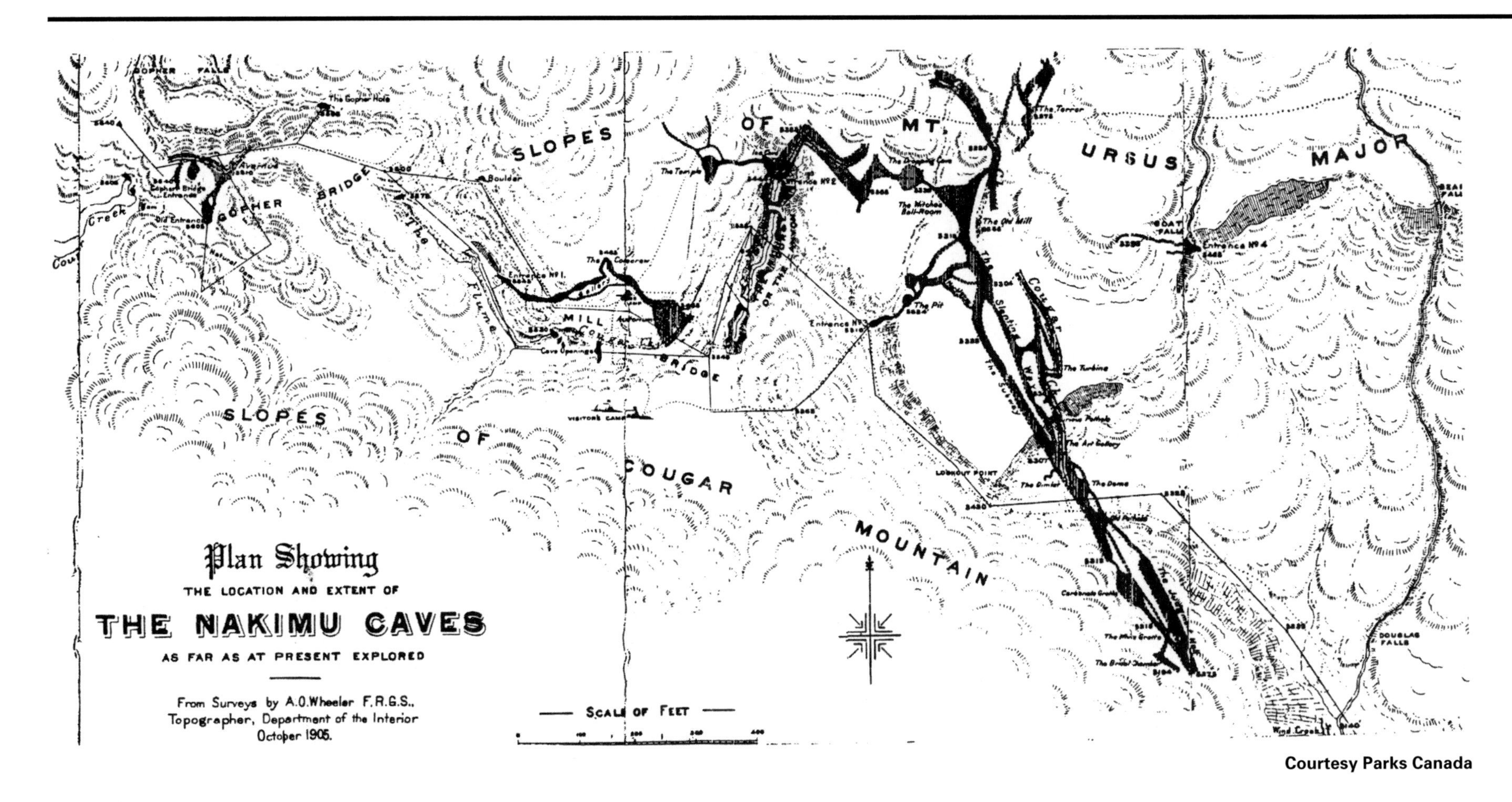

Courtesy Parks Canada

the explorers accessed The Gorge Series of Nakimu Caves, now known collectively as the Main Cave. A short ramp down the east side of the gorge accessed Entrance No. 2, and the large Entrance Chamber that contained a climb down "a rock face of some twelve feet with natural notches that would seem as though they had been cut with a cold-chisel for the special purpose" (still apparent today). Following this, they entered what they referred to as the Dropping Cave, "named from the fact that water drops from the roof in all directions," and by way of a narrow connecting passage, the Witches Ballroom, a large 15 m-high chamber "just the spot where a group of witch-hags might be expected to caper round the ghastly fumes of some hellish cauldron at a Sabbath meeting."

From the northeast extent of this chamber they entered The Terror and The Old Mill, and from the southeast, they located four ways on: to the left they could hear the muffled roar of the stream, to the right was a chamber called The Pit, which was later accessed from the surface via a 36 m shaft (Entrance No.3) in which a precarious stairway was later constructed (St. Peter's Stairway). In between lay the Subway and Slanting Way, beneath which was the active stream passage called The Turbine. The lower end of these passages in the vicinity of The Judgment Hall represented the extent of early exploration in the Main Cave, about 500 m short of where Cougar Creek returns to the surface. As Wheeler surmised: "It would appear that the largest accessible portion of the series is now on record."

Howard Douglas negotiated with Deutchman for the claim (reputedly a sum of $5,000 was agreed upon) and the Department of the Interior retained Deutchman as custodian of the caves, and commissioned him to take all steps necessary to open them to the public.

$1,200 was made available to build a trail to the cave, and $200 or $300 to build ladders and bridges. An early suggestion was to use Cougar Creek to generate electricity to light the caves, but this was never realized, and acetylene lights were used. In 1906 a bridle trail and a custodian's cabin was constructed. The Glacier House register indicates about 50 persons a week visited the caves during that first summer of 1906, with a total of 700 visitors in 1907, rising to over a thousand in 1909, dropping to around 600 by 1911.

Development of the caves proceeded slowly depending on Deutchman's spare time. In 1906, Entrance No. 1 was enlarged, a bridge was built over the creek in the Auditorium, and some pathway was enlarged by blasting in the Main Cave. In 1909, Deutchman and an assistant placed 80 m of iron railings in the caves.

The possibility of blasting another entrance into the lower part of Main Cave was looked at as early as 1911, as this would eliminate the need to retrace the route back to Entrance No. 2 or 3 from the larger passages at the southeast end of the cave. This work was finally carried out in 1916, and included double doors and a shed over the new entrance at a total cost of $400. In 1911, a coach road was started, and by 1914 ran for 8 km from Glacier Lodge to within 2 km of the caves, where steeper terrain prevented its completion. Nevertheless, in 1917 a regular stage visited the cave twice daily.

For Deutchman it was a constant struggle to develop the caves, and he regularly wrote to Ottawa requesting funding. He was only paid from June to September, and finally gave up his job as caretaker and guide in 1918 to take a mining job in Montana. When he died in Lime Rock, Connecticut, in 1962 at the age of 87, his obituary credited him as the discoverer of Nakimu Caves and with the death of 57 grizzlies!

Deutchman's long time assistant and successor, George Steventon, continued with the improvements that included the ladder up from the Pit to Entrance No. 3 (St. Peter's Stairway) and the route through to Entrance No. 2 that necessitated building a dam to deflect the water. In 1923, the C.P.R. built a teahouse at the entrance (their only direct investment).

With the closing of Glacier House in 1925, the viability of the caves as a tourist destination dwindled. 170 visitors registered in 1928, and 300 in 1930, when improvements were still being made to the caves with the addition of more wooden stairways and handrails. The caves were finally closed in 1935 with the death of George Steventon.

Nakimu Caves.
Photo Dave Thomson.

Interest in the Nakimu Caves was revived in the late 1950s with the routing of the Trans Canada Highway through the park (opened in 1962). In 1960, J. O. Wheeler (A.O. Wheeler's grandson) investigated the caves on behalf of the Engineering Services Division with a view to possible re-opening. Two visits by American cavers were made in 1957 and 1962. Further exploration was made downstream in the Main Cave, and the Bear Fall Series were partially explored.

In 1965 and 1966 Derek Ford and the McMaster cavers visited the caves and pushed a series of passages in the Cougar Creek Series, surveying the Dry and Wet Chutes, Micaschist Passage and the Alpine Grotto, finally connecting with the pioneer cave above. Attempts to pass a waterfall had to wait for the 1966 winter trip when lower water enabled a sump at –269 m to be reached.

Elsewhere in the cave, the McMaster cavers pushed beyond Terror Chamber to find Perseverance, Surprise Passage and The Crypt. The 1966 trip finally connected this part of the cave to the Great Slab in the Bear Falls Series where the American cavers' boot prints were seen in the mud. The Americans had entered via a small entrance alongside a waterfall, hence Bear Falls.

In 1975, 288 m of stream passage missed by the McMaster group was surveyed by Mike Boon, Martyn Farr (a cave diver from Wales) and Kitty Dunn and Wes Davis of the ASS. This followed the Streamway from the Turbine to below the Dry Chute. This completed exploration of the cave as we know it today.

In 1978, Parks Canada gated the artificial entrance to the Main Cave in order to enforce its new cave access policy (Section 34a Parks Act), and began removing sections of the old wooden walkways and ladders.

In 1980, a film was produced on the cave titled *Underground Rivers*. Together with a small cave exhibit, this can be viewed at the Rogers Pass Interpretive Centre.

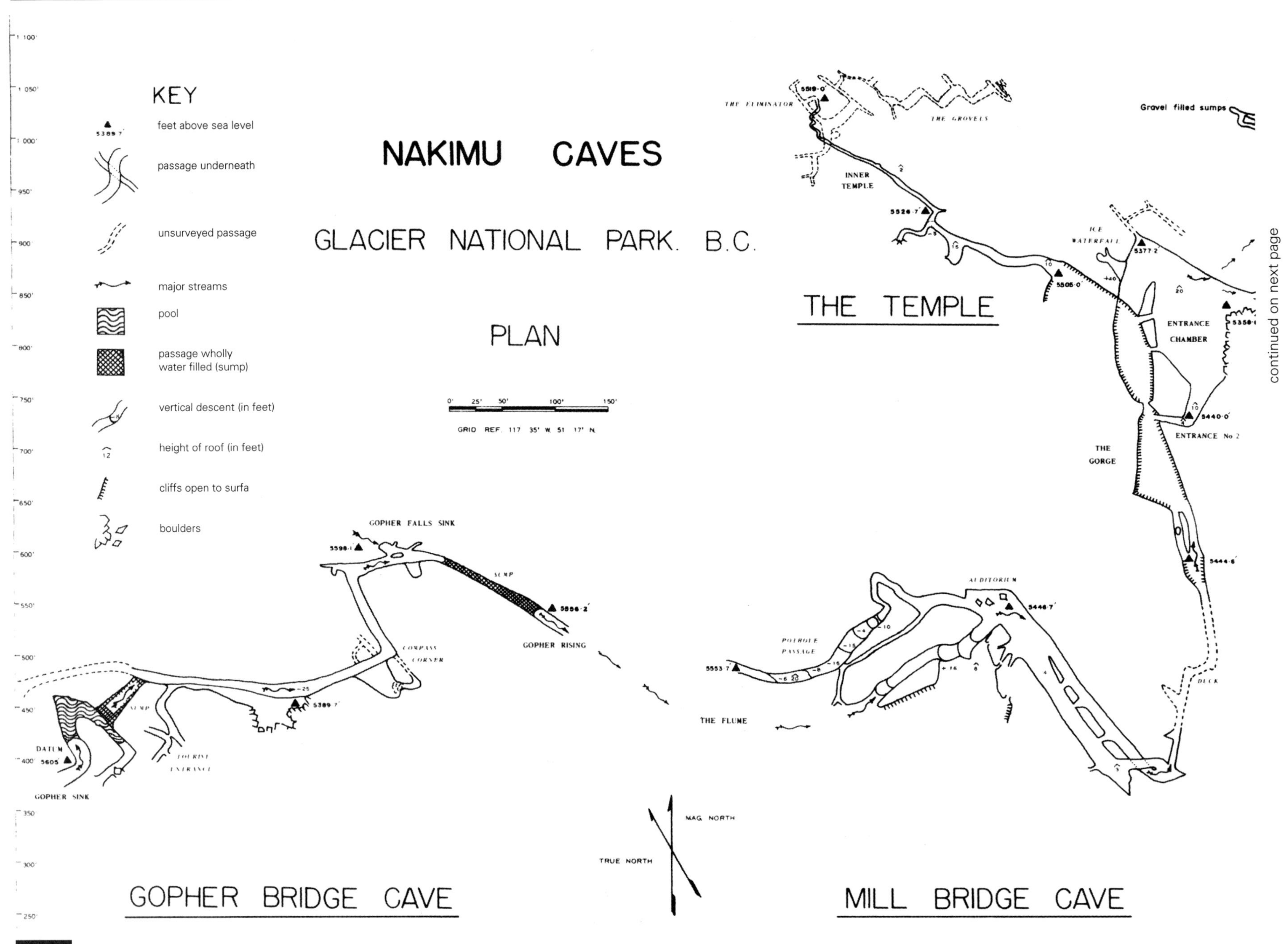

continued on next page

Geology The Nakimu Caves are formed in a small pocket of limestone, rare in the Selkirk Mountains. Identified as the Badshot Formation, the limestone beds containing the caves have been folded into a tight series of anticlines and synclines, in some places doubled in thickness. This is obviously a speleogenic limestone, because there have been reports of subterranean drainage in the Downie Creek area to the northwest of Rogers Pass, where the Badshot formation again outcrops. The limestone is shot through with distinctive white bands of calcite visible throughout Nakimu Caves.

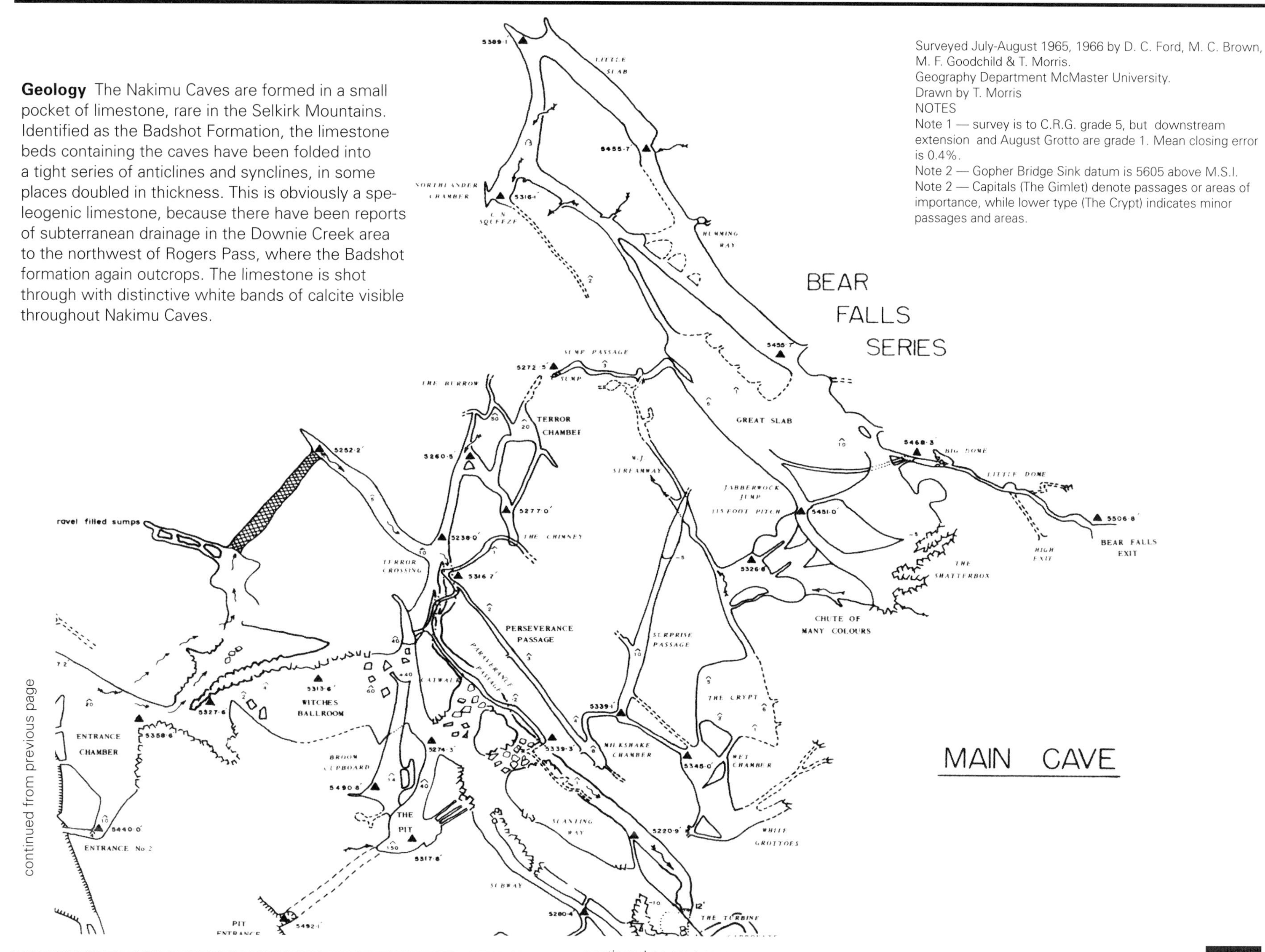

continued from previous page

continued on next page

continued from previous page

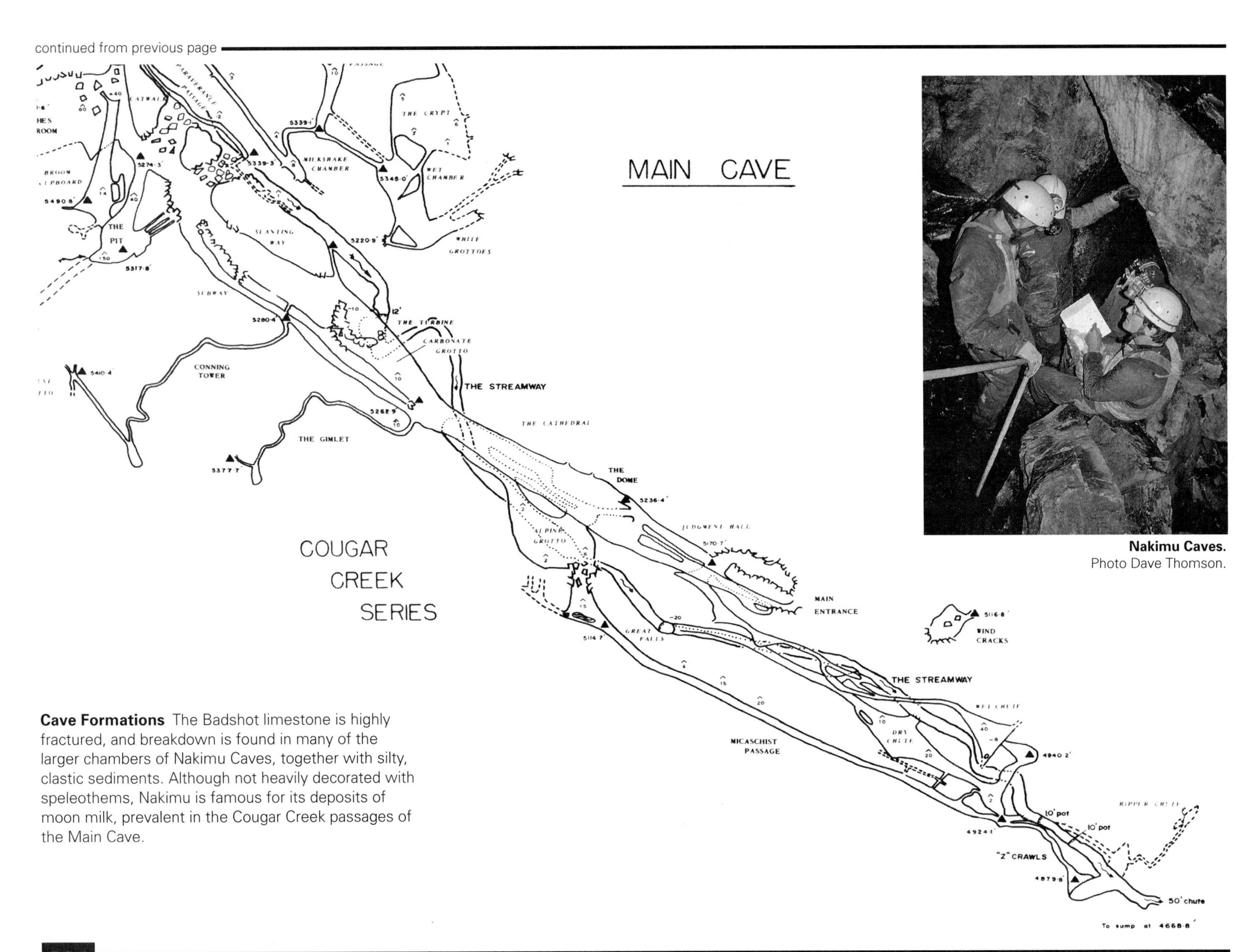

Nakimu Caves.
Photo Dave Thomson.

Cave Formations The Badshot limestone is highly fractured, and breakdown is found in many of the larger chambers of Nakimu Caves, together with silty, clastic sediments. Although not heavily decorated with speleothems, Nakimu is famous for its deposits of moon milk, prevalent in the Cougar Creek passages of the Main Cave.

The Tupper System

Unlike Nakimu Caves, the Mount Tupper sinks (Tupper Sink and Dyegone Sink) and spring (Raspberry Rising) are not located in limestone, but in a band of marble and dolomite sandwiched between slates and schists dipping at 45 degrees. The main hope for entering the system lies in diving and climbing in Raspberry Rising.

Tupper Sink

Map 82 N/6
Jurisdiction Glacier National Park
Entrance elevation 1862 m
Number of entrances 1
Length 30 m
Discovered 1966, by D. Ford & M. Goodchild

Access Located in the upper Tupper valley under East Tupper Glacier. Without a helicopter, locating the sink requires a major bushwack.

Exploration & Description A whirlpool fed by a melt water stream from the East Tupper Glacier was confirmed by dye test to connect with Raspberry Rising. A few meters of passage blocked with silt were found just above the sink; excavation produced more passage that closed down to a drafting fissure.

Dyegone Sink

Map 82 N/6
Jurisdiction Glacier National Park
Entrance elevation 1982 m
Number of entrances 1
Length 20 m
Discovered 1966, by Pete Fuller, Gary Pilkington & Tom Wrigley

Access Located in the upper Tupper valley under East Tupper Glacier. Without a helicopter, locating the sink requires a major bushwack.

Exploration & Description After Tupper Sink ended, the Tupper Valley was searched in an attempt to find an alternative entrance to the Tupper System. A thrash through devil's club and up small cliffs led to the location of Dyegone, which was dug to produce 20 m of good stream passage, ending in a low crawl. A dye trace again proved positive, the water taking eight hours to travel to Raspberry Rising.

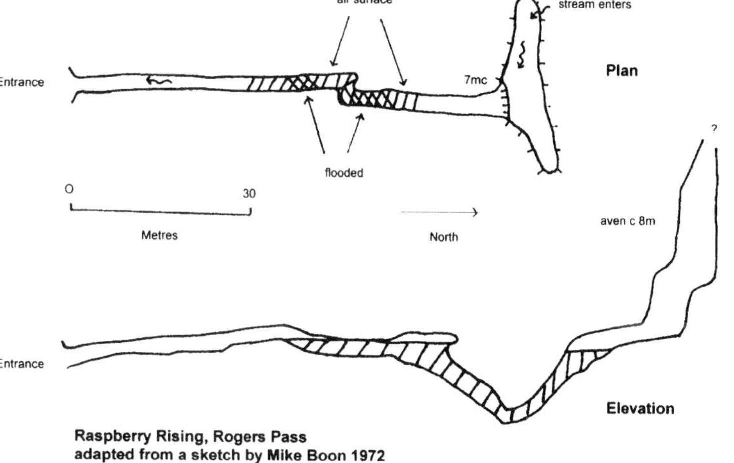

Raspberry Rising, Rogers Pass
adapted from a sketch by Mike Boon 1972

Raspberry Rising

Map 82 N/6
Jurisdiction Glacier National Park
Entrance elevation 1372 m
Number of entrances 1
Length 76 m (including sump dive)
Discovered Unknown

Access Heading west on Hwy. 1 (the Trans-Canada Hwy.) from Golden, park just before the first snowshed. Head north up a drainage under an old stone railway bridge. About 180 m higher, the water emanates from a spring known as Raspberry Rising.

Raspberry Rising is the resurgence point for Tupper and Dyegone Sinks. From Tupper Sink to resurgence, the water covers a horizontal distance of 2 km (485 m vertical) in only 53 minutes (summer), indicating the possibility of steep, air-filled passages.

Exploration & Description During Easter 1972, Mike Boon dived the spring. "The sump is in two sections: a canal-like section where the roof dips down below the surface for a few feet, and a steeply dipping gravel slope leading down to a slot several feet wide and about 2 feet high, beyond which another slope rises steeply to the far side. Here I took the kit off and explored a short section of cave to a climb at the foot of an impressive rift. The climb leads upwards about twenty feet to a kind of landing onto which the stream poured from a pitch thirty or more feet high."

During a later attempt at pushing the cave Mike Boon and Pete Lord dived through the sump with a Maypole! The results of this attempt were not recorded.

Warning Do not attempt to dive this spring unless you are an experienced cave diver!

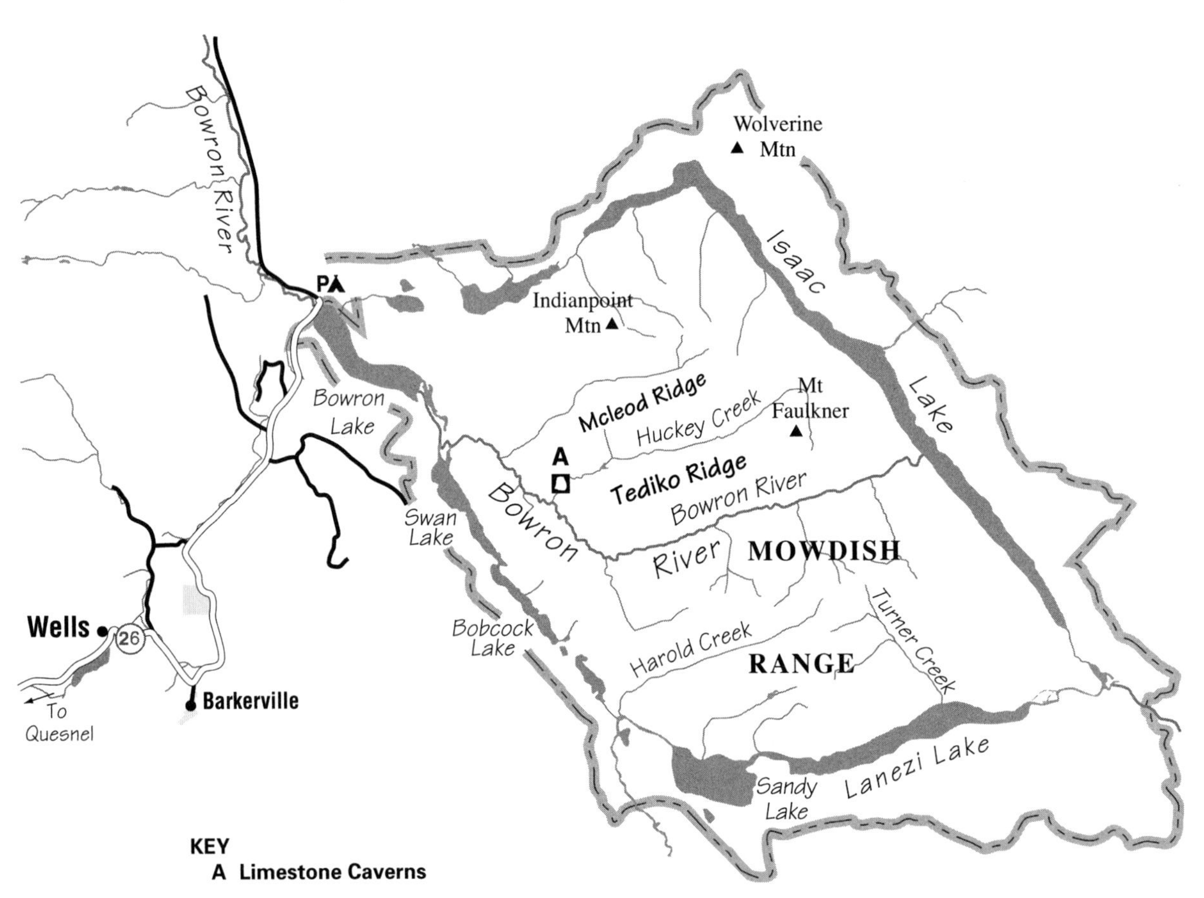

Limestone Caverns (also known as Grizzly Bear Cave, Huckey Creek Caves and Bowron Lakes Caves)

Map 93 H/3 198933
Jurisdiction Bowron Lakes Provincial Park
Entrance elevation 1020 m
Number of entrances Upper Series 2, Lower Series 1, Grizzly Bear Passage 2
Length, Depth Upper Series 118 m, 27 m; Lower Series 238 m, 16 m; Grizzly Bear Passage 24 m, 9 m
Discovered Probably by residents of Barkerville during the Cariboo gold rush of the 1860s. They were first noted on a map by BC Forestry in 1935.

Access Take Hwy. 97 to Quesnel, then Hwy. 26 east for 110 km, passing through Wells to Bowron Lakes Provincial Park. A drive-in campground (first-come, first-served) is open May 15 to September 30.

The popularity of the canoe circuit (reservations need to be made), and the presence of grizzlies in the area means the caves are probably best visited in winter.

Skiing is straightforward up Bowron Lake for 7.2 km, then up the interconnecting creek to the head of Swan Lake (4 km to a primitive cabin). From here ski 4 km up the Bowron River, then across meadows for 3 km to access the Huckey River and the caves that can be reached without too much difficulty.

In summer, the Bowron River is navigable to within 1 km of Huckey Creek. However, it is best to pull out 2 km before the confluence and hike east across meadows and through dense lodgepole pine forest (very boggy following rainfall).

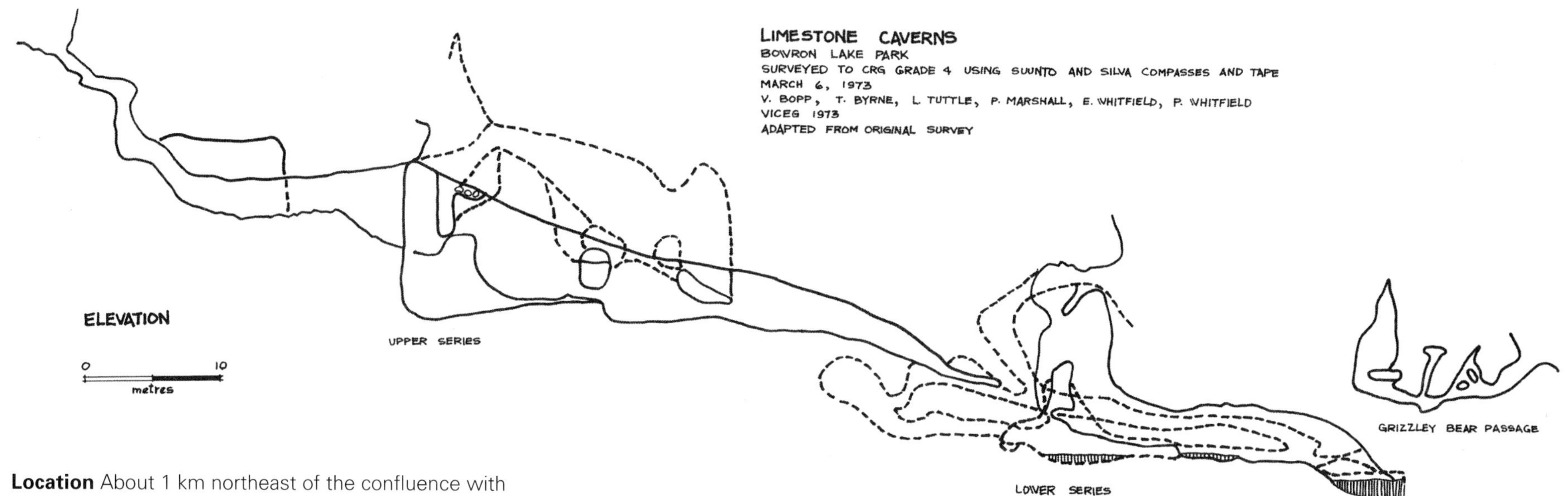

UPPER SERIES	LOWER SERIES
length 386.0 ft.	length 781.0 ft
depth 89.0 ft.	depth 51.5 ft.

Location About 1 km northeast of the confluence with the Bowron River, Huckey Creek passes through the caves via an opening on the north side of a narrow canyon.

Exploration Limestone Caverns were first visited by cavers in 1971, and then mapped by members of the VICEG at the request of BC Parks in March of 1973.

Description The caves occur where Huckey Creek encounters a small patch of Cunningham limestone. Between where the creek sinks and resurges are several sections of lower active and higher fossil passages, similar to the situation at Nakimu Caves near Rogers Pass. A dry surface canyon links the Upper Series and Lower Series entrances. Dye tracing has indicated little possibility of finding more passage.

The most impressive feature of Limestone Caverns is the large (10 m by 30 m) Upper Series sink entrance. The name of the highly enthusiastic neophyte caver who reported the wonders of gold, white frogs and sasquatch footprints is probably best left hidden in the dusty archives of the "Canadian Caver." OK, Paul.

"Preliminary explorations revealed that the cave contains: gold-bearing black sand; a hot spring; white frogs; miles of unexplored galleries with waterfalls; an underground lake; round metallic-coated rocks like cannon balls; super sized footprints of unknown age."
Vancouver Sun, August 19, 1972.

THE CARIBOO MOUNTAINS

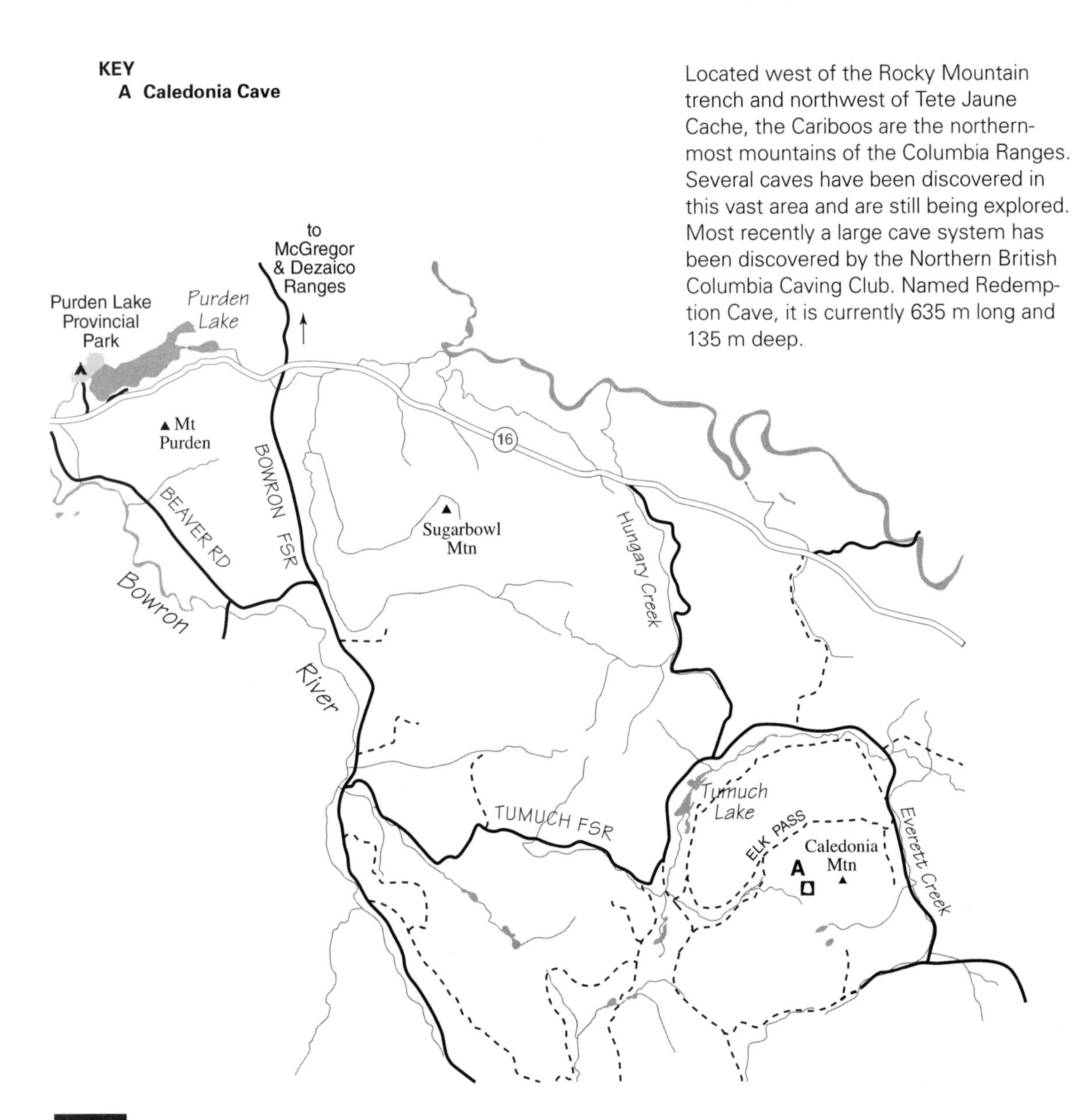

Located west of the Rocky Mountain trench and northwest of Tete Jaune Cache, the Cariboos are the northernmost mountains of the Columbia Ranges. Several caves have been discovered in this vast area and are still being explored. Most recently a large cave system has been discovered by the Northern British Columbia Caving Club. Named Redemption Cave, it is currently 635 m long and 135 m deep.

Caledonia Cave

Map 93 H/11 c.077480
Jurisdiction Cariboo Land District
Entrance elevation c. 1600 m
Number of entrances 1
Length, Depth 105.5 m, 74 m
Discovered 1987, by Dave King & Mike Nash of the Caledonia Ramblers

Access The best route starts from Hwy. 16, 8.5 km east of the Purden Lake Resort. Turn south onto the Bowron Forest Service Road (FSR). In 24 km turn left (east) onto the Tumuch FSR. Follow it for 54 km until just past the second crossing of Everett Creek, then turn right and up a short hill and continue for another 3 to 4 km to the west fork of Everett Creek (bridge has been removed).

The trail from here (overgrown and hard to find) ascends first on old roads through a cutblock and then up the mountainside through an old burn that extends all the way to the alpine. After cresting the ridge the route descends to a small alpine lake, then more steeply down through trees to the larger Emerald Lake.

From about halfway around the north shore climb north through a series of openings for about 300 m onto a ridge. Follow a small stream, trending west, down the other side for about 300 m and start looking for the cave a few hundred metres away from the stream on the north side. (Adapted from an email by Mike Nash.)

Location The cave is located on the forested southwest slope of Caledonia Mountain. Look for a small cliff. The entrance is hidden from view until you are standing on the lip of the large 25 m-wide entrance depression.

Exploration & Description The large entrance to Caledonia Cave was spotted in March 1987 from a helicopter during a winter caribou count. In June, Dave King and Mike Nash hiked in, locating the entrance about a kilometre down-slope from several sinkholes and a 50 m-long cave known as Caribou Pit. Lacking enough rope to proceed, they made a return trip in August when the entrance slope was descended 20 m to where it steepened. Thrown rocks and lots of counting established it as being about 60 m deep.

With the assistance of three ASS members, they revisited the cave in September. After descending 30 m and getting very wet, the team returned to the surface and successfully diverted the stream that enters the cave. The final visit two weeks later found the stream diversion still intact, and the cave was finally explored and surveyed.

From the large entrance depression a rubble-strewn slope led down to a steep drop which Rick Blak descended to the bottom of at -70m. Both the main pit and a side aven were explored, but were found to be choked. Ian McKenzie pendulumed into a small short passage off the pit through which another parallel shaft could be sounded, but not accessed. Before leaving the cave, Rick traversed over loose rock and ice above the main pit to where 50 m of boulder-filled rift ascended towards the surface.

Warning The pit is threatened by lots of loose rock. As a result of the change in stream flow, dangerous ice formations linger through much of the summer. (Mike Nash correspondence, 2003.)

Caledonia Cave entrance. Photo Ian McKenzie.

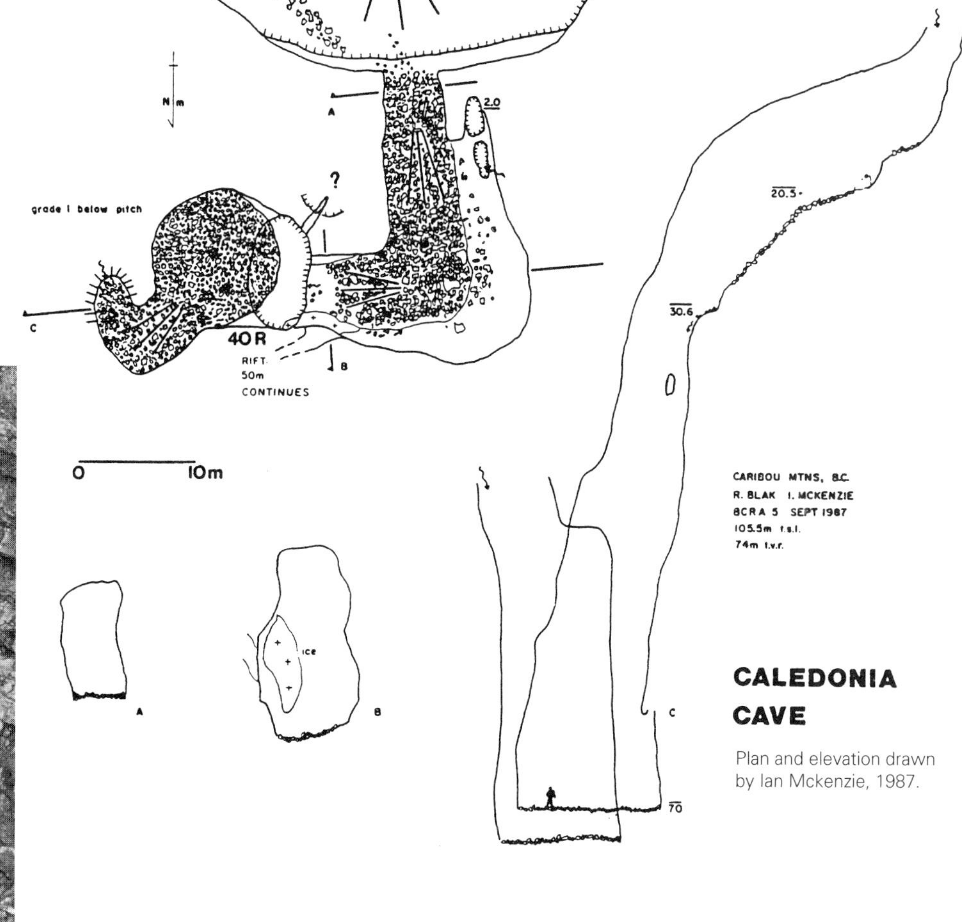

Plan and elevation drawn by Ian Mckenzie, 1987.

WOOD BUFFALO NATIONAL PARK

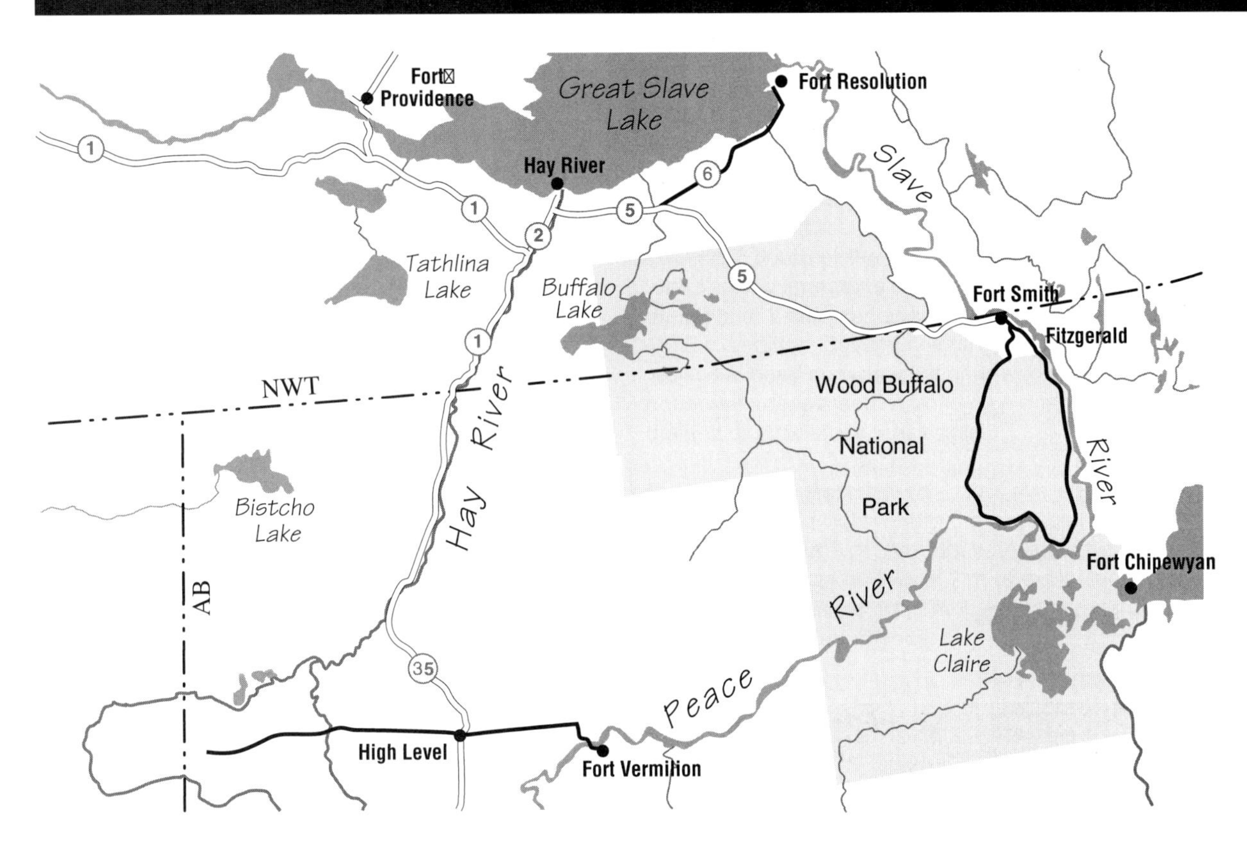

Straddling the border of northeastern Alberta and the Northwest Territories, Wood Buffalo National Park contains the largest area of gypsum karst in North America. Surface features include sinkholes, dolines, sinking and reappearing rivers, dry valleys, springs and caves.

Unlike the majority of limestone caves, gypsum caves are characterised by frequent collapse and are considered unsafe by parks staff. To date, two caves have been identified by park naturalist Rob Lewis. However, owing to their fragile nature and their importance as bat hibernacula, Parks Canada does not allow access.

Area Geology Gypsum (hydrated calcium sulphate, $CaSO_4 2H_2O$) is a soluble rock, belonging to a class known as evaporites (A. C. Kendall 1979). Large exposures of this gypsum are seen on the surface, and it has been estimated that more than 60% of the park has gypsum at depth (R. Stein, J. J. Drake and R. Lewis 1975). Interbedded with limestone, dolomites, anhydrites and shales, the gypsum is in places as much as 100 m thick.

Where the gypsum is capped by a thin layer of less soluble rock such as limestone, active ground water erosion takes place along the interface. Once no longer supported by the gypsum, the limestone capping collapses in, large blocks of it falling to the floor of the cavern below. By this process, known as stoping, the cavity migrates towards the surface forming sinkholes. Fissures present in the surface of the capping limestone direct water into the underlying gypsum, forming large cavities which become interconnected by the interface dissolution. With subsequent sinkhole collapse these cavities can become enterable caves. Due to low relief, and low hydraulic gradient, the potential for explorable caves above the water table is probably limited, but there are doubtless extensive underwater cavities. As indicated by the map, the sinkholes, the major feature of the park, lie in an arc from the Nyarling River in the north to Peace Point in the south. Some of these crater-like features are up to 60 m deep.

Springs Springs are a major feature of the Wood Buffalo karst, the most prominent being Neon Lake Spring, estimated to discharge at a rate of 8 m^3/sec. Springs also discharge in numerous lakes, as evidenced by deep blue holes. Algae and diverse chemical processes are thought to account for the spectacular range of colours of these lakes throughout the seasons, described as ranging from pink, yellow, and brown to deep blue. The Nyarling River is dry for about 16 km of its course, and has well defined sink and rising points.

Just west of the Slave River is a 15 to 60 m-high escarpment. Numerous briny springs resurge from the base of these cliffs and flow across a large area of salt flats near the salt river. These springs have deposited mounds of salt up to a metre high.

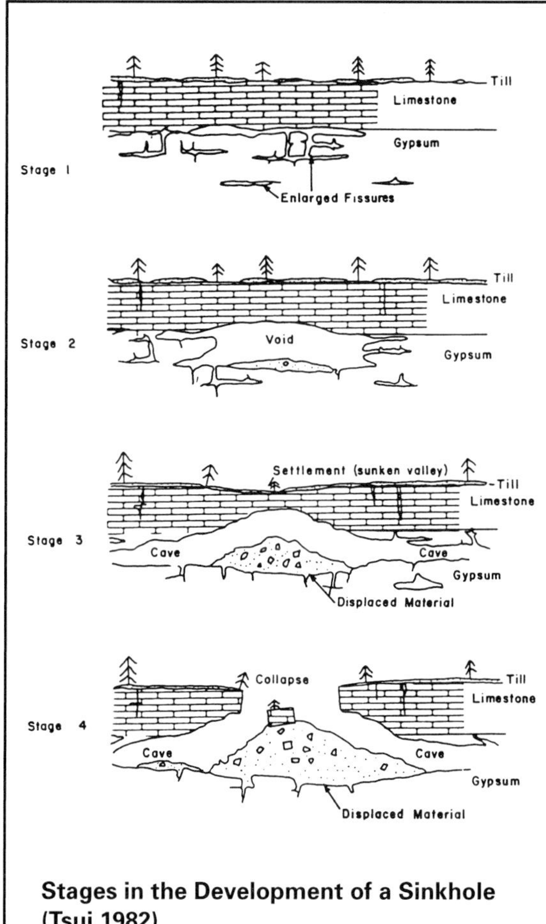

Stages in the Development of a Sinkhole (Tsui 1982).

Stage 1 Groundwaters dissolve gypsum along fissures and at the gypsum-limestone interface.
Stage 2 Enlarged fissures coalesce to form a void.
Stage 3 Void migrates toward the surface by stoping of the overlying limestone.
Stage 4 Void breaks through to surface to form a sinkhole.

The Gypsum Karst of Wood Buffalo National Park. (J. J. Drake et al. 1979.)

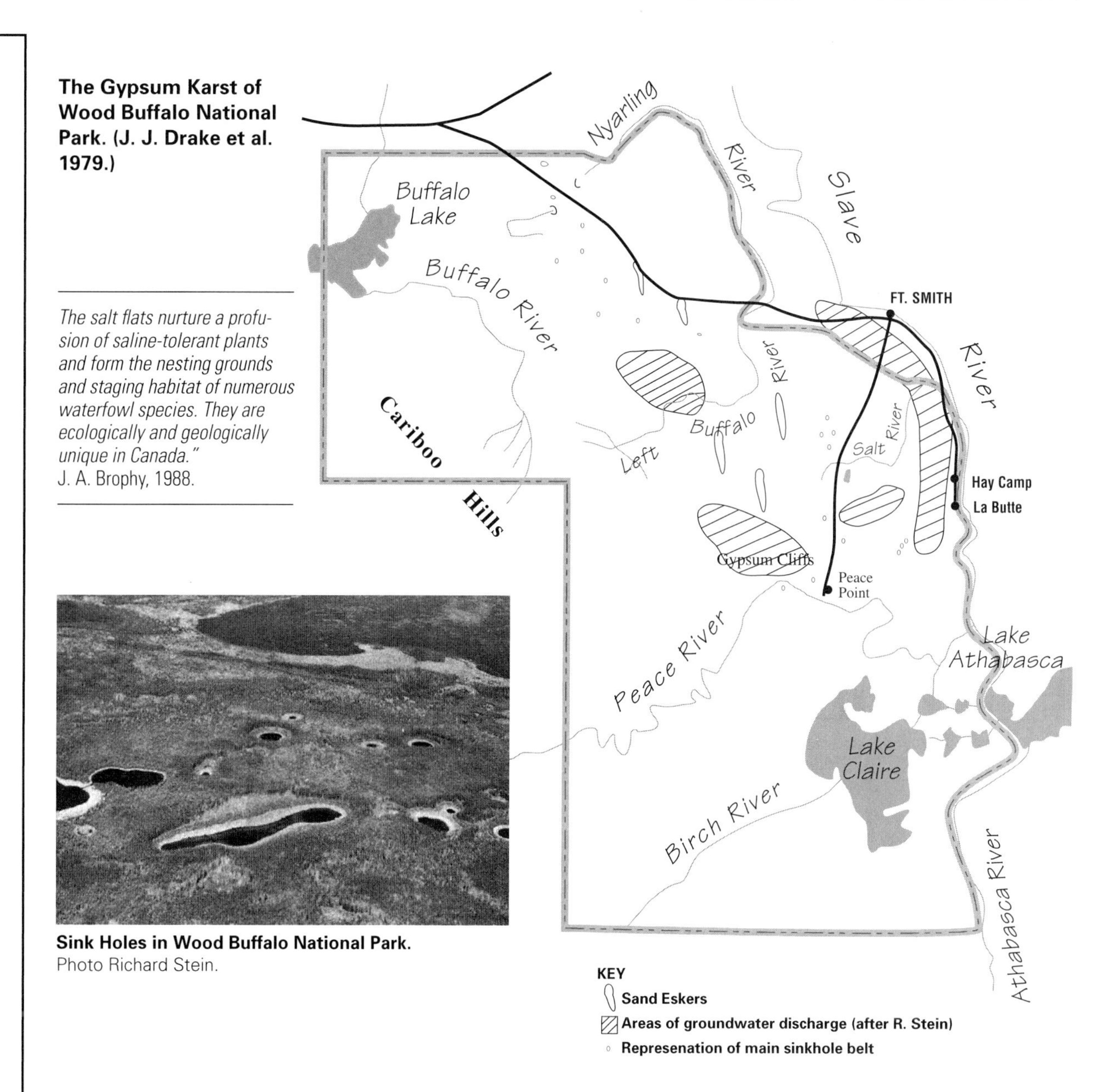

The salt flats nurture a profusion of saline-tolerant plants and form the nesting grounds and staging habitat of numerous waterfowl species. They are ecologically and geologically unique in Canada."
J. A. Brophy, 1988.

Sink Holes in Wood Buffalo National Park.
Photo Richard Stein.

Fauna The karst areas of Wood Buffalo National Park provide rare or unique habitats for many species of wildlife. The salt plains are important nesting habitat for numerous bird species. J. S. Nelson and M. J. Paetz (1974) have documented four species of fish in a sinkhole only accessible through an underground channel. Four species of bat hibernate during the winter in caves in the park, and the red-sided garter snake has adapted by hibernating beneath the frost line in karst formations (Parks Canada, 1985). These karst areas are obviously of tremendous ecological importance.

Ice Cave

Map Uncertain
Jurisdiction Wood Buffalo National Park
Entrance elevation Uncertain
Number of entrances 1
Length, Depth Length approx 150 m
Discovered Not known

Ice Cave.
Photo Richard Stein.

Location Located at the north end of a 20 m-deep sinkhole.

Access Unknown.

Description Ice cave is a cold trap cave where there is no air circulation. Consequently, water vapour freezes out on the wall in the form of ice crystals. Fallen ice crystals are in places 0.6 m deep, concealing a boulder-strewn floor. The entire cave roof and walls are covered in ice crystals. The entrance is literally one massive ice sheet, sloping steeply downward, with large slabs of rock cemented into place with a matrix of ice.

At the bottom of the cave is a stream. Behind a mound of fallen rock the cave system continues. It has been explored to its limit and terminates at a rockfall after approximately 130–180 m. The stream follows the same path and disappears under the rockfall. The average passage dimensions are 6 m wide and 3–6 m high. (Adapted from a report by Rob Lewis.)

Walkin Cave

Map Uncertain
Jurisdiction Wood Buffalo National Park
Entrance elevation Unknown
Number of entrances 1
Depth 70 m
Discovered Not known

Location Walkin Cave is believed to be located at the south end of the same sinkhole as Ice Cave, over a steep pile of fallen rock.

Access Unknown.

Exploration & Description Walkin Cave, the largest cave yet discovered in Wood Buffalo National Park, is a warm or dynamic cave in which the air is able to circulate through the chambers from a second opening. Large slabs of breakdown are everywhere in the cave.

At the base of the rocks, some 70 m below the ground surface, a small stream has created small lakes on the floors of the chambers. One quarter of the way down the main passage, another passage leads off to the right. Maintaining a width of 6–8 m and a height of 3–5 m, the passage continues until it opens up into a large room. The room is approximately 6 m high with a diameter of 30–35 m. A passage leads off this room and follows the stream through a narrow channel with a 2 ft-high roof.

A dinghy (small inflatable boat) was utilized to explore further chambers. The water in most places is less than 1 m deep, and in some places greater than 2 m deep. The channel empties into a very large room, 20 m wide, 120 m long and 15 m high above the rock pile on the floor which is itself 6–8 m above the water level. A faint ribbon of light enters the room from an adjacent sinkhole. A passage leading off this large room was explored to its length of 70 m. Approximately 150 bats were seen in this cave. (Adapted from a report by Rob Lewis in 1975.)

Natural History

Curtain formation in Rat's nest Cave.
Photo Dave Thomson.

GEOLOGY–THE ROCK THAT CONTAINS THE CAVES

All caves in the southern Canadian Rockies are contained in carbonate rocks, i.e., limestone and dolomite of the Paleozoic era (540–258 Ma). The most imporant are the Mural and Cathedral formations of the Cambrian Period, the Palliser Formation (Devonian) and the Rundle Group — Livingstone, Etherington and Mount Head formations of Mississippian Age. Apart from providing an absolute earliest date for the commencement of formation, the age of the cave-bearing formations themselves is no indicator of the age of the caves. Today's karst topography, with steeply-bedded limestones in a mountainous environment, does not provide the slow percolation rates needed for cave formation. We know that caves form mainly beneath the water table, which is currently in valley floors far below existing cave entrances.

Derek Ford writes that "Cavern entrances in the Rockies are often perched 500–1000 m above valley floors. Yet these caves formed beneath the water table, at a time when they were beneath the valley floors. The difference in elevation between perched cave passages and the present valley bottoms provides a rough measure of how much rock has been lost to erosion of all kinds since the caves formed. By doing some clever arithmetic and making some educated assumptions, geographers have estimated that the ridge lines we see now were at the level of the valley floors some 6–12 million years ago."

Enterable caves form best in massively-bedded carbonates such as the Cathedral dolomitized limestones that contain Castleguard Cave. In such fine-grained, relatively impermeable rocks, the passage of water is restricted to joints and bedding plane weaknesses, thus forming linear passages that travel for long distances without interruption. The Mural Formation is interbedded with shales and sandstones, caves having formed at the contact between the permeable and impermeable rocks, i.e. Arctomys and the Small River caves. The Palliser Formation, the most massively layered and homogenous limestone after the Eldon and Cathedral Formations, is prominent in the Rockies and forms such splendid moun-

"Even an untrained eye can look at this massive limestone formation, with its grid of fissures, and be fairly confident that it contains caves. A karst geomorphologist would simply say, Ah! Good rock."
Dalton Muir & Derek Ford, 1985.

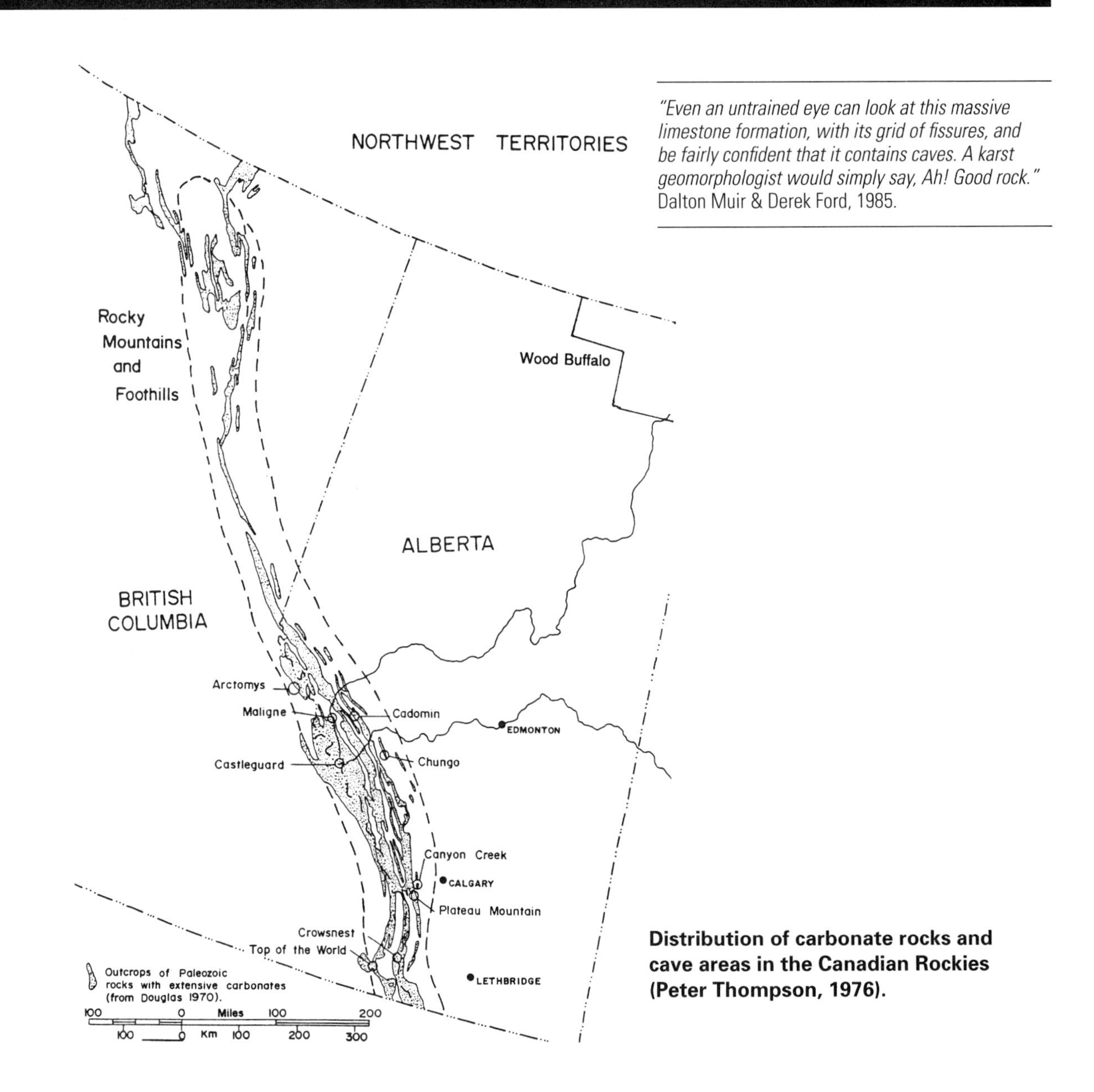

Distribution of carbonate rocks and cave areas in the Canadian Rockies (Peter Thompson, 1976).

tains as Windtower and The Three Sisters near Canmore, and Cascade Mountain near Banff. It is also a good cave-bearing limestone, as witnessed by Cadomin Cave and the Snaring karst. Palliser limestone is dark black with fine calcite veins, and is fairly fossiliferous, being composed of the remains of crinoids — a type of echinoderm from the same phylum as today's starfish. Limestone of the Livingstone Formation contains the extensive caves of the Andy Good and Ptolemy plateaus of Crowsnest Pass, and like the Palliser Formation, is a thick and resistant unit. It forms large grey cliffs such as those of the Livingstone Range, its namesake, which runs north from Crowsnest Pass to the Oldman River.

Surficial deposits, up to 300 m thick
Till, river and lake deposits, slides and debris, windblown material
Age: Quaternary, 1.9 Ma to the present
Found in all three regions (northern, central, southern)

Young clastic unit, 5 km thick
Mostly sandstone and shale. Age: Mesozoic and Tertiary, 245—1.9 Ma
Exposed in foothills and front ranges, all three regions,
and in Rocky Mountain Trench, Flathead and Elk valleys

Cave Bearing
Middle carbonate unit, 6.5 km thick
Mostly limestone/dolomite and shale
Age: Paleozoic, 540—258 Ma
Exposed in front ranges and main ranges in all three regions

fig. 12

Old clastic unit, 10 km thick
Mostly gritstone, shale and quartzite
Age: Hadrynian and Early Cambrian, 770—540 Ma
Exposed in main ranges of the central and northern regions

Purcell-type unit, 8.3 km thick
Mostly mudstone and limestone
Age: Helikian, 1500—1300 Ma
Exposed in southern front ranges and northern front/main ranges

North American Plate, 30—50 km thick
Granite and gneiss
Age: Aphebian, averaging 1700 Ma
Underlies all three regions

Basic geological column for the Canadian Rockies (Ben Gadd, 1986).

Part of the stratigraphy of the Canadian Rockies (Halliday and Mathewson, 1971).

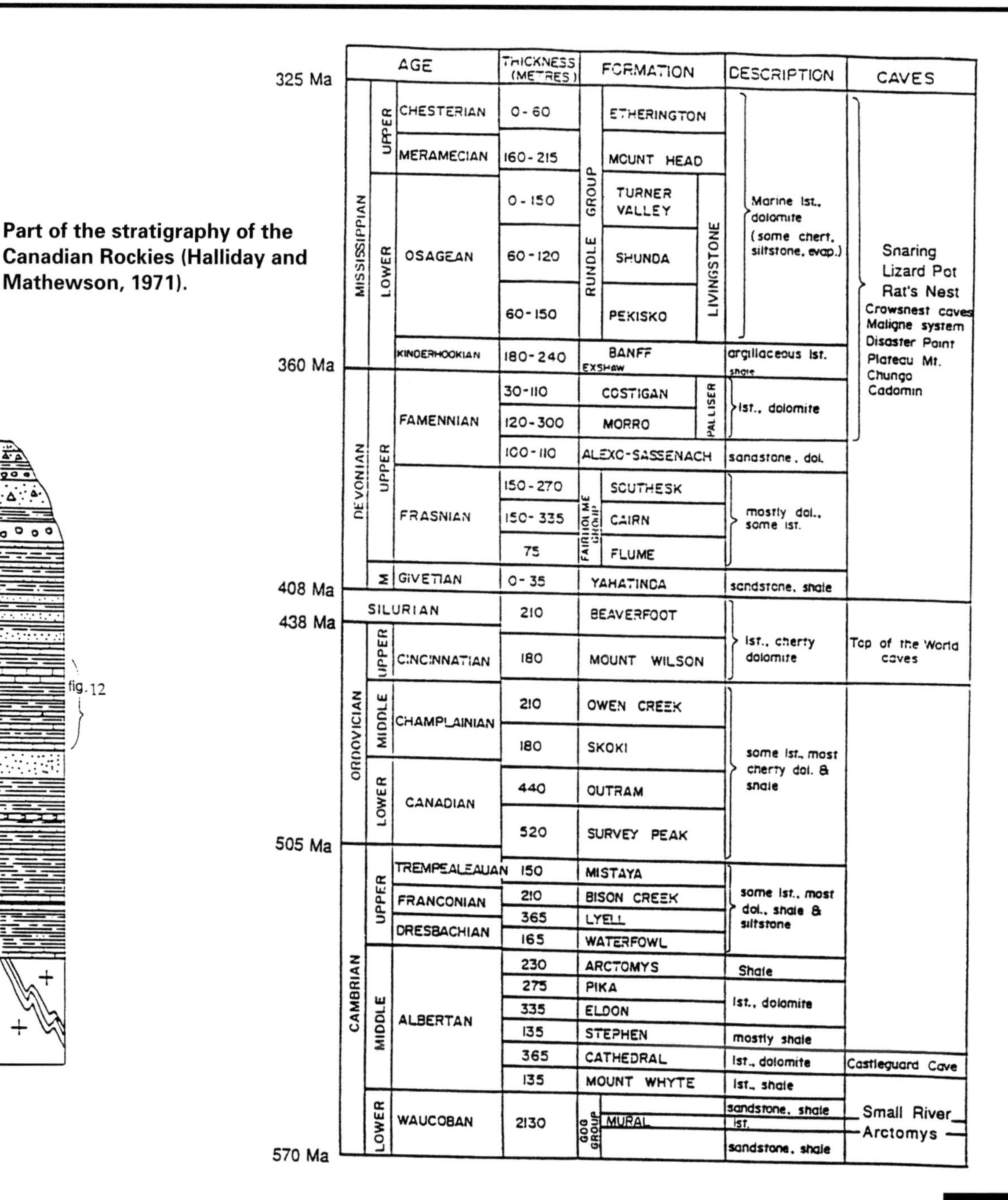

Ma	AGE			THICKNESS (METRES)	FORMATION			DESCRIPTION	CAVES
325 Ma	MISSISSIPPIAN	UPPER	CHESTERIAN	0-60	RUNDLE GROUP	ETHERINGTON		Marine lst., dolomite (some chert, siltstone, evap.)	Snaring, Lizard Pot, Rat's Nest, Crowsnest caves, Maligne system, Disaster Point, Plateau Mt., Chungo, Cadomin
			MERAMECIAN	160-215		MOUNT HEAD			
		LOWER	OSAGEAN	0-150		TURNER VALLEY	LIVINGSTONE		
				60-120		SHUNDA			
				60-150		PEKISKO			
			KINDERHOOKIAN	180-240	BANFF	EXSHAW		argillaceous lst. / shale	
360 Ma	DEVONIAN	UPPER	FAMENNIAN	30-110		COSTIGAN	PALLISER	lst., dolomite	
				120-300		MORRO			
				100-110		ALEXO-SASSENACH		sandstone, dol.	
			FRASNIAN	150-270	FAIRHOLME GROUP	SOUTHESK		mostly dol., some lst.	
				150-335		CAIRN			
				75		FLUME			
		M	GIVETIAN	0-35		YAHATINDA		sandstone, shale	
408 Ma	SILURIAN			210		BEAVERFOOT		lst., cherty dolomite	Top of the World caves
438 Ma	ORDOVICIAN	UPPER	CINCINNATIAN	180		MOUNT WILSON			
		MIDDLE	CHAMPLAINIAN	210		OWEN CREEK		some lst., most cherty dol. & shale	
				180		SKOKI			
		LOWER	CANADIAN	440		OUTRAM			
				520		SURVEY PEAK			
505 Ma	CAMBRIAN	UPPER	TREMPEALEAUAN	150		MISTAYA		some lst., most dol., shale & siltstone	
			FRANCONIAN	210		BISON CREEK			
				365		LYELL			
			DRESBACHIAN	165		WATERFOWL			
		MIDDLE	ALBERTAN	230		ARCTOMYS		Shale	
				275		PIKA		lst., dolomite	
				335		ELDON			
				135		STEPHEN		mostly shale	
				365		CATHEDRAL		lst., dolomite	Castleguard Cave
				135		MOUNT WHYTE		lst., shale	
		LOWER	WAUCOBAN	2130	GOG GROUP			sandstone, shale	Small River
						MURAL		lst.	Arctomys
570 Ma								sandstone, shale	

Features

Karst

Derived from the Slovene name for a cave area in former southern Yugoslavia, karst is "Any terrain where the topography has been formed chiefly by the dissolving of rock" (George Moore & Nicholas Sullivan, 1978).

Karst includes both surface and underground features, although karst areas do not necessarily contain enterable caves. Rockies karst has been glaciated, and is referred to as glaciokarst. By far the most distinctive features of glaciokarst are limestone pavements, where glaciers have removed the soil and weaker rocks, leaving extensive bare limestone surfaces punctuated by occasional erratics — glacially transported blocks. The Snaring karst, Hawk Creek and Castleguard Meadows all provide superb examples of limestone pavements. Joints and fractures in the limestone pavement become enlarged by solutional weathering and are known as grikes. The blocks of bedrock between the grikes are called clints. Smaller solutional features on the surface of the limestone pavement are collectively known by their German name as karren. Rinnenkarren are the runnels that form in the sides of clefts or on steep limestone faces, i.e., Burstall Slabs near Burstall Pots. Kluftkarren are clefts. Examples in the Upper Cathedral limestones at the north end of Castleguard Meadows are 10 cm wide and more than a metre deep (Dalton Muir & Derek Ford, 1985).

Freshly exposed glaciokarst often show striations left by rock fragments being dragged across pavement surfaces by ice. These features can be observed in the vicinity of Porcupine Cave below the Small River Glacier. In some instances, large cobbles of approximately 30 cm in diameter are still sitting at the end of the striations they cut.

Other large features common to all karst areas are depressions known as sinks, formed where streams sink underground. Some leave dry valleys and dolines, which are closed depressions formed by bedrock solution and collapse. Many sinks are quite small, just modified dolines, i.e. the numerous examples on the benches around Camp Caves below the Andy Good Col. Other

Rillenkarren, Hawk Creek karst.
Photo Mary Day.

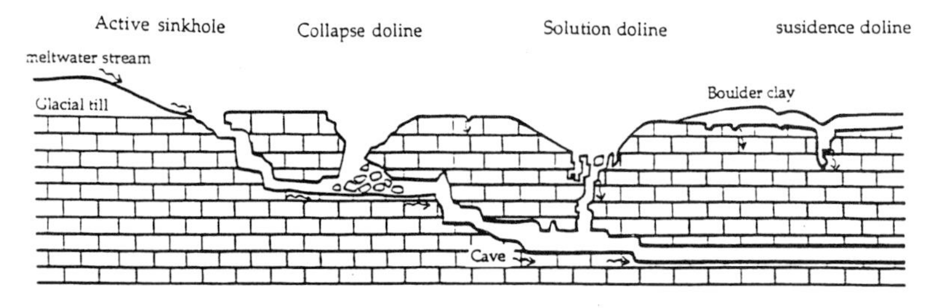

Combination of solutional activity, erosion and collapse. Subsidence dolines are formed where soil and boulder clay subside into fissures below (Tony Waltham, 1987).

sinks are more dramatic, with streams plunging into large shafts i.e. White Hole and Moon River, or pouring into cave entrances i.e. Fryatt Creek Cave.

Sinks are often geologically controlled by water running over impermeable rocks and sinking at the limestone contact i.e. Mad Dog and Arctomys caves. Many of the shafts found in the Rockies are fossil sink points that once took meltwater streams. Over 30 such shafts in the Hawk Creek karst mark stages in the recession of a glacier that has now retreated to the head of the valley just below Mt. Ball.

All karst areas in the Canadian Rockies have seen sequential periods of glaciation that have doubtless removed, deroofed, and buried many caves beneath tons of glacial debris. Examples of this destruction abound, with truncated cave passages heading out of mountainsides, and promising bedrock passages blocked by breakdown close to the surface. Camp Caves and Shattered Illusions provide solid evidence for glacial destruction of cave passages. There are some indications that sections of canyons, such as the Maligne Canyon near Jasper, may be cave passages laid bare by glacial erosion (Ben Gadd, 1986).

Relict, high caves may also be formed owing to glaciers creating high water tables. When the glaciers retreat, the water table falls rapidly, leaving the cave high and dry. Often this occurs before any vadose development of the cave passage has had time to take place, as is the case with Rat's Nest Cave.

Ben Gadd in *Handbook of the Canadian Rockies* makes an important distinction between surface karst and karst regions. "Any area underlain by caves can be called karst, in which case the Maligne system in eastern Jasper National Park is probably the biggest and certainly the longest karst region known in the Canadian Rockies. But the landscape of Maligne valley gives few clues at the surface as to what lies beneath. Aside from the sinks in Medicine Lake and the springs lower in the valley there is little karst expression there. The landscape in places such as Castleguard Meadows and the Snaring Karst, on the other hand, is marked by many sinks, fissures and shafts that show how and where water is going underground. This is surface karst — a rare phenomenon at our latitude, where millions of years of glaciation have carved up the landscape, removing entire cave systems along with relatively delicate surface karst features."

Springs

Many cave entrances are fossil springs abandoned as the water table dropped. Some still resurge occasionally, especially those having sumps close to the entrance, i.e. Castleguard Cave, Steaming Shoe Cave and Jaw Bone Cave. More often as the water table fell, new lower springs were created and they now drain that part of the cave system only enterable by cave divers. Springs occur in a number of situations, sometimes resurging high on mountainsides (Icefall Brook) and in the sides of canyons (Maligne), but more often they are hidden in riverbeds, buried beneath glacial till, or by sediments in the bottom of lakes (Crowsnest).

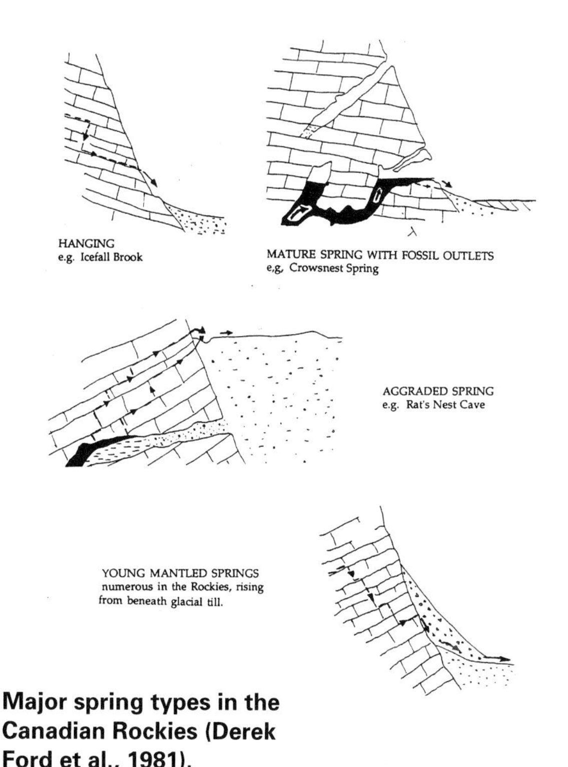

Major spring types in the Canadian Rockies (Derek Ford et al., 1981).

Two main types of springs can be recognized in the Canadian Rockies by temperature. Thermal springs — those with temperatures above the local mean-annual air temperature — only occur in mountainous areas where fault lines allow the deep circulation and geothermal heating of water. Thermal springs with temperatures greater than 37°C are commonly known as hot springs, i.e., Cave and Basin Hot Springs, Miette Hot Springs and Radium Hot Springs. Enterable caves are sometimes formed in travertine deposits at the resurgence sites of these springs, but are rarely of any extent i.e. Cave and Basin and the cave at Miette Hot Springs that collapsed many years ago. Hot springs show seasonal variations in temperature, tending to be hottest near the end of the winter and coolest following snowmelt or periods of heavy rainfall. However, many are invariable.

Cooler examples, where the water temperature is less than 37°C but more than 5°C above the local mean-annual air temperature, are known as warm springs, and include Crowsnest Spring and Turtle Mountain Spring (Crowsnest Pass), The Paint Pots (Kootenay Park), Mount Fortune (Spray Lakes) and the Cold Sulphur Spring east of Jasper.

Cold-water springs are numerous in the southern Rockies, and are a good indicator of possible enterable caves in the mountains above, especially where large rivers pass through major mountain breaks. Some of the larger cold springs by volume of discharge include Castleguard Springs below Castleguard Cave, Maligne Springs below Maligne Canyon, and Karst Spring near Watridge Lake in Kananaskis Country.

All karst areas in the Rockies have springs associated with them.

"Springs are points of natural, concentrated discharge of groundwater, at a rate high enough to maintain flow on the surface. The spring discharge commonly represents rain or snowmelt that has entered the ground a number of years earlier, at a higher elevation some distance away." R. van Everdingen, 1991.

How Caves Form

"When engineers drive a tunnel or open a mine, they work as rapidly as they can, amid a deafening clatter of machinery punctuated by the detonation of powerful explosives. Nature, by contrast, hollows out limestone caves in almost complete silence, usually with no other tool than slowly moving, weakly acidified water. However, nature has, so to speak, all the time in the world." George Moore & Nicholas Sullivan, 1978.

Most caves are initiated beneath the water table, which may be lowered into them as their reservoir capacity is enlarged. Later caves may then be formed in the drained "vadose" zone above them. This is especially common in alpine areas because glacial action diverts streams away from the original sinkpoints to create the younger, steep-to-vertical shafts that are typical of alpine karst globally.

Cave mineral formations are created later in depositional phases. Thus we know that larger caves are much older than the mineral formations they contain.

The solutional processes by which caves are formed consist of a number of chemical reactions between water and calcium carbonate, the dominant mineral which forms limestone. Simply put, "The carbon dioxide combines with water to produce carbonic acid, which in turn attacks the calcite and divides it into soluble ions. A cubic metre of water exposed to air containing 10% carbon dioxide, if kept in contact with limestone until the reaction ceases, can dissolve about 250 grams of calcite." (George Moore and Nicholas Sullivan, 1978.)

The limestone removed in producing a cave is the endpoint of a complex process. Carbon dioxide from the atmosphere, and more importantly the soil mantle, is the essential agent that combines with meteoric water (rainfall and snowmelt) to produce the weak carbonic acid that then goes on to dissolve the limestone. The dissolution process is also reversible under certain conditions. When this occurs, calcite is precipitated and cave calcite or aragonite formations result.

Although the production of carbonic acid provides a useful explanation for cave development under soil mantles, what of cave development in alpine regions where soil mantles are thin, or for the many karst areas in the Canadian Rockies where soil is totally absent? Recent research (Steve Worthington, 1991) indicates that many rivers draining karst areas in the Canadian Rockies are rich in sulphates. It is believed that sulfuric acid, produced by the dissolving of sulphate minerals in the limestone, or perhaps more significantly by the oxidation of sulphides in adjacent shales, may be a major dissolving agent of calcium carbonate. It is also important to note with respect to glacial conditions that cold water can dissolve greater amounts of CO_2 than warm water. The concentrations of H_2SO_4 in Rockies waters are never sufficient to explain cave enlargement. However, they may help in the critical initiation phase when caves are very small, when the acid makes a proportionally larger contribution.

Measured CO_2 concentrations in Rockies waters imply atmospheric concentrations of between 0.01 and 3.0% at their sources in the air, beneath glaciers or in soils. These are comparable to many warmer parts of the World. The 10% CO_2 that George Moore and Nicholas Sullivan give is realistic only in certain hot springs water. Most Rockies water has encountered atmospheric CO_2 in the range 0.03–0.3%.

Passage Formation

Notable caves are initially formed beneath the water table where a fine network of joints and partings in the bedrock are exploited by moving subterranean water. Preferential use of a route by groundwater may eventually create a cave-sized conduit, the form and direction of which is dependent on weaknesses in the bedrock. Some cave passages occur where limestone beds abut an impermeable rock, as is the case with Arctomys Cave, which is sandwiched between sandstones and shales. Groundwater will also often exploit bedding plane weaknesses, as with Pinto Lake Cave, or flow along joints, as with Lizard Pot. Faults, though less important in dictating cave formation than joints or bedding plane weaknesses, will sometimes redirect the course of a cave passage for a short distance, i.e. Rat's Nest Cave. If the limestone in which a cave is formed is homogeneous and thickly-bedded, a passage developed beneath the water table will take the form of a large pipe, solutional corrosion being equal on all surfaces. Such has been the case with The Subway in Castleguard Cave. More commonly, a joint or bedding plane causes differential enlargement to take place, creating an ellipse-shaped passage. A fine surface exposure of an elliptical tube lies just above the resurgence to Fryatt Creek Cave.

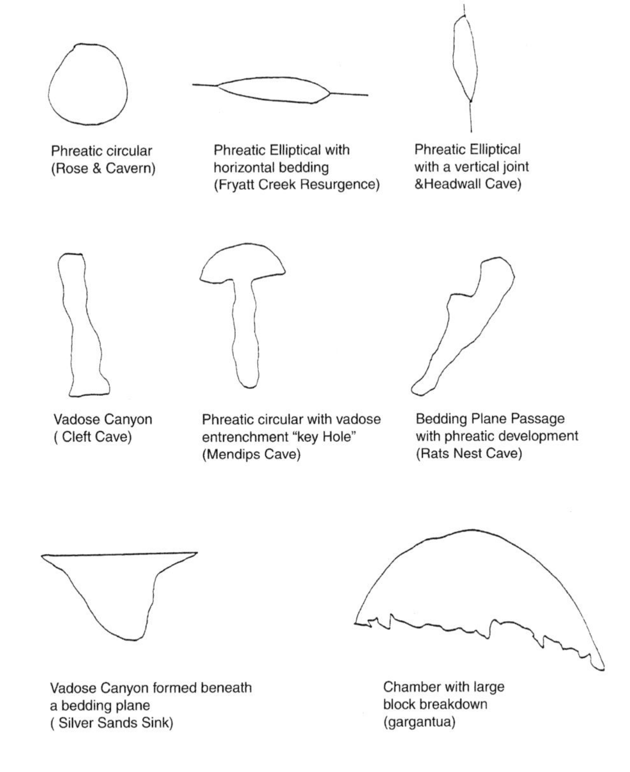

Common Cave Passage Shapes in the Canadian Rockies.

Passages formed above the water table develop in much the same manner as surface streams, with water down-cutting under the force of gravity. They tend to have a long and narrow canyon-like appearance. The First and Second fissures in Castleguard Cave provide good examples of canyon passages. A drop in the water table will often produce a combination of these two basic passage shapes: a distinctive keyhole design. Good examples of a keyhole passage can be found in many Rockies caves, notable examples being Mendips Cave and Rat's Hole Cave at Crowsnest Pass. Caves formed exclusively beneath the water table and having the characteristic round section are said to be phreatic; those formed above the water table and having a narrow rift-like section are said to be vadose. Caves such as Yorkshire Pot and Castleguard contain passages of both types, the position of the cave relative to the water table having changed over time.

Features of vadose passage development are shell-shaped depressions ranging from a few centimetres to 1 m across called solutional scallops. They are caused by moving water exploiting small irregularities in the rock surface. The steeper face of the scallop indicates the upstream direction, and the size of the scallop indicates the velocity of the water flow. The smaller the scallop, the faster the flow (on a logarithmic scale). Particularly fine examples of scallops can be seen in the Big Dipper and Roller Coaster passages of Gargantua and Yorkshire Pot.

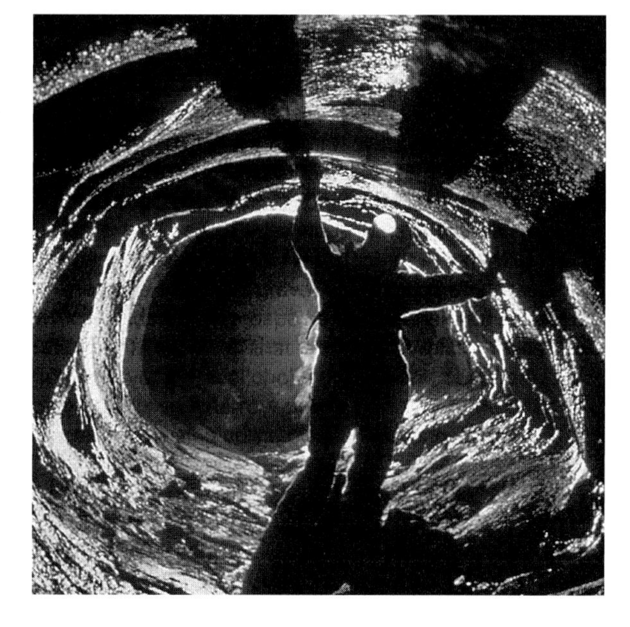

Top left: Preatic tube in Wapiabi Cave. Photo Ian McKenzie.

Top right: A keyhole passage in one of the terminus passages of Castleguard Cave. Photo Tom Miller.

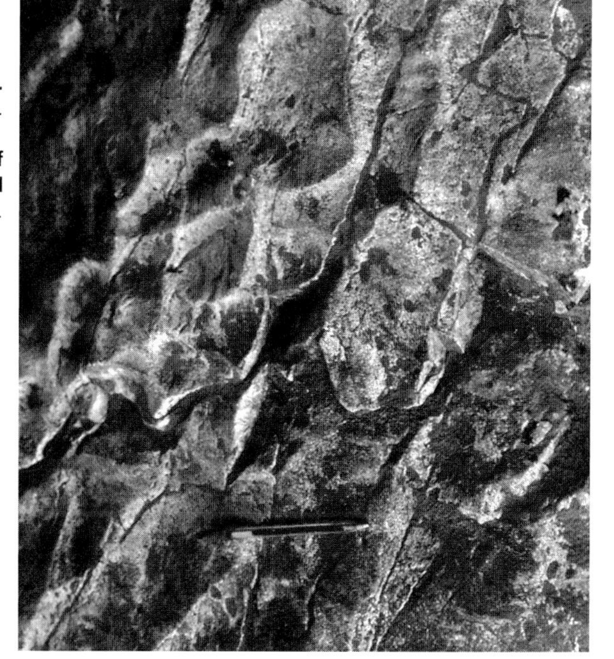

Right: Scallops in the entrance to Rat's Nest Cave. Photo Jon Rollins.

Breakdown Chambers

"The passage kept growing and growing in size. We passed a large lower entrance and still the passage grew. Finally, feeling very small, we stood among large boulder breakdown in a passage 30 m wide and 25 m high. It was mind-boggling, so we called it Boggle Alley." Tich Morris, 1970.

Especially large sections of passage, known as chambers, are formed through the collapse of rock from the walls and ceiling of a cave, often caused by a lowering of the water table and loss of buoyant support to the rock. The undercutting and removal of material may then continue the process by vadose water flow. Large amounts of collapse material, commonly called breakdown, litter the floor of Boggle Alley in Gargantua Cave, Canada's largest natural subterranean cavern. Other caves having large chambers are Fang, Cadomin Cave and Ice Chest. These features are not common in Rockies caves.

Shafts

"One steamboat, two steamboat, three steamboat, four steamboat, FIVE steamboat, and there it went again, the deep sonorous, rolling thunder. Suddenly, none of us wanted to stand close to the edge. They looked at me, envious yet relieved about not being first down this pit. How can I get out of this?" Ian MacKenzie, 1986, on the first attempt to bottom Close to the Edge.

Intense glacial entrenchment has left many cave entrances high above the current valley floors. Water entering these caves — precipitation and meltwater from snow pack and glaciers — seeks out steep bedding planes and joints as gravity and hydraulic pressure drives it on its path to the water table. Many of these bedrock weaknesses have been eroded out to produce some of the most impressive features of Rockies caves: vertical sections of passage known as shafts, pits or pitches. The term "Pot" (derived from Yorkshire) is also sometimes used for caves having round, shaft-type entrances.

Nearly all Canadian Rockies caves have a strong vertical component to their passages i.e. Yorkshire Pot has 130 m of entrance shafts that can only be descended using ropes. In contrast, Arctomys Cave, which descends over 500 m, has numerous sections of climbing, but only three pitches for which ropes are required. Close to the Edge in the Dezaiko Range has an entrance shaft consisting of a 244 m single drop. Many karst areas in the Rockies are dotted with pits and shafts, but unless they have some associated horizontal development or are more that 30 m deep, they are not usually recorded or surveyed. This is the case with the many shafts on the Dezaiko Plateau.

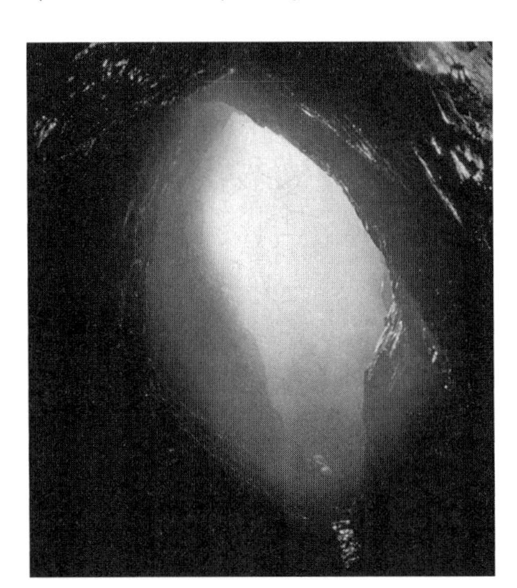

The shaft in Close to the Edge.
Photo Ian McKenzie.

Shafts are not always formed by water flowing down them. Phreatic lifting shafts, such as the 30 ft and 80 ft pots in Castleguard Cave, were formed by water being forced up a weakness under hydraulic pressure. Many of these shafts are magnificent examples of natural bedrock architecture, having fluted, solutional forms with arches and flying buttresses reminiscent of Gothic architecture.

Avens

Avens are vertical, domed features in the ceiling of a cave, often with no enterable passage leading off from them. They are sometimes formed by "mixing corrosion."

Water percolating toward the water table may quickly become saturated with calcium carbonate, or may move so quickly that much of the carbonic acid remains unneutralized. In either instance, when this water mixes with the calcium carbonate-saturated groundwater at the water table, the mixing of two waters that have different carbon dioxide contents can lead to an undersaturated mixture that has excess carbon dioxide, and thus is capable of dissolving more limestone. This explains why most cave formation tends to take place directly below the water table, with a tendency to form horizontal passage, even when the limestone containing the cave is folded or steeply dipping. It is thought that this process — mixing corrosion — plays a part in the formation of avens that sometimes punctuate cave passages, i.e., the 200 ft aven in Castleguard Cave, measured from the base using a hydrogen balloon, and the terminal aven in Larch Valley Cave.

Sediments

Most Rockies caves contain large amounts of glacial sediments ranging in size from large cobbles down to course gravels or silts and fine glacial flour. Material deposited at the snout of any glacier will give a fair indication of the range of sediments to be encountered in caves. In some caves, a degree of sorting has taken place (see Castleguard Cave), the larger material being deposited first, the smallest gently settling out of water left ponded in the cave. During a period of quiescence, calcite deposition may form a crust over the sediments, and a later resumption of water flow may bring in more material, thus adding another layer or sometimes re-eroding existing sediments. An examination of a section through cave sediments can give a good indication of the history of erosion activity on the surface. Sediments can sometimes fill cave passages to the roof, forcing further passage development to occur in the ceiling (paragenesis), or it can close off passages completely. Examples of both exist in the headward complex in Castleguard Cave. Memorable from a caver's perspective are the dry sandy sediments that make some crawl passages pleasant, and the wet thixotrophic clays that suck your boots off, such as in the highly unpleasant terminus passages of Rat's Nest Cave.

Perfectly rounded polished cobbles are sometimes found at the base of phreatic lifting shafts in Rockies caves, as in Castleguard Cave and Pinto Lake Cave. During times of high water these cobbles, which are too heavy for the water to eject, have been continually agitated by water under hydraulic pressure. Often they are found in a matrix of fine gravel that presumably enhances the polishing process.

Fine sediments deposited by glacial meltwater are sometimes laid down in cyclical layers reminiscent of lake varves, and are known as varved clays or rhythmites. These are deposited as the finer sediments settle out of ponded cave water. Examples have been examined in Castleguard, Nakimu and Rat's Nest caves.

Sediments are sometimes reworked by cave water into interesting shapes known as mud formations. Drip holes in Castleguard Cave and small domed "mud patties" in Cadomin Cave are good examples.

Mineral Formations

"Soon we were standing in a fossil roof passage admiring an incredible display of cave pearls. The weariness caused by seven hours of continuous caving evaporated as we scurried from pool to pool finding successively better displays. We were experiencing one of the rare rewards of caving — the discovery of a beautiful grotto after hours of toil through uninteresting passage. We left the pearls behind and promptly discovered a magnificent display of stalactites, helictites and a massive column." Peter Thompson, 1976.

Cave formations or speleothems are the extraordinary-shaped depositional features for which caves are famous. The process by which they are formed is opposite from that that forms the caves themselves. As the carbon dioxide-rich water percolates down through the limestone, it takes the calcium carbonate into solution. When it enters a cave where the carbon dioxide levels are similar to that of outside air, the carbon dioxide diffuses out of the water and calcium carbonate is precipitated out onto the walls. This process can be accelerated if there is effective evaporation, as in cave entrance zones. In fact, evaporates can form even in cave passages with typically high relative humidity (above 95%) if there is a vigorous draft. The Central Grottoes of Castleguard Cave are a good example of this.

The volume of speleothems in a cave depends on the amount of carbon dioxide lost by the waters that enter it, which in turn depends on the amount of organic activity in the soil above the cave. Thus, caves in the tropics, with large amounts of organic material and heavy precipitation, tend to contain more and larger speleothems than cold northern caves. In alpine karst, prevalent in the Rockies, the source of groundwater is often snow and ice-melt solutions that become warm on entering caves. Groundwater can carry increasing levels of calcium carbonate as water temperatures fall.

Because almost all caves are formed in limestones and dolomite, carbonate minerals form the majority of speleothems. Calcite is by far the most common constituent of Rockies cave formations, and occasionally aragonite (a common cave mineral in the tropics), which has the same composition but a different crystal structure called a polymorph. Small deposits of gypsum and hydromagnesite are also present in Castleguard Cave and are occasionally seen in other Rockies caves.

The dating of speleothems in Rockies caves indicates that many formations are at least several hundred thousand years old. Over such a period of time a cave undergoes many cycles of erosion, deposition and quiescence. This can be observed in formations that have undergone re-solution, when the chemical balance of the groundwater has changed from depositional (saturated with calcite), to aggressive (slightly acidic). Material carried in by floodwaters erodes speleothems and impregnates the remaining calcite with clay, often turning it brown. These processes can be observed in speleothems in many Rockies caves.

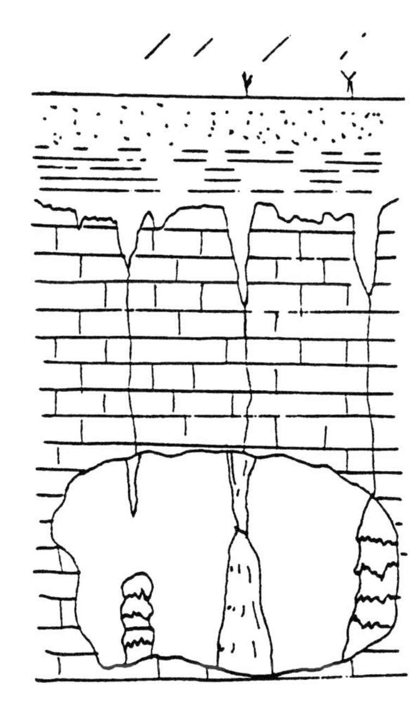

Precipitation of Calcite in a Cave (G. Holland et al., 1964).

The following are the major speleothem types to be found in Rockies caves in order of frequency of occurrence.

Coatings and Crusts

While the larger, more prominent formations such as stalagmites and stalactites tend to be the best known, the light-coloured coatings and crusts that line the ceilings, walls and floors of caves are far more common in the Rockies. In Time Traveler passage beyond the Crutch in Castleguard Cave, pebbles have been coated with gypsum, giving them a freshly washed, shiny appearance. Coatings and crusts tend to favour sloping surfaces common in bedding plane caves such as Rat's Nest, but can also be found in deep frost pockets and shelter caves. Quite detailed and complex, these coatings provide a base for subsequent speleothem growth.

They often form a layer over clastic sediments where the calcite acts as cement. In Provenance Cave, for example, a layer of calcite-cemented pebbles has been left suspended across a passage, the underlying glacial fill having been washed away. Similarly, in the Rose & Cavern Cave, remnants of false calcite floors have been left hanging following removal of clastic material. The collapse of such a false floor over a deep fissure in Castleguard nearly led to disaster for those cavers camping nearby.

Cave Coral (Popcorn)

Often found in cracks and fissures in drier sections of a cave, this coralloid consists of small knobby clusters with individual knobs typically 5–10 mm in diameter. Sometimes stalactites form, hanging from the coral, good examples of which can be seen near the entrance of Her Majesty's Cave. The lack of any central canal suggests that cave coral is formed by seepage followed by evaporation, rather than by dripping water.

Flowstone
Flowstone is probably the largest form of speleothem commonly found in Rockies caves i.e. the Lower Gallery in Cadomin Cave. It forms where thin films of water flow over large surfaces, not only over bedrock, but also over clastic sediments. If these sediments later become washed out, as might occur in a glacial recession, the flowstone can be left suspended, forming canopies that later may become decorated with stalactites hanging from the lip. Beautiful examples of these canopies can be seen in Rat's Nest Cave beyond the Birth Canal.

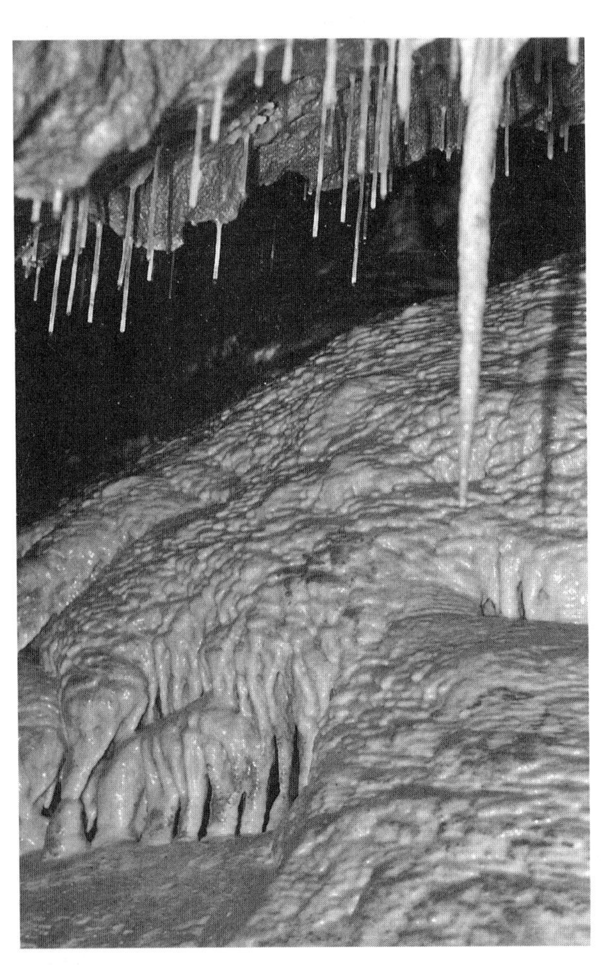

Curtains
Also known as draperies, these formations are created when water flows down the inclined ceiling of a chamber, depositing curtain-shaped formations. These are often translucent, and may display discolourations that correspond to trace element variations in the groundwater over time (streaky bacon). Undulations in the ceiling can cause the curtains to furl, and when the ceiling angle lessens, a stalactite often punctuates the formation. Rat's Nest Cave provides many good examples of curtains.

Stalagmites
These are formed by splashing water and thus have a larger diameter, and a more rounded appearance than stalactites. They occur in a puzzling array of forms, the most common being a flat-topped cone. Tiered forms are a product of a combination of drip formations caused by changes in deposition rates. When examined closely, many tiers are seen to be composed of tiny rimstone dams that have caused a succession of outward growths. If deposition is uniform over time, a stalagmite will have a cylindrical form. More often, however, owing to climatic changes, the deposition rates change. An increase in deposition can produce a stalagmite that thickens upward; a decrease can produce large stalagmites with a column on top. Often the base of a stalagmite will be thickened by the growth of other formations. Proto-stalagmites take the form of small discs of calcite, often seen on cave floors.

Stalactites
Stalactites are probably the best-known cave formations and are quite numerous in Rockies caves. In section, most stalactites display the central canal that shows they started as tubular stalactites or straws. The conical form evolves as calcite deposition occurs on the outside of the straw, crystal growth being perpendicular to the central tube, which in contrast is formed of longitudinally-aligned crystals. Stalactites can grow to spectacular sizes, some very fine specimens being located in the Grottoes and Holes-in-the-Floor passages of Castleguard Cave.

Columns are an advanced form of stalactite, created when a formation encounters its floor counterpart, a stalagmite. Floor-to-ceiling straws in the Subway passage of Castleguard form slender, delicate columns.

Far left: Flowstone and stalactites. Photo Jon Rollins.

Left: Curtains and stalagmites in Rat's Nest Cave. Photo Dave Thomson.

Redissolved stalactites in Cleft Cave. Photo Jon Rollins.

Soda Straws

A soda straw — the proto-form of the stalactite — appears quite different from its later robust conical-shaped form. Chas Yonge explains: "As seepage water falls from the cave ceiling, CO_2 is liberated from the water and an annulus of calcite is precipitated at the upstream portion of the drop. In time, successive joining of these calcite rings results in a hollow tube or 'soda straw.'"

It often forms in large groups or "clouds" such as those in the Straw Gallery of Arctomys Cave, and can grow to spectacular lengths. Often they break under their own weight, possibly at flaws in their crystal structure, and can be found littering the floor beneath actively growing straws, even in newly-discovered passages. Other possible explanations for natural breakage include passage invasion by meltwater streams and cataclysmic events such as earthquakes. Anomalies sometimes occur in collections of soda straws, such as when a straw has a carrot-shaped irregularity hanging from it. These are thought to form when the central canal of a soda straw becomes blocked, and calcite deposition occurs around the blockage.

Soda straws and stalactites in Rat's Nest Cave. Photo Dave Thomson.

Flags

Stalactites with sections of horizontal growth are known as "wind-flagged," the horizontal growth or flag occurring in the upwind direction owing to evaporation. Some fine examples occur in Castleguard Cave, where the same stalactite has flags growing in two directions, indicating possible long-term changes in the predominant wind direction.

Rimstone Dams

These are common micro-features of flowstone surfaces, each successive miniature dam impounding a small pool of water. Usually only a few centimetres high in Rockies caves, they are thought to form through carbon dioxide loss and associated calcium carbonate deposition caused by the agitation of water as it flows over rapids.

Spar

This crystal growth occurs in dogtooth or nailhead forms, and is a pseudomorph of calcite that forms in pools. The individual crystals take the form of combinations of flat rhombohedra and prisms that are visible to the naked eye.

Helictites

Usually passed unnoticed in Rockies caves owing to their small size — commonly only 1 mm in diametre — helictites have the appearance of a short convoluted thread, often with branches, emerging from the cave wall or ceiling. A fine central capillary tube conducts moisture to the extremities of the helictite. Often they can be observed growing from the sides of stalagmites. Sometimes they develop into soda straws, possibly in response to an increase in water flow.

Helictites have long fascinated speleologists. Their chaotic growth and the way they appear to defy gravity has tested all those who seek a logical explanation. Many theories have been proposed for their chaotic growth patterns. According to Chas Yonge, "The most popular theory at present involves a primary mechanism of hydrostatic pressure and capillary action combined with other factors such as evaporation, air flow, impurities in solution, water supply and intercrystalline seepage. A porous rock face, usually encrusted in carbonate, appears to be required for the initiation of helictite growth. Branching of helictites likely occurs where there is a blockage in the central canal forcing water out sideways and thus developing a new limb."

Helictites can be seen in Rat's Nest Cave, and deep in Castleguard and Arctomys caves.

Soda straw with helictite in Rat's Nest Cave. Photo Dave Thomson.

Cave Pearls

As the name indicates, these are small, loose, pearl-shaped formations, commonly found grouped together in "nests." In the Rockies they range in size from tiny seeds to 2 cm-diameter spheres. Like helictites, the formation of cave pearls has created much speculation amongst speleologists, especially when a nest of cubic cave pearls was discovered in Castleguard Cave. Similar to oyster pearls, they are thought to start with a single grain. Agitated in water over time, the grain becomes coated evenly with calcite, but remains a free body. The more rare cubic pearls are thought to have started as round (as indicated by cross-section analysis), but attained their cubic form from restricted movement in a crowded nest.

Moonmilk

Derived from the German for gnome (mon), thus gnome's milk, moonmilk is a white paste-like substance, powdery when dry, that is sometimes found on passage ceilings and walls. It is formed by precipitation from groundwater, but for some reason the mineral crystals never become large. In the Rockies, moonmilk is usually made of calcium carbonate, but in warmer caves can be composed of magnesium or dolomite. Its plasticity is due to its high water content — 40% to 80%.

Its occurrence near cave entrances in the Rockies suggests that in some instances bacteria, fungi or algae may be involved and that moonmilk may have biological origins. Most calcite moonmilks that have been studied (including those in Nakimu Caves for which the caves are famous) are fixed by microorganisms.

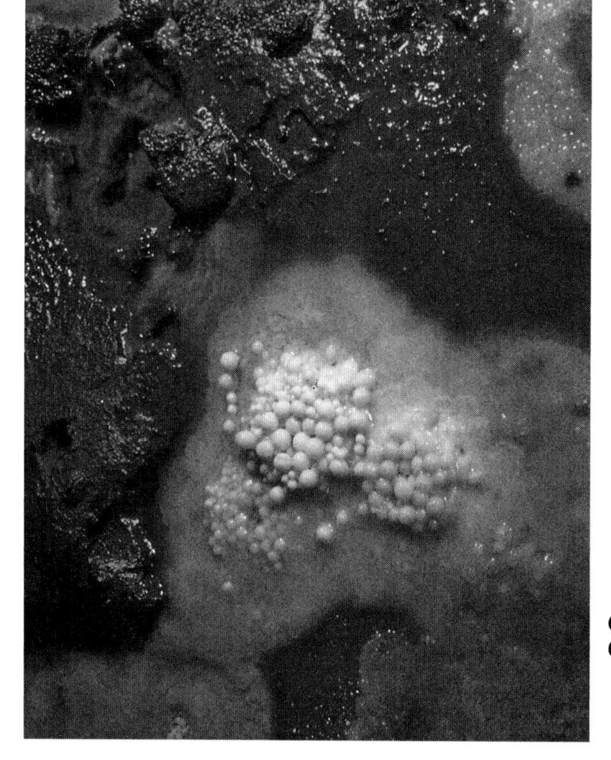

Cave pearls in Rat's Nest Cave. Photo Jon Rollins.

THE CAVE ENVIRONMENT

"The Sarcees tell of a monster wedged in a mountain cave. This creature was known as the Wind Maker. His ears were longer than the tallest spruce trees on the Sarcee Reservation, and when he was asleep, as he was most of the time, breathing was quiet and ears lay motionless along his back. But when the mountain monster was awakened, he would fan and beat his ears so violently that the hot breath from his nostrils was driven across the foothills and plains to send the temperature upward. Unfortunately, the beast would fall asleep again and his ears would cease to drive the warmth eastward, and the Chinook would end as suddenly as it had begun." EVDS, 1988.

Temperature

There is little doubt the originators of the tale described above had at one time visited a Rockies cave in the winter, and experienced its warm, moist breath. Caves retain the warm summer heat and receive varying degrees of geothermal heat, their rock walls providing excellent insulation. Beyond the entrance zone, caves of any extent provide year-round stable temperatures above freezing, even those located in an alpine environment where the annual surface temperature is below freezing. For example, the mean temperature for the Columbia Icefield is –2.1°C, yet the midsection of Castleguard Cave beneath Castleguard Mountain (entrance elevation 2000 m) has a temperature of 3.5°C. Lower caves experience warmer temperatures: Rat's Nest Cave (elevation 1480 m) with a mean surface temperature of 3°C has a temperature of 4.5°C, and is probably the warmest cave in the Rockies. Certainly, when you enter a cave such as Rat's Nest when the surface temperature is –25°C, the warm moist air with its heady scent of pack rat urine might be mistaken for the Wind Maker's breath.

An exception to this are ice caves (see below).

Humidity

As with temperature, the relative humidity increases in Rockies caves the farther in one goes until it approaches 100% and the air is saturated. Closer to the cave entrance the relative humidity falls, and caves such as Castleguard, which draft inward in the winter, can have extremely dry entrance zones. Rat's Nest Cave often drafts out in the winter, and the warm saturated air produces a thick layer of hoarfrost on the roof of the entrance alcove.

When warm air rises out of a cave entrance in the winter, it is often said to be "steaming," i.e. Steaming Pits near Ya-Ha-Tinda Ranch.

Air Movements in Caves

"In exploring virgin caves, the discovery of a draft is often greeted with joy and anticipation of extensive passages ahead. When digging out choked systems, a draft is often regarded as sure proof that the dig "must go," and used as bait to lure unwilling helpers to the site." Trevor Atkinson

Although the temperature is above freezing, many Rockies caves are made uncomfortable by the winds that blow through them. Most caves have a draft; in fact, a major clue to finding the way during cave exploration is to search for air movements. These can become quite strong where long cave systems have a constriction in their passages, as at the base of the 80 ft Pot before the Subway in Castleguard Cave, and just before entering Interprovincial Way in Gargantua.

At least three different mechanisms can cause air flow in caves. In many of the larger caves with complex patterns of passages at differing elevations, two or all three may operate simultaneously.

The first occurs in all caves and is caused by changes in barometric pressure between the surface and the passages inside the cave. This often occurs on a daily basis, resulting from differences in air temperature between day and night (which is also responsible for the anabatic and catabatic winds that characterize mountain valleys). During the day when the air warms and rises, causing a drop in air pressure, the air moves out of the cave. At night when the cooler air sinks and causes a rise in air pressure, the air flows into the cave entrance. Changes in barometric pressure owing to storms can be superimposed on this diurnal regime.

Chimney-effect winds are usually stronger and are caused by the vertical form of the cave combined with differences in the temperature and humidity of the air inside and outside of the cave. The creation of these winds requires a cave to have more than one entrance at different elevations, which is commonly the case in the Rockies, although other entrances may not be known or may be too small for human access. The cave functions like a chimney drawing the colder air in through the lower entrance in the winter, warming it, and causing it to rise out of the upper entrance. In the summer the wind is reversed, the cave temperature being colder than the outside air. In the spring and fall the average outside temperature may be close to the cave temperature, and the draft may reverse diurnally. Caves exhibiting these chimney effects include Rat's Nest and Castleguard, neither of which are known to have a second entrance. In the case of Rat's Nest Cave, it is probable that another entrance exists, the considerable overburden of rock above the explored end of the cave indicating much undiscovered passage. Castleguard Cave ends deep beneath the Columbia Icefield, and although an entrance in this area seems highly improbable, it is possible air may be exchanged via deep crevasses in the glacier above. There are probably many tiny entrances/exits from the cave into the base of the icefield that will permit air flow, except, possibly, when the ice base is saturated with water.

Some caves known as breathing caves have drafts that move in and out regularly, sometimes over periods of a few minutes or less. George Moore and Nicholas Sullivan suggest that this is because of air turbulence across the cave entrance that alternately compresses and expands the air inside the cave, much like that produced by blowing across the neck of a bottle.

Water

Caves with no running water are called fossil caves, and those with stream ways are called active caves. Although formed by water, most enterable Rockies caves are now high and dry, well above the water table. A few contain running water supplied by rainfall and snowmelt. These streams sometimes use existing passageways, forming "under fit streams" that, barely noticed, trickle along the bottom of rift passages, or meander in small channels cut in sediments on the floors of phreatic tubes. Others have cut new juvenile stream passages, sometimes targeting larger existing passages for short distances, but usually leaving them again, drainage basins and hydraulic gradients having been altered over time by glaciation. Examples of juvenile and misfit streamways can be seen in Gargantua and Yorkshire Pot, and in the fissures of Castleguard Cave.

A few caves at glacial margins still take large volumes of water, i.e., Porcupine Cave, and some juvenile stream sinks are enterable for a distance, the small passage size forcing cavers into close proximity with the water, i.e., Yohole, Silver Sands Sink, Assiniboine Sink, Mad Dog and Mistaya caves. Some stream-sink caves, though restricted in development by impermeable beds, still take significant volumes of water in their main passageways, as is the case with Arctomys Cave. Although not common in the Rockies, caves with significant amounts of water running in the main passageways are exciting to be in. The usually somber cave environment is enlivened by the roaring of waterfalls and the gurgling and splashing of rapids. Nakimu is a fine example of a dynamic cave: in July when snowmelt enters the cave, the passages shake with the intensity of the water flow. Even small stream ways can sound quite loud in enclosed passageways, and many explorers have been disappointed when the large roaring torrent they heard in the distance turned out to be an innocuous little thread of water running over a drop.

Many Rockies caves have sumps that usually take the form of a passage descending into a pool of water. These occur when cave passages encounter the local water table, or, more often in Rockies caves, where a U bend in a passage has become filled with water (perched sumps). The characteristic dip of the massive carbonate platforms in the Rockies allows water to circulate deeply under the water table and some sumps can be large and very deep, i.e., Crowsnest Spring. Sumps usually mean the termination of exploration of a cave, although occasionally they can be bypassed by dry passages, i.e. Yorkshire Pot, or one can return at a drier time to find the sump drained. Some caves, where the main passage has formed along the strike, contain several sumps terminating the down-dip passages i.e. Rat's Nest Cave. A few Rockies sumps have been dived and occasionally lead to an air-filled passage that in turn usually leads to another sump.

Ice

Although most caves have year-round temperatures above freezing, many alpine caves contain extensive ice formations in their entrance zones. Through isotopic analysis it has been ascertained that ice has formed from two main sources: water vapour that has desublimated and seepage water (Bill MacDonald, 1992). Work to date concludes that none of the ice masses in Rockies caves are old enough to be glacial remnants of the last ice age (Charlie Brown and Tom Wigley, 1969, Bill MacDonald, 1992).

Ice in Rockies caves can be separated into two main types. The first is ice that reforms in cave entrances each winter. Because air in caves is saturated with water vapour, when this saturated air meets the freezing surface air it can cause extensive hoarfrost and ice formations at cave entrances. Water in pools and water dripping from the roof of cave entrances also freezes, the resultant ice formations bearing a marked similarity to calcite stalactites and stalagmites found farther inside. Good examples of winter ice occur in Castleguard Cave, where residual water freezes on the passage floor for a considerable distance into the cave, forming the famous Ice Crawls. Just inside the entrance, where the passage walls restrict the formation of ice, volcano-shaped formations or pingos form where the ice is upheaved. Because winter air is so dry in this area, the ice is hard and superbly clear, and few cavers can resist pausing for a moment as they slide through to stare at the beautiful lines of elliptical bubbles trapped within the ice. Often the surface is covered by a fine film of glacial flour deposited by the outside breeze. Farther inside, local seepage forms floor-to-ceiling ice columns, which later in the spring have razor-sharp edges formed by evaporation from warm outside air blowing into the cave. Similar ice columns are found in Her Majesty's Cave, Fryatt Creek Cave and in many other Rockies cave entrances in the winter. Because Rat's Nest Cave is unusually warm (4.5°C), during cold snaps the entrance is heavily decorated with hoarfrost and numerous rounded ice "stalagmites and stalactites."

Whereas European scientists term any cave with seasonal or perennial ice an "ice cave," in North America, only caves that have ice in them all year round are known as ice caves. Stuart Harris of the University of Calgary defines an ice cave as "a cave where the rock temperature is partly or wholly below 0°C for more than one year."

Bill MacDonald in his 1989 master's project with the CRE program at the University of Calgary identified five different mechanisms for ice accumulation in Rockies caves:

Hexagonal ice crystals in Ice Chest. Photo Jon Rollins.

1. Permanent Ice A high-altitude cave, where the average annual temperature of the surface is less than 0°C, may have a temperature below freezing, and, depending on its size and configuration of entrances and chambers, may contain permanent ice. An example of such a cave is Serendipity at Crowsnest Pass.

2. Relict Permafrost Stuart Harris, who has carried out detailed work on the permafrost in Plateau Mountain Ice Cave, has found evidence that the cave lies in a zone of relict permafrost that was left as a result of an earlier cold period. It was found that in the Rocky Mountains of southern Alberta, areas of low winter precipitation have continuous permafrost at lower altitudes than those areas with a deep snow pack. Plateau Mountain Ice Cave is famous for the thick layers of ice crystals that cover the walls, some of them hand-sized hexagonal plates. Unfortunately, temperature rises recorded in the permafrost surrounding the cave is now causing the ice to melt (Bill MacDonald, 1991), although alterations in the ice plugging the rear of the grotto, prompting a vigorous through draft, is probably also responsible. Ice Chest Cave at Crowsnest Pass is also thought to be contained in relict permafrost.

3. Snow Trap Caves that have a shaft configuration may collect and hold winter snow. This snow builds up, and through partial melting and subsequent refreezing, becomes compacted into ice. Caves such as Disaster Point near Jasper and Provenance Cave in the Mount Ball karst are good examples of snow trap caves.

4. Cold Trap The phenomena of the cold trap cave is explained by the "Balch Effect," whereby cold dense air accumulating at the lowest point of a cave can maintain a temperature below freezing for over a year. This effect relies on an absence of air movement such as might occur in a blind pit cave. Coulthard, with its north-facing cold air "lake" provides an excellent example of this type of cave, coined "Cold Sock" by Tom Wigley.

5. Cold Zone Tom Wigley and Charlie Brown, cavers from the McMaster Karst Research Group, surmise that ice forms in caves owing to the creation of a cold zone a certain distance from the cave entrance. The explanation for the cold zone lies in the unequal cooling and warming of air in this section of a cave. In the winter the outside air and the dewpoint temperatures are below the cave temperature, and the cold zone air is cooled by evaporation and by direct contact with the colder air. In the summer when the cave is above the dewpoint but below the outside air temperature, the cold zone air is cooled by evaporation and warmed by direct contact with the outer air. The imbalance arises from the greater cooling effects of evaporation. Once ice forms in a cave, the rock temperatures stay below freezing until all the ice has melted (Charlie Brown and Tom Wigley, 1969).

The cold zone phenomena and the effects of evaporative cooling combine to produce the enormous iceberg-like formation that fills Cathedral Chamber in Serendipity. Even in caves at low altitudes, such as Canyon Creek Ice Cave, large volumes of ice are maintained. Other caves in the southern Rockies containing large ice masses include Provenance and Larch Valley caves (Mount Ball Karst), Upper Sentry Cave (Crowsnest Pass), Wedge Cave (Top of the World) and Disaster Point Cave near Jasper.

Through casual observation by cavers it has been noticed that the so-called "permanent" ice in Rockies caves is currently melting fast, whether this is because of cyclical trends in temperature or overall climatic warming is not yet clear. The repeated opening and closing of the squeeze in Canyon Creek Ice Cave suggests a 20 to 30 year cycle of ice build up and melting. However, only time will tell how long the current recession of ice in Rockies' caves will continue.

Ice mass in Serendipity. Photo Jon Rollins.

CAVE FAUNA & FLORA

Cave Fauna

Cave fauna is divided into three distinct types according to its relationship and degree of adaptation to the underground environment. Trogloxenes use the cave entrance zones for temporary shelter or permanent nests, but get all their food from outside the cave. Troglophiles may live their entire life cycle in a cave, while other individuals of the same species live outside. Troglobites live permanently in the dark zone and are found exclusively in caves.

All the mammals, herptiles and birds described here are trogloxenes. Some of the insects are thought to be troglophiles, possibly trogloxenes that have become trapped in a cave. Only four troglobitic life forms have been discovered in Rockies caves — a species of isopod, two species of amphipod and a mite.

Mammals and Birds

Pack Rats *(Neomata cinerea)* This delightful nocturnal creature is also known as the bushy-tailed wood rat, and by mountaineers in the past as a mountain-rat or snafflehound. Its elaborate nests are found in the majority of Rockies caves, even in those that are highly visited such as Rat's Nest.

Ben Gadd in "Handbook of the Canadian Rockies" writes that, "Pack rats breed in February, gestation taking 27 to 32 days and normally produce two litters of three to four young each year. Weaning takes 26 to 30 days and maturity is reached in nearly a year, a long time by rodent standards. Maximum longevity in the wild is about four years."

Pack rats are excellent climbers and will go considerable distances into caves using scent trails to find their way in total darkness. Exactly why they occasionally venture deep into caves is not understood; perhaps they enjoy caving?

In the summer, small bundles of vegetation can often be seen on ledges around cave entrances, having been gathered and left to dry there by the pack rats before being stored away for winter forage. In the winter, a silent early morning visitor to a cave entrance may be rewarded by the site of a pack rat, delicately balanced on its hind feet, drinking water as it drips off the tip of an icicle.

Cave entrances that have been inhabited by pack rats for long periods of time often have large middens that can fill passages almost to the ceiling as in Brazeau and Rat's Hole caves. The organic material contained in pack rat middens would be a good source of information regarding past vegetation types and thus climatic change, the more so because these accumulations are common and widespread in the Rockies. In the Grand Canyon, pack rat middens in caves have yielded pollen up to 4,000 years old (Park Science, 1991).

Bushy-tailed Pack Rat. Photo Dave Thomson.

Bats, mainly Little Brown Bats *(Myotis lucifugus)* The little brown — near the northern limit of its distributional range — is by far the most numerous bat seen in the Canadian Rockies. The colony in Cadomin Cave was thought to have originally numbered over a thousand, although it has been severely depleted in recent years by thoughtless human visitors who sometimes light fires in the cave. The little brown bats average 8 cm in length (tail 3.5 cm), weigh 7.5 g and have dark-brown backs with a lighter underside. They are thought to live up to 30 years.

In the Canadian Rockies bats spend between five and seven months in hibernation, seldom spending winter and summer in the same roost. Large colonies are known to use buildings as summer roosts, usually close to water where there are abundant insect populations. Where they migrate to in winter remains uncertain, although a good number probably head south to the northern United States.

Of the 1 to 1.5 million little brown bats estimated to populate Alberta (Margo Pybus, 1988) only a small proportion are thought to over-winter in caves. All the bats from a summer nursery colony do not necessarily occupy the same hibernaculum cave, although individuals return year after year to the same hibernaculum. This means that many bats die if a cave hibernacula is degraded or destroyed (S. R. Humphrey, 1978). Interestingly, males predominate in these winter roosts, the location of most females during the winter being unknown (Margo Pybus, 1988).

Several caves in the southern Canadian Rockies and Wood Buffalo National Park are known to provide winter hibernacula, but only in Cadomin Cave has any attempt been made to record sightings. Here, apart from a resident population of little brown bats, big brown bats, hoary bats and a long-legged bat have also been identified (Ben Gadd, 1986). Other caves in which bat sightings have been made include Guano Sauna, Procrastination Pot near Jasper, Chungo Cave in the Front Ranges, Rat's

Nest Cave in the Bow Corridor and Bisaro But Beautiful near Fernie. Bats have also been noted in many mines in southeastern B.C. (Wayne Lamphier, 1989 personal communication). Only Guano Sauna, the caves in Wood Buffalo National Park, Cadomin Cave and Procrastination Pot are thought to have roosts of any size, but as bat bones have been found in other caves and bats can fit through incredibly small apertures, it is possible that roosts exist in unexplored or inaccessible sections of many known caves.

Observations of little brown bats roosts (J. J. McManus, 1970) indicate that the optimum temperature for hibernation sites is 3°C. Bats exposed to higher or lower temperatures have faster rates of metabolism, use more energy and accumulate waste products, chiefly urine, more rapidly. Frequent arousal uses up valuable fat reserves required for the bat to survive the winter until food (aerial insects) again becomes available. It is not warmth that bats seek in caves, but the lowest stable temperature short of freezing.

"Hibernating bats seek an environment in which the chemical processes involved in metabolism transform fat into energy at the minimum rate. Not many caves provide such environmental stability, and conditions that satisfy the needs of one species may be totally unsuitable for a closely related species" (C. E. Mohr, 1976). The cool stable temperatures are best found in caves with a good influx of cold air. Such caves tend to have one or more large entrances, together with voluminous passages. The specific site chosen in a cave involves humidity, little browns apparently seeking areas where the humidity approaches 90% or more. Many dormant bats observed in Procrastination Pot had moisture condensed on their backs. Bats don't always go directly to their mid-winter roosts; they will move farther into a cave as it gets colder.

"If a healthy little brown bat finds a favourable environment for hibernation — with no disturbance — it can survive a long winter. At an optimum temperature of 36°F it would expend energy at the rate of twenty calories per day, a total of 3,000 calories of energy for an average 150-day period of hibernation. The metabolic energy yield of one gram of fat is 9,500 calories. Since most individual bats burn up two grams during the winter, it is apparent that much of the time the bat's metabolism rises far above the state of complete dormancy. If it didn't a bat could theoretically survive at that level for six years. It doesn't have to do that, but its whole life style is so regulated — its lifetime energy use so budgeted — that it far outlives similar-sized insectivores, the shrews. A tiny shrew, with its furious living pace, may expend as much calories in a sixteen-month life span as a little brown bat does in twenty years" (C. E. Mohr, 1976).

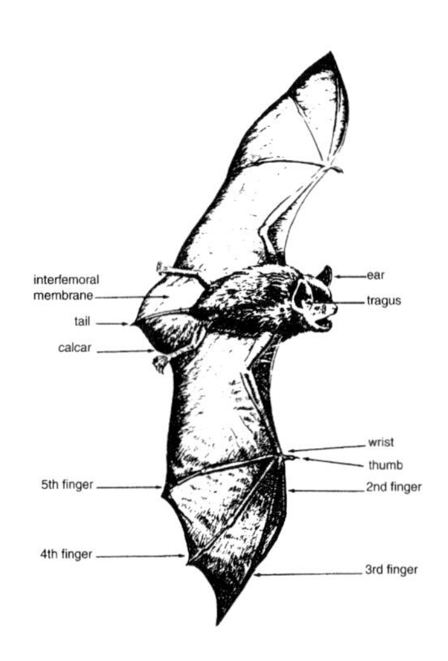

Little brown bat *(Myotis lucifugus).* Drawing by Margo Pybus, 1988.

Little brown bats roosting in Procrastination Pot. Photo Dave Thomson.

Much has been made in recent years of the bat's ability to navigate and catch insects using echolocation. The experiences of cavers are that when flying undergound, bats often do not use this ability. Rockies caves used as bat roosting sites often contain large amounts of bat bones. In some areas of Procrastination Pot bat bones litter the floor like pine needles in a forest, and some are even embedded in calcite. It is thought that in such well-used hibernaculums bats fly on memory and that they fly into foreign objects such as cavers who block their path. It is also possible that bats recently awakened from hiberbation are not yet fully operational. Although the role that bats play in the food chain has not been examined in any detail, bat guano and dead bats are obviously important inputs in the ecosystem of a cave, providing nutrients for cave insects, salamanders and cave isopods.

Metabolic economy is the key to bat survival in Rockies caves and the intrusion of humans into this environment can be a major threat. In accessible caves that provide bat hibernacula like Wapiabi and Cadomin, heavy visitation has doubtless led to the depletion of bats. It is strongly recommended that visitation be regulated during the winter months.

In early spring, writes Ben Gadd, "Little brown bats emerge from hibernation ... when small flying insects, especially mosquitoes, become active. Mating occurs in autumn in the Rockies, just before hibernation begins, but fertilization is delayed until after hibernation. Gestation takes 50 to 60 days and delivery of the single young occurs from mid-May to July. Carried about in its first few days, the young bat gains weight quickly and soon flies; it matures during the first annual hibernation, which begins with the disappearance of flying insects in fall, usually by mid-September in the Canadian Rockies."

Warning: All species of animals are believed capable of contracting and transmitting rabies. Handling of bats should be avoided.

Long-toed Salamanders *(Ambystoma macrodactylum)* Only two recorded sightings have been made in Rockies caves, one in the Green Pool Series of Yorkshire Pot (providing strong evidence for an entrance in the vicinity, since discovered), and the other in Disaster Point Caves. The salamander feeds on small arthropods. It may hibernate in caves, or be an accidental trogloxene.

Swallows *(Hirundo pyrrhonota)* Swallows, likely cliff swallows, and their elaborate nests can be seen in the entrances of a number of Rockies caves, notably Cliffside Cave at 2350 m in the Snaring Karst, and Gargantua at 2501 m at Crowsnest Pass.

Humans *(Primates)* The discovery of two prehistoric Indian arrowheads (Pelican Lake style, 2000 to 3000 years old) in Rat's Nest Cave and petroglyphs in the entrance to Crowsnest Spring and Devona Cave near Jasper, indicate that humans have visited some of the caves in this guide. However, Rockies caves were probably too cold and damp to be suitable for long term habituation.

Today, humans can penetrate long distances into cave systems relying on artificial light sources and using ropes to descend vertical sections of passage. Humans apparently no longer use caves for shelter, but enter them for a number of complex reasons probably symptomatic of over-populated living conditions in cities, and the need to escape sedentary lifestyles.

Insects & Spiders

Very few studies of cave invertebrates have been undertaken in the Canadian Rockies. The following information is based on collecting that took place in Castleguard Cave in 1977, 1978 and 1980 by John Mort and Anneliese Recklies, and in 1984 by Miloslav Zacharda and Chris Pugsley. More recently, in 2001, collecting was done by Heidi Macklin in Rat's Nest Cave, Crowsnest Spring and Low Sentry Cave, and in 2002 by Pat Shaw in Cadomin Cave. All collections were made in the winter and were partial surveys (a full survey would include sampling in more than one period of the year). The most abundant and diverse order collected was Diptera (flies), followed by Opiliones (harvesters), Acari (mites), Collembola (springtails), Coleoptera (beetles) and Lepidoptera (moths). Less numerous were Notoptera (ice insects), Plecoptera (stoneflies), Siphonaptera (fleas), Orthoptera (crickets) and Araneae (spiders).

Flies *(Diptera)* Most common were fungus gnats (family Mycetophilidae), followed by dark-winged fungus gnats (family Sciaridae). Less numerous were winter crane flies (family Trichoceridae) and dance flies (family Empididae). Also collected were moth flies (family Psychodidae), gall midges (family Cecidomyiidae) and mosquitos (family Culicidae).

The fungus gnats, as the name suggests, feed on fungi commonly found growing on decaying organic material in cave entrances. They have the appearance of a small mosquito-shaped fly. I have observed in many Rockies caves the filamentous feeding strands of larval Mycetophilids.

Winter crane flies are frequent inhabitants of cave entrances where they swarm in late winter and early spring. They were also found in the dark zone of Rat's Nest Cave, suggesting they may adopt a troglophilic lifestyle. Also mosquito-shaped, they are usually larger than gnats, and have long fragile legs and smoky-coloured wings.

Female mosquitos are thought to over-winter in some caves, as the author can affirm, having been bitten during his winter guiding activities in Rat's Nest Cave!

Harvesters *(Opiliones)* "Caves and abandoned mines are common in most parts of the mountainous regions of western Canada but to find one with a piece of 'moose hide' hanging on a vertical face might seem a bit unusual — especially when close inspection of the 'moose hide' revealed that it consisted of hundreds or thousands of dangling legs that were quite capable of propelling their owners very quickly away. As inhabitants of the above ground parts of the earth, we first thought that this phenomenon was very unusual. However, after talking with several cavers, we have found that many have had similar experiences." (Nello Angerilli & Robert Holmberg, 1984.)

What is described here are over-wintering clusters of harvesters, commonly known as daddy long-legs. Harvesters belong to a group of arthropods called Opiliones, and are related to spiders in that they have eight legs and a similar body construction. However, harvesters have a

Long-toed Salamander.
Drawing by Robert Stebbins.

Fungus gnats.
Drawing by K. Rollins.

Springtail.
Courtesy Ben Gadd, *Handbook of the Canadian Rockies.*

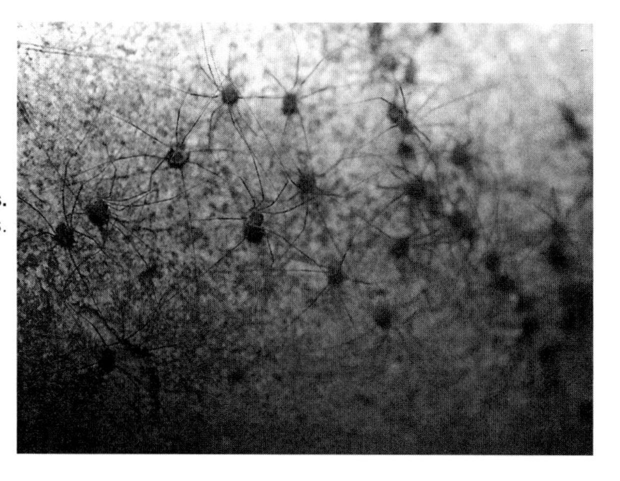

Harvesters.
Photo Jon Rollins.

one-piece body and only two eyes, while spiders have two major body divisions and usually eight eyes. Harvesters also lack the spinning organs and poison glands characteristic of spiders. On the other hand, harvesters have a distinctive pair of scent glands known as repugnatory glands that are used as a defense mechanism at times of disturbance.

Rarely found beyond the twilight zone, harvesters can be seen in the entrances of many lower elevation Rockies caves, usually clustered in solution pockets in the ceiling. Consisting of roughly equal numbers of males and females, the numbers in a particular site vary. Nello Angerilli and Robert Holmberg classified the aggregations as "loose" or "dense." Dense aggregations contain up to 2.6 individuals per square centimetre. These types of aggregations consist of an inner layer of individuals that cling to the rock's surface with their mouthparts as well as several outer layers that hang onto those underneath by means of legs and other appendages. Most members of these aggregations tend to let their legs hang straight down, thereby creating a fur-like appearance reminiscent of moose hide. Individuals in loose aggregations hang onto the rock by means of their legs and do not form layers. When disturbed, the individuals on the outside of the aggregations start bobbing up and down, this followed by the release of a strong odour and then scattering. The aggregating behaviour possibly stems from the stronger defence mechanism provided by collective odour release, although warmth generation and improved mating success probably also plays a part.

Mites *(Acari)* Predatory mites (unidentified) were found in the twilight zone of Rat's Nest Cave and Rhagiidae (family Prostigmata) in the dark zone of Cadomin Cave. A new troglobitic mite, *Robustocheles occulta* (family Prostigmata) was found in Castleguard Cave in 1984 near the Ice Plug. Originally thought to be an example of subglacial relict fauna (surviving underground from the late Wisconsian glaciation), the same species was later found in caves in Washington and Iowa.

Mites are tiny and require magnification for viewing. *Robustocheles occulta* has a body just 768–970 µm long.

Springtails *(Collembola)* These primitive wingless insects, otherwise known as snow fleas, have been seen on several occasions by the author in Rat's Nest Cave on the wet walls of the head of the 18 m pitch. It is assumed that they are visitors and not permanent residents. Springtails are 2–6 mm long and are commonly found on the surface above treeline, on the snow and under rocks and tundra plants. Springtails can jump using a mechanism in their tails, hence the name. Ben Gadd decribes what happens: "The furcula, an abdominal plate, cocks forward, held by a tiny catch. When a springtail decides to leap, the catch lets go and the furcula snaps back, flipping the little chap 10–15 cm into the air. Listen closely to an airborne collembolan; you may hear it saying 'Wheeeeee!'"

Beetles *(Coleoptera)* All beetles were collected in the twilight zone of Rat's Nest Cave with one exception, and are believed to be troglophiles. Most common was a morphospecies of the feather-winged beetle (family Ptiliidae). Rove beetles (family Staphylinidae) were found throughout Rat's Nest Cave, and were the only insect other than Diptera found in the dark zone.

Moths *(Lepidoptera)* Two types of moths were collected, those of the family Geometridae being found exclusively in the twilight zone of Crowsnest Spring. The larvae of this family are the well-known inch-worms. Owlet moths (family Noctuidae) were found in the twilight zone of Rat's Nest Cave and Crowsnest Spring.

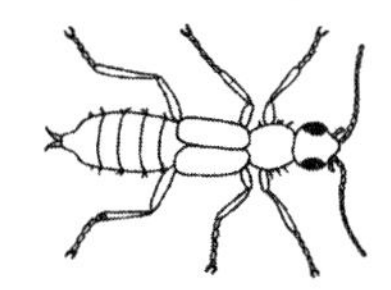

Beetles (Staphylinidae). Drawing by K. Rollins.

Ice Insects *(Notoptera)* Collected in Cadomin, Rat's Nest Cave and Low Sentry Cave, ice insects (family Grylloblattidae) are rarely found on the surface, being intolerant of high temperatures. They are believed to be troglophiles. Happiest at temperatures near 0°C, they develop slowly and live for up to 7 years. Their major food source in caves appears to be the wingless crane fly *(Chionea obtusa).*

Stoneflies *(Plecoptera)* All stoneflies were found in the twilight zone of Crowsnest Spring, close to the water. From the family Capniidae, they are our smallest stoneflies (5-10 mm long). It is believed the stable temperature and the water remaining unfrozen all winter makes this an ideal site for the final growth stages of the stonefly nymphs and larvae.

Fleas *(Siphonaptera)* Found on mice and rats the world over, these fleas (family Hystrichopsyllidae) would be parasiting the pack rats. All were collected in the twilight zone of Rat's Nest Cave.

Crickets *(Orthoptera)* Crickets have been observed occasionally by the author in Rat's Nest, where they insert themselves neatly into tiny vugs in the wall near the entrance. Those collected in Rat's Nest and Low Sentry caves are from the family Gryllacrididae.

Spiders *(Araneae)* Pirate spiders (family Mimetidae) and sheet-web spiders (family Linyphidae) were found in the entrance of Rat's Nest Cave. It is not known whether they were over-wintering or were accidental trogloxenes. The pirate spiders are so called because they predate on other spiders. The sheet-web spiders are tiny (less than 2 mm) and white.

Moths (Geometridae). Dra wing by K. Rollins.

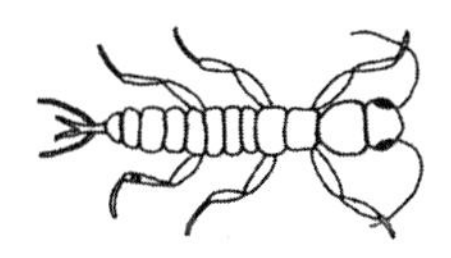

Ice Insect (Grylloblatta). Drawing by K. Rollins.

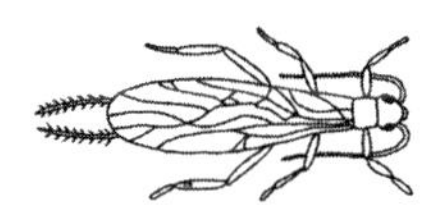

Stoneflies (Plecoptera capniidae). Drawing by K. Rollins.

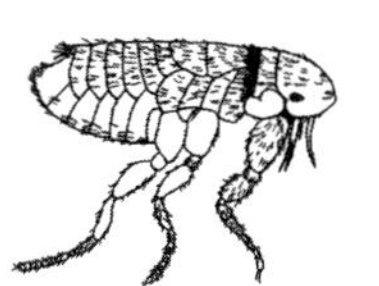

Fleas (Hystrichopsyllidae). Drawing by K. Rollins.

Crustaceans

"Look closely at the next cave pool you drink from. You might accidentally suck up a new addition to the Canadian cave fauna!" Pat Shaw, 1987.

These eyeless, unpigmented troglobitic crustaceans include the isopod *Salmasellus steganothrix* Bowman and the amphipods *Stygobromus canadensis* Holsinger and *Stygobromus secundus* Bousfield and Holsinger.

Stewart Peck believes that "the presence of these crustaceans, assumed to have limited dispersal abilities, suggests that they survived in situ in sub-glacial groundwater refugia during times of Pleistocene glaciation, and that they may be relics of older, preglacial distributions."

Isopods resemble a small horizontally-compressed shrimp, while amphipods look like a small, laterally-compressed shrimp. The isopod *Salmasellus steganothrix* was found in numerous pools in Castleguard Cave, from just beyond the Ice Crawls to the end of the Second Fissure, their tracks evident in the undisturbed sediments in the bottom of the pools. They were first identified by T. E. Bowman in 1975 from one collected in the gut of a trout from Horseshoe Lake in Jasper National Park, and have since then been identified in several springs, including Cadomin Spring.

The amphipod *Stygobromus canadensis* Holsinger was located in only one location in Castleguard Cave, in a series of pools between the end of The Subway and the start of First Fissure. Fast moving, the amphipods were tricky to collect. The discovery of a previously undescribed species caused some excitement!

The amphipod *Stygobromus secundus* Bousfield and Holsinger was found in a spring near Rocky Mountain House.

***Salmasellus steganothrix* isopod, Castleguard Cave. Length 6–10 mm.** Drawing by Pat Shaw.

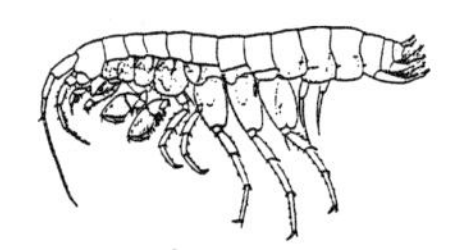

Cave Flora

"Demonstrations in English caves have shown that the mere passage of explorers into and out of a cave can cause extreme contamination by certain types of bacteria that had not previously existed in the cave" George Moore & Nicholas Sullivan.

Without sunlight to provide energy for photosynthesis, the surface flora we are familiar with does not grow in caves. The main observable plant life consists of fungi, mould and mushrooms that grow on dead animals, bat and pack rat feces, organic debris brought or washed into caves, and human trash or waste. The fungi and mould grow in a surprising variety of forms, and the moulds sometimes have bright colours. These forms of cave flora are known as heterotrophs, because they are dependent on introduced nutrient sources. Autotrophs are compounds that can derive energy directly from inorganic raw materials. A subgroup of autotrophs known as chemoautotrophs obtain all the energy they need by transforming certain minerals to different ones, and are thought to form the basis of the food chain for entire caves. A chemoautotrophic iron bacteria present in many Rockies' caves uses moisture and iron compounds to live, and turns cave sediments a dark brown colour.

***Stygobromus secundus*, a troglobitic amphipodcrustacean found in the Rockies. Length 5 mm.** Drawing by Pat Shaw.

Opposite: Sink in Hawk Creek karst. Photo Mary Day.

Appendices

Top Ten Caves (2003)

Longest		metres	Rank in Canada
1	Castleguard Cave	20122	1
2	Yorkshire System	13812	2
3	Gargantua	6001	9
4	Nakimu Caves	4500	10
5	Rat's NestCave	4003	12
6	Arctomys Cave	3496	14
7	Fang Cave	3342	15
8	Cadomin Cave	2791	16
9	Porcupine Cave	1794	28
10	White Hole	1320	33

Deepest			
1	Arctomys Cave	536	1
2	Close to the Edge	472	2
3	Castleguard Cave	390	4
4	Yorkshire Pot	389	5
5	Ptarmigan Cave	318	8
6	Gargantua	286	11
7	Pachidream	282	12
8	Nakimu Caves	257	13
9	White Hole	253	14
9	Dezaiko Cave	253	14
10	Fang Cave	247	16

The World's longest cave, Mammoth in Kentucky, is 571,317 m long. The deepest, Voronja-Krubera Cave in the west Caucasus, is 1,710 m deep.

The list for Canada's longest and deepest caves is maintained by Rob Countess, countess@pacificcoast.net. Please forward information and copies of data and maps if possible, to Rob whenever new surveying is done in any cave in Canada over 100 m deep or 1000 m long.

Caves Suitable for Novices

Although parts of the following caves are easy, getting to them may be more challenging than the caving itself. Good hiking and basic mountaineering skills may be needed. Once in the cave, be aware of unpredictable hazards such as rockfall, slippery surfaces and unexpected drops.

Make sure you have adequate caving equipment. A short rope may be handy for assisting weaker members of your party, but could also, unwittingly, encourage you into trouble. Make sure all novices stay with the group leader who should be an experienced caver. Never attempt a climb you cannot comfortably reverse. Only go as far as the weakest member of your party feels able to.

Cadomin Cave
Only the upper section of the cave is suitable for novices. The approach to the entrance is steep; take an ice axe if you expect to encounter snow.

Camp Caves
Although you can barely get out of daylight, Camp Caves contain some superb examples of glacially de-roofed canyon passages. Early in the season snow may barely cover the numerous entrances, so use caution!

Rat's Nest Cave
You will require a guide to visit this cave. Contact Canadian Rockies Cave Guiding at 1-888-450CAVE.

Canyon Creek Ice Cave
Loose rock is a hazard at the cave entrance. To minimize this hazard, use the trail that approaches the cave from the east, wear your helmet during the final approach and don't hang around below the cave entrance.

Cleft Cave
While all the cave is accessible to novices, be careful at the two lookouts at the back of the cave. Owing to loose rock, it is best if one person at a time attempts the scramble into the cave entrance. Wear your helmet during the approach.

Fang Cave
Only the upper and lower entrances are suitable for novices. Watch for high water levels in both entrances and don't attempt any drops as some become steeper and very slippery as you descend. Significant avalanche danger may exist in winter.

Gargantua
Only the upper section of the cave (Boggle Alley) is suitable for novice cavers. Do not attempt this cave early in the season — usually before mid-July — as the approach is avalanche prone. Take an ice axe to negotiate snow patches.

Hoodoo Creek Cave
Only the short entrance series of this modest cave are suitable for novices. Do not attempt to climb into the steep upper areas.

GLOSSARY OF CAVING TERMS

These are not intended to be complete explanations of these terms, but they should provide sufficient information to assist in comprehension of the text for those not versed in caving or geological terminology. Many of the terms are explained in more detail in the natural history section.

Ablation — The removal of rock or ice by wind action.

Active — Carrying water.

Alpine — For the purpose of this book, alpine refers to the "alpine zone," which is defined as land above the treeline. This is not dependent on absolute elevation since it varies with latitude, nature of the topography and aspect of the slope. For the Canadian Rockies treeline lies at approximately 1300 m. Rather than being a definite "line," it involves transition in vegetation type over several hundred metres of elevation.

Amphipod — An amphipod is a tiny crustacean resembling a small, laterally compressed shrimp. Several species of troglobitic (cave-limited) crustaceans (amphipods and isopods) have been found in Alberta caves.

Argillaceous — Containing clay.

Aven — Also "dome pit," an aven is a dome-shaped shaft that forms when sinking surface water intersects existing cave passages.

Bedding Plane — The plane along which bedrock splits. Exploited by groundwater, it thus forms a focal point for cave development.

Boon — A Canadian caver of some notoriety, after whom features in several Canadian caves are named (see Castleguard Cave).

Breakdown — Large pieces of bedrock that have collapsed from the ceiling of a cave passage, sometimes forming a "Breakdown Chamber."

Calcite — A white mineral, calcium carbonate ($CaCO_3$), that constitutes the majority of mineral formations found in Canadian Rockies caves.

Carbide Lamp — Also known as a "cap lamp," this is a small brass miner's lamp of early 20th century design that is still used by Rockies cavers. It consists of two compartments, one containing water and the other carbide chips. When the two are combined, acetylene gas is formed, which burns with a bright white light. A small container of carbide chips can provide a caver with several days of light, water being readily available in most Rockies caves.

Carbonates — Rocks which consist predominantly of carbonate minerals, limestone and dolomite being the cave-bearing carbonates in the Rockies.

Cave Coral — Small globular calcite formations, formed by water seepage, often found on passage ceilings.

Cave Diving — The use of scuba equipment to pass through sumped passages to hopefully discover more air-filled passage beyond. With a fatality rate similar to Himalayan climbing, taking up this sub sport of caving is a serious decision.

Cave Pearls — Pearl-shaped and sized, these calcite cave formations are formed by water dripping in pools.

Cave Radio — This "radio" uses electromagnetic induction, enabling communication to be made through bedrock.

Chert — A silica-based deposit that forms sharp-edged brittle protrusions sometimes seen on cave passage walls and known as "cherty flags."

Choke — When a surface cave feature or subterranean passage is impenetrably blocked with snow, ice or glacial deposits, it is said to be choked. Caves are sometimes extended by digging through chokes.

Cirque — A glaciated hanging valley with a characteristic rounded shape, also known as a "Corrie" or "Cwm."

Clastic — Material produced by the weathering of other rocks, often transported and deposited in caves by water.

Clints — The ridges in bare limestone rock surfaces.

Closed Depression — Also known as "Polje," a circular valley where water sinks underground. These features, which can be quite large, are common in karst landscapes.

Column — A floor to ceiling cave formation. Formed of calcite, they are a joining of a stalactite and stalagmite. Ice columns can sometimes be seen close to the entrances of caves in the winter.

Coralloids — A collective name for cave mineral formations having a rounded, globular, botryoidal or nodular appearance.

Cross-passage — A passage formed at a tangent to another cave passage, either crossing below or above it. Such features indicate a change in the direction of drainage during the formation of a cave system.

Cross-rift — A cross-passage having a rift configuration.

Descender — A metal friction device used for sliding down ropes.

Desublimation — The opposite of sublimation — the conversion of water vapour directly into ice.

Dike/Dyke — A discordant intrusion that cuts across bedding planes, and sometimes interrupts the course of cave passages (see Castleguard Cave). Usually formed of igneous rock.

Dip — Dip and "strike" (passages) are terms used to describe the location of cave passages in relation to the bedding structure of the rock in which they are contained.

Dip Tube — A phreatic cave passage running down dip.

Dissolution — The process by which rock is chemically dissolved, forming a solution.

Doline — Funnel-shaped depression on the land surface where water sinks.

Dolomite — Also "magnesium limestone," rocks containing more than 15% magnesium carbonate.

Draperies — Also "curtains." Thin, translucent sheets of calcite that hang down from cave ceilings.

Duck — A section of passage almost flooded to the ceiling. Usually there is just enough air space to breath.

Dye Tracing — The use of concentrated dye, usually rhodamine or fluorocene, to discover possible drainage routes taken by subterranean streams.

Evaporites — Formed of sediments following the evaporation of water, evaporites such as gypsum and anhydrite can form karst, as in Wood Buffalo National Park or northern Manitoba.

False-Floor — Following the removal of sediments and clastic material by invasive water, calcite deposits on cave passage floors (see flowstone) can be left intact, suspended in space (see Castleguard Cave and The Rose & Cavern).

Fault — A fracture in the rock where there has been an observable amount of displacement. Faults create weaknesses in the bedrock that are sometimes followed by caves. Characteristics of cave passage formed on a fault are "slickensides" where the two rock surfaces become polished, with linear grooves and ridges running parallel to the direction of movement. Such features are present at the beginning of the First Fissure in Castleguard Cave.

Felsenmeer — Meaning "Sea of Rocks," are large areas of angular frost-shattered debris, often found bordering glaciated karst.

Flags — Arrow-shaped protrusions on sodastraws thought to be caused by constant drafts in caves (see Castleguard Cave).

Flowstone — Layers of calcite deposited by water flowing over bedrock, sediments and breakdown in caves.

Formations — Also "speleothems" or "mineral formations." The collective name for mineral deposits which are usually composed of calcite.

Foraminifer — Phytoplankton that secrete calcite skeletal structures during growth.

Fossil — Opposite of active; no longer taking water.

Free Hang — A rope rigged in a cave in order to descend and ascend a shaft that hangs free and does not touch the walls.

Free Dive — A dive made without diving equipment, usually through a short, easy sump.

Frost-Pocket A joint or bedding plane surface exposure that has been enlarged by weathering, especially freeze-thaw action. Frost-pockets take on the appearance of cave entrances, but rarely go back far enough to lose daylight.

Glacial Flour A white powder, the finest material found in glacial deposits.

Glacial Till Also "boulder clay" and "drift," these are unsorted glacial deposits.

Grikes Also "grykes," the gaping fissures in limestone pavements.

Helictite A mineral formation consisting of a fine, chaotically twisting tube found on passage ceilings and walls.

Hibernaculum Place of hibernation. Caves often provide the constant low temperatures necessary for successful insect and bat hibernation.

Hydraulic Gradient Differing groundwater pressures that dictate which direction water will flow in.

Hydrology The movement of water underground. The subterranean drainage systems provided by caves are very ancient, and often bare no relationship to surface drainage basins.

Hydrostatic Pressure The load pressure on rock from the material above.

Hydrothermal Water heated by igneous activity, as with hot springs.

Isopod Tiny crustacean resembling a small horizontally compressed shrimp.

Isotope One of a set of chemically identical species of atom that have the same atomic number but different mass number.

Joint A rock fracture with no discernible movement.

Jumar A mechanical prusicking device used to climb ropes, as in SRT (single rope technique).

Juvenile Karst in which the subterranean drainage routes have not been enlarged enough to produce enterable cave passages.

Karren Collective name for the variety of solution grooves found on bedrock surfaces in karst areas, including kluftkarren, meanderkarren, rillenkarren, rinnenkarren and trittkarren.

Karst (karstic) A distortion of the Slovenian word "kras," a region straddling the Yugoslavia-Italy border characterized by thin soils, white limestone and an absence of surface water. Karst is defined as a landscape that has been developed by water solution of the rocks.

Keyhole Passage having a rounded phreatic ceiling with a slot-like vadose trench in the floor is known by its characteristic cross-section as "keyhole passage."

Krummholtz The small, windblown, stunted trees growing at the highest extent of treeline.

Lamination The development of thin, discrete layers of sediment or rock.

Lava Tubes Also "volcano karst," these are tubular passages in volcanic rock that are formed by the surface of the lava cooling faster than the still molten lava that runs out from beneath.

Lead A part of a cave where it is believed there is a good possibility of finding more passage.

Lifting Shaft Also "lifting chimney." Applies to vertical sections of cave passage where water is driven up under hydraulic pressure.

Meteoric Water Water that penetrates the rock from above, i.e., rain, dew, hail, snow and the water of rivers and streams.

Misfit Stream A stream flowing along a cave passage that it did not form. Fossil cave passages often act as a hydrological focus for active streamways, which may utilize them for part of their course.

Mixing Corrosion The mixing of two waters of different carbon dioxide content leading to an undersaturated mixture that is capable of dissolving more limestone.

Moonmilk A white paste-like deposit found on cave walls.

Moulin Also "mill well," a hole in a glacier down which water sinks.

Nick-Point Also "knick point," an abrupt change in a passage level often characterized by a waterfall, where the old and new profiles of a stream way interface because of land uplifting or a drop in the water table (rejuvenation).

Null	The point above a cave radio transmitter where there is no reception, indicating the receiver is directly above the transmitter.
Paleo-magnetism	When the Earth's magnetic poles change over time, these reversals are sometimes recorded in the alignment of mineral grains in cave deposits.
Paragenesis	In terms of caves, the resumption of passage formation following the plugging of an existing cave passage by sediments or clastic fill.
Perched Sump	A sump above the water table, the water being trapped in a U-shaped section of passage.
Phreatic	Cave passage formed below the water table, usually having a rounded cross-section.
Phreatic Lift	See "lifting-shaft."
Pictograph	Aboriginal rock paintings.
Pingo	Ice "volcano" formed when ice formation is constricted, as on the floor of a cave passage, causing it to heave upward.
Pit	See "pot" or "pitch."
Pitch	Vertical section of cave passage requiring a rope to descend and ascend.
Plug	See "choke."
Popcorn	See "cave coral."
Pot	A narrow circular shaft, characteristic of caves in Yorkshire, where the sport of caving is often referred to as "pot-holing" (pronounced "pot'oling").
Pseudomorph	One mineral occurring in the crystal form of another.
Pushing	Cave passages are often partially blocked by breakdown or sediments, and thus get very small, or require difficult climbing manoeuvres to traverse. To continue to explore such passages is known as "pushing" them.
Rappel	To slide down a rope using a descender.
Rebelay	The reattachment of a rope to a cave wall below the primary anchor point, usually to prevent damage to the rope through abrasion. Passing rebelays during descent and ascent of a rope requires practice.
Redissolved	Cave mineral formations are sometimes eroded by aggressive acidic percolation water, or by the subsequent flooding of cave passages. This produces an amorphous or "redissolved" appearance.
Resurgence	A cave where water flows out onto the surface. These caves may take the form of a spring, and if full to the roof with water, may not be enterable except by diving. Some Rockies resurgences have spectacular torrents of water pouring out of them in the spring and summer, but may be enterable in the winter. The partner to a resurgence is a sink or swallet cave, where the water sinks into the ground.
Rift Passage	A fault-guided cave passage, usually high and narrow in section.
Rigging	The use of ropes and natural or artificial attachment points (threads, bollards, pitons, bolts and chocks) to progress through a vertical cave system.
Rimstone	Also "rimstone dams," these are located on cave floors. The dam-shaped mineral formations that impound small pools are thought to be formed by an increase in CO_2 release and a corresponding increase in calcium carbonate deposition as water flows over small rapids and is agitated.
Scallops	Small dish-shaped features on the walls of phreatic passage.
Scree	Also "talus," these are areas of shattered angular rock fragments on mountain slopes, often believed to be covering the entrances of caves.
Shaft	A large vertical section of cave (see pot or pitch).
Sinkholes (Sinks)	Points at which water goes under ground (see dolines).
Siphon	See "sump."
Slickensides	See "fault."
Soda Straws	Also "straw," a fragile tubular stalactite mineral formation common in Canadian Rockies caves.
Solifluction	The slow downhill movement of soil, enhanced by water absorption.
Speleogenesis	The history of cave formation.

Speleogenic Conducive to cave formation.

Speleology The science of caves.

Spelunking North American slang for the sport of caving.

Sporting Dangerous, but character forming.

Squeeze The act of forcing one's body through a small aperture.

Stalactite Usually a straw, curtain or cone-shaped mineral formation that hangs from the cave ceiling.

Stalagmite A cone-shaped mineral formation (usually more rounded than a stalactite) that forms on the cave floor.

Straw See "soda-straw."

Strike See "dip."

Sublimation The conversion of ice directly into water vapour.

Sump Also "siphon." When a cave passage encounters the water table, or more commonly a U-shaped section of passage filled with water, it is said to sump. Unless very short, sumps require diving equipment to pass through.

Tackle See "rigging."

Talus See "scree."

Thixotrophic Materials that are effectively solid when stationary, but that become mobile liquids when subjected to shearing stresses. Clays having thixotrophic qualities are sometimes found in caves, and can make movement through passages very unpleasant.

Thread A hole in the wall or ceiling of a cave passage that provides a natural attachment point for ropes (see rigging).

Till See "glacial till."

Travertine Mineral deposits found at hot springs.

Tricounis Nailed boots.

Troglobites Life forms that live permanently in the dark zone and are found exclusively in caves.

Troglophiles Life forms that may complete their entire life cycle in caves. Species of the same type are found on the surface.

Trogloxenes Life forms that visit caves, sometimes for hibernation purposes.

Underfit See "misfit."

Vadose Passage formed above the water table, often with a canyon-shaped cross-section.

Varve Some Rockies caves contain large quantities of sediments that have in some cases been carried in by glacial meltwater. When deposited in layers, these are said to be "varved." In its broadest sense the term varve refers to a layer of sediments deposited in a single year, the coarser layer representing summer deposition, and the finer layer, winter deposition.

Vertical Range Difference in elevation of the cave passages between the highest and lowest points. Many Canadian Rockies caves have passages higher than the cave entrance, thus the cave "depth" is inaccurate.

SOURCES AND RESOURCES

Caving Organizations

For up to date information on caving organizations across Canada go to www.cancaver.ca and follow links.

Caving Guides

- (Jon Rollins) Canadian Rockies Cave Guiding 1-888-450-CAVE, www.caveguiding.com
- (Chas Yonge) Canmore Caverns 1-877-317-1178, www.canadianrockies.net/WildCaveTours
- (Jerry Fochler) Inroads Mountain Sports in Stony Plain at (780) 817-1512 or at the field office in Hinton at (780) 817-1512, jerry@inroadsmountainsports.ab.ca.
- (Reg Banks) Alpine Ventures (Nordegg) at (403) 721-2171.

Caving Courses

- Calgary Parks and Recreation: Introduction to Caving (403) 268-3800
- University of Calgary Outdoor Programs: Introduction to Caves and Caving (403) 220-5038

Equipment Suppliers

Some Canadian cavers make good quality caving equipment for sale, and some act as agents for caving equipment suppliers such as Petzl. Contact the Alberta Speleological Society for the latest information.

Purpose-made caving equipment can also be ordered from:

- Bob and Bob's (Lewisburg, WV) (304) 772-5049
- Bat Products (Wells, UK) 01144-1749-676771
- Caving Supplies (Buxton, UK) 01144-0129-871707

Currently, caving equipment is not usually available through retail outlets in Canada. However, many outdoor stores stock climbing gear that may be suitable.

Videos

- *The Longest Cave* (Castleguard). National Film Board,1974. Order at www.nfb.ca. ID# 113C0174540.
- *Underground Rivers* (Nakimu Caves). Parks Canada/ Dream Machine, 1981. Available for viewing at Rogers Pass Interpretive Centre.

Caving Publications

The Canadian Caver, the bible for Canadian caves is published twice a year. To subcribe follow links from www.cancaver.ca.

For other Canadian publications including club newsletters go to www.cancaver.ca and follow links.

Books

Check out the Alberta Speleological Society Library, available free to members, and the National Speleological Society website at www.caves.org.

On Canadian Caves:

- *Under Grotto Mountain Rat's Nest Cave.* Charles J Yonge. 144pp Rocky Mountain Books, 2001.
- *Castleguard.* Dalton Muir & Derek Ford. Canadian Government Publishing Centre, 1985.
- *Cave Exploration in Canada,* a special Issue of *The Canadian Caver.* Peter Thompson. 183pp, 1976 (out of print but available at some libraries).
- *Caves and Karst in Manitoba's Interlake Region.* W.D. McRitchie and Kim Monson (Ed.) 181pp, Speleological Society of Manitoba, 2000. Available from the University of Winnipeg bookstore.

On Caving Techniques

- *On Rope.* Allen Padgett and Bruce Smith. 341pp, NSS publication, 1987.
- *Caving Practice and Equipment.* David Judson (Ed.). 296 pp, 1991. British Cave Research Association. Available from British caving stores (see above).
- *Caving Basics.* 128 pp. 1987. NSS publication
- *An Introduction to Cave Surveying.* Bryan Ellis. 40 pp 1988. Available from British caving stores (see above).
- *Cave Photography A Practical Guide.* Chris Howes. 67pp, Caving Supplies, Buxton, 1987. Available from British caving stores (see above).

Natural History of Caves

- *Speleology The Study of Caves.* George W. Moore and G. Nicholas Sullivan, 150pp, Cave Books, St. Louis, 1981.
- *Karst and Caves.* T Waltham. 32 pp, 1987. British Cave Research Association. Available from British caving stores (see above).
- *Karst Geomorphology and Hydrology.* D.C.F. Ford and P. Williams. 601 pp, Unwin and Hyman, 1989.
- *Bats of Alberta.* Margo Pybus. 16 pp, 1988. Available from Alberta Fish and Wildlife.
- *Handbook of the Canadian Rockies.* Ben Gadd. 831 pp, 1995. Corax Press.

Books For Kids

- *One Small Square Cave.* Donald M. Silver. 48pp, W. H. Freeman and Co,. New York, 1993.
- *Les Caverns.* Pauline Gravel, 31pp, Societe quebecoise de speleologie, 1989.
- *Caves: Facts, Stories, Activities* Jenny Wood, 32pp, Scholastic Canada, 1990.

Acknowledgments

Because caving is not a popular activity in the Canadian Rockies, most of the information and cave surveys in this guide are the result of countless hours of toil by about sixty dedicated cavers, members of the Alberta Speleological Society.

All cave discoveries continue to be documented in *The Canadian Caver*. This periodical together with ASS newsletters, expedition reports, and twenty years of active caving with the ASS has provided me with the bulk of the information.

Special thanks to:
Ian McKenzie, not only for his encouragement and his wealth of information on Rockies caves, but also for the use of many surveys and photos used in this guide. To Derek Ford for his encouragement and for input into the geology and natural history of caves, and for allowing me to use cave surveys produced under the auspices of the McMaster Karst Research Group. To Peter Thompson for allowing me to use diagrams and information from *Cave Exploration in Canada*. To Dave Thomson and Ian Drummond for their great caving photos, many taken in very difficult situations. To Chas Yonge for introducing me to caving in the Canadian Rockies, for his input into their natural history and for numerous cave surveys and photos. To Ben Gadd for information on the Snaring Karst, and for writing his *Handbook of the Canadian Rockies* on which I have drawn heavily for natural history information. To Tom Barton — discoverer of many caves, to Steve Worthington and Mike Evans for their cave location maps for Crowsnest Pass and their information on cave hydrology in the Crowsnest Area. To Bill MacDonald for the Bastille Karst maps and information and for his work on ice in caves. To Heidi Macklin and Pat Shaw for their work on cave biota, Margo Pybus for her work on bats and Jim Burns for information on bones in caves. To Henry Bruns for information on the Ptarmigan area and for his fine photo gracing the cover of this book. To Tom Miller for taking me caving in Belize.

Thanks are also due to:
The McMaster cavers who under Derek Ford left an incredible legacy of cave discoveries and groundbreaking cave science in the 1960s and 70s: Charlie Brown, Mike Shawcross, Mike Goodchild, Pete Smart, Chris Smart, Tom Wigley, Alf Latham and John Drake.

To the cavers from British Columbia who have organized expeditions and explored caves in the Rockies including Phil Whitfield, Steve and Olivia Grundy, John Pollack, Martin Davies, Dayle Gilliat, Paul Hadfield, Eric von Vorkampff and Clive Keen.

To the various caving expeditions that have visited the Canadian Rockies and invited myself and other Canadian cavers along, including the Imperial College Caving Club (ICCC) from London, members of Lancaster University Caving Club (LUSS), and the Sheffield Cavers of the ACRMSE expeditions who provided information and cave surveys for the Small River and Goat Valley karst areas. Especial thanks are due to Deej Lowe for geological and speleogenesis of these areas.

To Phil Whitfield for making us feel a little bit safer by undertaking an impossible task: the organization of cave rescue in the Canadian Rockies.

To all ASS cavers including Ron Lacelle for his continuing support and keen sense of humor, Taco van Ieperen for his Crowsnest Pass cave surveys, Dennis Weeks for Burstall Pass information, Tim Auger for information on the Vroom Closet, Tich Morris who almost had me convinced he discovered most of the caves in the Rockies, Jason Morgan whose energy leaves us all breathless and to those Alberta cavers who first formed the ASS: Dave Doze, Harvey Gardner, Kitty and Bruce Dunn and Mary-Helen Posey. To the many ASS members whom I have had the honor to go caving with over the years: Marg Saul, Julian Coward, Randal Spahl, John Donovan, Pam Burns, Kim Smallwood, Don Prosser, Dave Chase, Keith Sawatsky, Yves Bellemare, Gille Roy, Eric Neilsen, Art Peters, Wes Davies, Pierre Lebbel, Will Gadd, Jim McPhail, Don Rumpel, Ian Phillips and Maria Cashin.

To Garry Pilkington, Bugs McKeith, Linda Hastie and Rick Blak whose exploits live on in all our memories, and particularly to Mike Boon, caver extraordinaire, whose amazing underground exploits and cave writing motivated me and who will not approve of this guide book.

To the Faculty of Environmental Design, University of Calgary, that allowed me to do my masters defense project on Caves of the Canadian Rockies — the genesis for this guide, specifically professors Dixon Thompson, David Henry and Steve Herrero.

To the Banff Centre for Mountain Culture for their financial support and to the Whyte Museum of the Canadian Rockies for the use of images. To my wife Karen for her many hours of toil over a "hot" drawing board.

Lastly to Gillean Daffern of Rocky Mountain Books without whom this guidebook would never have seen the light of day.

INDEX OF CAVES & KARST AREAS

The Author

Jon Rollins. Photo Dave Thomson.

Jon Rollins was born in London, England. From an early age he was taken hiking by his father to the Peak District and North Wales, these regular visits instilling in him a love of the outdoors which was to became a dominant force in his life.

After moving to Canada, Jon lived for two years in Vancouver where he discovered the delights of caving on Vancouver Island. Since then he has lived in Edmonton, Calgary and now Canmore with his wife Karen (herself a keen caver) and their two daughters.

Most of Jon's cave exploration is with members of the Alberta Speleleological Society (ASS), with whom he has been on caving expeditions to Mexico and Central America. His other passions — rock climbing and back-country skiing — are, as any Rockies caver will tell you, essential skills for getting to the caves.

Jon currently works as an independent environmental consultant specializing in karst management, and as Mountain Activities Co-ordinator for the Alpine Club of Canada. For the last 15 years Jon has operated Canadian Rockies Cave Guiding.

Other Guides to the CANADIAN ROCKIES AND COLUMBIA MOUNTAINS available from Rocky Mountain Books

Under Grotto Mountain: Rat's Nest Cave
Charles. J. Yonge
Pull out the map and follow cavers as they explore 4 km of subterranean passages in this popular cave near Canmore. 144 pp.

Exploring Prince George
An outdoor guide to North Central B. C.
Mike Nash
Explore Prince George and the surrounding backcountry of the McGregor and Dezaiko ranges with an experienced outdoorsman. 240 pp.

Scrambles in the Canadian Rockies
Alan Kane
Award-winning guide to over 150 non-rechnical peaks for mountain scramblers. Route-marked photos.
336 pp. 4th revised ed.

Ghost Rock
Andy Genereux
Front Range rock climbs in the Ghost valley area near Calgary. 336 pp. 3rd ed.

Waterfall Ice *Climbs in the Canadian Rockies*
Joe Josephson
A wide selection of climbs ranging from practise areas to grade 6 routes. Covers the area from the US border to Grande Cache. 400 pp. 4th ed.

Sport Climbs *in the Canadian Rockies*
John Martin and Jon Jones
Guide to one of the fastest growing adventure sports. Topos. 336 pp. 5th ed.

Bow Valley Rock
Chris Perry and Joe Josephson
Authoritative guide to multi-pitch rock climbs in the Bow valley from the prairies to the Banff National Park boundary. Includes Yamnuska. 432 pp.

Mixed Climbs *in the Canadian Rockies*
Sean Isaac
Details some the World's most difficult climbs from time-honoured classics to modern sport-style testpieces. 208 pp. 2nd ed.

Selected Alpine Climbs *in the Canadian Rockies*
Sean Dougherty
A modern up-to-date guide to the best mountaineering routes on the best mountains. Route-marked photos. 320 pp.

Backcountry Biking *in the Canadian Rockies*
Doug Eastcott.
Over 200 trails in the Rocky Mountains of Alberta and B. C. 408 pp. 3rd ed.

Hiking Canada's Great Divide Trail
Dustin Lynx
A guide to Canada's 1400 m-long Great Divide Trail from the US border at Waterton Lakes National Park to Kakwa Lake in northern B.C. 192 pp.

Summits & Icefields
Canadian Rockies
Chic Scott
Alpine ski tours, ski ascents and the grand traverses on both sides of the Great Divide. 224 pp.

Summits & Icefields
Columbia Mountains
Chic Scott
Alpine ski tours, ski asents and the grand traverses of the Purcell and Selkirk ranges. 208 pp

Ski Trails *in the Canadian Rockies*
Chic Scott
Groomed and backcountry trails in national and provincial parks. Includes established circuits such as the Wapta Icefields traverse. 224 pp. 2nd ed.

Hiking the Historic Crowsnest Pass
Jane Ross & William Tracy
Hikes in the Flathead and High Rock Ranges which feature some of Alberta's largest "windows' and deepest caves. 176 pp. 2nd ed.

Exploring the Historic Coal Branch
Jane Ross & Daniel Kyba
Hikes and rides around Cadomin on the border with Jasper National Park. 176 pp.

David Thompson Highway: A Hiking Guide
Jane Ross & Daniel Kyba
69 hikes between Nordegg and Banff National Park. 256 pp.

Kananaskis Country Trail Guides (2 vol.)
Gillean Daffern
Describes a total of 340 hikes in Kananaskis Country with photos and detailed trail maps. Volume 1, 272 pp., covers Canmore and the Kananaskis Valley. Volume 2, 320 pp.,covers the Highwood, Sheep and Elbow areas. 3rd ed.